Understanding the Cultural Landscape

Understanding the Cultural Landscape

Bret Wallach

THE GUILFORD PRESS
New York London

A Division of Guilford Publications, Inc.
72 Spring Street, New York, NY 10012
www.guilford.com

Printed in the United States of America

This book is printed on acid-free paper.

Last digit is print number: 9 8 7 6 5 4 3 2 1

Library of Congress Cataloging-in-Publication Data

Wallach, Bret, 1943–
Understanding the cultural landscape / by Bret Wallach.
p. cm.
Includes bibliographical references and index.
ISBN 1-59385-119-7 (pbk.) — ISBN 1-59385-120-0 (hardcover)
1. Human geography. 2. Human ecology. 3. Nature—Effect of human beings on. I. Title.
GF41.W354 2005
304.2—dc22

2004024793

Maps by Geoff Maas.

For JJP and WZ

CONTENTS

PHOTO SOURCES

All of the photos in the text also appear on the website *greatmirror.com*. They can be viewed there, larger, in color, and accompanied by related photos and extended captions. This list gives the website location of each picture.

	Country	Chapter	Photo #
Cover (upper)	Spain	Granada: Palaces	39
Cover (lower left)	Austria	Bramberg	1
Cover (lower middle)	China	Beijing: Temple	16
Cover (lower right)	Singapore	Modern	5

Book page	Country	Chapter	Photo #
32	Sudan	Darfur and Kordofan	12
40	India Themes	Farming Technology	3
41	Poland	Countryside	9
42	Pakistan	Karimabad	4
43	India Themes	Irrigation One	7
44	Sudan	Darfur and Kordofan	4
45	Sudan	Darfur and Kordofan	1
54	Syria		8
57	Egypt	Icons	2
67	China	Xi'an	9
68	China	Dali and Lijiang	4
69	China	Beijing: Symbolizing Political Power	30
71	China	Yunnan Village	8
76	Northern India	Khajuraho	8
80	Southern India	Sanchi	8
82	Northern India	Kutb Minar	1
85	Austrian Alps	Stubachtal	13
90	India Themes	Irrigation Four	7
92	Italy	Florence: Churches	7
94	U.S. East	D.C.	1
104	India Themes	Plaques and Statues	6
111	Sudan	Khartoum	1

Book page	Country	Chapter	Photo #
112	China	Beijing: Hutongs, Siheyuan, and Highrises	33
123	Kenya		2
130	China	Weihe Plain	7
135	China	Baisha and Longcheng	17
139	U.S. West	Pioneer Oil Fields	1
144	China	Yunnan Village	12
175	Northern India	Noida	9
183	Jordan		11
188	China	Chongqing	10
197	U.S. Oklahoma	Small Towns	29
200	China	New Suzhou	22
216	U.S. West	Los Angeles Two	27
219	U.S. Oklahoma	OU Campus Two	2
220	U.S. Oklahoma	OU Campus Two	11
221	Belgium	Bruges Revived	6
223	U.S. East	Manhattan	9
228	U.S. West	Laredo	6
256	Burma	Rangoon One	19
257	Trinidad	Port-of-Spain to Toco	9
261	U.S. West	Northwest	8
264	U.S. West	Pioneer Oil Fields	15
266	U.S. West	Great Plains	8
275	U.S. West	Pioneer Oil Fields	18
309	U.S. East	D.C.	14
310	U.S. East	D.C.	21
311	U.S. East	Manhattan	6
314	U.S. East	D.C.	40
319	U.S. West	Laredo	29
326	U.S. East	Santa Fe	10
334	U.S. Oklahoma	Small Towns	4
340	U.S. West	Santa Fe	8
347	U.S. West	Sierra Nevada	2
349	Northern India	Calcutta Walk	51
353	China	Shanghai: Shikumen to Xintiandi	1
354	China	Shanghai: Shikumen to Xintiandi	9
355	China	Shanghai: New Chinese City	17
356	Southern India	Hyderabad Cyberabad	3
358	Northern India	Calcutta Walk	35
359	Burma	Rangoon One	26
360	Northern India	Taj and Fatehpur Sikri	4
361	Jerusalem	Walls, Gates, Streets	1
362	West Bank	Hebron One	10
363	Singapore	Colonial Singapore	21
364	China	Beijing Hutongs, Siheyuan, and Highrises	28
368	Bangladesh	Tagore Country	9
370	China	New Suzhou	20
371	Sri Lanka	Town and Country	16
373	China	Hong Kong Four	6
375	West Bank	Battir	2

CHAPTER 1

INTRODUCTION

Geography is a strange subject. You may think you know what it's about, but stick around.

To begin with, geography is primitive. It starts with every child looking around and making sense of the world. What's this? What's that? Why is it this way? Why here? Even when an adult takes it up, geography remains terminally childish. How do a dozen farmers in India share the water that runs periodically in the tiny ditch that irrigates their fields? Why do suburban streets in America curve? Why, wherever you look, are old airport terminals painted the same color?

Primitive, geographers begin and end with tangibles. This makes them a disgrace to the sophisticated branches of learning, which grapple first and last with intangibles: the historian's documents, the chemist's formulas, the mathematician's equations. Pity the object-driven geographer. Muddy boots never won anybody a Nobel.

And that's not all. Geographers are magpies. Even when they limit themselves to the cultural landscape, by which I mean the elements of landscape created by people, geographers trespass over a multitude of well-groomed disciplinary turfs. The Eskimo belong to anthropology, but human geographers are interested in them because the Eskimo are outside that mental window that geographers—all of us, really—have on the world. The Great Pyramids of Giza belong to history, but they're also in human geography because those stone-block mountains are outside that window. Fast-food restaurants are a topic for business school analysis, but they're in human geography, and you know why.

Look at a map legend. There are symbols for natural features: rivers, mountains, deserts, swamps. Call them the stuff of physical geography. Then there are symbols for cultural features: boundaries, roads, buildings, parks, monuments. They're the stuff of human geography. Look out an airplane window on a clear day and you see the same division. Natural features slide by: Niagara, the Grand Canyon, Cape Comorin at the southern tip of India. Overlaying them are the cultural ones: bridges, roads, villages, cities. Fly over the American West on a good day and with a good seat, and you can hardly miss the overlay of the cultural landscape atop the natural one. The High Plains are an immense natural table set with cultural center-pivot irrigation systems, neatly gridded towns, and lonely roads. It passes so slowly that it seems inconceivable you're going 500 miles an hour.

So much to look at, whether you're a physical or a human geographer. And you likely suffered at the hands of an embittered physical education teacher who was forced to teach geography on the side. Thanks to him, you habitually associate this grandly primitive, tactile, unrestricted inquiry with the dumb memorization of state capitals. Not likely.

Geography means earth description or, to put it less etymologically, studying the character of the earth's surface. That's a lot more than mnemonics to recall Augusta and Tallahassee.

Just think of the neglected stacks of *National Geographic* magazines that pile up in offices and attics coast to coast. Those yellow covers promise articles about one place after another. Nobody seems to read the articles, but the pictures talk. What do they say? More broadly, because I don't really want to talk about that magazine—in its bland way, another degrader of this discipline—how shall we handle a subject as vast as the cultural landscape, the world we have made?

One way is to attack it regionally. Begin anywhere—say, Russia. Then drift south to India or east to China. Eventually jump the Pacific and do the United States. This can be interesting, especially if the teacher is the rare bird with firsthand experience of these places, but it gets repetitive, because over and over you're talking about fields, factories, highways, and markets.

An alternative is to organize things topically, with agriculture one day, energy resources the next, cities a third, religion a fourth. Between these two approaches, I much prefer the topical, because it avoids repetition. Still, how should the topics be sequenced?

My answer is anything but original. I'm following the teacher who taught human geography to me almost 40 years ago. His approach was simple: Make human geography the story of how human beings have transformed the earth. It wasn't an original approach with him, either. He learned it from his teacher, Berkeley's Carl Sauer, who for several decades towered over American academic geography.

There is at least one big difference between Sauer's approach and mine. Sauer's geography was basically a German geography of the 19th century. This meant that it was a scholarly undertaking and that anyone who sought to make geography practically useful was suspect. German geography was very different, in other words, from the British geography of the period, which explicitly served the needs of empire. Look at the list of officers of the Royal Geographical Society a century ago, and you see a parade of military officers and colonial administrators, with no less than the king as patron. Not so with the imperially deprived German geographers. But though, in his maturity at least, Sauer had almost no interest in applied geography, he was not a disinterested scholar. He thought of himself as a child of late 19th-century rural Missouri and believed—this is hardly an exaggeration—that human history had peaked sometime before 1900. That was when people were wealthy enough to have food and books. It was before they had lost the ability to distinguish between freedom and the freedom to shop.

Naively, I once asked Sauer if he'd serve on a doctoral committee for a dissertation I hoped to write on some California oil fields. I was back in the hall before I had a chance to sit down. If I had proposed looking at fish poisons in Southeast Asia, things would have been different. If I had proposed a study of the diffusion of corn across Africa, Sauer would have said yes. He would have signed up straight away if I had wanted to study aboriginal burning of grasslands. But oil fields? No thanks. There wasn't even a thanks, just an "I don't think so," spoken in a tone suggesting that the topic was depressing.

This aversion to the modern world explains why Sauer's own courses in human geography (he called it cultural geography) stopped somewhere about 1850. Even as a student, I thought this was absurd. We'd learn a lot of interesting things—say, that pastoral nomadism was widespread across the Old World but unknown in the New—but we'd learn nothing about skyscrapers, shopping centers, or railroad networks. I shared Sauer's romantic sympathies—what American doesn't, bombarded as we are with advertising for home cooking, pies like grandma used to make, and spaghetti sauce from an old Italian recipe?—but it just didn't make sense to me for geographers to close their eyes to the world outside today's window. That's why I've rebalanced things here. Perhaps I've gone too far the other way and treated the distant past so cursorily that I insult the Sauer tradition. Truth to tell, I have to work to get interested in the things I tackle up front, under the heading "Anthropological Foundations." It's probably because they're mostly things I can't see.

Throughout, whether discussing the distant past or the here and now, my presentation is both advanced and elementary. It's advanced in the sense that it's factually very dense, a real jungle. More about this in a moment. At the same time, however, I try to write the way I would like to talk. You can't find a better, surer way to raise sophisticated eyebrows. What I've written is elementary in another way, too, because I rarely use words such as hypothesis, data, and model. This may sound innocent, even charming. I like it myself, having never had the least interest in being a scientist. Almost without exception, however, human geographers today—especially the more celebrated among them—want more than anything else to be scientists. Accordingly, they shun description and plunge deeply into the production of theory. But, then, they've analyzed the production of so many things. Space, place, memory, love: you name it, they've investigated its production. For them, Chopin perfected the production of nocturnes.

It doesn't take a genius to see whose side I'm on. For me, human geography belongs with the humanities, and contemporary efforts to make a science out of it are just another chapter in the dismal story of the social sciences seeking respect. But now you see why high-flying insiders will dismiss what I've written as not merely elementary but regressive. It's also why a clever title for this book would be *An Old Geography of a New World.* Or maybe just *A New Old Geography.* My own working title was *The Great Mirror,* a title suggesting that the cultural landscape is useful as a way to see and understand ourselves. As well as any book or conversation, the cultural landscape shows us that Western civilization, uniquely powerful and uniquely secular, is also uniquely successful and uniquely destructive, uniquely rewarding and uniquely devastating. These are paradoxical pairings, but they're no odder than the fact that unhealthy food can taste good. What *is* odd is that geography's contribution to our understanding of the world continues to be ignored by academics who trust and perhaps enjoy words more than reality. The truly afflicted will say that words *are* the only reality.

Another warning: The evolutionary approach can be initially confusing. At the start, you'll think you somehow picked up a mislabeled book on anthropology. After all, I begin with human origins and our diffusion over the earth. We'll soon look at the beginnings of agriculture and the emergence of civilization. Sounds like archaeology to me. Then we'll get to Europe and the rise of what's become a global culture. We'll look at its roots and its expansion—and you'll look at the spine of this book and squint to find the word history. We'll spend a lot of time looking at how people make a living. By then you'll think this must be a book for novice investors trying to decide where to put their money. We'll get to social upheavals and environmental problems, and you'll think that this must be a study in political science or environmental studies. At some point, you'll ask where the geography is in all the topics that come flashing by—and flash they do in a survey such as this. But everything we touch is either there to be seen in the cultural landscape or needs to be understood to make sense of the cultural landscape.

You may also feel that you're drowning in facts, which are certainly here in abundance. Which brings me back to the advanced aspect of this book, its factual density. I like facts a lot more than generalizations—yes, yes, including this one. But I also want you to be able to follow things up, be able to dig more deeply. The old way of doing this was by taking a list of further readings to the library. I'm not shy about offering bibliographical pointers—I've included enough to fill a big bookcase—but a lot of the information here is too current to be amplified from books. No problem. Nowadays you can dig amazingly fast on the internet. By supplying very specific names I'm hoping to give you search terms to ease your life on the prowl.

At some point, and however interesting you find the details, I hope you'll realize that my real interest is in their interconnections. I want you to see one thing—any thing—and an instant later see the strands leading off from it in every direction. "Everything in Every Thing" might be my mystical (or mystifying) motto. Odd for a geographer to have feelings: We're supposed to be supremely pedestrian, measuring the exact dimensions of rivers and mountains and cities and exports. But human geography finally produces, at least in a loose sense, a philosophy of human ecology. That's what Sauer had, as he shuffled up and down Berkeley's corridors, pipe in hand, smoke curling behind. It wasn't the fish poisons and crop diffusions and aboriginal burning that kept him going: it was our changing relationship with this planet. That's a long way from the wretched test handed to my seventh grader last year. Her social studies teacher wanted her to name all the states starting with "O."

PART I

ANTHROPOLOGICAL FOUNDATIONS

CHAPTER 2

HUMAN EVOLUTION, DIFFUSION, AND CULTURE

The mucky shores of Lake Texoma, on the Texas–Oklahoma border, are scattered with 30-pound snails a foot across. They're dead. They're more than dead: they're fossils. And they're not really snails: They're cephalopods, not gastropods. Still, they're about 100 million years old. That's old enough to get your attention, but these pseudo-snails are latecomers in the story of evolution. Worlds of multicellular life forms had come and gone by the time these creatures lived and died at the edge of a now-vanished shallow sea.

Thirty-five million years passed after the extinction of the snails. All the nonavian dinosaurs had gone with them. Then—it's hard not imagining this happening on a fine Spring day—the mammals began their own evolution. Among the emergent groups were the apes, once so widespread that they lived even in North America. Eventually, the quicker-breeding monkeys pushed them to refuges, primarily in Africa. There, some 6 million years ago, another group of primates appeared.

AFRICAN ENVIRONMENTS

Before we turn to the hominids, stop and look at a map of Africa. Looking at a map is an old-fashioned thing to do, but human geography begins with the physical earth, and you can't find a bigger chunk of ancient rock than Africa. There are recent mountains in the Northwest: they're the Atlas, part of the Alpine system that rims Asia all the way east to Burma. To the south, the continental bedrock is frequently masked by sedimentary debris, for example, desert sands. Patches of volcanic rock poke through, too. Some, like the Hoggar and Tibesti mountains of Algeria and Chad, lie deep in the Sahara. Another patch forms Mt. Cameroon, just inland from the angle of the Gulf of Guinea.

Dig through Africa's surficial sediment, however, and you'll almost always find precambrian basement, older even than the dinosaurs, let alone the mammals. Almost the whole continent is a precambrian block, lower in the north, higher in the south. The most dramatic edge is in South Africa's Drakensberg, or the dragon's mountain. Ocean-going ships that might head many hundreds of miles up the Amazon or the Yangtze can't get far up the Congo. About 100 miles from the sea, they're stopped by a waterfall over the

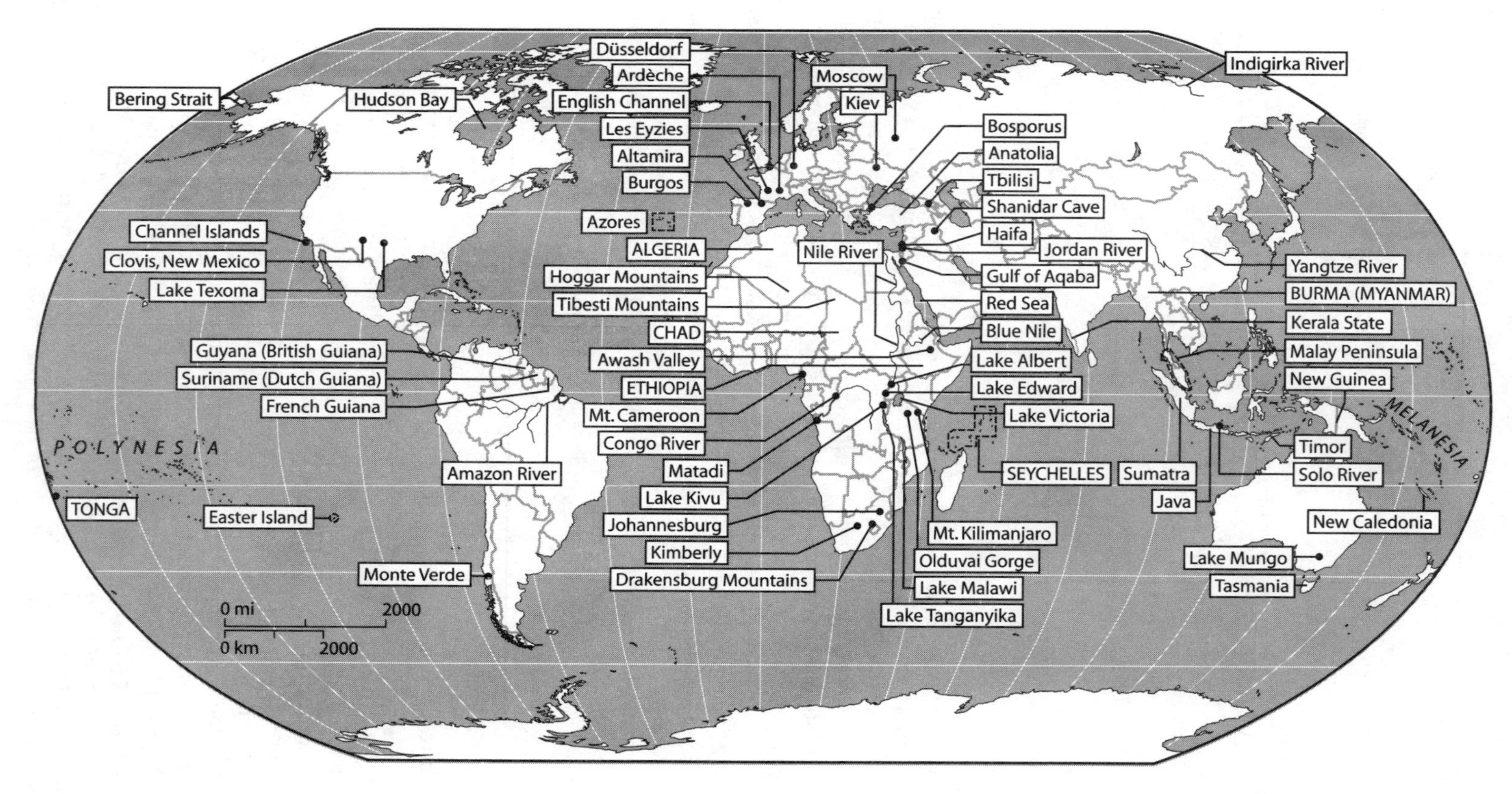

Note. No substitute for an atlas (for its wealth of thematic maps I recommend *Goode's World Atlas*), this map attempts, like those that follow, to locate all but the most obvious places mentioned in its chapter.

plateau's edge. Such freight as there is—the Congo's economy is a wreck—must be carried by rail or road from the port of Matadi upstream to the river port at Kinshasa.

About 5 million years ago, Africa began to split. The separation was never completed, but the process left behind a scar, which can be traced from Turkey down the line of the Jordan River, Dead Sea, Gulf of Aqaba, Red Sea, and then inland across Africa. There the scar is a trench partly filled by narrow but very deep lakes: Albert, Edward, Kivu, Tanganyika, and Malawi, formerly called Nyasa. Lake Victoria sits in a bowl between two arms of the trench, and though its surface area is large, the lake is never more than a few hundred feet deep. Tanganyika, in contrast, is as much as 4,500 feet deep.

This geological rifting was filled by upwelling lava, most dramatically in Mount Kilimanjaro but also in the broad and rugged upland of Ethiopia. There the volcanic rock was eventually trenched by rivers like the Blue Nile. Great canyons were formed, which is why the country has sometimes been called Abyssinia.

The formation of the rift uplands changed the climate of East Africa. West of the rift, equatorial Africa is rain forest, moistened by winds drawn in from the Atlantic by the heat of the equatorial lowlands. Rifting and uplifting produced mountains that drained those winds of their moisture, so the equatorial plains and coast east of the rift became dependent on eastern winds. Normally, east coasts at this latitude receive generous amounts of rain. Think of the Guianas and the Malay Peninsula. Those places have thousands of miles of ocean to their east, however. East Africa's winds often originate over dry Arabia. Winds from the Indian Ocean do bring summer moisture to East Africa: Without them, there would be no annual flood on the Nile. But the water vapor in these winds does not condense until forced to rise by the inland mountains.

HUMAN ORIGINS

East of the rift, in short, Africa dried out, creating a new ecological niche. It was soon occupied by the new group of primates, the hominids. They were bipedal.

That's a very short sentence but a very big idea. There are 4,000 other mammals on the earth, including about 200 primates. None except the hominids regularly walk on two feet.

Why the hominids did so isn't clear. (Such is life. The most important questions never have indisputable answers.) In *Upright* (2003), Craig Stanford argues that the trait emerged in the forest, where it was advantageous to stand upright on branches and reach fruit with one hand, while the other held another branch for balance or support. Out in the hot savanna, or crossing patches of savanna between islands of forest, bipedalism remained helpful because the sun doesn't bake an upright animal quite as much as it does one on all fours. To live in the hot sun, the early hominids also lost their hairy coats and began to sweat over the whole of their bodies.

Like chimps, they ate meat. Stanford observes that they were probably better hunters than the chimps, whose knuckle-walking gait is inefficient for sustained hunting. That's why chimps will kill and eat monkeys in their path but won't spend a day looking for game. The bipedal hominids, however, could devote themselves to the chase or—more commonly than we'd probably like to admit—to finding a carcass to scavenge. It's tempting to link their reliance on meat to increasing brain size, but Stanford points out that hominids were on the savanna for 5 million years before brain size, relative to body size, began growing.

Apart from upright posture, sweating, hairlessness, and a probable tendency to undertake long hunts for meat, these first hominids—the australopithecines, to give them their proper name—were more apelike than human. There's no evidence that they regularly used tools, for example, even though chimpanzees do. Nor were their brains larger than those of contemporaneous apes.

From one australopithecine species, probably *Australopithicus afarensis*, the larger and tool-using genus *Homo* emerged about 2.5 million years ago in the species *Homo habilis*. The line of human descent may pass about 1.8 million years ago to what some experts recognize as the species *Homo ergaster*, which possessed for the first time a distinctively modern body shape, with the long legs that indicate a completely terrestrial rather than at least partly arboreal habit. Other experts consider *Homo ergaster* simply the earliest example of *Homo erectus*. (The significance given by the name *erectus* to upright posture is misleading, because the australopithicines also stood upright. Still, the name *Homo erectus* has stuck ever since it was coined in the 1940s by Ernst Mayr, who was consolidating the finds popularly known as Java Man and Peking Man.)

Homo erectus lived for well over 1 million years before disappearing about 200,000 years ago. Then, half a million years ago and for reasons that remain unknown, brain size began increasing rapidly. *Homo sapiens* now appeared, first in archaic forms like the Neanderthals. (The name is sometimes spelled without the "h," as Neandertals.) Modern *sapiens,* indistinguishable from ourselves, appeared only about 150,000 years ago. There are experts (the best known is Milford Wolpoff) who believe that *Homo sapiens* emerged separately in several parts of the Old World from *erectus* stocks. Call it the multiregional hypothesis. The more popular theory—it's summarized in Chris Stringer's *African Exodus: the Origins of Modern Humanity* (1997)—holds that *Homo sapiens* emerged once, in Africa, then radiated outward about 100,000 years ago to the Middle East and beyond.

Experts in this field have a weakness for cute names. Stanford, for example, calls bipedalism the "heavenly gait." Stringer, too, indulges himself and calls the African-origin story the Out-of-Africa Hypothesis. Cutely named or not, the hypothesis rests on the fact that of all the people now living on Earth, the ones genetically closest to the earliest modern *Homo sapiens* are Africans, specifically some of the !Kung San Bushmen and Biaha Pygmies. The matter's still hotly debated, however, and there's a middle position, too, combining an African origin with genetic blending as the African population spread across Eurasia. It's tempting to hope that these alternatives are true, because one brutal implication of Out of Africa is that our ancestors killed the Neanderthals.

So much for a fiercely compressed summary of a complex subject. Let's turn to some details. A nice one is the gap in the ridge brow, or torus. You can feel it between your eyebrows, where the skull goes flat. It's the defining mark of modern *sapiens*.

We could also venture into the saga of the history of knowledge in this field. It's been a detective story of epic proportions, surrounded by controversy since 1856, when the first Neanderthal bones were found near Düsseldorf, Germany. They were dismissed as pathological by Rudolf Virchow, the outstanding anatomist of the day. The publication of Darwin's *Origin of Species* (1859), however, inspired a search for so-called missing links. In 1891 a Dutch physician, Eugene Dubois, made a spectacular discovery on the banks of the Solo River near Trinil, in East Java. It was just a skull and femur, but it was the first recovered remains of Java Man. (Dubois's story is breathlessly recounted in Pat Shipman's *The Man Who Found the Missing Link* [2002].) Dubois himself was embittered for the rest of his long life by the scientific establishment's rejection of his claims. Several decades passed before the anatomically similar Peking Man was found near Beijing. This discovery, along with Java Man, was what Mayr in the 1940s would designate *Homo erectus*.

In Taung, north of Kimberley, South Africa, Raymond Dart in 1925 had meanwhile found the first australopithecine. Like Dubois, his work was rejected by Europe's experts, who balked at tracing human origins to Africa. The importance of the australopithecines—and of Africa in the story of human evolution—did not become paradigmatic until the 1960s, when many fossils were dug from Tanzania's Olduvai Gorge. It's an erosional slice through an ancient lake bed, where sediments were deposited over a period of 2 million years. They have been famously worked by the Leakey family—Louis, wife Mary, and son Richard. Mary was the one who made the first big discovery, of an australopithicene the Leakeys called *Zinjanthropus boisei.* They reported their discovery in the *National Geographic Magazine* in 1960, but the real shocker came later, when potassium-argon dating revealed that their find was not, as the Leakeys thought, 600,000 years old, but 1.75 million years old.

Discoveries continue to be made, and some of them will no doubt upset the tidy history presented here. In 1998, an apparently very complete australopithecine skeleton was found at Sterkfontein, near Johannesburg and dated to nearly 3.5 million years ago. Early in 2001, another Leakey—Meave—discovered what she described as a new genus of the same age; she called it *Kenyanthropus platyops.* A few months later, a graduate student named Yohannes Haile-Selassie found some fossils in Ethiopia's Awash Valley. They were dated to more than 5 million years ago and apparently came from a creature close to the dividing point, 6 million years ago, between chimpanzees and hominids. Judging from its toe joint, the creature walked upright. Strikingly, however, the environment in which the creature lived was forest, not grassland. This fact threatens the conventional wisdom that human evolution is a story of the African grasslands. In the summer of 2002, parts of a chimp-like skull with human-like teeth were found in Chad and dated to 6–7 million years ago. Although the skull, classified as a new species nicknamed Toumai, may be from a chimp ancestor, it may also prove to be from a human one. If so, it jeopardizes what Stringer, still in good form, calls the East Side Story. Potentially, it reduces East Africa to a sideshow. In the summer of 2003, however, Tim White announced the discovery in Ethiopia of *Homo sapiens idaltu* or Herto humans, from the village where these anatomically almost modern humans were found and dated to 160,000 years ago. Multiregionalists were not persuaded, but Out-of-Africa supporters took this discovery as more evidence in favor of African origins.

There's no slacking to this tide. Early in 2004 a group of medical researchers led by Hansell Stedman published a paper in *Nature* in which they discussed a mutation to the gene MYH16, which builds the strong jaw muscles characteristic of primates. The mutation makes the gene inactive, is found in all humans, and can be dated to approximately 2.4 million years ago. They speculated that this reduction of jaw musculature might be linked to the change in skull shape and the enlargement of the hominid brain that also occurred at this time with the first species of the genus *Homo*.

We could venture, finally, into the controversy surrounding the idea of human races. There is more genetic variation between members of a given race than there is between members of different races, which is why anthropologists repudiate race as a tool to categorize people. Yet the idea of race, resting ultimately on skin color, continues to be rooted extremely deeply in most societies, despite its ruinous social consequences. It's not an American or European specialty, either. The name Sudan, for example, means land of the blacks and was used as a term of contempt by Arabs who encountered the Nilotic tribes upstream from Egypt.

I'm not going to dwell on any of this. Instead, I want to look at the spread of humanity across the earth and then take up the question of culture.

DIFFUSION

We're wanderers—always have been. It goes back to upright posture and our ability to outwalk our relatives.

Homo erectus walked to the Solo River in Java 1.8 million years ago; to Dmanisi, near Tbilisi, 1.7 million years ago; to Beijing about 1.3 million years ago; and to Europe about 800,000 years ago. The European date comes from the Gran Dolina, near Burgos, Spain, but the path taken there must have been indirect, because the Mediterranean—even the relatively narrow but deep channel at Gibraltar—was an insuperable obstacle until boats were invented. The ice ages, however, lowered sea levels by about 600 feet, and *Homo erectus* could walk into Asia through a now-flooded land passage across the narrow strait at the southern end of the Red Sea. (The strait is called the Bab al Mandeb, in Arabic literally the gate of tears.) The mountains of Anatolia probably blocked the route north and west into Europe, leaving the earliest hominids to enter Europe by backtracking across the flat lands of Central Asia. Once in Europe, they probably became the ancestors of the Neanderthals, who arose about 250,000 years ago only to disappear perhaps because they were less adaptable or competitive or belligerent than modern *Homo sapiens.*

The diffusion of modern *sapiens* is more important to us. There's good evidence from the Skhul cave near Haifa of this species living 100,000 years ago in present-day Israel, which was reached by diffusion across North Africa toward the Mediterranean at a time when the Sahara was not particularly dry. Puzzlingly, these modern *sapiens* seem to have coexisted with Neanderthals for perhaps 50,000 years. Then, starting more than 50,000 years ago, modern *sapiens* spread east along the Asian rim to Southeast Asia. There's not much evidence for this, but you wouldn't expect there to be, because the easiest path of diffusion would have been along coastlines now under water. Lower sea levels allowed early modern *sapiens* to walk not only across the Red Sea but also from the Malay Peninsula to Sumatra and Java—to walk, in other words, across the former land mass known as Sunda.

Australia is a different story. During periods of lowered sea level, Australia was part of Sahul, a continent including New Guinea and Tasmania. Sahul was separated from Sunda by about 50 miles of deep water south of Timor and its eastern neighbor, Tanimbar. Reaching Australia took boats, in other words, and that's why *Homo erectus* never got to Australia. It's also why 1 million years separate the first hominids on Sunda from the first on Sahul. We've had boats for about 60,000 years though, which is also the earliest firm date of Australian settlement. Another date comes from Lake Mungo in New South Wales, where a grave has been dated at more than 40,000 years ago. (Lake Mungo is also the site of the world's oldest known cremation, 25,000 years old.) Once in Australia, people could walk across Sahul to New Guinea and Tasmania, but boats were needed once again to get to more distant islands like New Zealand. The knowledge of how to build boats, or the ability to do so, sometimes disappeared, leaving people stranded. A very late example is Easter Island, only reached about 1,500 years ago. Descendants of the settlers apparently destroyed the island's forests. With the decay of the settlers' boats and the absence of wood on the island, the settlers were stuck.

There was diffusion west, too, bringing *Homo sapiens* at least 45,000 years ago to Europe, where they are familiarly known as the Cro-Magnons. That name comes from the name of a rock ledge near the French village of Les Eyzies, in the Dordogne Valley. Ancient bones were unearthed here during railway construction in 1868.

Once again, land bridges were important, both at the Bosporus and the English Channel. But don't imagine that there was only one wave rolling north into Europe.

Genetic studies suggest that about 5% of the ancestry of modern Europeans dates back to a pioneer wave of settlement that took place about 45,000 years ago. About 80% dates to migrations occurring 20,000–30,000 years ago. The remainder comes from still later migrations.

Settlement eventually curved northeasterly to Russia, with its wealth of meat in the form of mammoths. The Ice Age climate required the development of specialized tools, such as needles to sew clothing, but these people were inventive. At Mezhirich, near Kiev, archaelogists have found houses framed about 25,000 years ago with mammoth tusks. Even the brutal climate of eastern Siberia was no bar to settlement: by 12,000 years ago there were people living at Berelekh, which is at the mouth of the Indigirka River and well above the Arctic Circle. It may just be that these people were genetically adapted to the climate. That's the finding of Douglas Wallace, who believes that the mitochondria of northerners generate unusual amounts of heat.

The most famous of the land bridges was the one across the Bering Strait, crossed by people drifting northeasterly from Siberia. It's not clear whether they came from the west or the south or both, but the idea that the first Americans came via the Strait goes all the way back to José de Acosta in 1589. When did the migration occur? In 1932, fluted points dated to about 13,000 years ago were found near Clovis, New Mexico. Since then, the conventional wisdom among archaeologists has been that the Americas were settled at about that time, when a gap opened in Canada between the great ice sheet that spread west from Hudson Bay and the smaller glaciers descending the eastern slope of the Rockies. It suddenly became possible for the Clovis Hunters to stream through the gap and spread over the Western Hemisphere.

Lately, this picture has been called into question. First, there is general agreement that a site called Monte Verde in southern Chile was settled 15,000 years ago. The site includes not only remnants of food and hides but familiar objects such as cords and tent pegs. The important implication is that the settlement of the Western Hemisphere had to begin when there was no passage between the ice lobes. Is there an alternative route? Coastal migration has been proposed, but again it's difficult to document because here, too, ancient campsites are now mostly under water. There are a few exceptions, though. Peru lacks a broad continental shelf, for example, and the remains of coastal settlements have been found there and dated to about 11,000 years ago. More support comes from a skeleton dated to 13,000 years ago from the Channel Islands off the southern California coast.

The astonishing thing is that people by 10,000 years ago had occupied just about the entire ecumene, or inhabited world. Only some islands and very high latitudes remained unoccupied. Even most of those empty places soon had people. The islands of the Mediterranean and Caribbean were settled about 4,000 years ago. The Arctic came later, although recent claims suggest that there were settlers in at least the European part of Arctic Russia as long as 35,000 years ago. Last of all was Polynesia, whose settlement required not only advanced navigation skills but also the possession of domesticated plants that could provide a food supply. It seems that Polynesia was settled very quickly by people from Southeast Asia who had settled for a time in Melanesia. A group of those people, almost unmixed with the local Melanesians, moved east to Tonga about 900 B.C. (The evidence for the Melanesia–Tonga connection is the so-called Lapita pottery, which is found both at Lapita, in Melanesian New Caledonia, and on Tonga.) From Tonga, groups radiated across Polynesia in outrigger canoes loaded with dozens of people and their domesticated plants and pigs. They sailed southwestward to New Zealand, eastward to Easter Island, and northeastward to Hawaii.

Unoccupied areas? Antarctica is the most obvious. Others include the Azores and Seychelles.

CULTURE

The final thing I want to look at here is culture in the anthropological sense. It's a word that's played a pivotal role in anthropology ever since Edward Tylor brought it into common use in the sense of the totality of a society's knowledge and beliefs, growing through innovation and intergenerational transmission. Tylor was Oxford's first professor of anthropology, and his *Primitive Cultures* (1871) is a classic in the field. He was also the first person to use the word "prehistoric," at least in print.

Culture in this sense isn't unique to us—chimps and even the near solitary orangutan are cultured in this anthropological sense—but we use innovation and transmission not just to open spiny fruits but to create and move into a world almost wholly of our own making. Culture in this anthropological sense extends from science and technology and fashion and amusement all the way down to the bedrock beliefs that make us part of the group or groups to which we belong. (I use the plural because we're all part, these days at least, of several cultures, including some at the local level and others at a national or even supranational one.) Notice how different this sense of culture is from the other, perhaps commoner use of the word, which suggests a concern with literature and the arts. "Isn't the Waldstein in C Major?" "No, no, the 'View of Delft' is in the Hague." "Dostoyevsky made a huge splash in England because the Victorians needed some Slavic passion." Most people aren't cultured in this sense. In the anthropological sense, everybody is.

Consider the case of a young man born and raised in England who goes home to India. It's home because his father is Indian, though his mother is Irish. It's his first trip, and he stays with his maternal grandmother in her village in Kerala, down in the southwest. She's a woman of high status in her community, but in formal social situations she always appears bare-breasted. She thinks that only a loose woman would cover her breasts in public. What kind of shock does the young man feel? The case I'm recounting is a real one, that of Aubrey Menen, who recounts it in *Dead Man in the Silver Market* (1954). Menen goes on to say that his grandmother was disgusted by the thought that Europeans ate together—actually saw each other masticate. It was scandalous. Everyone of taste and breeding knew that eating was strictly private. There was no talking her out of her foolishness—or Menen out of his.

Culture in the anthropological sense, in short, includes the fundamental values guiding human behavior. These values are held with a confidence that not only transcends doubt but almost precludes the possibility of doubt. Sometimes, as with Menen's examples, these values appear as curiosities, but they can quickly become politically charged. Many African societies hold that when a man dies, his property—including his widow—belongs to one of the deceased man's brothers. It's an old practice but one coming under attack nowadays as an unjust denial of women's rights. So is female circumcision, another very widespread practice. The controversies over the continuation of these practices are intense precisely because they strike at core values, which people don't want to think about.

The same thing happens closer to home. European and American society was torn in the 19th century by slavery and in the 20th by woman's suffrage. In the 21st, it seems, the issue will be gay and lesbian marriage. The United States has also seen bitter fights over the right of a child in the public schools to be taught in his or her native language. Americans also consider it axiomatic that people should be judged without regard to race, religion, or ethnicity, but France carries this idea much further, with socially polarizing results. In the course of conducting an official census, for example, the French government does not count race, religion, or ethnic origin. There's color-blindness for you. But France also prides itself on its historic revolution, which created a secular state from the wreckage of both the monarchy and religious zealotry. With an increasing number of Muslim girls attending public schools and wearing head scarves, the government in 2003 decided that students would not be allowed to wear religious insignia in state-supported schools.

Suddenly there's a huge controversy, with Muslim girls facing suspension for insisting on the right to wear the *hijab*. (The *Qur'an* says that women should be modest. From learned interpretations of this simple injunction spring a vast array of garments, culminating in the famous *burqas* of Afghanistan, whose advocates consider women in head scarves lewd.) Which side is right? One says that the state is interfering with freedom of religion. True. The other says that such insignia encourage students to identify with a religious group at the very time and place when they should be identifying themselves as citizens of the French republic. True, too. There's no satisfy-everyone solution, precisely because questions such as these rest on beliefs that neither side is prepared to question, let alone relinquish.

Language alone sometimes reveals how deep feelings run in such matters. Female circumcision has been illegal in Britain since 1985, for example, but the newspapers refer to it matter-of-factly as "genital mutilation," a label that by itself condemns the practice. In 2004, Britain made it a criminal offense to take a girl out of the country for the purpose of circumcising her. The maximum penalty is 5 to 14 years. The British home secretary explained that "no cultural, medical or other reason can ever justify a practice that causes so much pain and suffering." Justify to whom? From many cultural viewpoints, the home secretary is expressing a humane sentiment, but millions of Egyptians, Sudanese, Ethiopians, Eritreans, Somalis, and Yemenis think that female circumcision is perfectly justified, even obligatory. Think you can persuade either side to change its mind?

The most vital tool for the transmission of culture in the anthropological sense is language. Despite its importance, we know very little about its origin. That's unlikely to change soon: The first chapter of a recent book on the subject begins by suggesting that the evolution of language may be the "hardest problem in science." Some people link language to the human hand, with its elegant coordination and the brainpower required to control it. Others talk about the lowering of the voice box, which facilitates production of a wide variety of sounds at the expense of leaving us the only mammal that can't drink and breathe simultaneously. Others point to the hypoglossal canal, a hole in the base of the skull through which nerve fibers lead to the tongue: The enlargement of that canal in *Homo erectus* suggests that delicate control of tongue movement goes back a long way. Still others look to a mammalian gene called FOXP2, which changed in *Homo sapiens* about 120,000 years ago. Almost all of us have the altered version, which improves fine motor control of the muscles needed for speech. The rare individuals who lack it have trouble both speaking and, for unknown reasons, writing.

Anatomically based discussion leads to speculations about human beings having had language for the whole course of the existence of our species, say, 150,000 years. Students pursuing the FOXP2 connection shorten that period. At a minimum, if one assumes that language is a prerequisite to religion and art, then language must be at least 40,000 years old.

Is there an original human language? If so, we know nothing about it. We do know of at least 5,000 different languages, many of which can be grouped into families, each derived from a common language. There are bold theorists (the recently deceased Joseph Greenberg is the best known) who believe that all but a handful of these languages belong to a few superfamilies including Euroasiatic, four African groups (Afro-Asiatic, Nilo-Saharan, Khoisan, and Niger-Kordofanian), and a single Amerindian group, but such groupings are highly controversial.

What did we talk about? Part of the answer must be that we talked about technology—better ways to make and use tools. Another way of saying this is that we talked about material culture. It's the first answer that comes to mind because stone tools have been the daily bread of archaeologists for a long time, not just because our society is technologically oriented but because bits of rock, rather than ideas, are what's survived. By now there's an overwhelmingly elaborate record of increasingly sophisticated tools, starting with core tools, evolving through flakes, and culminating in the elongated flakes known as blades. These blades appeared about 50,000 years ago, and we know them from arrows and spears. They were also the tools of the Clovis Hunters, who tackled mammoths presumably with as much aplomb as Pygmies today, at 4′ 9″, hunt elephants. Along with tools came other technologies. Fired pottery and weaving, for example, can be traced back at least 25,000 years.

This focus on technology, however, can easily blind us to other aspects of our ancestral character. The highest paid people in our society aren't technologists but people who play games, sing, or act. It's very strange from a practical perspective—downright frivolous—but it's good evidence of the importance of nonmaterial culture.

The surviving evidence of nonmaterial culture in prehistory, however, is extremely fragmentary. We have statuettes traditionally but speculatively treated as fertility objects. The most famous is the tiny—only 4 inches high—Venus of Willendorf, discovered in Austria in 1908 and made of a limestone not found locally. Another is the Venus of Lespuge, about twice as large and carved from mammoth ivory. Both are about 25,000 years old.

There is cave art, too, found mostly in Spain and southwestern France. When first found, at Altamira in 1879, these cave paintings were dismissed as a hoax. How could there be bison in Spain? (Good thing that the discoverers didn't wander into the Chauvet cave in France's southern Ardèche: It includes paintings made 30,000 years ago of rhinoceroses, lions, and mammoths.) More striking than the species portrayed is the vitality of their portrayal. No still lifes here. Henri Breuil argued in *Four Hundred Centuries of Cave Art* (1952) that this was hunting magic, but there are other interpretations.

Then there's death. The Neanderthals buried their dead, rather than leaving them for scavengers. A famous case is the Shanidar cave of northern Iraq, where burials date to 60,000 years ago. They include one person who could not have survived without lifelong care. They also include a body that, judging from the extraordinary pollen count around it, may have been buried under a heap of wildflowers. If you distrust the imputation of sentiment to Neanderthals, perhaps you'll accept it for early modern *sapiens* in the case of a child buried 15,000 years ago with thousands of ivory beads at Sungi, 100 miles from Moscow. Jewelry can probably be pushed back further still, for example, to eggshell beads from Bulgaria and Turkey 40,000 years ago. A recent find described as orange and black seashell beads, found at South Africa's Blombos Cave, has been tentatively dated to 75,000 years ago and described as evidence "of selfhood, of aesthetics." If you're adventurous, on the other hand, you may attribute humanity to the hominids who 350,000 years ago left an unused, pink-quartz axe amid the bones of 27 hominids found buried in 1998 at the Sima de los Huesos, or Bone Pit, near Burgos, Spain. Does the axe hint at human qualities antedating both Neanderthal and modern *Homo sapiens?* Could it have been an offering? Many other tools from this time have been found in the region, but all have been of common rocks such as limestone and sandstone.

We know nothing certainly about what such distant ancestors believed or felt, but when you think of the lure of a campfire, or even a fireplace, you're probably echoing the value they placed on warmth and security. (Did *erectus* sit before a campfire? It's debatable, but the Neanderthals certainly did.) When you stop to take in an arresting sunrise, or when you smell some especially fresh morning air or smile at the spreading branches of a tree, you're probably echoing the attachment these people had to the world around them. When you respond to a smiling infant (or a crying one), you're behaving in a way shaped in our most distant past. When you pet your dog or stroke a cat, you're not alone. When the hair on the back of your neck stands on end during a ghost story, even though you don't believe in ghosts, you're a child of our collective past. We haven't changed very much in the last 100,000 years. The biologist E. O. Wilson has made this point nicely in his autobiography, *Naturalist* (1994). Our ancient fear of spiders and snakes is perfectly intact, he writes, but we have no fear at all of much greater new dangers, like cruising down a highway at 70 miles an hour.

CHAPTER 3

FORAGERS

We've been around for 100,000 years, more or less. That's a hard number to visualize; maybe it's better to say 5,000 generations. That seems like a good long time, especially if you figure that Julius Caesar lived only 100 generations ago—say, 2% of the way back.

For 90% of our collective existence, we lived without agriculture. The common term for our economy during this, the overwhelmingly greater part of our existence, is "hunting and gathering." A better term is "foraging," because it includes scavenging, which almost certainly has been a major source of food over the many millennia.

Given the duration of this foraging period, it's surprising how little we know about it. But of course we didn't write and didn't build monuments, and little of our handiwork has survived except stone tools and fragments of bone or ash. That's why in trying to understand our early selves it's tempting to look at vestigial groups, the few foragers still around. Well, *were* around until the last century or two, as nobody today is unaffected by the modern world.

I'm going to try a quick survey of these vestigial groups because I think we can infer from them things about faraway times. That's not the same thing, however, as saying that these vestigial peoples lived as our ancestors had always lived, and between my survey and my inferences I'll point out some of the ways in which these vestigial groups were probably very different from those of the mists.

FORAGERS OF THE RECENT PAST

At the time of Lewis and Clark, there were plenty of foragers in North America. Not the peoples on the East Coast and parts of the Southwest, who had agriculture, but those on the Great Plains and farther west and north. Here, people survived by hunting, fishing, and gathering wild plants. We all know of the salmon-based societies of the Northwest, with their longhouses and totem poles. We know of the bison hunters of the Plains, with their teepees. We know of the Eskimo, with their famous igloos. Each of these peoples consisted of several distinct subgroups. The Eskimo, for example, can be initially subdivided into those who speak Inuit Eskimo (they stretch east from the Bering Strait to Greenland) and those of southern Alaska and far eastern Russia, who speak Yupik Eskimo. The North American Eskimo today often live in American-style houses, drive snowmobiles, and hunt with rifles. The culture of the Siberian Yupik has been more brutally uprooted. Two

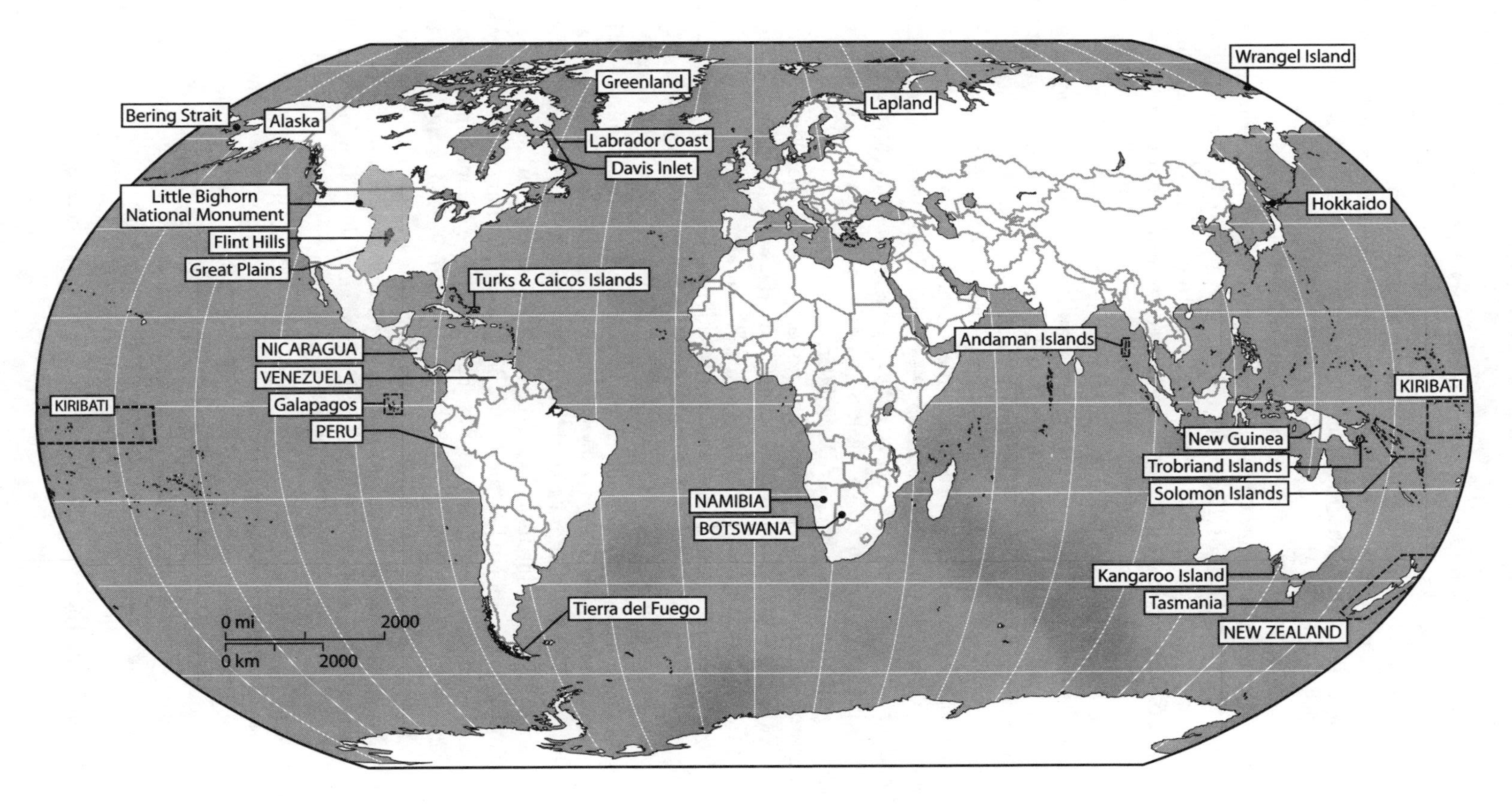
Wrangel Island
Greenland
Lapland
Bering Strait
Alaska
Labrador Coast
Davis Inlet
Little Bighorn National Monument
Hokkaido
Flint Hills
Great Plains
Turks & Caicos Islands
NICARAGUA
Andaman Islands
VENEZUELA
KIRIBATI
Galapagos
KIRIBATI
PERU
New Guinea
Trobriand Islands
Solomon Islands
NAMIBIA
BOTSWANA
Kangaroo Island
Tierra del Fuego
Tasmania
0 mi
2000
0 km
2000
NEW ZEALAND

centuries ago, the Yupik were whalers, but 19th-century Russian and American whalers introduced them to money, jobs, and imported foods. The Soviets outlawed native whaling, and the Yupik were reduced to chopping whale meat for Russian fox farms.

Odd as it may seem, there were few foraging groups in Central and South America, but that's because agriculture was the rule there. Exceptions include the Miskito Indians of Nicaragua and the Yanomami of southern Venezuela, made famous in Napoleon Chagnon's *Yanomamo: The Fierce People* (1968). Peru's Mashco Piro and Yora are another exception, so isolated that even today they are known as the uncontacted tribes. Far to the south, there were also the Ona and Yahgan of Tierra del Fuego. They shocked Charles Darwin back in the 1830s, when he saw them almost naked yet enduring blustering winds and icy cold. Only a handful of Yahgan remain today. They're known now, from their language, as the Yamana, and they're concentrated on Isla Navarino, just south of Tierra del Fuego.

In Africa there were the famous Bushmen. That name brings us to the battle of nomenclature. We almost bumped into this subject two paragraphs ago, when I blithely used the word "Eskimo." You won't find that word used much in Canada these days, because it translates disparagingly as "eater of raw flesh." The Canadians now say Inuit, which means the people. Similarly, the neighboring Naskaupi and Montagnais Indians of eastern Canada are now the Innu. At the end of 2002, the Naskaupi of Davis Inlet on the Labrador coast moved 10 miles to a new community, Natuashish. They were happy—these were their first homes with running water—but Canada's *Globe and Mail* reported the story as an Innu one. It was an upbeat story, too, although within a year 40 Innu youngsters at Natuashish were sniffing gasoline. Many had already been through detox back at Davis Inlet.

In the United States, the use of Indians as mascots is passing quickly from schools and universities, though at the level of professional sports we still have the Washington Redskins, Kansas City Chiefs, and Atlanta Braves. There aren't many hard-and-fast rules here, apart from the general precept that it's best to call people what they want to be called. Which brings us back to the Bushmen. Is that name derogatory? It certainly isn't the name they use for themselves. No single self-applied name, however, covers the dozen or so peoples—the best-known call themselves the !Kung—whom Europeans dismissed as Bushmen. One alternative is San, which comes from the language group they share, but apparently there are Bushmen who don't mind the old, outsider-applied name. Sometimes the labels applied by outsiders are so intrinsically derogatory, however, that they almost certainly must be discarded. Many Indians of the Great Basin and adjoining parts of the American West, for example, survived partly by digging roots and so were contemptuously dismissed in the 19th century as Diggers. (Mark Twain uses that term in *Roughing It.*) That name is gone now. So is Lapp, referring to the reindeer herders of Lapland. The word translates as "good for nothing" and is giving way to the self-applied name, Saame.

Today there are about 50,000 San or Bushmen divided primarily between Botswana and Namibia. There are also about 200,000 Pygmies, another hard-to-label group. Their name can be used in a derogatory way, but what other name can we substitute for these people? If we follow the rule of using the names applied by the people themselves, we will have to speak of half a dozen groups including the Aka, Baka, and Mbuti, each of whom now numbers about 25,000. There's another difficulty, too, because all these groups live in a mutual dependency with Bantu villagers. (The Bantu treat the Pygmies as inferiors. They will marry Pygmy women, for example, but they won't tolerate Pygmy men marrying Bantu women.) Half the year, the Pygmies work as farm laborers for the Bantu, in exchange for simple goods and food. Then they disappear into the forest, where for the other half of the year they hunt and forage. It may have been that way since the Pygmies first entered these forests. In other words, the Pygmies may never have lived solely as rain forest foragers.

Asia in the last few centuries has had few nonagricultural groups, though reindeer herders, such as the Iukagir, survived in Siberia. On Hokkaido, there were the famous Ainu, still extant today with a population of several hundred thousand, though none live in the traditional way, dependent on river fishing. On the islands of Southeast Asia there were the Andamanese and the Semang. I almost said "are." Two centuries ago, there were 5,000 Great Andamanese; now there are perhaps 40. Then, there were a dozen languages spoken on the Andaman Islands; today, there are four. Only two of the Andamanese groups still live in approximately the old way. They are the Jarawa of South and Middle Great Andaman and the Sentinelese of North Sentinel Island. Both are among the most isolated remnants of the once numerous groups known in India as tribals.

The Jarawa, in particular, have received a good deal of recent attention. By the late 19th century, the British saw them as hostile. A superintendent of the Andamans, however, J. C. Houghton, defended them in 1860 by saying that they "have had hitherto but too good cause to look upon the rest of the human race as their enemies. . . . I have determined to use my best endeavours to avoid all aggression upon them. . . . " A reservation was set up for the Jarawa in 1932, and a sympathetic chief commissioner named Cosgrave wrote a few years later, in 1938, that "the Jarawas are the original inhabitants of the island and we have some obligation not to interfere with the occupation of their ancient habitat more than they themselves [by attacks on Europeans and Indian settlers] make it necessary." The first prime minister of independent India shared these sentiments. Jawaharlal Nehru said that the government should encourage the preservation of traditional tribal cultures.

The Jarawa reserve, 764 square kilometers for about 200 people, was rimmed with guard posts manned by armed police. They were there not to protect the Jarawa from the settlers but the settlers from the Jarawa. (Some of the settlers were Bengali refugees. Others were former convicts at the prison the British had established on the Andamans.) Beginning in 1974, the Andaman authorities began paying monthly visits to the Jarawa and giving them bananas, rice, coconuts, cloth, and medical services. Meanwhile, the government built the Andaman Trunk Road through the reserve. Contact was suddenly easy. The American Anthropological Association's president elect, Louise Lamphere, wrote to New Delhi in 1999 to say that "the state has no right to attempt to force a tribal people to abandon its culture or take on an alien mode of production, whether in the name of progress, modernization or assimiliation into the local or national society." The Supreme Court of India seems to have agreed, because in 2002 it ordered the closure of the road and an end to logging in the reserve. Contacts are likely to continue, however. They bring with them not only the dependency established by regular gifts but the risk of devastating disease. A measles outbreak in 1999, for example, killed a tenth of the Jarawa.

Finally there were the Aborigines of Australia, the only continent that, as of 1800, lacked agriculture. (Oddly, there was agriculture on New Guinea, which was settled through Australia and is the place where sugarcane was domesticated.) There are about 425,000 Aborigines today, reduced from about 1 million in 1788 but up sharply from a nadir of 93,000 in 1901. At European contact, the Aborigines spoke about 200 languages. All but 50 are lost, and of those 50 only a handful have as many as 2,000 speakers. There were also the neighboring Tasmanians, exterminated by British settlers seeking sheep pastures. Eight years after the death of the last Tasmanian in 1869, the so-called Last Tasmanian died, but Truganini was actually an Australian Aboriginal woman born in Tasmania. Before her death, she asked to be buried. Instead, in a fine display of European ethnocentrism, her skeleton was put on museum display for many years.

APPRAISING THESE PEOPLES

If you want to learn more about these peoples, a simple place to start is the *Cambridge Encyclopedia of Hunters and Gatherers* (1999). I want to turn, however, to the point I made at the beginning of this chapter about the fallacy of thinking that ancient peoples lived like those I've just surveyed.

A fairer statement is that ancient foragers, compared to more recent ones, had both easier and harder lives. Ancient foragers, after all, lived during the Ice Age, which was an appallingly tough time even in lower latitudes. Its end some 10,000 years ago, however, brought comparatively easy living to many parts of the world. People became less mobile, more sedentary. The West Coast of Canada is a good example, with Indians by 4000 B.C. living in permanent settlements and supported by locally rich marine resources. That's a famous example, but there are plenty of less familiar analogs, including Japan's Jomon culture, Scandinavia's Ertebolle, and Southeast Asia's Hoabinhian.

That's one swing of the pendulum: from hard times to easy living. The reverse swing occurred with the expansion of agriculture, which pushed foragers into refuges, tough environments that farmers didn't want.

These pendulum swings were accompanied by social changes. Fired pottery and basketry, for example, can be traced back at least 25,000 years, well before the great melting, but the retreat of the ice and the development of sedentary life magnified their importance. Foragers of the deep past had been too nomadic to store food. Instead, they cooperated in getting it, then sharing it quickly before it spoiled. With sedentary life and storage, however, families could be much more self-sufficient, and there was less need for cooperation and sharing. The flexible associations characteristic of nomadic bands tended to be replaced by much more rigid ideas of who belonged—and who didn't—to the family. Ancestry became very important, a fact indicated by increasing attention to burial practice. There was as yet probably no social hierarchy, but inequalities began to emerge between richer and poorer families.

These distinctions are developed at length in Peter Bogucki's *The Origins of Human Society* (1999). In general, they pull ancient peoples farther away not only from us but from the foragers we know from the recent past. It's all the more reason to be cautious about interpreting the meaning of cave art and, more broadly, the character of ancient cultures.

Still, we're left with the question: What shall we make of our ancestors?

The intractable fact is that for thousands of years we haven't been able to make up our minds on this point. The Greeks wrote of a Golden Age, when life was simpler. At the same time, however, they spoke dismissively of barbarians, who were precisely the people living those simpler lives. Closer to our own time, Thomas Hobbes wrote in *Leviathan* (1651) that life in the state of nature was "solitary, poor, nasty, brutish, and short." Fifty years later, John Dryden wrote a play, "The Conquest of Granada," in which a character says: "I am free as nature first made me / Ere the base law of servitude began. / When wild in the woods the noble savage ran." Which is it? Certainly Dryden's view is the more popular one today, at least in Europe and America. A movie like "Dances with Wolves" opposes good Indians and bad whites, but you can hear an echo of the opposing view in the United States today when somebody who is habitually late is described as running on Indian time. You don't have to hunt very hard in Brazil or China or Indonesia today to hear government officials refer to the foragers in their jurisdiction as savages. The Tasmanians were brutally killed, but that doesn't mean that every British settler in Tasmania applauded. Among the voices in opposition was George Augustus Robinson, who wrote to his wife in 1830: "What can be more revolting to humanity than to see persons going forth in battle array against the people whose land we have usurped and upon whom we have heaped every kind of misery?"

This collective ambivalence has ramifications that go far beyond the way we judge distant peoples. If we're ambivalent about them, after all, we're necessarily ambivalent about ourselves. And so, indeed, we are. We like cities; we hate cities. We're fascinated by machinery; we're repelled by machinery. We love nature; we hate bugs. Contradictions like these are not just embedded in our behavior. They're visible in our skyscrapers and parks, our factories and craft fairs. We'll bump into them repeatedly when we consider the cultural landscapes of our own time, which reflect one side or the other of our divided cultural personality. Ambivalence about our own culture is a fundamental and recurrent theme in human geography. I don't like rankings, but I sometimes think this ambivalence is the most important theme in contemporary human geography.

Even anthropologists can't make up their minds. James Frazer's *The Golden Bough* (1890) was written at the apogee of Victorian arrogance, and it set out to show that magic was nothing but an inferior form of science. Fifty years later, Ruth Benedict's *Patterns of Culture* (1934) swung in the opposite direction and sensitized readers to the hazards of ethnocentrism. From time to time since then, anthropologists have gone still further and implied that prehistoric peoples were superior to us. At the peak of the so-called counterculture, in 1972, an important conference was organized at which foraging societies were said to have more leisure than we do. (The proceedings were published as *Man the Hunter* [1972], edited by Richard B. Lee and Irven Devore.) Marshal Sahlins in the same year published *The Original Affluent Society,* a book whose title plays on John Kenneth Galbraith's *The Affluent Society*, a 1969 portrait of contemporary America.

These opposing expert views can be unsettling. Look at the posthumously published field diaries of the eminent anthropologist Bronislaw Malinowski. During World War I, Malinowski worked in the Trobriand Islands, near New Guinea. There he almost invented participant observation, a commonplace practice among anthropologists today. *A Diary in the Strict Sense of the Term* (1967), however, is full of disparaging remarks of the sort that anthropologists today would never admit to thinking. In 1915, for example, Malinowski writes that "on the whole my feelings toward the natives are decidedly tending to '*Exterminate the brutes.*' "

Some anthropologists were angry that Malinowski's diaries were even published, but at about the same time that Malinowski was in the Trobriands the pioneering nature photographer Martin Johnson was in the nearby Solomon Islands. His wife Osa, who accompanied him almost everywhere, was horrified to see the treatment meted out to a runaway wife. In *I Married Adventure* (1940), she describes how a large rock was heated in a fire, then placed on the back of one of the woman's knees. The leg was bent back and tied, so that the rock deeply burned the leg muscles and tendons. Permanently crippled, the wife limped away, while male villagers taunted her to try running now. Square that with the naive, uncritically admiring view that is the norm in Western societies today. And it's no isolated case. In a controversial passage of *The Voyage of the Beagle* (1860), Darwin writes of cannibalism among the Yahgan of Tierra del Fuego. He says he was told that in times of famine old women were forcibly held over a campfire until they died of smoke inhalation. Then they were eaten. Some writers have argued that Darwin was deceived by his translator, but Darwin himself had no doubts. "We are told," he continues, "that they often run away into the mountains, but that they are pursued by the men and brought back to the slaughter-house of their own firesides."

University students ride the pendulum when, searching for a society better than their own, they elect to major in anthropology. Still another sign of the times is the Native American Graves Repatriation Act of 1990, under which museums have been shipping their collections back to many Indian tribes. A spectacular case was the shipment of 2,000 skeletons from Harvard to the Taos and Jemez Pueblo Indians near Santa Fe. Museums everywhere are trying to present their surviving collections more empathetically. In 1991, meanwhile, the Custer Battlefield National Monument became the Little Bighorn National Monument.

If foragers were paragons of spiritual wisdom, would they have crippled runaway wives? For that matter, would they have exterminated the megafauna around them, the animals weighing over 100 pounds? Certainly the saber-toothed tigers, along with all the horses, mammoths, camels, and giant bears of North America, disappeared shortly after the arrival of humans. Extinction was the fate of the mammoths and lions and bison of Europe, too, with the last mammoths killed about 2000 B.C. on the remote Siberian refuge of Wrangel Island.

The chronological correspondence between human arrival on a continent and megafaunal extinctions there is very close. Australian extinctions preceded American ones, for example, and American ones preceded those of the Pacific Islands. You can even show that people eliminated the larger animals first and then the smaller ones. There are still a few so-called climatists who discount our role as killers, but they're in the minority. There's just too much contrary evidence, including not only the chronological correspondence but the survival of nearly all very small animals and our minimal impact on marine life of the time. You don't have to imagine, by the way, that we personally exterminated the great predators; we merely had to exterminate their herbivorous prey—or possibly introduce diseases that did the job.

At a minimum, we've been responsible for some well-known exterminations. New Zealand once possessed a family of flightless birds called the moas. Some were small; others, bigger than ostriches. They had never encountered human beings before the Polynesians arrived about 1300. By the time Captain Cook came in 1769, the moas were gone. We keep learning of such cases. Scientists report, for example, that after 800 A.D. there was a sharp decline in the population of sea turtles in the Turks and Caicos Islands. The animals, which had formed a local staple, were apparently hunted to near extinction. And Africa? Why should megafauna have survived comparatively well there? Perhaps it's because we evolved there with them. Certainly animals there are wary of us, which was demonstrably not the case when people in recent times first landed on the Galapagos.

Experts seeking to describe these events sometimes use clinical language, like first-contact extinctions and the overkill hypothesis. Others have turned to metaphor and envisaged a bow-wave model of extinctions accompanying our diffusion across the planet. More dramatic metaphors have been used, too, including Blitzkrieg. For more on this subject, see *Extinctions in Near Time*, edited by Ross MacPhee (1999).

Foraging societies had a less emotive but equally momentous impact on vegetation, because they are probably responsible for the world's grasslands. Early Europeans in Australia, for example, were appalled by the carelessness with which Aborigines let campfires escape. That's why much of Australia at contact was grass and scrub, while uninhabited Kangaroo Island, near Adelaide, was forested. (This difference has been masked by European modification of Kangaroo Island, most of which is now as barren as the mainland.)

How many of the world's other grasslands are similarly anthropogenic? This is a debated subject, but the savannas of tropical Africa and South America do seem to be the result of aboriginal burning. The temperate grasslands are more controversial, with a vocal minority insisting that they are natural. Their staunchest spokesman is probably still James C. Malin, whose *The Grasslands of North America* was published in 1947. Yet the ecotone or contact between trees and grass is undeniably moving west on the Great Plains, apparently because we've done a good job suppressing fire. Certainly the ranchers in the Flint Hills, east of Wichita, Kansas, are convinced that they'll lose their pastures to woodland if they don't deliberately burn them annually.

In *The Ecological Indian: Myth and History* (1999), Shepard Krech demolishes the contemporary Western tendency to ascribe an innate environmentalism toward North America's indigenous peoples. Similar books could be written for the other continents. But once we quit glorifying these peoples, what shall we say about them—apart from the fact that we can't decide how to judge them?

For starters, they survived. Think of the Eskimo who wants to catch a seal swimming under the ice floe on which he stands. He somehow finds a minute blowhole, builds a little shelter around the upwind side of it, then places a feather at the opening of the hole and squats to wait motionless for hours until the feather flutters from a seal's breath. He quickly drives the harpoon down through the hole, and the ice turns bloody. Or think of the Bushman who draws water from a sip well where, seeing nothing, I would die of thirst. Think of basketry, which we stupidly dismiss with jokes about courses in remedial basket weaving. Baskets can be extraordinary things: fibers woven so tightly that the basket will hold liquid and yet with fibers so handsomely decorated that beauty seems as important to the weaver as utility. We're now bumping against jewelry and body decoration, including scarification and body piercing. Those last words are another reminder of how much like us these people were.

More than survive, these people also had imaginative lives. A hundred years ago, we dismissed their myths as, well, myths, by which we meant irrational nonsense. We don't do that anymore, thanks largely to the French anthropologist Claude Lévi-Strauss, whose many books (the most readable relevant one is *The Savage Mind*, published in 1966) rest on the thesis that myths explain the world, make sense of it, order it. That's something we tend to recognize now as a diagnostic human trait, and so we've grown respectful. On the other hand we still balk at the belief in magic that often accompanied these explanatory stories. We dismiss it as superstition, a word that almost by definition is pejorative.

There are good reasons for debunking magic—New England's experience with witchcraft has been a powerful teacher—but maybe we're too dogmatic in our rejection of it. John Neihardt in *Black Elk Speaks* (1932) writes of Black Elk inducing a light rain. Laurens van der Post in *The Lost World of the Kalahari* (1958) writes of cameras jamming in places where the Bushmen said they should not be used. Should we be as skeptical of such stories as we are of the once popular books of Carlos Castaneda? But consider Arthur Grimble, a British civil servant stationed for many years in the Gilbert Islands, now part of Kiribati. It's not so easy discounting an English career civil servant as a quack, and in his autobiography, published in the United States as *We Chose the Islands* (1952), Grimble tells a strange story.

He writes of being scolded by the villagers for being too thin. They say he should eat porpoise meat. A villager tells him that it can be arranged; in fact, he says, there is a hereditary class of porpoise callers. Grimble chooses a date some weeks in the future and journeys to the distant island where all this will happen. The porpoise caller greets him and asks him to wait. Hours pass, and Grimble grows bored during the man's trance. Eventually the porpoise caller bursts from his hut and cries "Arise! Arise! They come, they come! Our friends from the west." A thousand villagers rush into the water. Nothing. Then Grimble makes out a line of approaching porpoises. "They were moving toward us in extended order with spaces of two or three yards between them, as far as my eye could reach." The animals come aground and allow the villagers to lift them from the water. The villagers then fall upon the animals and cut them to pieces. Grimble says he found this so savage that he couldn't eat the meat, but he reports the whole episode matter of factly. He says that he cannot explain it, merely saw it. Shall we consider him a credible witness or merely credulous—or perhaps dishonest? Scientists are likely to say that nobody can call a porpoise—and to say it with a completely unscientific dogmatism.

LEGACY

It's worth repeating from the previous chapter that, as far away from us as these cultures now are—and more or less destroyed—remnants of them are still very much part of us. Think of the importance that fishing and hunting have to many of us. Think of Norman Maclean's *A River Runs Through It* (1989), where the "it" of the title has less to do with real estate than with the author's life. Think of Christmas trees, an echo of ancient beliefs in a world of spirits. Desmond Tutu has written that we only live as human beings if we live together. That's another bit of the ancient legacy. It's a bit that we must at least sometimes regret as we live increasingly atomized lives—jobs terminated (30 minutes to clear your desk and be walked to the door by security guards), serial marriages, kids bounced from parent to parent, and children growing up and moving far away. The truth is that the things we hold dearest are those we inherited from our remotest ancestors.

CHAPTER 4

DOMESTICATION

Over the last 10,000 years, forgotten peoples have transformed hundreds of plants and animals into the domesticated species which we absolutely depend on today. "Absolutely" is none too strong a word. It's even an understatement, because although there are about 1,500 of these genetically transformed plants, a mere handful sustain the human race. Jack Harlan, until his recent death a leading student of the subject, wrote in *The Living Fields* (1990) that 61% of the edible dry matter of the world's crops comes from just three plants: wheat, corn, and rice, in that order. The runners-up, in descending order, were barley, soybeans, cane sugar, sorghum, potatoes ("Irish" potatoes), oats, manioc (a tropical staple best known to Americans as the source of tapioca), sweet potatoes, sugarbeets, rye, and millet. No other crop accounted for as much as 1% of the total. In *The Origin of Cultivated Plants* (1967), Franz Schwanitz emphasized our reliance on domesticates in another way. Without agriculture, he asserted, the world's population would not exceed 30 million people.

DOMESTICATES

What is a domesticate? What distinguishes it from a wild plant or animal? Sometimes the differences are obvious. Domesticated sheep, goats, pigs, and cattle are smaller than their wild ancestors, for example, presumably because this makes them more manageable. Wild plants, on the other hand, have been transformed by repeated selection so that their edible parts are bigger and tastier. The wild tomato, for example, has fruits half the size of a cherry tomato, and wild lettuce is very bitter. In the wild, the carrot has many branching roots instead of one massive one. Many domesticates have lost their natural defenses, too: ancestral citrus, for example, was thorny, while the reproductive eyes of white or "Irish" potatoes have been slowly bred away from their original location. Originally they were deep inside the tuber; now they are close to the surface, where they are vulnerable but easy for us to remove. Carried to its extreme, domestication produces plants that are not only competitively disadvantaged but sterile, unable to reproduce without human help. Bananas and pineapples are examples; so are seedless grapes and navel oranges.

The grains which we rely on so heavily have been subjected to several less visible changes. Like grasses generally, the panicles of the ancestral grains shattered when their seed was ripe. This is a valuable trait in a wild plant, which is more likely to reproduce if its seed

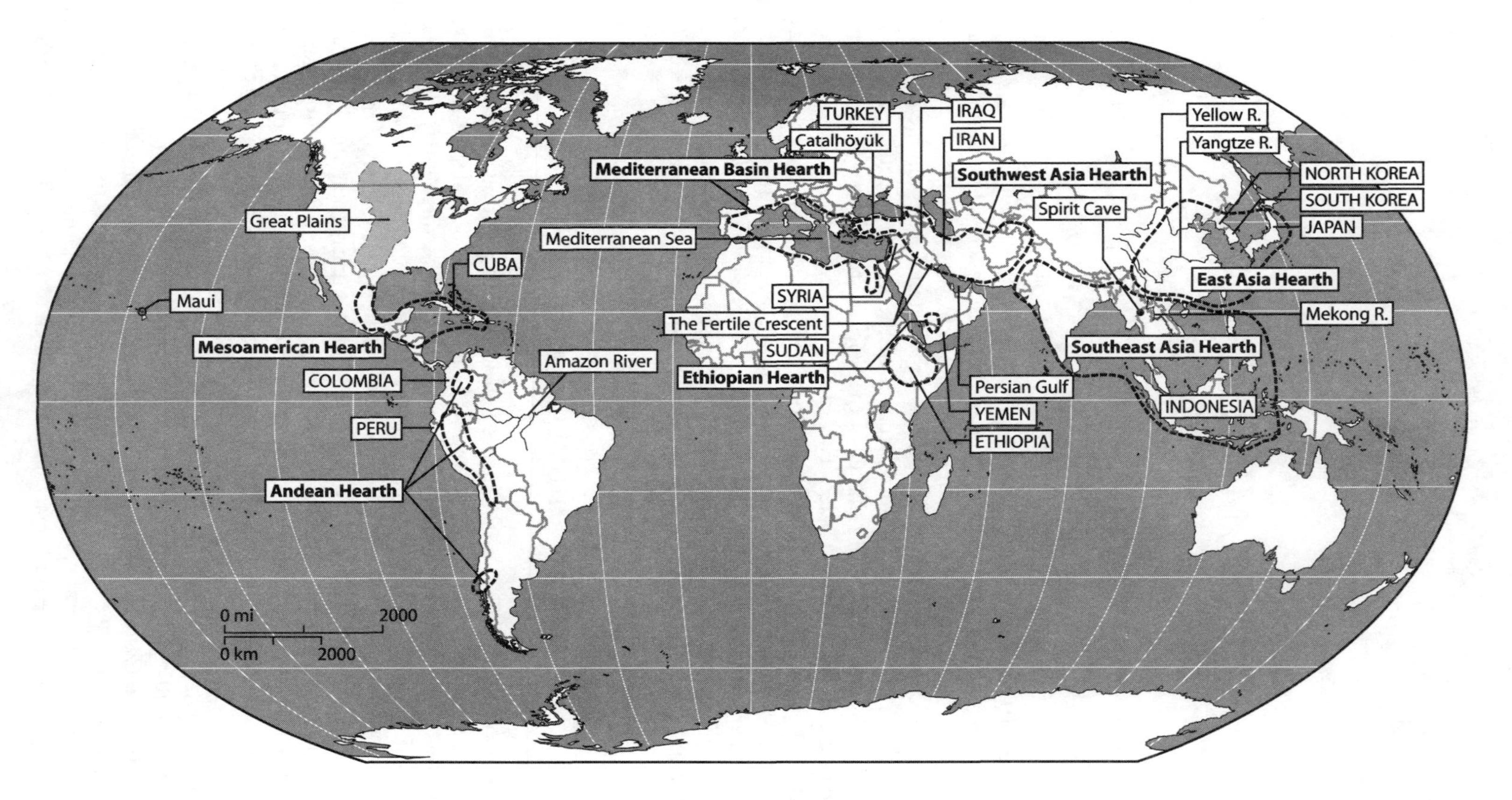

TURKEY
Çatalhöyük
IRAQ
IRAN
Yellow R.
Yangtze R.
Mediterranean Basin Hearth
Southwest Asia Hearth
NORTH KOREA
SOUTH KOREA
Spirit Cave
JAPAN
Great Plains
Mediterranean Sea
CUBA
East Asia Hearth
Maui
SYRIA
Mekong R.
The Fertile Crescent
Mesoamerican Hearth
SUDAN
Southeast Asia Hearth
Amazon River
COLOMBIA
Ethiopian Hearth
Persian Gulf
INDONESIA
YEMEN
PERU
ETHIOPIA
Andean Hearth
0 mi 2000
0 km 2000

is widely scattered, but it is ruinous to a farmer. Domestication has therefore tended to make these grasses nonshattering. Corn is a particularly dramatic case. Domesticated in Mexico about 5,000 or 6,000 years ago from a grass called teosinte, the domestic plant holds its kernels so tightly within the husk that they cannot escape at all. Peas in their pods are another illustration of an entombed seed, very different from a wild progenitor that sprang open when ripe.

Less visibly still, the seed coats of the grains have tended to become thinner, because while a thick coat protects seeds in the wild, farmers want seeds to sprout quickly and shade out the competition. A thinner coat implies a larger grain, too, which means more food for the farmer.

One more change for the grains is that their ancestors matured serially, requiring repeated harvests. Domesticated grains, on the other hand, tend to concentrate their seed in a single, compact mass. Consider sorghum, which as a domesticate has only a single flower. Corn has only a few.

HEARTHS OF DOMESTICATION AND AGRICULTURE

So much for the "what" of domestication; I want to consider now the where, how, and why of it. These questions will take us into increasingly speculative territory.

In the 19th century, Alphonse de Candolle, a Swiss botanist, culled written records to study the distribution of plants over the last several thousand years. His *The Origin of Cultivated Plants* (1884) is still of interest. In the 1920s and 1930s, however, the problem was attacked by a legendary Russian biologist, Nikolai Vavilov. Although he was the director of the All-Union Institute of Plant Industry, which should have kept him busy shuffling papers, Vavilov was a hyperenergetic field collector who traveled to over 50 countries searching for the hearths of plant domestication. In a series of publications beginning with *Centers of Origin of Cultivated Plants* (1926), he postulated that a domesticated plant likely originated where the greatest range of varieties of the plant is found. He writes, for example, of Soviet scientists finding not just 1 species of white potato in the Andes but 18. There, he surmised, the potato had been domesticated.

Sixty years after Vavilov starved to death in a Soviet prison camp in 1943—he refused to recant his belief in Darwin's theory of natural selection—Vavilov's seven hearths still form the basis for discussions of the geography of plant domestication. The seven are Southeast Asia (including India and Indonesia), China (including Korea and Japan), Southwest Asia (including Turkey, Iran, and Afghanistan), the Mediterranean Basin, Ethiopia (with Yemen), Mesoamerica (with Cuba), and the Andes, including Colombia and Peru.

These are generally warm and mountainous places, presumably because such environments stimulate mutation. Each appears to have given us several crops. A sample would include rice, soybeans, rhubarb, apricots, and perhaps tea from southern China; sugarcane, coconuts, nutmeg, and clove from the East Indies; eggplants, mangoes, jute, and indigo from India; wheat, onions, turnips, apples, figs, melons, and alfalfa from Southwest Asia; olives, dates, lettuce, and sugarbeets from the Mediterranean Basin; coffee and okra from Ethiopia; corn and avocados from Mesoamerica; and "Irish" potatoes and strawberries from the Andes.

A few plants originated outside these hearths. The sunflower is native to the American Great Plains and was domesticated there or near there, even though it was later cultivated primarily in the forested east. Farther south, peanuts, pineapples, cashews, and tobacco come from coastal Brazil. Manioc and sweet potato seem to have come from Amazonia. Some crops, meanwhile, cannot be pinned down to a single hearth. Citrus and bananas seem to have been domesticated independently in both China and Southeast Asia, while beans were domesticated independently in both the Andes and Mesoamerica.

Domesticated animals arose in only a few hearths, too. One stretches from India eastward to southern China: it is the home of the water buffalo and zebu cattle, as well as one home of the pig, dog, cat, and chicken. A second hearth is central Asia, from the Ukraine through to the Gobi: This is the home of Taurus cattle (the parent of the breeds raised today) and the yak, plus one home of goats, sheep, horses, and camels. A third is the Middle East, one home of pigs, cats, dogs, horses, goats, and sheep. The last and least important hearth is Mesoamerica and the Andes, from which come llamas, guinea pigs, and, most important to world agriculture today, turkeys.

Agriculture conceivably might have been invented only once, subsequently diffusing to these hearths of domestication. Since the 1950s, however, when Robert Braidwood worked in Iraq and Richard MacNeish worked in Mexico, archaeologists have looked closely for evidence of this transition, and it now appears that agriculture arose independently at least seven times. Specifically, it seems that agriculture arose three times in Asia (once in the Fertile Crescent, which arcs from Lebanon into Syria and Iraq, and twice in China), once in subsaharan Africa, twice in North America, and once in South America. Arranged chronologically, agriculture arose in the Fertile Crescent about 10,000 years ago, in Mexico about 9,000 years ago, in China about 8,000 years ago, in South America about 7,000 years ago, and in eastern North America and subsaharan Africa about 4,000 years ago.

The earliest agriculture in the oldest of these areas, the Fertile Crescent, was based on archaic domesticated wheats (emmer and einkorn, respectively diploid and tetraploid forms in the genus *Triticum*, which includes the hexaploid bread or modern wheats) and barley. All these crops evolved in present-day Syria. Proof comes from archaeological sites such as Abu Hureyra, where domesticated emmer and einkorn were grown 9,500 years ago. (A very important site, Abu Hureyra was excavated in the early 1970s shortly before it was flooded by a reservoir on the Euphrates.) It was there or nearby that the use of sickles probably triggered the rapid transformation of wild grasses into these grains. The plants then diffused eastward to the foothills of the Zagros Mountains of Iran. There, on the eastern side of the Fertile Crescent, they were taken up by herders who had recently domesticated goats. Instead of spending their time killing animals, in other words, they began protecting them, keeping them alive by seasonally migrating up and down the Zagros in search of grass. These transhumants—they're sometimes called vertical nomads—were the ones who pioneered mixed farming, the hugely important integration of crops and animals in which animals feed off stubbles and in turn fertilize the fields.

A lot of work in recent years has gone into untangling the process of animal domestication in the Middle East. One line of inquiry focuses on the ratio of male to female animals. Find a shift away from balance, the argument goes, and you're probably dealing with domesticated animals because then, as now, people probably slaughtered all but a few superior males. Which species was domesticated first? Excluding pets, domestication seems to have happened first with pigs in southeastern Turkey. For the moment, the evidence suggests that pigs were followed by goats, sheep, and cattle.

Mexican agriculture meanwhile arose around the famous trio of corn, beans, and squash. Bruce Smith speculates in *The Emergence of Agriculture* (1998) that the transition may have occurred near Guadalajara or farther south, in the Balsas River Basin.

Agriculture along the Yangtze arose around rice, whose wild ancestors remain unknown today because the area is so intensively cultivated. Smith argues that the process may have begun when foragers extended the ancestral plant's range by seeding it in diked and flooded areas, then harvesting it and repeating the process annually. Farther north, in the colder areas along the Yellow River, the core domesticate was millet, especially foxtail millet.

Agriculture in South America included not only lowland crops such as corn, beans, and squash but highland specialties, chiefly the "Irish" potato. (I keep inserting the quotation marks to emphasize that although these white potatoes became a staple in Ireland and are commonly called Irish potatoes, they are in fact South American.) Here the llama was bred from the guanaco and the smaller alpaca from the vicuna. The shift to domesticated animals in South America is indicated archaeologically by a declining percentage of deer bones. Why weren't deer domesticated? That question brings us back to the placidity required of domesticated animals.

Farmers in eastern North America relied on sunflowers for thousands of years before corn came their way. In fact, they treated corn as a secondary crop for a thousand years until a short-season variety evolved for higher latitudes. They also appear to have independently domesticated green and yellow squashes, though orange ones from Mexico eventually reached them.

Farmers in subsaharan Africa relied heavily on sorghum, which they domesticated and which remains a staple today across the grassland belt that separates the Sahara from Africa's tropical forests. It's a versatile plant. In the Sudan, for example, thin sorghum pancakes—*kisra* in Arabic—are a staple, dipped in a spicy stew called *mulah*. The Sudanese frequently eat *asida*, too, a stiff sorghum mush. They drink *abri*, made by soaking *kisra* in water. They make *marisa,* which ranges from a thin, barely fermented sorghum porridge to a dark beer, and they distill *marisa* into *araki*, a high-proof liquor.

METHOD

Plant breeding can be a very deliberate process, and today it usually is. Just consider the case of the gold pineapple that dominates the fresh pineapple market in the United States today. The gold started out in the 1970s as #73-114 at the Pineapple Research Institute on Maui. Del Monte Fresh established, or at least claimed, exclusive right to it. National marketing began in 1996, and today Del Monte Fresh has 70% of the world's billion-dollar fresh pineapple market. There are lots of similar cases. The Swiss company Syngenta, for example, has produced the tiny and ultrasweet pure-heart watermelon and the almost black but very sweet and juicy tomato it calls the kumato.

The earliest and most important transformations, however, were probably unintended. Consider the major changes that occurred to the grains. Seed retention and concentration probably occurred quickly—within a few centuries or less—once collectors began harvesting with sickles. There's no mystery about it: With a sickle, scatter-prone grains are lost, while plants with compactly bunched and retained seeds are collected and saved for the next season. Similarly, the tendency toward thinner seed coats probably emerged when natural selection acting on the seeds saved and planted by the early farmers: Thinner-coated seeds

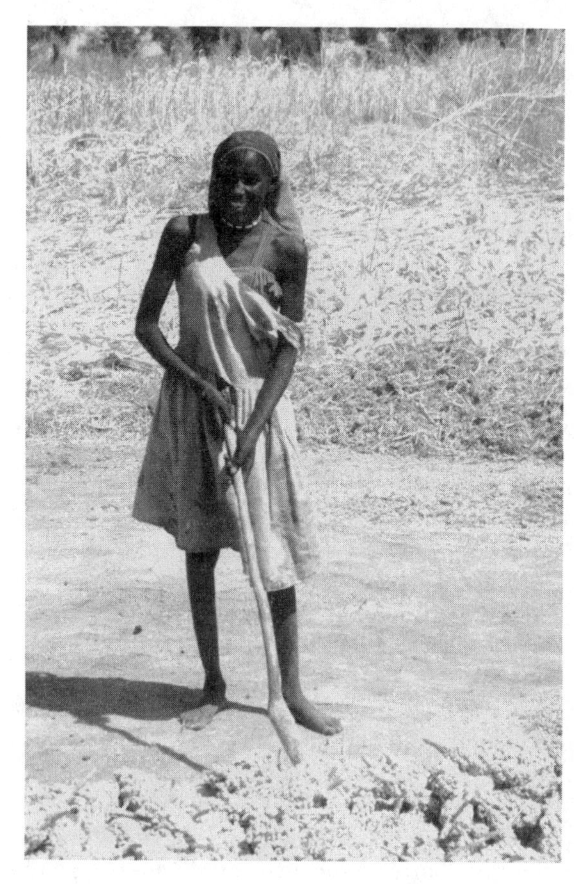

Sorghum, or *dura* in Arabic, is the traditional staple of the grasslands fringing the south side of the Sahara. Many traditional varieties are grown. Some are irrigated. Others—as in this case, near Geneina, in the western province of Sudan known as Darfur—are rainfed. Here the crop is manually threshed.

had a competitive advantage in germinating and growing more quickly than their thick-coated neighbors.

MOTIVATION

Strikingly, all the hearths were ecologically rich. This is an important point, because it conflicts with the so-called stress models that we almost instinctively turn to when we try to explain the rise of agriculture. Stress models rest on the deceptively attractive assumption that agriculture arose because people were hungry. The grandaddy among this family of explanations was proposed by the archaeologist V. Gordon Childe, whose much-reprinted *New Light on the Most Ancient East* (1928) proposed that hunters turned to farming when the postglacial climate became drier, forcing them to adapt or die. (In 1925, by the way, Childe coined the term "Neolithic Revolution." The phrase emphasizes the importance of the change from the paleolithic world of hunters to a New Stone Age of farmers.)

Certainly the transition was hugely important, but archaeologists to this day have found no evidence for the desiccation that Childe's theory demands. They've looked for alternatives. Kent Flannery proposed in 1973 that agriculture arose because forager populations grew until those foragers had to move to areas where the grasses they relied upon were scarce. It's plausible, but the earliest agricultural sites occur within the core areas of those grasses, not peripheral ones. There are more flamboyant theories, too, such as a rise in sea level crowding foragers at the head of the Persian Gulf onto higher ground, where they had to farm to survive.

We like stress models: After all, we're comfortable with the idea that necessity is the mother of invention. But what shall we do if stress models lack archaeological support? A good place to begin the search for other explanations is with the realization that the knowledge of seed reproduction is much, much older than agriculture itself. Paleolithic peoples had a deep knowledge of plants, and there are many stories of botanists being amazed at the precision with which foragers classify plants and understand their habits. There are plenty of nonagricultural peoples, for that matter, who came very close to farming. The Aborigines dug wild yams with such thoroughness that early European visitors thought the Aborigines were harvesting fields of potatoes. The Paiutes (the 19th century's Diggers) of Nevada not only harvested wild grasses but planted them in ground they had burned; they then diverted irrigation streams onto the fields so that the grasses would grow better.

But if agriculture wasn't the result of environmental necessity, and if the reproductive facts on which it rests were common knowledge among foragers, then the "why" question behind the Neolithic Revolution becomes still more perplexing. It's a question that was implicitly raised in Carl Sauer's *Agricultural Origins and Dispersals* (1952). Sauer proposed that the first farmers lived comparatively comfortable lives in permanent settlements, most likely along tropical coasts like those of Southeast Asia. For a time it seemed that the archaeologists would back him up. In 1972, for example, excavations at Spirit Cave in Northwest Thailand led one anthropologist to write that he was now convinced "that somewhere among the forest-clad mountains of the region man's first tentative efforts to exploit wild plants and animals opened the way first to horticulture and then to full-scale agriculture and animal husbandry." Since then, the seeds that were found at Spirit Cave have been evaluated as probably wild, and a recent review concludes that Spirit Cave's "early plant remains . . . are too few, and so tentatively identified, that a more cautious interpretation is necessary." Sauer's more general point, however, about the affluence of the first farmers has slowly won widespread acceptance. Archaeologists recognize that foragers in the Middle East, for example, collected such substantial quantities of wild emmer and einkorn that they could live in permanent settlements.

The plants that these and other pioneer farmers domesticated—emmer, rice, sorghum, corn, potatoes—had been collected for a long time. Nor did the development of domesticated plants mean that early farmers stopped collecting the wild plants. In Syria, for example, wild rye was collected long after domesticated rye began to be grown.

What kind of explanation can we offer for the development of agriculture under such circumstances? One approach is that of Peter Bogucki, an archaeologist I mentioned in Chapter 3 in connection with the links between sedentary life, food storage, and the decline of egalitarian band structures. He suggests that farming was adopted when it made sense, namely, when sedentary life and storage facilities allowed families to prosper through innovation and hard work. Until then, by implication, it did not make sense—and so was not adopted.

There's a lot of conceptual elegance in this approach, and it certainly doesn't strain the imagination. Maybe it's true, too: After all, we're squarely in the realm of speculative archaeology now. But there are other possibilities. An intriguing one labors under the ponderous name of postprocessual anthropology. It is based not in economics but in nonmaterial culture. People might have turned to agriculture, for example, because they needed mountains of food for ritual feasting, or for dowries.

Another postprocessual explanation is that the rise of agriculture is linked to a revolution in religion. Consider Çatalhöyük, a famous archaeological site near Konya, in south central Turkey. Ten thousand years ago, Çatalhöyük was an immensely crowded village of perhaps 10,000 people, with 40 houses to the acre. It was so congested that there was no room for streets and doorways. Instead, homes were entered by stairs descending into roofs.

Çatalhöyük's wall murals show a mother goddess and bull god. These are very different from the subjects chosen for representation in paleolithic societies, where game animals such as gazelles dominate. For that matter, they're very different from the sheep, goats, and pigs on which the people of Çatalhöyük actually depended—but which they almost never portrayed in art. Ian Hodder, who has worked at Çatalhöyük, suggests that agriculture was conceived as a way to demonstrate or exercise human domination of nature. This idea has been linked to the goddess and bulls through the work of Jacques Cauvin, who argues that a cultural revolution preceded the shift to farming. In a book significantly titled *The Birth of the Gods and the Origins of Agriculture* (2000), Cauvin correlates the beginnings of agriculture with what he calls a "revolution of symbols." He's thinking again of the mother goddess and bulls. He's thinking, also, of art apparently showing human figures in positions of supplication. These images, he writes, introduce "an entirely new relationship of subordination between god and man." It's only a small step from here, he continues, to humanity "striving towards this perfect, transcendent being." Cauvin notes that it was only in the Neolithic that houses began to be made square rather than round. Although there are technical reasons for this, he also ascribes symbolic value to the change, with the circle as the transcendent sun and the square a product of human design. Thinking both of agriculture and housing, Cauvin writes that "man could not completely transform the way he exploited his natural environment, his own settlements as much as his means of subsistence, without showing at the same time a different conception of the world and of himself in that world." That conception was one of self-confident assertion, shaping the world.

Farfetched? We think of swinging as child's play, but it can be traced to old religious practices where swinging was done very seriously as a way of sympathetically encouraging crops to rise from the ground. It's one of many kinds of farming magic. In Laos, canoes traditionally opened a path through the water of the Mekong so mythical serpents could come forth and bring rain to the rice paddies. Or think of kissing under the mistletoe, a plant considered sacred in many societies because it's green in the winter, the season of death. So? So think of people deeply fearful of endless winter and looking for ways to encourage the return of spring. That's why there are so many rituals aimed at ensuring seed germination. Such ritual was apparently at the core of the Eleusinian Mysteries of ancient Greece, but you can find similar practices in the midwinter rituals of the Hopi, in the American Southwest. Shall we dismiss such ritual as ornamentation of the hard economic facts of agriculture? Or is the agriculture of Kansas and California today merely a vulgar, secularized variant of ancient religious practice? The beauty of speculative archaeology is that whichever side you take, nobody can prove you wrong.

CHAPTER 5

THE DIFFUSION AND EARLY DEVELOPMENT OF AGRICULTURE

Domesticated plants are incredible travelers, as footloose as their creators. Wheat originated in the Middle East, for example, but the biggest producers today are China, India, and the United States. Corn originated in Mexico, but Mexico's crop is only a third the size of Brazil's, a sixth the size of China's, and a tenth the size of America's. Potatoes began in the Andes, but the leading producers today are China, Russia, and Poland. Rice is unusual, because about a third of the world's crop still comes from China, where rice originated.

It's precisely the mobility of domesticates that has made us so dependent on just a few of them, but diffusion isn't restricted to those few staples on which our survival depends. Sugarcane originated in Southeast Asia, but the leading producers today are India and Brazil. The leading coffee producer is Brazil, not Ethiopia. The leading cocoa producer is Ghana, not Mexico. The leading peanut and tobacco producer is China, not Brazil.

You can read diffusion stories from language, too. The biggest apple producers in the world today are China (overwhelmingly), followed by the United States, France, and Italy. Hunt down the origin of apples, however, and you'll wind up in the Tian Shan, the "heavenly mountains" of Kazakhstan. The capital of that country is Almaty (in Soviet times, Alma Ata); either way, the name means father of apples. (A sad story: the apple forest stretching into the mountains south of Almaty has been reduced from 125,000 acres in the 1940s to about 10,000 today. The reduction worries apple breeders around the world, who hope to find trees here to broaden the narrow, disease-prone gene pool of commercial varieties, almost all of which come from two parent trees.) Another example: if you're a gardener or foodie you know about Damsons, or Damson plums, but you probably don't know that the word Damson comes from Damascene, which is to say from Damascus. The English word "orange," to take one more example, comes from the Persian *narang*, which is Persian for the color orange. The Persian (and Arabic) name for the fruit, on the other hand, is *portucal*, because it was Portuguese traders who first brought oranges from the Orient to the Middle East. On your own you can figure out the staging point for tangerines as they came to Europe. If you're unsure, just find the biggest town on the African side of Gibraltar.

Or, consider the Romans and barley. The Roman staple was emmer wheat, which they called *far*. It's the root of the English word "farina." North of the Alps, the "f" became a "b," so *far* became *bar*. From this change, it's no great distance to our word "barn," the

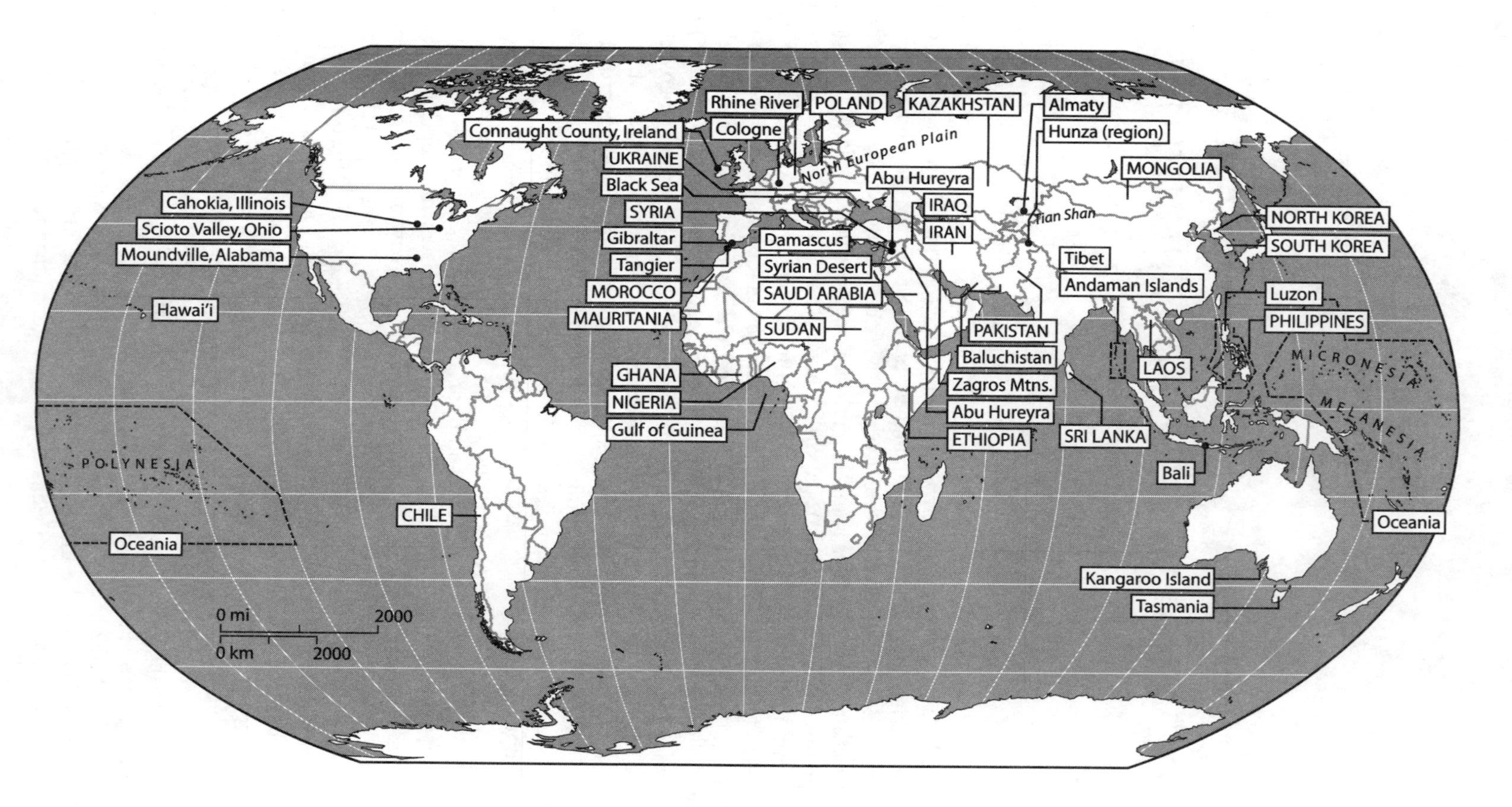

Rhine River
POLAND
KAZAKHSTAN
Almaty
Connaught County, Ireland
Cologne
Hunza (region)
North European Plain
UKRAINE
MONGOLIA
Abu Hureyra
Black Sea
Cahokia, Illinois
IRAQ
SYRIA
Tian Shan
NORTH KOREA
Scioto Valley, Ohio
IRAN
Gibraltar
Damascus
SOUTH KOREA
Moundville, Alabama
Tibet
Tangier
Syrian Desert
Andaman Islands
MOROCCO
SAUDI ARABIA
Luzon
Hawai'i
PHILIPPINES
MAURITANIA
SUDAN
PAKISTAN
Baluchistan
MICRONESIA
GHANA
LAOS
Zagros Mtns.
NIGERIA
MELANESIA
Abu Hureyra
Gulf of Guinea
ETHIOPIA
SRI LANKA
POLYNESIA
Bali
CHILE
Oceania
Oceania
Kangaroo Island
0 mi 2000
0 km 2000
Tasmania

building used to store the *bar*. The only change was the grain itself. Emmer didn't do well north of the Alps, and so another Middle Eastern domesticate was substituted. Naturally, it was the grain we call barley. The Romans were also responsible for the first German planting of peaches, which came from China. Radishes, on the other hand, were a staple of the Egyptians who labored on the pyramids, but radishes did not come to Europe until the 16th century.

Livestock moved too. The beef and dairy breeds we know today come from Central Asian ancestors that were brought west, while the fierce native European cattle, the aurochs, were driven to extinction. The turkey, meanwhile, is an American bird, but in the confusion surrounding its introduction to Europe the idea took hold that it had come from the mysterious East, which is why we call it the turkey. The Turks knew it wasn't Turkish, but they didn't know what it was, so they called it *hindi,* "Indian." The French followed the Turks, perhaps because they mistakenly assumed that the Turks knew what they were talking about. Then there's the South American guinea pig, mistakenly named as though it came from West Africa.

DIFFUSION

We've already touched on how grain from the western side of the Fertile Crescent moved to the eastern side and was incorporated into mixed farming. Not long afterward, those same plants and animals began moving to Europe, where agriculture arrived in a wave that swept across the continent between about 9000 and 4000 B.C. Perhaps "swept" is too dramatic a word, because the new technology actually diffused at an average rate of about 1 kilometer a year. It started in western Anatolia and moved westerly and northerly across the continent, mostly along the coasts and major rivers. Farmers soon discovered, however, that Europe's best farmland isn't in those places. Instead, it's a belt of wind-blown silt called loess. This material was winnowed from glacial debris and blown south, where it settled like a blanket along the line separating the North European Plain from the hills to its south. Stretching from near Cologne eastward through southwestern Poland to the southern Ukraine, the loess belt was discovered about 6,000 years ago by a surprisingly homogeneous group of peoples known as the LBK farmers. (The name comes from *Linearband-keramik,* the German name for the distinctive, line-banded pottery they produced.) The LBK farmers relied primarily on emmer and lived in longhouses scattered across this broad domain of easily worked soil. (Loess isn't unique to Europe, by the way: It's common in periglacial environments, including North China and the American Midwest.)

One late case of diffusion: Recall that the settlement of Oceania depended not only on seafaring canoes but on the possession of crop plants that could be packed along. The end of the line was Hawaii, occupied about 500 A.D.

Plants and animals weren't the only things that moved. So did the people who were carrying the planting technology and intermarrying along the way with local people. This insight has been developed by Luca Cavalli-Sforza, a Stanford geneticist and coauthor of *The Great Human Diasporas* (1996). Cavalli-Sforza has taken blood samples, then done principle-components analysis of genetic variation in those samples. He concludes that the Basques, with a remarkably high incidence of Rh negative genes, are a remnant of the preagricultural Europeans.

Other workers have taken up this line of research. A group of researchers at Trinity College, Dublin, reported in *Nature* in 2000 that almost all the men in western Ireland's

Connaught County are, like the Basques, descended from Ireland's preagricultural peoples. In other words, they possess a chromosome marker almost never found among Turks but which rises to about 63% in England, 78% in Ireland, and 98% in Connaught. The authors speculate that the Irish with this marker are descended from the wave of pioneer settlers who reoccupied Europe after the last Ice Age. It's a wave that may have begun in Spain, which was a refuge during that period for many plants and animals, as well as people.

Agriculture swept eastward, too, from the Middle East to India, and genetic studies are beginning to untangle the story. The residents of the Andaman Islands, for example, carry a Y-chromosome mutation, Marker 174, which is found only in remote parts of Asia, such as Japan and Tibet. The locations of the peoples with this mutation suggest that they belong to the preagricultural Asian population, forced into refuges by agriculturalists advancing across the continent about 6000 B.C.

Language has been used as another tool to study agricultural diffusion, but these are treacherous waters. South Indian languages, for example, are Dravidian, not Indo-European. That's because India's Dravidian population was pushed south by an advancing tide of Indo-Europeans, the famous Aryans. Gordon Childe in 1926 hypothesized that the Aryans began as horse nomads in central Asia. According to him, they invaded Europe, too, which is why English, like so many other European languages, has many similarities to Hindi, the dominant language of North India. Cavalli-Sforza's studies reinforce this interpretation by suggesting that the migration of Indo-Europeans into Europe from the steppes north of the Black Sea occurred about 5,000 years ago.

Other researchers, however, believe that the parent language of all the Indo-European languages was spoken by early farmers in Anatolia, who subsequently migrated east and west. By the same logic, it has been argued that the Semitic languages arose among Levantine farmers. Similarities between the Dravidian languages and Elamite meanwhile support the theory that Mesopotamian farmers slowly diffused east to India. The whole Mesopotamian–Indian link remains open, however, and one recent discussion concludes by saying simply that "our present knowledge does not permit any definitive conclusions." The same can be said for the even more daring speculation that the Indo-European, Semitic, and Dravidian languages all belong to a hypothesized Nostratic superfamily whose original language was spoken by Middle Eastern farmers before their great migrations.

Joseph Greenberg has suggested that agriculture moved with the Bantu peoples from an origin in modern Nigeria south and east to cover all the farming areas of the continent, except those with a summer-dry, or Mediterranean, climate. (A Mediterranean climate is also found not only around the Mediterranean but in South Africa and parts of Chile, California, and Western Australia. All these places lie on the poleward side of west-coast deserts, which in summer extend to higher latitudes. The Mediterranean climate is a difficult environment for farmers, who have adapted to it by using irrigation where possible, by growing deep-rooted crops such as grapes, oranges, and figs, and by planting wheat in the relatively mild winter and harvesting it before the arrival of the summer drought.) The Pygmies may have been overrun by this advance, after which they worked out the symbiotic relationship they have today with the Bantu.

Jared Diamond and Peter Bellwood, writing in *Science* in 2003, turn such considerations upside down: instead of looking at language as a way to understand agricultural diffusion, they look at agricultural diffusion as a way to understand the pattern of languages. They suggest, in other words, that the world's major language families originated in agricultural hearths. Echoing Greenberg, they suggest that the 1,400 or so languages of the Bantu family evolved as yam cultivators began diffusing east and south from West Africa. Japanese, they suggest, may have replaced the languages of Japanese foragers when rice cultivators came from Korea. (Japanese isn't similar to Korean, but they suggest that the Koreans who came to Japan were part of the Koguryo kingdom, which was defeated in Korea by the Silla kings, whose language became modern Korean). The Austronesian family of languages, they suggest, originated among Chinese rice cultivators who introduced rice to Taiwan and whose descendants settled the Philippines and Polynesia. The Indo-European languages, they say, may have originated with the diffusion of Middle Eastern farmers, but they hesitate and recognize that it may have originated, after all, with Childe's famous horse nomads of the steppe.

ADAPTATIONS

Farmers without animals or some other fertilizer find that yields decline after a few years. To maintain yields, early farmers shifted to new fields, which in most cases were actually old fields that had been fallow for a generation or more. Often labeled pejoratively as slash-and-burn agriculture, shifting cultivation is still common in the tropics. The farmer burns a patch of secondary forest or grassland and uses a hoe or digging stick to plant several crops in the resulting ash. They're fertilized by that ash and yield good harvests for several years. The actual practice of shifting cultivation, however, is much more complicated than these summary sentences suggest. If you want an intimidatingly detailed picture, take a look at William Allan's *The African Husbandman,* (1965), which is a study of Zambia, and Marvin P. Miracle's *Agriculture in the Congo Basin* (1967).

Provided that the fallow period is long enough to yield a thick layer of ash, shifting cultivation can be sustainable, and prehistoric farmers probably managed the system well. In the last century or so, however, population growth and the expansion of modern economies have reduced the habitat available to shifting cultivators. Crowded, they are obliged to plant steep slopes or insufficiently rested fields. The system collapses, confirming the negative stereotype.

Shifting cultivators found several ways to increase productivity. One, which we touched on with the folding of goats onto fields of barley stubble in the foothills of the Zagros, was to combine farm animals with crop husbandry. With large enough herds farmers could cultivate a field perhaps 1 year in 3.

With the domestication of cattle came the plow, another technology of intensification. Originating about 5,000 years ago in the Middle East, plows spread east and west across Europe, North Africa, and Asia. Strikingly, the plow was not used either in subsaharan Africa or the New World before the arrival of Europeans. Farmers in these places relied on backbreaking hoes or mattocks.

At first, the plow was a relatively simple tool that did no more than scratch the soil as an ox pulled a crooked and pointed stick through the earth. Such plows are still used in poorer parts of Asia; a man can carry one on his back. In the drier parts of India, camels still pull the same kind of plow. By Roman times, however, wheeled plows had been developed with soil-inverting moldboards. Such plows, pulled by gangs of oxen, could break northern Europe's heavy soils.

With manure and plows, a whole new farming economy emerged. In medieval England, each village's cultivated lands were commonly divided into three large fields. Each was subdivided into thin strips cultivated by individual villagers, and each field was cultivated approximately 1 year in 3. Although each peasant had his own strips in each of the fields,

In Haryana, the state immediately west of Delhi, a farmer works hard to position his plow.

each field was planted homogeneously. It was not a system that welcomed innovation, but it was stable. It survived until fallows were bypassed with the introduction of soil-enriching legumes and imported or manufactured fertilizer.

The Chinese added the use of human waste—the standard euphemism is nightsoil. Even today, as China modernizes as an extraordinary rate, men collect buckets of soupy human excreta from public latrines and carry it to the fields.

Irrigation is a third way to intensify agriculture. It can be a complex affair. Until replaced in the 19th century by perennial or year-round irrigation, much of the Nile flood in August and September, for example, was drawn away from the river in short canals that were too high to capture water the rest of the year. The water was spread to a depth of about 40 inches in about 200 earth-rimmed basins averaging 8,000 acres. The water saturated the soil and deposited fertilizing silt. Forty days later, the Nile would normally have subsided, and the remaining floodwater could be drained from the basins back to the river, which would carry it down to the delta, where it would be put to work again. Crops were grown on the saturated soil, which without further irrigation sustained one crop per year. (The classic discussion of this subject is by William Willcocks in *Egyptian Irrigation* [1899].)

Irrigation can also take much simpler forms that don't require engineers and the disciplinary apparatus of government. The most spectacular examples of these village-based systems occur in the steeply terraced hills of Asia, for example, in the Hunza country of northern Pakistan, in Sri Lanka's central upland, in the mountains of northern Luzon, and on the volcanic slopes of Bali. Often photographed for their beauty, these irrigation systems are so intricate that they are hard to describe or illustrate with precision. See, however, Harold Conklin's unbelievably meticulous description of the paddy lands of northern Luzon, published as the oversized *Ethnographic Atlas of Ifugao* (1980). Water trickles through these terraces the way bearings roll through a wire-maze toy, top to bottom but with countless about-faces, steps, and faster and slower stretches.

Farmers also developed a wide range of simple machines to help them lift water. The most arduous was probably the two-man swing basket (*natali* in Egypt, *boka* in parts of India). At each swing, it raises a few gallons of water a backbreaking 3 feet. A more efficient device is the Arabic *shaduf* or Hindi *picota,* which consists of a beam on a post. There's a bucket on one end and a weight on the other; the operator raises the water by pulling down on the weighted end of the beam or by walking back and forth atop the beam. Another single-person device is the *don* of eastern India; it looks like an oar-sized boat suspended from a tripod. A similar device, resembling a spoon, is used in China. There's the Indian *mhot*, too, which relies on animal power to hoist a large leather bucket as much as 30 feet, as bullocks or a camel march down an artificial slope at the side of a well. At the base, the water is dumped into a ditch, and the animal walks backward up the slope to start again. Still another animal-powered device is the Persian wheel, or *saqiya*. A bullock is harnessed to a large horizontal gear that meshes with a vertical cage wheel, around which a chain of buckets is draped and hangs into a pool from which water is raised perhaps 30 feet to an irrigation channel. Unlike the other devices, the Persian Wheel can run for hours with an unsupervised animal walking in circles. Like the Islamic call to prayer, the creaking sound of the Persian Wheel rises over many dawn mists.

The most sophisticated of these simple irrigation technologies is the *qanat*, which consists of a shaft dug at the foot of a mountain through the alluvial mantle as far as bedrock, where groundwater may be encountered. A tunnel is then dug at a low gradient out for miles from the mountain base until it emerges at the surface, bringing for use at this artificial oasis the water that would otherwise run along the bedrock to depths impossible to reach

The open fields of medieval Europe are a thing of the past, but traces of them remain, as in these narrow strips in a field in southeastern Poland.

Karimabad, in northern Pakistan, is apricot country, but in February the trees aren't yet in leaf and so the massive terraces on which they grow are naked.

without modern technology. There are thousands of *qanats* from Morocco to western China. (The name I'm using here is Farsi. Outside Iran, these structures have different names, such as *karez* in Baluchistan and *foggara* in Mauretania.) Many are still in use, maintained by families that pass on the needed skills of surveying and excavating. On the ground, they're recognizable by the line of access shafts dug to remove excavated tunnel material. (For more on *qanats*, see Anthony Smith's *Blind White Fish in Persia* (1953); its odd title refers to Smith's fruitless quest for a fish said to inhabit *qanats*.)

Another famous agricultural adaptation is pastoral nomadism, the endless desert-fringe quest for grass. For lack of appropriate animals, pastoral nomadism never existed in the New World or subsaharan Africa, but it was widespread across Eurasia and North Africa, where people lived with herds or flocks of goats, sheep, cattle, camels, horses, or reindeer. Where and when did nomadism originate? One line of inquiry leads to Mongolia about 1000 B.C., when a colder climate forced farmers onto horseback. The newly nomadic Mongols traveled widely and adopted many aspects of the cultures they disrupted; their descendants may include the historic Scythians and Huns. There's not much left to indicate their homeland except hundreds of stone monoliths—deer stones—that still exist in northern Mongolia and southern Siberia. Found near ancient human and horse burial mounds, the columns show chiefs whose bodies are apparently tattooed with images of deer. Another line of inquiry suggests an origin in the Syrian Desert around

7000 B.C. These people appear, however, to have been seminomads, staying put while a grain crop grew each year.

Why these Syrians turned to nomadism is unclear, because they appear to have left more humid, coastal areas to take up this difficult life. Jacques Cauvin, mentioned at the end of Chapter 4 in connection with the possible role of religion in the development of agriculture, has suggested that mobility may have had an intrinsic or cultural appeal for these people. He writes: "poorer in terms of material goods than the villager or city-dweller whose enervating comforts he scorns, the nomad values his freedom of movement and prides himself greatly on his mastery of space, eminently positive values that should be understood as the matters of choice." However you explain it, pastoral nomadism comes after agriculture, not before it. The Bible has the sequence correct in its story of Abel and Cain. Cain becomes a nomad, in other words, only after he kills his brother: "When thou tillest the ground, it shall not henceforth yield unto thee her strength; a fugitive and a vagabond shalt thou be in the earth."

Despite their aura of independence, pastoral nomads have always depended on agriculture, whether they grew crops, bought them, or stole them. They haven't even been drifters in the unrestricted sense in which we often use the word nomad. Instead, they always have a turf and follow a regular annual route through it. The grasslands of Mongolia, for example, though unfenced and seemingly open to the horizon, were for centuries divided and subdivided into grazing grounds known as *hoshou* and *sum* (banner and arrow in Chinese), within which certain families had pasture rights. It was a complex arrangement but simple com-

A labor-saving device, at least for the man. Once going, the bullock may be left alone to circle for hours. It's dawn on the Ganges Plain.

Camels around a well in a dry streambed, Darfur, Sudan.

pared to the Sudan, where camel nomads at least until the 1960s followed annual loops that overlapped another set of loops followed by cattle nomads. The camel loops were north of the cattle ones and extended into desert fatal to cattle. Movement through the two loops was synchronized, so that in winter the camels returned south to ground vacated by the summer-resident cattle. It was a kind of rural time-share, with each group having occupancy rights at different times of the year. (For more on Mongolia, see David Sneath's *Changing Inner Mongolia* [2000]; for Sudan, see J. H. G. Lebon's *Land-Use in Sudan* [1965].)

Like shifting cultivation, pastoral nomadism can be stable, but in the last century it too has come under a lot of pressure, largely prompted by the fact that urban societies generally regard nomads as barbarians or, if mounted, as terrors. That's why Reza Shah, the father of the last Shah, brutally sedentarized the nomads of the Zagros. (For more on this, see the profusely illustrated *Nomadic Peoples of Iran*, edited by Richard Tapper [2002].) Stalin did about the same, starving 1 million Kazakh camel and sheep herders to death in the 1930s. A generation later, the Soviet Union forced the nomads of Mongolia to collectivize. In the 1990s the collectives were broken up. Some members chose to move out of town and become herders once again, but they had forgotten the old ways and no longer took their flocks on fattening autumn treks. When winter came, the sheep died and the would-be herders were reduced to accepting Red Cross handouts. Every now and then one hears about other nomads being forced into cities. A recent case is the million or so Kuchi of Afghanistan. Devastated by war and drought, they live in camps and are likely to settle in cities.

Arabia's Bedouin have fared poorly, too. One of the first tasks of the newly organized government of Saudi Arabia in the 1930s was to persuade Arabian nomads to become villagers. It did so by stressing the importance in Islam of regular attendance at a Friday or congregational mosque. Across greater Arabia, nomads today often live in houses, have gardens, work at city jobs, but have a flock of animals that one or more sons tends. Perhaps the animals follow the old grazing routines. More likely, the animals are trucked to seasonal pastures or simply kept near the family home, where they are fed purchased fodder. (For more on nomads in the Arabian peninsula, see *The Transformation of Nomadic Society in the Arab East*, edited by Marthy Mundy and Basil Musallam [2000].)

CONSEQUENCES

So much for diffusion and adaptation. It's time for my favorite question: so what?

One clear consequence of agriculture was the deterioration of diet. Foragers did not survive on endless bowls of cereal or endless loaves of bread. Farmers did—and in many parts of the world still do. A diet based on corn, for example, accounted for severe nutritional deficiencies among the poorer populations of the American South well into the 20th century. It's ironic, given our casual assumption that the adoption of agriculture is a good thing. Once again, however, the Bible has it right. Adam is expelled from Eden. No more

Fifty miles to the south of the camels in the preceding picture, a cattle herder drives his animals.

gathering food. He and Eve are told: "in the sweat of thy face shalt thou eat bread." Not a bad description of the traditional farmer's life.

Along with poorer diet came new kinds of physical stress. Theya Molleson has studied the skeletons found at Abu Hureyra, in Syria. She reports that the people were short—5′ 1″ for women, 5′ 4″ for men. The women's bones in particular reveal how hard their lives were, and why. Molleson writes of one woman whose "metatarsals of the big toes were severely deformed by the effects of gross degenerative disease." The woman's "injuries, especially to the back and pelvis, were such that she would have had difficulty walking without the aid of a staff." What was the cause of these problems? The simple answer is that the women of Abu Hureyra spent hours every day kneeling and rocking back and forth to mill grain into flour. Their toes were in this way permanently bent backward, while their tibias were enlarged. Clinically, Molleson writes that "individuals with degenerative joint disease and spinal injuries doubtless owed their condition to injuries incurred through operating a saddle quern."

I've saved for last the most important consequence. Unlike poor diet and physical stress, this consequence is still with us. At the end of Chapter 3, I mentioned the emergence, with sedentary life, of relatively rich and poor families. That's not necessarily the same thing as social inequality, even though we tend to equate the two. What's missing is the idea of hierarchy, of superior and inferior. Agriculture isn't essential to the appearance of hierarchy—the foraging societies of the Pacific Northwest had slaves and chiefs—but agriculture provides the economic base for most hierarchical societies. It proved an absolute prerequisite to the development of urban ones, which is perhaps why we assume that agriculture is a good thing.

The emergence of hierarchy—more broadly, the idea of cultural evolution—has periodically interested anthropologists since the time of Edward Tylor, the pioneering Oxford anthropologist. In an echo of Darwinian theory, anthropologists of his day tended to see societies evolving from forager savagery through agricultural barbarism to socially hierarchical civilization. Early in the 20th century, there was a reaction to this sequencing and its implication that modern peoples are better than earlier ones. You'll recognize in this reaction the ambivalence we've already encountered in our contradictory attitudes toward prehistoric peoples. Indeed, the leader of this reaction was Franz Boas, among whose students was Ruth Benedict, the author of *Patterns of Culture* (1934).

Since then, there has been another wave of interest in cultural evolution. One of its leaders was Elman Service, the author of *Primitive Social Organization* (1962). This book contains a simple typology captured in the sequence: band, tribe, chiefdom, and state. Service's band is egalitarian, with related families living in groups of 30–100 people and possessing only a modest degree of informal authority. The tribe is a larger organization and rare before agriculture; it lacks hierarchy but recognizes a charismatic leader. The chiefdom adds hierarchy, because it has a hereditary leader and ranks its members by their genealogical proximity to him. (Hawaii is the classic example, with a complex structure culminating in a paramount chief.) The state adds legal structures to enforce the chief's will, and it contains social classes, including an elite. Always found in states, agriculture is nearly universal in chiefdoms. Exceptions occur in very rich foraging environments, such as those of the Pacific Northwest.

Service's typology is beguilingly simple, and with the exception of the insertion of chiefdoms, it is hardly more than a euphemized equivalent of the older sequence of savagery and barbarism as civilization's precursors. Like those terms, Service's explain nothing. They merely lay out categories. Worse, they imply sharp divisions between categories that actually shade from one to another, which is why one anthropologist's tribe can be another's chiefdom.

Other typological terms have been introduced for these transitional states. Anthropologists working on the frontier between tribe and chiefdom speak mellifluously of transegalitarianism, for example. They point to Mesopotamia's Ubaid culture as an example—more of which very shortly. They also point to the Hopewell or Middle Woodland Culture that existed in Ohio's Scioto Valley. This culture left behind very large enclosures and mounds within whose tombs archaeologists have found grizzly teeth from Wyoming, alligator teeth from Florida, and copper from Lake Superior. The Hopewell people obviously had extensive trading contacts, and they farmed although they did not have corn. Yet there's no evidence of government. So, too, with the Hohokam and Anasazi peoples of the Southwest, who had elaborate irrigation systems, large buildings, and roads, but no evident hierarchy. For true chiefdoms, archaeologists will point to the Mississippian culture, supported by corn and leaving behind the ruins found today at Cahokia, Illinois, and Moundville, Alabama.

Prehistoric Europe presents analogous societies. Compare the farmers of late Neolithic Europe with those of the succeeding Bronze Age. Neolithic farmers used livestock to plow fields and pull carts: Both were technologies that produced widespread social inequalities. They also used copper, too soft for tools but fine for ornamentation of both the living and the dead. They left behind thousands of megaliths, massive stone constructions about whose symbolic purpose we know very little. The Bronze Age came to these people about 3400 B.C. It implied long-distance trading in metal, and it made possible the concentration of wealth. Megaliths were replaced by tumuli ("barrows" in English usage), which contained a single grave instead of many. Some of these burials were accompanied by treasure—breastplates and daggers with Irish gold and Baltic amber—and allowed powerful families to create a landscape memorializing their ancestors. In *Rethinking the Neolithic* (1991), Julian Thomas has argued that this change may correspond to a shift from a predominantly foraging economy to a predominantly farming one, with the economic unit switching from kin groups to households. He argues that the great stone monuments of the Neolithic—Stonehenge and the like—were built by people without any great economic surplus. He puts it this way: "the population of Neolithic Britain did not live in major timber-framed buildings, quite probably did not reside in the same place year-round, did not go out to labour in great walled fields of waving corn, were not smitten by over-population or soil decline, and much of their day-to-day food may have been provided by wild crops." What, then, does he make of the megaliths? In an echo of Jacques Cauvin's approach to agriculture, Thomas argues that these structures were "a means of imposing a certain conceptual scheme upon the world." He's thinking of an "interlocking series of binary oppositions: us/them, in/out, culture/nature, tame/wild."

We still haven't explained anything, of course—only described changes. I'll come back to an explanation in Chapter 6, although as you can anticipate I won't have hard answers. For now, I want only to say that agriculture proved to be prerequisite to the development of urban societies but did not automatically lead to it. One might imagine otherwise, because a logical consequence of agriculture is the growth of larger human populations. But world population did not rise as fast as the technology of food production allowed. There are Indians in the Amazon, for example, who practice shifting cultivation based on yams. They farm for 3 years, then leave the land fallow for 25. They choose to cultivate no land more than 3 miles from their village. These people could produce enough yams to support villages of 2,000 people. In fact, the villages have 100–300. The implication is that farming peoples can maintain their numbers by choice and are not checked only by starvation.

CHAPTER 6

THE EMERGENCE OF CIVILIZATION

I want to turn now to the class-divided societies that Elman Service called states. I don't want to use that word, however. I want one elastic enough to cover societies that are larger than states, such as Western Civilization. The word "civilization," in fact, will do the job.

Like the word "culture," however, the word "civilization" is used in two ways. Etymologically, it's akin to citizen and city: all three words come from the Latin *civis*, or citizen, and they all suggest that there's a relationship between class-based societies and urban ones. In fact, although they may sound like apples and oranges, they're one and the same, because urban societies are class-based, and class-based societies with at most minor exceptions always have cities. In another sense of the word, meanwhile, civilization suggests an ethically superior, esthetically refined way of life. In this second sense, uncivilized peoples, like uncultured ones, are inferior to civilized ones. Rather than run the risk of appearing pejorative when talking about societies without class divisions and cities, anthropologists today often avoid using the term civilization. Instead, they speak of complex societies. It isn't much of an improvement, because it implies that bands, tribes, and chiefdoms are simple, which they aren't. That's why I'll stick with "civilization" and emphasize that I'm using the word in only its etymological sense.

By civilizations, then, I mean class-divided societies or societies with cities. Take your pick: you'll have a hard time finding one without the other. Civilizations also invariably possess specialists: an administrative corps, soldiers, priests, and craftsmen. (In some cases, like the Maya, the craftsmen are themselves members of the elite.) They possess monumental architecture that symbolizes the power and in this way fortifies the rule of the elite. They have elaborate trading systems and a supporting infrastructure such as roads and canals. They produce standardized goods for the masses, while the elite have luxury goods: fine textiles, vessels, jewelry. Almost always (the exceptions are precontact Peru and subsaharan Africa), civilizations employ writing as a means of record keeping and long-distance communication. Add it up, and you can see why anthropologists call these societies complex.

(On the other hand, the label "complex" may be exactly the wrong word. Just as agriculture simplifies natural vegetation, so civilized life tends to become standardized, in our own case with identical pants, identical beverages, identical houses, identical towns, identical forms of work and entertainment, and so forth *ad nauseam*. Civilization in this light is an exercise in simplification, the better to exercise centralized control over both people and nature. It's a sobering thought.)

I want to look at the earliest civilizations and see them hatch from chiefdoms. After all,

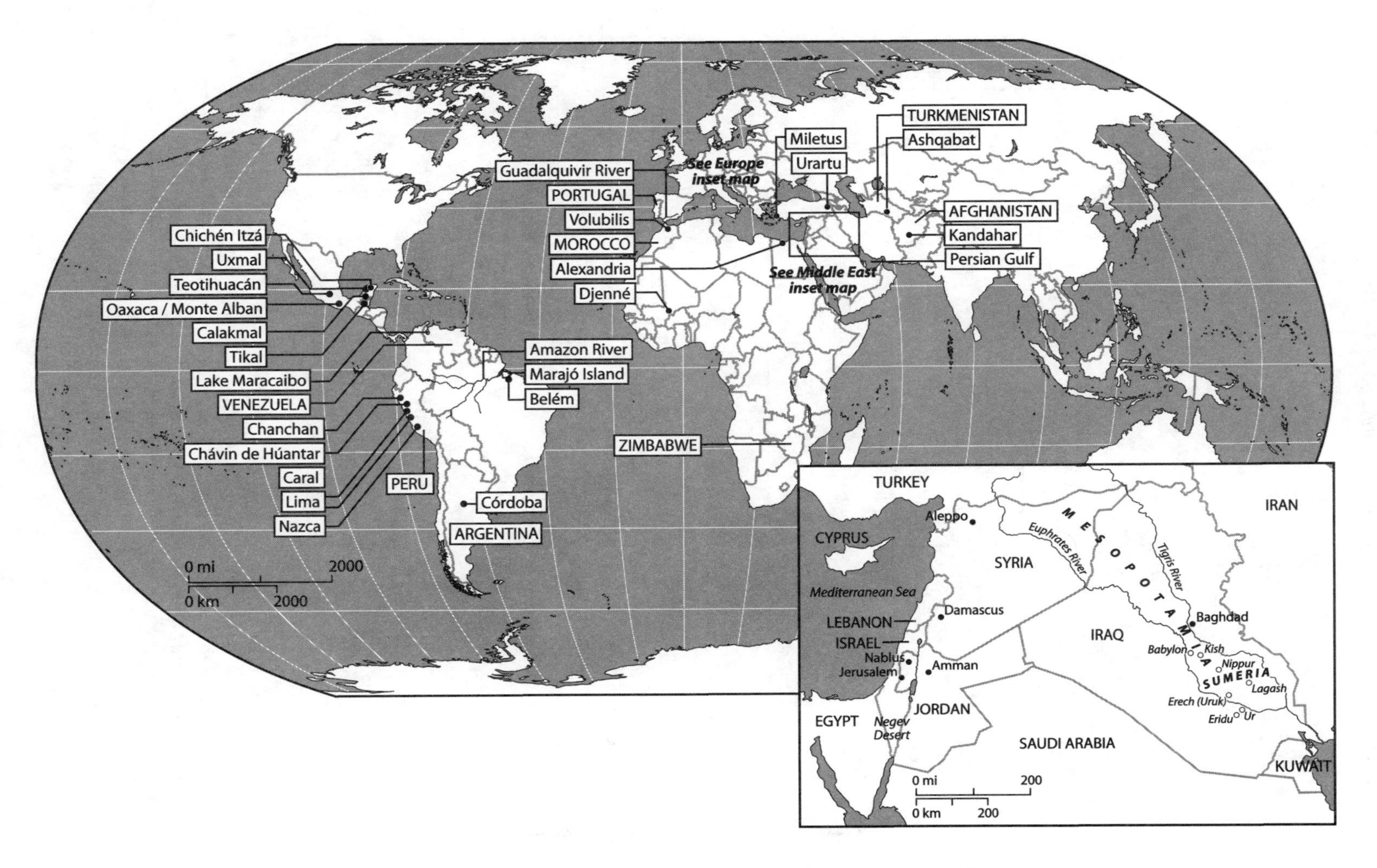

TURKMENISTAN
Ashgabat
Miletus
Urartu
See Europe inset map
Guadalquivir River
PORTUGAL
Volubilis
MOROCCO
Alexandria
Djenné
AFGHANISTAN
Kandahar
Persian Gulf
See Middle East inset map
Chichén Itzá
Uxmal
Teotihuacán
Oaxaca / Monte Alban
Calakmal
Tikal
Lake Maracaibo
VENEZUELA
Chanchan
Chávin de Húantar
Caral
Lima
Nazca
PERU
Amazon River
Marajó Island
Belém
Córdoba
ARGENTINA
ZIMBABWE
0 mi 2000
0 km 2000
TURKEY
IRAN
CYPRUS
Aleppo
SYRIA
MESOPOTAMIA
Euphrates River
Tigris River
Mediterranean Sea
Damascus
Baghdad
LEBANON
ISRAEL
Nablus
Jerusalem
Amman
IRAQ
Babylon
Kish
Nippur
SUMERIA
Lagash
Erech (Uruk)
Eridu
Ur
JORDAN
EGYPT
Negev Desert
SAUDI ARABIA
KUWAIT
0 mi 200
0 km 200

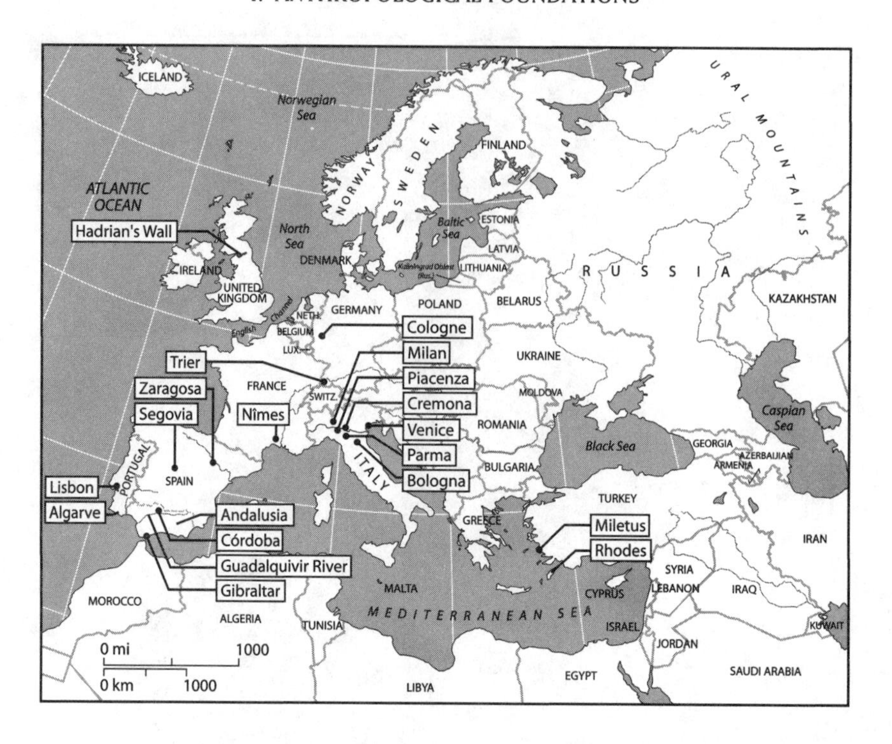

the appearance of urban societies is a major stepping-stone between human origins and the contemporary world. Then I want to indulge once again in speculative history and ask why people, time and again, have shown their willingness to surrender their autonomy. That's what the social order of civilization involves. It's another meta-question, and like the motivation behind the adoption of agriculture, it can't be confidently answered. Air-conditioning, nice clothes, and vacations in Hawaii may explain our willingness to participate in our own civilization, but they haven't been even a remote prospect for the great majority of people in the world's other civilizations. Something else is going on.

SURVEYING THESE SOCIETIES

Civilization seems to have arisen first about 4000 B.C. in Sumeria, the southern part of Mesopotamia. The evidence comes from several Sumerian cities, including Ur, Eridu, Nippur, Kish, and—best-known of all—Uruk. It's called Erech in the Bible, and it's Warka in modern Arabic.

German excavations begun during the Ottoman period revealed a village established at Uruk in the fifth millennium, in the so-called Ubaid period. (I mentioned Ubaid in Chapter 5's discussion of transegalitarian societies.) The settlement was in a rich environment at the head of the Persian Gulf, and the Ubaid people grew wheat, barley, and flax, as well as figs and grapes. They spun wool from domestic sheep. At the same time, they continued to gather wild plants and to fish. There was a village temple, though it appears to have wielded no political authority. Some specialists think that the village was self-sufficient and egalitarian, while others see emerging social hierarchy. For evidence, they point to the houses of the period, which vary from small to immense.

Around 4000 B.C., during the transition from the Ubaid to the Uruk periods, Uruk developed a class system—became civilized in the anthropological sense. Economic production remained in family hands, but tribute was demanded by an elite not only to meet its own requirements but to support craftsmen and the laborers engaged in the construction of monumental temples and walls. By the late fourth millennium, Uruk had become the largest city in lower Mesopotamia. It covered about 250 acres. Scribes kept track of things. Originally they did so with tokens, but protocuneiform soon emerged. True cuneiform, a mixture of ideograms and phonetic signs, would not appear until about 2500 B.C.

About 3000 B.C. Uruk entered the Early Dynastic period. A majority of the Sumerians were now living in cities and towns. The tributary economy was replaced by an *oikos*, or household economy, in which people were organized in hugely extended households, some of which (to judge from the case of Gu'abba in the state of Lagash) had as many as 4,000 adults. Temples themselves constituted *oikos,* although many other *oikos* were dominated by leading families. No longer self-sufficient, the common people contributed their labor in exchange for support. The elite now controlled not only the distribution but the production of goods.

By 2500 B.C. Uruk covered almost 1,000 acres and had a population of 40,000, a record that would not be broken until the expansion of Rome and Teotihuacán millennia later. The first of the famous ziggurats had been built. It was a sacred mountain where the rulers, including the famous Gilgamesh, could communicate with the city's goddess Inanna, later known as Ishtar. Artisans supplied the elite with luxury goods, while for the common people the decorated pottery of older Uruk was replaced by crude, mass-produced ware. A huge diversity of grave goods indicate the gulf between the privileged few and the many. Judging from the nearby Royal Tombs of Ur, dated to about 2400 B.C., those privileges included the right of kings to expect that at their deaths their servants would be sacrificed and buried with them.

The first empire, that of Sargon, was about to emerge, along with the idea of the king as a military hero, rather than an administrator acting on behalf of a god. Defensively, the rulers of the old city-states had invented the tradition that their cities were ruled by—were the property of—their god. It was a myth that much later fooled the German archaeologists who excavated Uruk. They mistakenly thought that the rulers of these states actually believed the myth. They didn't, but the myth remains a significant point in any discussion about why transegalitarian villagers should choose to become social inferiors. I'll return to it.

Such was the hatching of civilization in Mesopotamia—the what of it, though not the why. Agriculture had already diffused from the Fertile Crescent to North Africa, but it had bypassed the difficult environment of the annually flooded Nile Valley. Shortly before 5000 B.C., however, predynastic Egyptians began cultivating the land along the Nile. They grew wheat and barley, raised cattle, and fished. They lived in huts, and their burial goods suggest little social differentiation. During the fourth millennium, however, trade goods began appearing from the Levant—the Mediterranean's east coast—including copper from what is now Israel's Negev Desert. After 3500 B.C. a contrast appears between simple pit burials and burials in mud-brick tombs, plastered and painted. Sometime after 3200 B.C., upper and lower Egypt were unified, though details and chronology are so poorly understood that experts speak of a "Dynasty 0." Significantly, the building style of the earliest dynasties is so similar to that of Mesopotamia, with recessed brickwork, that the question of cultural diffusion inevitably arises. Many experts infer that the rise of Egyptian civilization owes much to Mesopotamia, although others disagree.

It takes an industrial-strength diffusionist to believe that civilization did not arise independently in the New World. In Mesoamerica, squash, beans, and corn were domesticated by the fourth millennium B.C. As late as 2000 B.C., however, which is the beginning of the formative or preclassic period, people still lived only in villages. Thereafter, social stratification quickly developed, first on the coast, then inland, especially in the valley of Oaxaca. The dominant coastal group was the Olmec, who in the first millennium B.C. built an elaborate center at San Lorenzo, with a complex religion and public buildings displaying gods as sharks, jaguars, eagles, and other predators. The Olmec possibly employed a system of writing as early as 650 B.C., but the glyphs, if that is what they are, have not been deciphered. Inland, the Zapotec city at Monte Albán, in Oaxaca, had a population of 17,000 by 200 B.C.; by then, it had a writing system. Power shifted during the subsequent Early Classic Period (200 B.C.–200 A.D.) to Teotihuacán, with its pyramids of the sun and moon, and to the coastal lowlands, where the Maya built such well-known centers as Tikal and Chichén Itzá. Nobody knows why all these civilizations collapsed, but translations of recently discovered Mayan texts suggest that warfare between the leading centers of Tikal and Calakmal weakened the Maya. There appears to have been at the same time a possibly damaging drought. In the case of Teotihuacán, hierarchy reappeared with the Toltecs, followed by the Aztecs.

In South America, too, civilization has surprisingly deep roots. Agriculture came to the Andes during the third millennium B.C. Potatoes were grown, along with squash and beans. Ceremonial centers were established almost immediately. A spectacular example is Caral, in the arid Supe Valley about 120 miles north of Lima and about 15 miles inland. There, by 2600 B.C., irrigation systems were producing a surplus of squash, sweet potatoes, and beans; workers also apparently wove local cotton into nets traded to coastal fishermen in exchange for dried fish. The spectacular features at Caral are its enormous raised mounds, 50 feet high and 500 feet on a side, as well as sunken circular plazas 150 feet in diameter. Their use remains unknown, and the site appears to have been abandoned after 1600 B.C.

During the first millennium B.C. (the so-called early horizon), a new South American civilization appeared in the highlands at Chávin de Huántar, not far from Caral. A temple shows the probable deity of the place as human but fanged and with snakes for hair. About 200 B.C. Chávin was abandoned. A couple of centuries later, however, another center appeared nearby with the Moche, whose capital of Cerro Blanco possessed an immense pyramid of the sun and a companion pyramid of the moon. To the south, another coastal center appeared at Nasca, which evolved from foothill villages into large towns. A subsequent middle horizon is dominated by the Wari Empire, replaced by a late intermediate phase dominated by the Chimu state, with its capital, Chanchan, back in the Moche valley. In the wake of Chanchan's decline, the Inca arose. Like the Aztec, they were the last in a line of indigenous civilizations, and we know them better than their predecessors partly because they are closest to us and partly because they had the misfortune to meet the Spanish.

DIFFUSION

These are not the only cases of civilizations arising from transegalitarian roots: I haven't even mentioned China or India, for example. But even the cases I've reviewed suggest the fragility of this new social order, which over and over in different places at different times emerged, expanded, retracted, and collapsed. In Egypt, for example, the Old Kingdom pharoahs built the pyramids, but their power disintegrated into an intermediate period, a name which sounds neutral but which designates a period of chaos. In Mesopotamia, Uruk faded into insignificance before 2000 B.C. Its ziggurat had just been rebuilt by the ruler of Ur, which lay downstream, at the head of the Persian Gulf. It was a good location, but Ur was sacked in 2004 B.C. The Euphrates soon shifted course, and Ur was never rebuilt. Power shifted to the Babylon of Hammurabi, whose unified empire was briefly established over southern Mesopotamia. Hammurabi's city vanished. Much later, after 600 B.C., Babylon was rebuilt by Nebuchadnezzar II, not only with the famous hanging gardens but with new walls and palaces of baked brick and imported paving stone. The political history of independent Mesopotamia ended a few years later with the invasion of Cyrus, a Persian king.

A convenient place to learn more about these early civilizations is the *Oxford Companion to Archaeology* (1996), but I want to steer away from archaeological detail. I'm headed for a meta-question, after all. For a moment, though, I want to look at the diffusion of civilization and at its traces in landscapes and place names.

It's possible to trace a line of descent from these early civilizations to the 191 members of the United Nations. The Roman Empire from this perspective is a way station between Mesopotamia and Europe. Greece lies between Mesopotamia and Rome.

These are famous cases, but there are less well-known ones. In South America a poorly known civilization existed on Marajó, an island which is twice the size of Massachusetts and which lies at the mouth of the Amazon. Was this civilization the product of diffusion? If so, from where? From whom did Southern Africa's Great Zimbabwe borrow? From whom, old Djenné on the Niger? Archaeologists have only recently recognized what, for lack of a better name, is called the BMAC, or Bactria Margiana Archaeological Complex. At Annau, near present-day Ashqabat, Turkmenistan, a BMAC city appears to have existed about 2300 B.C. It lay astride a trade route that in later days would become famous as the Silk Road. The residents appear to have had a unique method of writing, but their cultural ancestry remains unclear. If history is any guide, some experts will declare that the BMAC derives from another civilization. Others, especially local ones, will insist on the creativity of the people on the spot.

It's also possible to trace a line of descent focusing less on states than on broader attributes. During the first millennium B.C., for example, there was a small state around Jerusalem. It was unique because it was anchored by the world's first monotheistic religion. Viewed territorially this state was trivial—almost too small to consider alongside the great empires of the time. Judaism, however, gave rise to two other religions. One of them, Christianity, anchored European society for almost 2,000 years. The other, established by the prophet Muhammad in the seventh century, shows no sign of flagging.

Both Christianity and Islam can be seen as civilizations, as in the phrases "Christian Civilization" or "Islamic Civilization." They aren't states, although at various times in the past they have claimed to be, for example in the Holy Roman Empire and various caliphates. Even so, they embrace states and have ideologies that organize and sustain social hierarchies.

You don't have to rely on books to discern these historical sequences. North of Jerusalem, the streets of old Nablus have a gridded street pattern, with one set of parallel east–west streets intersected by another set of north–south ones. Farther north, in Syria, there's a similar pattern in the core of Aleppo. There are even fragments of a grid in Jerusalem's old city. Think it's coincidence? Think again. The grid pattern is said to have been invented by Hippodamus of Miletus. Miletus itself certainly was gridded, and the pattern followed in the wake of Alexander. Alexandria itself was platted about 330 B.C. by Deinocrates of Rhodes, which is another gridded town.

The dark opening at the bottom-center leads into a long covered street, which can be traced by its skylights and which is the main north-south street of the grid running through the old part of Aleppo, Syria.

In this as in so much else, the Romans emulated the Greeks. When laying out a city, they constructed a north–south street called the *cardo* (from which we get the word "cardinal"), an east–west one called the *decumanus*, and a lattice-work of streets paralleling one of those guidelines. Italy today is full of such towns: they include Cremona, Piacenza, Milan, Bologna, and Parma, all established about 200 B.C. The results can be seen farther afield, too, very stylishly in the prosperous (but heavily realigned) streets of Germany's Trier, established by the Romans as Augusta Trevirorum. The Romans left towns across North Africa, too, from Morocco's Volubilis all the way east to Jerash, near Amman. They left other relics, of course, like the great drywall aqueducts at Segovia and the Pont du Gard, near Nîmes. There's Hadrian's Wall, too, 73 miles long and 15 feet high as it marches across what is now northern England. Much of it is gone, but the entire route can be hiked along the new Wall Path.

Place names often provide corroborating evidence of vanished civilizations. The name Nablus is a corruption of Neapolis, Greek for new city. Spain's Zaragosa is corrupted from Caesar Augustus. Germany's Cologne comes from Colonia Augusta Agrippinorum. Afghanistan's Kandahar comes from Alexander the Great. Or, consider the great river of southwestern Spain. It's the Guadalquivir. Spanish as that sounds, it is a corruption of the Arabic Wadi al-Kabir, Big River. The resort region of southern Portugal is the Algarve, a corruption of al-Maghreb, Arabic for the West. An even better-known Arabic example is Gibraltar, which is a corruption of Jebel-al-Tarik, the mountain of Tarik, who was the warrior who led the Muslim forces that conquered the peninsula in 711. Those forces took southern Spain from its rulers of the time, the Visigoths, who had just replaced the Vandals. Drop the "v" from that name, and you have the Arabic name of the region, Al-Andalus, from which we get today the Spanish region of Andalusia.

Place names trace the diffusion of surviving civilizations, too. When the Europeans crossed the Atlantic, for example, they immediately introduced familiar place names. Some are obvious, like New York. Some are a bit less so. With over 1 million people, for example, Córdoba, Argentina, is almost four times the size of its namesake in Spain. Others are obscure. Belém is a city of over 1 million people at the mouth of the Amazon, but it is also the old port of Lisbon, the place to which Vasco da Gama returned from India. A few years later, Amerigo Vespucci explored Lake Maracaibo. The houses in the shallow waters reminded him of a little Venice, so he called the place Venezuela.

MOTIVATION

Such details may not be to everyone's taste—hard to believe—and so I'd better get to the puzzle at the heart of this great transition. Why have the majority of people in all these societies acquiesced in becoming near slaves in the service of a ruling elite? Why should they continue to acquiesce, generation after generation?

This question is yet another expression of the ambivalence we've repeatedly touched on. In raising it, I'm raking the coals of a fire that probably started the day after the birth of the first civilization. Recall Jean Jacques Rousseau, whose *Discourse on the Origin of Inequality among Men* (1754) ends with this flourish: "it is plainly contrary to the law of nature, however defined, that children should command old men, fools wise men, and that the privileged few should gorge themselves with superfluities, while the starving multitude are in want of the bare necessities of life." You could add the vitriol of Pierre-Joseph Proudhon, who wrote that "to be governed is to be at every operation, at every transaction, noted, registered, enrolled, taxed, stamped, measured, numbered, assessed, licensed, authorized, admonished, forbidden, reformed, corrected, punished." There's a beeline from those words to worlds of political upheaval.

Marx picked up the idea and applied it not only to the European proletariat but to the realm of what he termed "Oriental despotism." There's nothing particularly Oriental about it, but you know what Marx had in mind when you think of the palaces built by the many for the few. Thirty years ago, Lewis Mumford wrote that civilizations were the product of what he called the megamachine. The term didn't catch on, but its meaning is plain enough: you hardly have to read Mumford's *The Myth of the Machine* (1967) to get his point that civilizations are giant machines—inhuman, though of human parts. A generation later, Jared Diamond would write in *Guns, Germs, and Steel* (1997) that the rise of civilization amounts to a kleptocracy seizing control. Even a sober archaeologist like Susan Pollock, author of *Ancient Mesopotamia* (1999), can't resist a sardonic subtitle: "The Eden that Never Was." Wander down the long corridors of London's Victoria and Albert Museum and you'll get a sense of the treasures produced for elites and collected by them today in great museums. Porcelain vases, silk carpets, golden chalices, bronze vessels: The list is endless. The objects are magnificent—the kind of stuff that sustains Sotheby's—but the supporting social structure mocks ideals such as liberty and justice. It's enough to make you glad we have Wal-Mart.

Rousseau not only touched the nerve of social injustice but laid the groundwork for contemporary anthropological approaches to this great puzzle. Without calling his work anthropology, that is, he organized the *Discourse* as if it was a universal history, based on inference from ethnographic data. Anthropologists today continue to explore the mystery. Guillermo Algaze, for example, argues in *The Uruk World System* (1993) that the site of Uruk was naturally fertile and well-watered, which meant that it had the capacity to produce a food surplus. He's thinking of recessional irrigation, at least at first. The land would flood. Crops were then seeded and brought to maturity on residual soil moisture, as they were in ancient Egypt. Only later, when the Persian Gulf receded with geological uplift did the declining river levels force the development of extensive canal systems. Algaze argues further that Uruk was centrally located on an extensive river network, which allowed it to became a transportation center. Industries therefore arose at the site, and the city began exporting its own manufactures, especially textiles. The weavers were supported by the agricultural surplus, and the income from the textiles paid for imports, everything from roof timbers to luxury goods. Such growth stimulated more economic diversification, much as it is does in cities today. Administrative technologies developed, including writing for record keeping.

Algaze faces plenty of opposition from archaeologists who doubt that Uruk had the power to establish such dominance. But there's something else missing. The Uruk weavers were encumbered laborers, which is to say little more than slaves. To make these societies work, therefore, people had to accept positions of inferiority. What would make people do such a thing, particularly in a society where there was no material reward?

Robert Carneiro, writing in *Science* in 1970, proposed that population growth in Mesopotamia led to conflict, which led to winners and losers who constituted the beginnings of a class structure. If that were so, of course, then every conflict between foragers since the beginning of time would have led to civilizations. It didn't.

Another line of thought is that civilizations arise when a society becomes strong enough to create a safe haven in a dangerous world, a haven powerful enough not only to offer physical protection but food from stockpiles created for times of famine. Call civilization, then, the "primordial protection racket," in Adam T. Smith's neat phrase. But this explanation, like so many others, begs the question of how such a society becomes strong enough to offer those benefits.

Aidan Southall speculates that Mesopotamia's early egalitarian communities had ritual leaders. Some of the communities grew large, and their ritual leaders attracted visitors from smaller places. Gradually those leaders became "more separated from agricultural and materially productive tasks, acquiring the status of divine kings, and also acquiring subordinate staff to assist them, all willingly supported by the people in recognition of their shared success." But why "willingly supported"?

Perhaps we should answer the question by going back to the idea that all cultures include beliefs that everyone in the culture accepts without question. The motto of Spain's Guardia Civil catches it nicely: "*Todo por la patria.*" For such blind loyalty, you need to embed an ethos, a set of controlling ideas that explain the world in such a way that even a discussion of the merits and demerits of hierarchy becomes subversive. You need, in short, an ideology.

We're used to the ideologies of capitalism and socialism, but they're latecomers to this party. The party stalwarts—the specialists who develop, preserve, and defend ideologies—have generally been priests. Like shamans, they claim to know the answers to ultimate questions, but unlike shamans they're professionals, spiritual craftsmen who hold office. Not surprisingly, priests habitually defend the centralized society that supports them in that office.

They do it much as priests do today, with a combination of rhetoric and visible symbols. Call if reification if you must. Max Weber once said that there were three kinds of power: traditional, charismatic, and rational–legal. They all acquire symbols, and it is the idea behind those symbols that finally becomes real enough that people comply. The state becomes real because people believe it's real.

The people of ancient Mesopotamia, by this reading, lived in cities that claimed to be the property of a god. Once that idea was accepted, weavers were very unlikely to rock the boat. That's why elaborate rituals were designed to make the god as real as possible. People grew food for the god, offered food to cult statues, dressed the statues, and even carried them each night to a bedroom in a temple atop a ziggurat.

The Egyptians similarly believed that there was no guarantee that the sun would rise the next morning, that spring would follow winter, or that the Nile would flood the next summer. All these rhythms depended on gods, especially the sun god Re. Each night he died, faced a midnight struggle against forces of evil, and was reborn at dawn. Pharaoh was equated with Re and therefore stood as the Egyptians' protector, not only in the practical sense of providing justice but in a cosmological sense. This was *ma'at,* or things as they are supposed to be. Take Pharaoh away and life would be overwhelmed by *izfet,* disorder. Nothing worse could be imagined, and so the Egyptians accepted a social order that placed nearly all of them at the bottom of a heap. (For vivid pictures of Ra's journey, as depicted

The preeminent icon and metaphor of civilization: The fine limestone casing is long gone from the pyramid of Khufu (Cheops), and the coarse underlying blocks of limestone form a tempting but prohibited flight of onerous steps.

on the four walls of Amenhotep II's burial chamber, see *The Quest For Immortality: Treasures of Ancient Egypt*, edited by Erik Hornung and Betsy M. Bryan [2002].)

All the early civilizations became wonderfully adept at telling stories and building walls, temples, and palaces that strengthened the power of the elite. It's true even of small civilizations that most people have never heard of. "The earth was a wilderness; nothing was built there; out of the river I built four canals, vineyards, and I planted the orchards, I accomplished many heroic deeds there. Argishti, son of Menua, powerful king, great king, King of the lands of Bianili, ruler of the city of Tushpa." It sounds like a Hollywood script writer cutting loose, but it's not. It's a rock inscription from the kingdom of Urartu in eastern Turkey about 1000 A.D.

It's a strategy with many contemporary echoes. When the U.S. Army in the 1880s established a cypress and eucalyptus forest on the sand dunes of the San Francisco Presidio, the major in charge wrote that the trees would "indirectly accentuate the idea of the power of government." How modest this seems in comparison to Saddam Hussein. Baghdad's Umm al-Mahare Mosque—the Mother of All Battles Mosque—was fitted with a copy of the *Qur'an* supposedly written in his blood. It also had an ornamental pool with an island that reproduced Saddam's thumb-print. He built a city near Wadi Tharthar, too. Every brick of Saddam City was embossed with his name. Want to bet that it didn't "accentuate the idea of the power of government"?

PART II

HISTORICAL DEVELOPMENTS

CHAPTER 7

CHINA

Many though not all of the pyramids still stand, and many Egyptians today have ancestors who lived in pharaonic Egypt. Still, Egypt today rests primarily on a comparatively recent Islamic base. Gone, too, are the pristine civilizations of Mesopotamia and of the New World, however much the occasional dictator thinks of himself as the embodiment of ancient kings.

China and India, however, are different. Like those places, they possess relics of the distant past, and their populations are genetically contiguous with the founders of ancient civilizations. Despite evolution, revolution, and the huge changes occurring in both countries today, the ancient civilizations of China and India survive, changed but unbroken, culturally intact after thousands of years. That's why I want to focus on them in the next two chapters. Besides, they're demographic heavyweights and, increasingly, economic and political ones elbowing their way into a world long dominated by people of European ancestry. If ever China and India could be patronized as the exotic homes of rice growers, tea pickers, and craft producers, that day is past.

GETTING SOME BEARINGS

I want to start, as I did with Africa at the start of Chapter 2, with pointer-to-board geography, that lost art. Why was it lost? Was it because the teachers doing it had next to no knowledge of the places they were naming? Was it because their ignorance made the places as meaningless as a listing of lunar features? Perhaps it was just that geography disappeared into the maw of social studies, always dominated by history teachers. Or perhaps I'm being unfair to the moon.

In any case, the map of China shows a far western province of Xinjiang. (If you're using an old atlas, it probably shows this province as Sinkiang. That's because other methods of transliterating Chinese were used in the past. For more on these systems, see the note at the end of this chapter.) Xinjiang comprises two arid basins separated by the Tian Shan, or Heavenly Mountains. The basins are the more northerly Dzungaria, through which the old Silk Road passed, and the more southerly Kashgaria, also known as the Tarim Basin. Most of Kashgaria is a desert—the Takla Makan by name—and agriculture consists of scattered oases where streams descending the surrounding mountains are diverted before they sink under the desert sands. Xinjiang was once called Eastern Turkestan, an obsolete name but

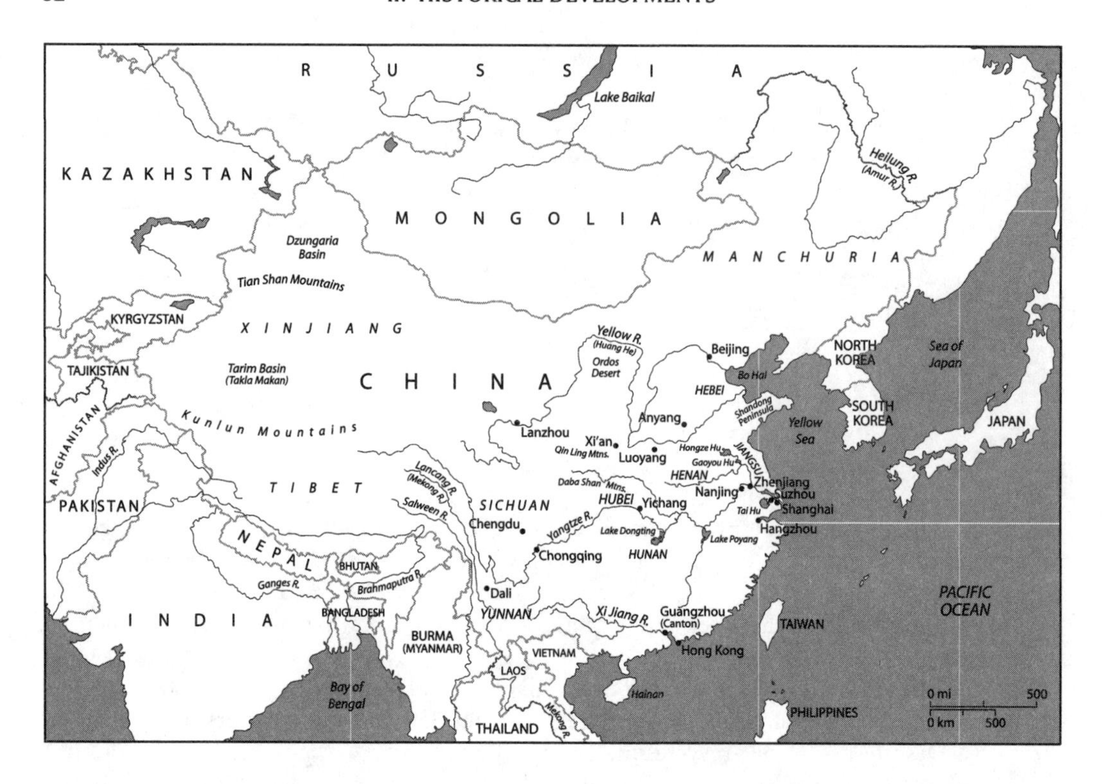

one that is still useful as a reminder of the province's predominantly Islamic population. The name Xinjiang itself, which means New Frontier, hints at the fact that Xinjiang became part of China only about three centuries ago. (Chinese place names are exceedingly literal. This is very handy for foreigners trying to get a grasp of a place whose names they find difficult, but the Chinese themselves pay no more attention to these meanings than Americans do when they hear names such as Long Beach, Twin Falls, or New Haven.)

An even more recent territorial acquisition lies south of the Kunlun, which are the mountains on the south side of the Tarim Basin. Tibet is the only place on earth where you can lay out a square 500 miles on a side and within it find no spot below 13,000 feet above sea level. It's amazing, but it's no geological mystery. A fragment of an ancient continent called Gondwanaland long ago broke free from its ancient position wedged between Africa and Antarctica. Having drifted for a long time at about the speed of a growing fingernail, for the last 50 million years it has been slowly ramming into Asia. (Gondwanaland takes its name from Gondwana, a region of central India that is home to the Gonds, Indian tribals.) The leading edge of this fragment has forced its way under Asia's southern rim and lifted it to form Tibet.

Although it has an average precipitation of less than 10 inches a year—Tibet's too cold for its atmosphere to hold much water vapor—Tibet is the world's most concentrated source of great rivers. The Yellow River and the Yangtze rise on its eastern edge. The Mekong of Vietnam, the Salween of Burma, and the Brahmaputra of Bangladesh all rise at its southern and southeastern edge. The Indus of Pakistan flows from its southwestern corner. Of all these great rivers, only the Ganges originates on the south side of the Himalaya, which form Tibet's (and now China's) border with India. Tibet's 3 million people cluster primarily in the south, especially in the valley of the upper Brahmaputra, here called the Tsangpo, which follows a trench along the northern edge of the Himalaya.

The map also shows a projection of Chinese territory north from Beijing to the Heilung Jiang, or Black Dragon River. Bordered on the west by the mountains of Inner Mongolia, this is the region formerly known as Manchuria, the home of the Manchu people. Like Xinjiang, it came under Chinese rule about 1700. It was briefly lost during the first half of the 20th century, when it was administered first by Russia, then by Japan. Chinese once again, it's now colorlessly called Northeastern China.

Set aside these three domains—Xinjiang, Tibet, the Northeast—and you have China proper. Its almost circular coast extends from near Beijing on the north (the name Beijing in fact means Northern Capital), on the east to Shanghai (the name means Near the Sea), and on the south to Hong Kong. On the west it extends to Xinjiang and Tibet.

The dominant physical feature of China proper, apart from its fanlike coast, is the Yangtze River. (The river's name in China is Chang Jiang, or Long River, and recent English-language atlases often show it that way.) Coming east from Tibet, the Yangtze flows through the agriculturally rich and densely populated Red Basin, which occupies the center of Sichuan, a province with more than 80 million people. (The old transliteration was Szechuan, which survives in the regional cuisine, heavy on the red pepper introduced by the Spanish.) The river then enters gorges extending to Yichang. Emerging from them, it crosses the closely cultivated Tangho Basin around Lake Dongting, a natural backwater of the Yangtze in flood. It then flows east past another large backwater, Lake Poyang, before arriving at Nanjing (the name means Southern Capital) and the river's mouth, just north of Shanghai. The provinces north and south of the river as it passes through the Tangho Basin are Hubei (North of the Lake) and Hunan (South of the Lake). Both have populations exceeding 70 million.

Shanghai lies at the southern end of a lowland that extends north 700 miles to Beijing. This is the North China Plain, China's largest tract of arable land, and it was built by the Yellow River, or Huang He. At Lanzhou, the Yellow River emerges from Tibet. Rather than flowing directly eastward to its plain, the river makes a great northward loop around the loess plateau of the Ordos Desert, where it picks up the yellow silt that gives the river its name. Returning southward, it makes a right-angle turn near Xi'an, an ancient Chinese capital. Shortly below this elbow, the river enters the Plain. Its course here has changed many times over the centuries and has included routes that brought the river to the sea south of the Shandong (Eastern Mountains) Peninsula. The provinces through which it passes as it crosses the plain include Hebei (North of the River) and Henan (South of the River). Both have populations of about 70 million.

Apart from the Red Basin and North China Plain, China proper is hilly or mountainous. Travelers going south from Xi'an, for example, cross the formidable Qin Ling, the eastern extension of the Kunlun. Even from the air, their ruggedness is exhausting. Just to the south and parallel to the Qin Ling are the Daba Shan, which the Yangtze crosses in its gorges. South of the Yangtze, the Nan Shan (Southern Mountains) mark the hydrographic divide separating rivers flowing north to the Yangtze from those flowing south, chiefly to the Xi Jiang (West River), at whose mouth lies Guangzhou, first called Canton by the Portuguese. The western extension of the Nan Shan lies in the historically remote province of Yunnan (South of the Clouds), which borders Burma. The famous Burma Road, which went west from Yunnan, had to cross the profound gorges of the upper Yangtze, Mekong, and Salween. All three rivers are tightly bunched together here, though they soon go separate ways.

THE ORIGIN OF CHINESE CIVILIZATION

Academic histories of China for many years began with the Shang Dynasty (1523–1027 B.C.), which ruled from a series of capital cities all based along the Yellow River. (The most famous of these is the late Shang capital of Anyang, roughly equidistant from Beijing, Xi'an, and Nanjing.) Chinese sources which spoke of earlier dynasties were dismissed as mythic for lack of archaeological confirmation.

Now, although matters have not been fully worked out, it is clear that the transition to civilization occurred long before the Shang. Agriculture can now be traced back to about 6000 B.C.. The classic Chinese crop division was soon established, with rice to the warmer south and millet and wheat in the colder and drier north. Rice was then being cultivated near Tai Hu (Tai Lake), not far west of present-day Shanghai, while on the North China Plain farmers were cultivating millet and raising chickens and pigs. Social stratification appears by the fourth millennium Hongshan Period: The evidence is jade jewelry and tombs differentiating rich from poor. By the Lungshan Period (2500–2000 B.C.), pottery was supplemented by cast bronzes, and villages had grown as large as 600 houses. West of Luoyang, which is roughly midway between Anyang and Xi'an, archaeologists have found at Erlitou a city with a walled, rammed-earth platform 100 yards square. They have found jade ornaments and bronze wine vessels, as well as palaces whose wooden columns were set atop human sacrifices. Dated to 2000 B.C., Erlitou is attributed to the formerly mythic Xia. This sequence implies, but of course does not explain, the now-familiar transition to a hierarchical, urban society.

DYNASTIES AND THE EXPANDING ECUMENE

From the Shang onward, dynastic histories extend over the next 3,000 years, highlighted by the Zhou or Chou (1027 B.C.–B.C.), the Han (202 B.C.–220 A.D.), the Tang or T'ang (618 A.D.–907 A.D.), the Song or Sung (960–1279), the Mongol Yuan (1271–1368), the Ming (1368 to 1644), and the Manchu Qing or Chi'ing (1644–1912). From the Han onward but with occasional interruptions, imperial power extended over China proper, that roughly circular core domain. The brutal but very brief Ch'in Dynasty (221 B.C.–206 B.C.) first established control over all this area except Yunnan. The Han added Yunnan, as well as substantial parts of present-day Vietnam and Korea.

"Control" here must be taken with a grain of salt. The southern area, in particular, was the home of many unassimilated peoples who, like the Tibetans today, were slowly overrun by the Han. The great lake basins of Dongting and Po'yang, for example, were not reclaimed for cultivation by the Chinese until after the 16th century; Peter Perdue tells the story in *Exhausting the Earth: State and Peasant in Hunan, 1500–1850* (1987). Their agricultural status today is precarious, because the levees along both the Yangtze and the rivers feeding the lakes from the south frequently fail. There's no feasible remedy, except to evacuate large areas during floods.

New crops, as well as people, moved over the centuries into these new areas, which is why China's southern uplands are planted not only to indigenous crops such as rice and wheat but also to American imports such as corn, potatoes, and tobacco. With an expanded ecumene and a more diverse range of crops, China's population rose from about 100 million people in Han times to 400 million in 1900. Since then, it has more than tripled.

IDEOLOGY

Once again, we have the puzzle of enduring and deep inequality. One side of the polarity is represented by the cultural treasures on display in Beijing's Forbidden City, where the Ming and Qing emperors lived in grand seclusion. The society producing these treasures consisted of a tiny elite, a small group of administrators, a somewhat larger group of landowners and businessmen, and a mass of peasants who comprised 90% of the population. These peasants lived in tiny villages clustered around periodic market towns where, once a week, peasants could buy a few necessities—matches, oil, cloth—from traveling merchants. The peasant standard of living was so low and the peasants' oppression so great that China's most important crop has been described as people, grown by the elite.

Ancestor worship has been strong throughout Chinese history, and by teaching deference it provided the seed for the ideology needed by this civilization. That seed matured in Confucianism and Taoism, which both arose in the sixth century B.C. The one was concerned primarily with social behavior; the other, with the relationship between humanity and the natural world. Both became state cults. Taoism did so especially during the Tang Dynasty, whose founder claimed to be a descendent of the founder of Taoism. Confucianism did so during the Han Dynasty and again for a very long period from the 14th century until the removal of the last Chinese emperor, early in the 20th century.

Confucius advocated good behavior (*jen*) that was not only outwardly correct but inwardly true to one's nature. He did not think in Western terms of ego satisfaction or accomplishment for the sake of personal development and recognition. Instead, he emphasized family and social responsibility, along with respect for elders and social superiors. Mencius gave this moral teaching a political flavor in the fourth century B.C. by saying that rulers are obliged to act morally in the interest of their subjects. They should do so not by wielding power arbitrarily but by demonstrating virtue, which leads their subjects to reciprocate. This principal of reciprocity has echoes in Taoism, whose central text, the Tao-de Ching, teaches rulers to act imperceptibly. Such was the path followed by those who understood the *Tao,* or Way. They understood the perpetual flux of the alternating forces that comprise the world, the two breaths (*ch'i*) of the dark side (*yin*) and the bright (*yang*). They understood that what appears to be tumult is in fact unity, and this understanding keeps the world in harmonious balance. (A surprisingly good source on Chinese as well as other religions is *Merriam-Webster's Encyclopedia of World Religions* [2000], which in addition to short entries has extended essays on the world's major religions.)

By the 14th century, when Confucianism began its long reign as the state cult, a canon of Confucian texts had been organized. It remained at the core of Chinese primary education and civil service examinations until 1905. Almost 100 years have since passed and the emperor is gone, but loyalty and cooperation are highly prized. Deference both within the family and upward through the social pyramid remains very powerful, too, which is why Mao was followed blindly during the ruinous Cultural Revolution. Even intellectuals are kept firmly in check; in the words of one Beijing University professor, they are "supposed to act like children who never talk back to their parents."

SYMBOLS AND EXPRESSIONS

Like every other civilization, China reinforced its ideology with monumental architecture. A Zhou ode ascribes limitless power to the Son of Heaven: "all [lands] under Heaven are the soil of the King; [all people] within the boundary of that soil are the subjects of the King." No wonder that the emperor, as Jeffrey Meyer writes in *The Dragons of Tiananmen: Beijing as a Sacred City* (1991), "was commissioned by Heaven to rule China, pacify the outlying territories, and ultimately to set such a shining example of perfect government that the whole world would come to the foot of his throne and offer submission."

What kind of city suited such a ruler?

For nearly 2,000 years, the designers of Chinese cities followed to greater or lesser degrees the Zhouli Kaogongji. This was a Han text that, to increase its own authority, purported to be from the Zhou. It stipulated that a "ruler city" should be located in the center of the kingdom, at the place "where earth and sky meet, where the four seasons merge, where wind and rain are gathered in, and where *yin* and *yang* are in harmony. Therefore the myriad things are at peace." To stress the centrality of the city, surveyors according to the Zhouli "demarcated it as a square with a side of nine *li* [one *li* is approximately half a kilometer], each side having three gateways."

This formula produces a city that is square and approximately 3 miles on a side. The Zhouli called for a wall around cities, although these walls were probably more symbolic than defensive. In the words of Rhoads Murphey, the walls were "as imposing as the rank and size of the city dictated, but in every case were designed to awe and affirm, only secondarily to defend, although of course they might be useful in troubled times."

The Zhouli go on to say that "within the capital there were nine meridional and nine latitudinal avenues, each of the former being nine chariot tracks wide." There are different renditions of this street grid (*jing wei-tu*), some showing the streets clustered rather than evenly spaced, but the city in any case was divided into nine blocks by the streets that ran from gate to gate, and the central block was reserved for the Son of Heaven, whose throne faced south. The Zhouli explain that "in the front is the Imperial Court, while at the back lies the market." Residential neighborhoods are treated almost as an afterthought. Until the Song Dynasty, they were grouped in walled and gated quarters called *lu-li* and segregated by occupation. Later, these ward walls disappeared, and commercial activities began to appear along the city streets, with merchants living above their shops. Yinong Xu writes in *The Chinese City in Space and Time* (2000), that "commercial and residential activities, and thus shops and houses, were no longer segregated from each other, but rather mingled in the web of city streets and alleys."

Many writers, including Gideon Golany in *Urban Design Ethics in Ancient China* (2001), explain the basic nine-cell grid as an urban variation of *jingtian*, which was a Zhou practice of subdividing farmland blocks into nine squares. Those at the periphery were cultivated by peasants for their own use, but the square at the center was cultivated by the peasants for the benefit of the landlord. The same nine-block geometry is echoed in the Chinese conception of China itself as the Middle Kingdom, or Zhongguo.

There are other interpretations, however. Victor Sit, in *Beijing: The Nature and Planning of a Chinese Capital City* (1995), calls the *jingtian* analogy superficial and traces the nine-cell form to the ancient Ming T'ang, or Hall of Light. This is a structure whose origins go back to pre-Shang times, when the Ming T'ang was a single room, octagonal but thatched, that served as the seat of government. By the Zhou Dynasty, the Ming T'ang was a more complex structure, with a central room and eight surrounding ones. The Ming T'ang declined in significance during the Han, who moved their seat of power to a nearby throne room, but the Han maintained the old center. One Ming T'ang from the Han period was excavated in the 1950s at Xi'an. It consists of a circular moat several hundred yards in diameter around a square wall with four central gates entering a precinct at whose center is the Hall of Light itself.

An ancient text explains that the Ming T'ang was "round on top and square at the base. . . . The rounded upper part takes its form after Heaven and the square foundation symbolizes Earth." Sit himself says simply that the Ming T'ang "evolved to become the basic concept of the Chinese national capital, or, for that matter, the concept of the Chinese city in general."

No Chinese city ever followed the Zhouli exactly, but an amazing number of them followed it broadly, particularly with regard to walls, gates, cardinal axiality, and a central palace. Shanghai and Guangzhou (Canton) are European creations that ignore the pattern, but Suzhou still exhibits the ancient form, even though the city's 5 million people sprawl far beyond the remaining bits of the old wall. The pattern is evident in Xi'an, the Ch'in capital, although a Ming wall rims it today. Five miles to the northwest is the site of Chang'an, the famous and still unexcavated Han capital: its square wall was reputedly 5 miles on a side. Chengdu, in Sichuan, has worked hard to obliterate all trace of its old form, but smaller southern towns, such as Yunnan's Dali, have worked hard to preserve it.

The grandest and most faithful manifestation of the Zhouli, however, is Beijing. This is ironic, because the city was built as the capital of foreign invaders, the Mongols. For that very reason, however, Kublai Khan sought to legitimate his rule by building a grand capital in the purest Chinese tradition. He called the place Da Du, Grand Capital, and it was built

We're standing at Nanmen, the "south gate" of Xi'an. The south side of the city wall runs to our right and left, while before us is Nanda Jie, the main street running north to the city's Bell Tower, the circular structure in the distance. The tower is the crossing of the city's main north-south and east-west streets.

A much smaller example of the classic Chinese city: This is Dali, in western Yunnan. The wall and its entrance gate have been reconstructed in the name of historic preservation and tourism promotion. It's a useful strategy; as the picture suggests, many thousands of Chinese tourists come to Dali to stroll its gridded streets and browse its shops.

late in the 13th century with three nested walls. The outer one, confusingly, enclosed the Inner City. The middle one enclosed the Imperial City. The inner one enclosed the Ze Jin Cheng, known in English as the Forbidden City.

When the Yuan Dynasty fell to the Ming, the new rulers chose Da Du as their capital, though they renamed it Beiping (Capital of Peace), then Beijing (Northern Capital). They abandoned the northern half of the Mongol city but in 1553 added to its south an abutting Outer City. These names—Inner, Outer—hint at the importance of the southern approach. The Chinese excavated lakes on the western side of the Imperial City and heaped up the quarried material to form Coal Hill, which now rises behind the Forbidden City on the spot that was once the center of Mongol Da Du. All this earth moving was no accident: It was a way of ensuring that the Mongols would not return. For good measure, the Ming city had no central gates on its north side.

A mile south of Beijing's Forbidden City, and slightly to the east of the city's grand north–south axis, is the triple-roofed Hall of Annual Prayers, popularly called the Temple of Heaven. Today the roofs are all black, but the Ming covered the top one in blue tile, to signify heaven, the middle one in yellow to signify crops, and the bottom one in green to signify the earth. (Its location relative to the Forbidden City is the same as that of the Ming T'ang to the Han palace at Xian 2,000 years ago.) The Ming and Qing emperors came annually to pray for a prosperous year. Commoners had to hide indoors during the procession. They were forbidden even to look upon the emperor. August visitors might do so—might even have an imperial audience—but their approach into the Forbidden City was through a set of walls and gates on such a scale and of such opulence that the visitor was reduced to physical insignificance. That was the idea, after all.

Americans engage in a mild form of this kind of intimidation themselves, sequestering their leaders—political and corporate—in impressive buildings and screening them from public view except under highly controlled circumstances. In Washington, Americans have built their own symbols of majesty, but Americans are comparative novices at this business. For all its columns, Washington lacks the decorative magnificence of the Forbidden City. A more important failing of Washington, judged strictly from the viewpoint of strengthening the elite, is that it makes no effort to root political power in religion or cosmology. The Chinese city was not so modest. As Paul Wheatley wrote in *The Pivot of the Four Quarters* (1971), the Chinese capital city was on the cosmic axis of the world so the emperor could better communicate with the heavens and, through the city's geometry, the four quarters of his empire.

That geometry diffused eastward and is visible in the imperial cities of both Korea and Japan. It even appears in Mandalay, laid out by the Burmese in 1857 and in its gridness looking as though the British had platted it, as, with a different logic, they would soon plat Rangoon (Yangon).

Meanwhile, China developed the infrastructure to control its human crop. This infrastructure included the Grand Canal, or the Yunliang He—literally, the Grain Transport River. The canal was begun in the fourth century B.C., when it linked the Yangtze delta near Zhenjiang to Qingjiang on the Huai River, farther north. The canal was fed by reservoirs

The grand axis of Beijing's Forbidden City, seen here from Coal Hill, to the north. The ceremonial approach was always from the south, and even today taxi drivers are very reluctant to drop tourists at the perfectly legal north entrance.

like Hongze Hu and Gaoyou Hu. Both are large enough to be called lakes and to show up on good atlas maps of China today. A thousand years later, the canal was extended south to Hangzhou, then the capital of the Southern Song dynasty, and north to Loyang, the Sui capital. By 1300 the Mongols were relying on the canal for grain in Beijing. By then the canal was 1,100 miles long, with major river crossings and locks that brought the canal to a maximum elevation of 138 feet. Pairs of water buffalo might drag a set of 40 boats, lashed together. In case of flood or low water, the grain could be put into emergency granaries along the line of the canal.

The imperial Chinese built roads, too, though the high cost of road transport meant that they were used more for political and military than commercial purposes. The Ch'in designed their roads to be precisely wide enough to handle military chariots, which were built on a standardized axle length. By the time of Kublai Khan, a dozen imperial highways crossed China. The most important were the Ambassador's Road, which ran south from Beijing to Guangzhou (Canton), and the Silk Road, which headed west into Dzungaria and Central Asia. Across soft lowlands, the roads were paved with large stone blocks. Through the Yangtze Gorges, at the other extreme, a highway ran along sheer cliff faces, sometimes on wooden platforms anchored in the rock and sometimes in half tunnels dug into the cliffs. (For pictures and details, see Joseph Needham's *Science and Civilization in China*, volume IV:3 [1971].)

What is one to make of it? The rapid economic development of China today is often attributed to Confucian values, including the acceptance of one's position within a hierarchy and a determination to work incredibly hard out of a sense of responsibility, both to one's superiors and inferiors. The prodigious changes that have come over China since it broke free from the stranglehold of Communism probably would not have occurred so quickly without this sense of responsibility.

At the same time, however, civilizations are always vulnerable to rebellions arising from the deep awareness of the injustice at their heart. That's why civilizations always rely on force as well as persuasion. It's no accident that Chinese citizens all have a *dang'an*, a dossier that follows them through their lives with information about their schools, jobs, personal life, and politics. People cannot see their own file, which is kept under close control by the Communist Party. Consistent with this distinctively Chinese kind of totalitarianism, 70% of criminal defendants in China have no legal representation. Defendants are nearly always convicted, and about 14 are executed every day, many after confessions induced by torture. Being a defense lawyer is itself a risky business, because lawyers are routinely subjected to harassment and worse. No wonder that in 1989 the Chinese leadership was willing to open fire on students in Tiananmen Square, even though students are far more respected in China than they are in the United States or Europe. Deng Xiaoping's final instructions were "stability above everything else." He might as well have said hierarchy and an elite above everything else.

On the one hand, the Forbidden City hints at a vanished world of beautiful textiles, jewelry, furniture, bronzes, pottery, and more. North of town one can visit the Ming tombs, some of which are still in ruins, others restored. The neglected ones—usually locked but sometimes accessible with a well-timed bribe—have a majesty that the products of American factories cannot touch. On the other hand, old photos show that the price of these luxuries was crushing poverty imposed on the great mass of the Chinese people, along with a penal system specializing in refined instruments of torture. Even today, there are parts of rural China where malnutrition, if not starvation, prevails. There are still plenty of people shivering in houses poorly heated with unvented coal stoves that induce lung disease. It was pre-

The valley outside this house in Hengdi, in Yunnan's Zhenxiong County, is meticulously cultivated in wheat and corn and tobacco, but the profits are too slim to afford anything beyond the barest furnishings. At 5,000 feet above sea level, the coal brazier is necessary but very unhealthy.

cisely the sense of injustice arising from such conditions that fueled the storm of revolution that swept across 20th-century China, as though to say that the powers of ideology and compulsion are finite.

Note on transliteration. Chinese names exist in three major transliterations. The one developed by the Chinese themselves is called Pinyin and is striking for its use of consonants in a way that seems bizarre to native English speakers confronting a name like Xinjiang. In fact, it's not as impossible as it looks, because the letter "x" in Pinyin is pronounced "sh." The two other and older systems are the Postal Atlas and Wade-Giles systems. The great province around Chongqing, for example, is Sichuan in Pinyin, Szechuan in the Postal system, and Ssuch'uan in Wade-Giles. Beijing in the Postal system is Peking; in Wade-Giles, it's Pei-ching. Guangzhou is Canton in the Postal system and Kuang-chou in Wade-Giles. Pinyin is rapidly replacing the older systems. They linger in restaurants, with Peking Duck and Chungking chicken. Airline bags are still tagged CAN for Guangzhou and PEK for Beijing. Hong Kong is still Hong Kong, thankfully, and not Xianggang.

CHAPTER 8

INDIA

For a century or so until 1947, the part of Asia south of Tibet was the British Indian Empire. Now it's chiefly India, Pakistan, and Bangladesh, and it has about as many people as China. Do the arithmetic: India with 1 billion, Pakistan with 140 million, Bangladesh with 120 million. That's about the same as China's 1.3 billion people, and we're forgetting about 80 million people in nearby Burma (Myanmar), Nepal, and Sri Lanka. Surprisingly, perhaps, South Asia has lots of almost empty areas, especially in the Great Indian Desert, the Thar, which sheaths the lower Indus River and extends eastward into India. Parts of peninsular India are empty, too, including the rugged mountains now in the new state of Chhattisgarh (the name means Thirty-six Forts). Kipling set his *Jungle Book* here, although he never visited the area—just heard about it. There are also some fantastically crowded areas—and not just in the huge cities of the region. Bangladesh is the most extreme case. It packs 120 million mostly rural people into 55,000 square miles. Oklahoma, in contrast, has 3 million mostly urban people clumped over 70,000 square miles.

Is India, the heart of contemporary South Asia, a civilization? Before the British came, it had only sporadically been a unified state. Still, measured by an enduring social hierarchy, monumental architecture, and literacy, India is indeed an ancient civilization, coherent and self-sustaining.

THE SETTING

Start with the Himalaya. They're part of the Alpine System, that set of chains created in the geologically recent past by the drift of Africa, Arabia, and India into Eurasia. The Alpine System includes not only the Alps proper but the Atlas and Pyrenees to their west and the Carpathians, the Pontic and Taurus mountains of Turkey, the Elburz and Zagros of Iran, and the Hindu Kush of Afghanistan to their east. The Himalaya—literally, the abode of snow—are farther east still. They hang like a drapery swag from a western knot called the Pamirs, mostly in Tajikistan today, and from another knot north of Burma. There, they change direction and head south in a rippling washboard of parallel crests to form the Arakan Yoma, or Arakan Mountains, of western Burma. The Arakan Yoma march straight into the Indian Ocean, from which some of their higher hills emerge as the Andaman Islands.

North of the Himalaya, the Indus and the Brahmaputra flow away from each other

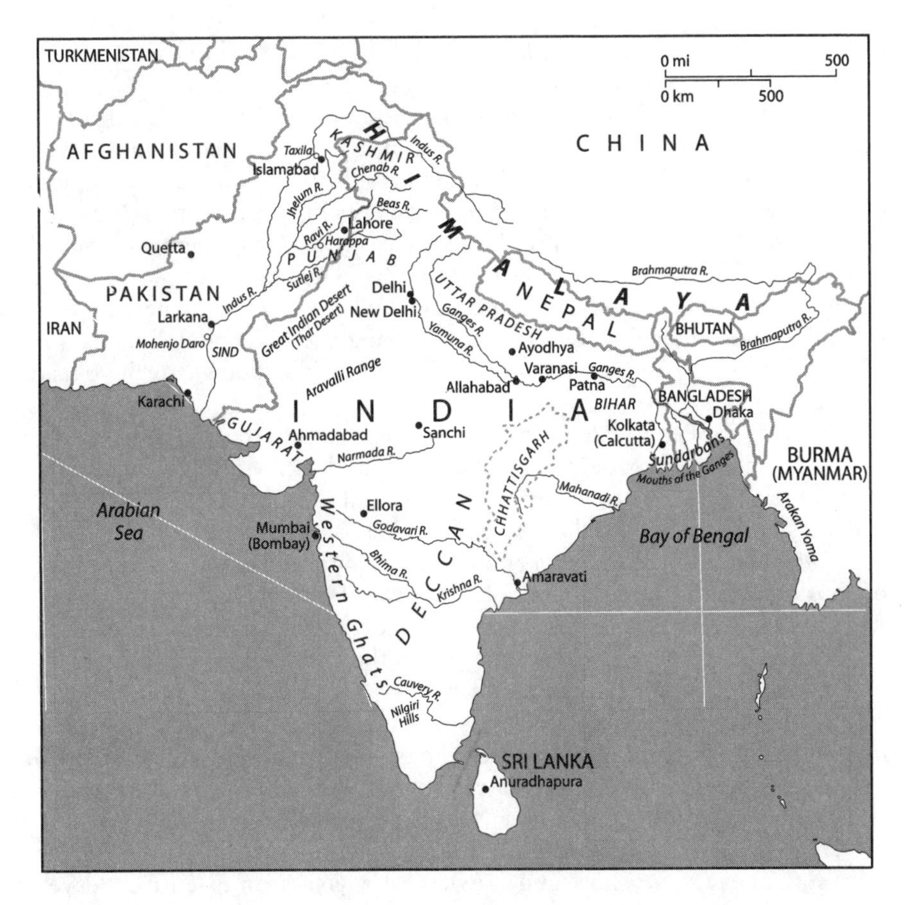

along the mountain wall before penetrating it and heading south. The Ganges, despite its religious importance, originates more modestly on the south-facing Himalaya slope. (Truth to be told, there's less water in it than in its western tributary the Yamuna, or Jumna, which flows past Delhi and joins the Ganges at Allahabad.) All these streams flow across and have blanketed with sediment the great Indo-Gangetic Plain, which stretches 1,000 miles from Peshawar on the west to the Ganges delta. The imperceptible divide separating the waters of the Indus from those of the Ganges lies on these plains a bit west of Delhi and is less than 1,000 feet above sea level.

The western section of the Indo-Gangetic Plains is the Punjab, literally Five Rivers. The name comes from five Indus tributaries—the Jhelum, Chenab, Ravi, Sutlej, and Beas. Since 1947, the Punjab has been divided into a Pakistani Punjab and an Indian one. The Pakistani Punjab is the breadbasket of Pakistan. It is more than that, too, because it contains the old city of Lahore, the country's cultural center. Compared to it, Karachi and Islamabad are vulgar upstarts, with nothing to compare to Lahore's fort, mosques and tombs. Farther south, in Pakistan's southern state of Sind, the Indus is a solitary stream flowing through the Thar. The British built huge irrigation works here in the 20th century, particularly the Sukkur Barrage, and they sheathed the river in green that obscures its historic character, which was to flow through a desert like the Nile, losing water along the way. The Indus in its natural state must have made a huge impression on early invaders, because the name Sind means flood. The Greeks later dropped the first letter and referred to everything east of the river as Ind. From them, we get the name India.

To the east of the Pakistani Punjab lies the second Punjab. It's India's most agriculturally progressive state, one where fields are often cultivated by tractors and harvested with combines. East of it lies U.P., Uttar Pradesh, literally Northern State. U.P. occupies the bulk of the fertile Gangetic Plain and not surprisingly ranks as India's most populous state, with a population approaching 170 million. Still farther east, the Ganges enters Bangladesh; the name means Land of the Bengalis. The Ganges there is joined by the great Brahmaputra, and the delta of the combined rivers is enormous. Kolkata (Calcutta) lies on one Ganges distributary, the sizable Hooghly. Dhaka (Dacca), Bangladesh's capital, lies on another. The coast is the region of the Sundarbans, forested swamps best known to the outside world today for tigers.

The Indo-Gangetic Plains extend south from Delhi to the Aravalli Hills, which with the Narmada River and Satpura Mountains to its south marks the northern edge of the Deccan Plateau. The Deccan (literally, the South) is the triangular peninsula that most people think of when they visualize a map of India. Most of it is bleak semidesert, largely covered by mesquite, a plant introduced by the British and which, despite wicked thorns, furnishes a major source of cooking fuel for the rural poor. The Deccan is the most geologically exposed part of the block of Gondwanaland that drifted slowly north, wedging itself under Asia and lifting Tibet. The line of collision or the seam between the two land masses is deeply buried by the river-laid sediment of the Ganges plains.

The Deccan is slightly tilted to the east, so the rivers draining it flow east, usually from headwaters near the Arabian Sea. The escarpment there is called the Western Ghats, or Steps. If it weren't perpetually hazy, from the summit you could see the Arabian Sea. You might even catch Mumbai (Bombay), India's largest industrial center. Such greenery as there is on the Deccan occurs largely on the Ghats and along the plateau's rivers, which include the Mahanadi, Godavary, Krishna, and Cauvery. All have been extensively developed for irrigation, especially at their deltas. (For much more about the geography of India—physical and human—see O. H. K. Spate's *India* [1954]. Updated in later editions only through 1972, it's badly out of date, but at least among English geographies it's probably the only one with a sense of humor.)

THE EMERGENCE OF INDIC CIVILIZATION

Civilization in South Asia can be traced to the Indus Valley about 4,500 years ago. Not surprisingly, its appearance remains fundamentally inexplicable, with archaeologists divided on the importance of Mesopotamian influence. The Neolithic farming village of Mehrgarh, near Quetta, shows that barley and other crops were cultivated as much as 8,000 years ago. Ceramics and copper appear later, but there is no indication of social stratification. Then, out of the apparent blue, comes the Indus Valley Civilization, which thrived over a very large area for about 500 years, until suddenly disappearing without apparent cause.

This civilization remained completely forgotten until the British realized that some rocky mounds they were quarrying for railway material were ancient ruins. Archaeologists went to work in the 1920s and began excavating two soon famous cities, Harappa and Mohenjo-Daro, the Mound of the Dead. The first lies on the Ravi River downstream from Lahore; the second is another 400 miles downstream, at a point about 20 miles south of

Larkana. Both cities used writing, of a kind not yet deciphered; there is also clear evidence of trade with Mesopotamia.

The two cities are famous for their gridded streets, modular housing, and drainage systems fitted with manholes. This is striking, because the rainfall at the time the cities were occupied was about 4 inches per year. Why build drains? Water seems to have had a special importance in these cities, because special attention was given to buildings tentatively identified as ritual baths. The building at Mohenjo-Daro called by its excavator the Great Bath, for example, stands apart from the rest of the city. It is bitumen-sealed and appears to have been the center of an elite residential district. Conceivably, there is a link between this bath and the profound attachment felt by Hindus today for certain sacred rivers, especially the Ganges. Early in 2001, some 20–30 million people came to Allahabad and the confluence there of the Ganges and Yamuna rivers. In the *Purna Kumbh Mela*, a festival recurring every 12 years, they bathed in the frigid water and blissfully washed away their sins.

IDEOLOGY

After the fall of the Indus Civilization, 1,000 years passed before a new urban tradition arose. In the meantime, the Aryans (literally, the Nobles) had arrived, first as a pastoral people in the Punjab. Wherever the Aryans came from, the Iranians claim them as their own. In fact the name Iran, adopted by Persia in the 1930s, means "[land of] the Aryans." The newcomers gradually moved into the rice-growing areas of the lower Ganges plains and south into the drier, sorghum- and millet-based country of the Deccan.

Under the Aryans, city-states eventually stretched across the Indo-Gangetic lowlands from Taxila in the west to near the present city of Patna. During this period and in ways that remain unknown, the *sanatana dharma,* or eternal law, developed. This is the body of belief that Westerners since about 1800 have called Hinduism. It spread far to the east, which is why there are Hindu monuments today at Angkor in Cambodia and on Bali in Indonesia. Buddhism and Islam eventually replaced Hinduism in these countries but not in India.

Because of the Western association of religion with deity, foreigners often approach Hinduism by studying its polytheistic structure. They read about Brahma the Creator, Vishnu the Sustainer, and Shiva the Destroyer. They learn about the divine *avatars*, literally the descents from heaven of these gods. Krishna and Rama, for example, are two very popular avatars of Vishnu. Foreigners learn also about the symbols of the gods, preeminently Shiva's phallic *lingam,* or sign. They learn about the freedom of Hindus to worship whichever of these gods, or many others, they like. Hindus can even worship none of them but still be considered good Hindus.

Some of that worship occurs in the home, where a room or space is usually set aside for an image of the family's god of choice. There, in the domestic equivalent of a temple, there is a daily *puja*—literally, praising. Food and prayers are offered to an image of the family's chosen god, and family members are purified and marked on the forehead with a daub of color, a *tilak,* its shape identifying the god worshipped.

There are public temples, too. They are everywhere, perhaps even more numerous than the churches and mosques of Europe and the Middle East. Like them, the Indian temple dominates the space around it partly by its size, partly by its lavish ornamentation, and partly by the expensive and permanent material—usually stone—of which it is made. Unlike churches and mosques, however, which strive inside and out for a sense of height or movement into a higher realm, the Hindu temple makes no effort to rise above the earth. Instead of making polar opposites of the material and spiritual worlds, the Hindu temple suggests that the material world is merely a curtain obscuring spiritual reality—"real" reality—from the eyes of the ignorant. There is no point in rising above the curtain. The task is to see through it.

The temple exterior often takes the form of a massive tower profusely decorated with images of animals, human beings, and gods. Despite its height, it is static, not dynamic, and its height suggests the importance of the deity within but not spiritual yearning. Unlike the church or mosque, the Hindu temple suggests that what is needed is insight, not effort.

Inside, even in the grandest temples, several rooms or porches lead under the towering mass of masonry to a small, dark chamber. There rarely is room for more than a dozen people in that room at a time, and in a popular temple turnover is rapid. The worshippers crowd before the statue of the temple god. There is no communal service. Instead, priests help supplicants make an offering or perhaps wash the image or decorate it with flowers. The goal is *darshan*, a moment of visual contact in which worshippers believe that not only are they seeing the deity but the deity is seeing them. It is no hollow ritual: Hundreds of millions of Hindus take it very seriously, some at face value and some as a way to become aware once again of the reality behind the visible but illusory world.

The gods of Indian temples today are not the gods who figure in the oldest Hindu texts, the Vedas, literally Knowledge. Written 2,000–3,000 years ago and concerned largely with

Khajuraho, between New Delhi and Allahabad, is one of the most famous Indian temple complexes. The two temples shown—Mahadeva and Jagadambi—are among the largest and most ornate, but neither has any congregational space. Prayer is personal, not social.

the ritual minutia of fire sacrifices, the chief Vedic gods are Indra and Varuna. About 2,000 years ago, when the epic Mahabharata was written, these gods were displaced. Shiva and Vishnu assumed primary importance, along with their avatars and the Great Goddess. She is known by many names, among them Durga and Kali.

Apart from this complex and changing mosaic of deities and practice, Hinduism also generated an immensely powerful ideology that sustains what is probably the most elaborate hierarchy of any civilization. This ideology is clear in later parts of the Vedas, specifically in the Upanishads, which move beyond rules of sacrifice to philosophical concepts that remain of central importance to Hinduism today. Among these concepts is *samsara*, or transmigration, which imagines the soul cycling endlessly between births. A second crucial concept is the division of society into four hierarchical *varnas,* literally colors. Ranked from top to bottom, they consist of the *brahmins* or priests, *kshatriya*s or warriors, *vaishyas* or merchants, and *sudras* or laborers. A third and unifying concept is *karma* (action), according to which one's actions in life determine the *varna* into which one is subsequently reborn. These beliefs were codified in the Laws of Manu, one of many *dharma sastra*s, or manuals of religious law, and they justify what Westerners have long called the caste system. The word "caste," however, is not of Indian origin. It was introduced by the Portuguese and means race or breed.

The caste system is much more complex than the fourfold division of the *varnas* suggests. It actually consists of perhaps 3,000 *jati* (literally, births). Traditionally, these *jati* corresponded to an occupation, but this does not imply 3,000 occupations. There is a great deal of duplication, with many *jatis* sharing the same vocation but found in different parts of the country. A Hindu, then, is born into a farming *jati* and traditionally at least has been destined to be a farmer; another *jati*, and the Hindu was destined to be a tanner. Tanners were an especially low caste, because they worked with cow hides in a country where cows are sacred. (It's probably more dangerous in India today for a motorist to hit and kill a cow than a pedestrian—certainly more likely to produce a threatening crowd. McDonald's, you may recall, got into trouble with Hindus around the world in 2002 because it adds a minute trace of beef flavoring to the french fries it sells outside India.) The *jatis* of the lower *sudras* constitute the group, 160 million strong today, called untouchables, people whose touch—even whose shadow—was (and to some extent still is) considered ritually polluting, especially by Brahmins.

How can a person's caste be identified? In villages, it's an easy matter, because occupations there still correspond well with caste. The cleavage is visible in the village landscape, too. On one side of the main road there will be *pukka*, or well-built, brick houses. These will be for high-caste villagers, some of whose houses have satellite dishes on the roof and cars parked outside. Over on the other side of the road, there will be *kutcha*, or poorly made, huts of mud and thatch, with no electricity in sight. Safe bet, that's the neighborhood for untouchables. (Residential segregation among the different *varnas* and of non-Hindu groups is also strong but less visible than the chasm between untouchables and everyone else.)

In cities, the task is more challenging. Family names are a powerful indicator but are also very complicated. The single group of land-owning sudras called the *kayasthas,* for example, is associated with 100 family names—half a dozen or more in each Indian state. Non-Indians are unlikely to find their way through this thicket. For them, one indicator of caste arises from the fact that strict Brahmins will not eat with non-Brahmins. Indeed, they will not eat at all, unless they know the food has been prepared by a Brahmin. It's called commensalism. Come lunch, everyone will sit down except one or two people who slip away. "They're Brahmin" will be the explanation.

Caste can be defended as a wonderfully enduring way of organizing a society and making everyone in it mutually interdependent. The defense of caste goes beyond economic logic, however. To the hundreds of millions of Hindus who accept it unconditionally, caste provides tremendous emotional support. People who, by Western lights, live unbelievably wretched lives will feel no bitterness, will say that God's will must be obeyed, will say that their lifetime of suffering is the inevitable and just result of their actions—their *karma*—in a previous life. Besides, if they live ethically in this life, they will come back blessed in the next. The taxi driver and rich passenger might just change places.

There's another side to the story, though, because caste provides an excuse to disregard human suffering. See a double amputee crawling along the street with a begging bowl? No problem: "It's his *karma*." See a child with a begging bowl, a plastic bag of urine tied to his side and hoses penetrating his abdomen? No need for concern: He deserves it. See the destitute widow, cast off by her family since her husband's death? Feel no pity. Caste, in short, provides an impregnable rationale for ignoring pain. That's why it often strikes Westerners as appallingly callous. They've been taught to believe, at least, that people should help each other. That's why beggars in India invariably zero in on Westerners as easy marks.

Indians with a Western education often share the Western reaction. Gandhi returned to India after years abroad and tried among other things to remove the stigma of untouchability. His tactic was to call untouchables *harijans*, people of God. The Indian constitution, written after his death, took a more legalistic approach and simply declared untouchability illegal. It also tried to replace that name with the less pejorative "scheduled castes." As an American might guess, discrimination persists despite the law, which is why militant untouchables now—and there are many, some highly educated—call themselves *dalits*, the oppressed.

A more ingenious ideology for sustaining a hierarchical society can hardly be imagined. About 80% of India's judges and 60% of its senior administrators (members of the elite Indian Administrative Service) are Brahmins, even though Brahmins constitute only 5% of India's population. Low-caste villagers, on the other hand, are still beaten for trying to enter a temple or draw water from a village well. Even in India's booming cities, glitzy new buildings while under construction are juxtaposed against the crude plastic and burlap pup tents that are the homes of communities of low-caste or tribal construction workers. Husbands, wives, children spend their lives breaking rock and doing other brutally hard tasks. Often their mood is far happier than that of Westerners of wealth. "They are resigned to their fate," an educated Indian says, and he is right. Thank *karma*.

CHALLENGES TO THE IDEOLOGY

Pick up the *Sunday Times* of India. It's in English, which means it's for the country's elite. Then turn to the matrimonials section. It's neatly bifurcated: wanted, brides; wanted, grooms. It's then meticulously subdivided. The listings begin with "Cosmopolitan," where ads typically contain the words "caste no bar," even though the caste of the family placing the ad is usually specified. This section runs two columns or so and is followed by "Scheduled Caste/Scheduled Tribe," which runs all of 6 inches and is for people at the bottom of the hierarchy. Then, neatly walled off from them, comes the caste parade. Advertisements of available Brahmin men fill four columns. There are 5 inches advertising *kshatriya* men. There's a column of *vaishya* and three columns of *sudra*, specified not as such but by more flattering specific names, such as the land-owning *kayasthas*. (In fairness, at the end of these listings by caste, there are also listings by language, profession, religion—Buddhist, Christian, Muslim, Sikh—and by nationality, meaning nonresident Indians, especially permanent residents of the United States.)

With caste so powerful even among India's Western-educated elite, it's easy to see why attacks on the caste system have failed. Perhaps the most famous took place in the third century B.C., when the Emperor Asoka expanded the Mauryan Empire until it covered almost all of India, Pakistan, and Afghanistan. Appalled by the violence he witnessed in the course of his military campaigns, Asoka adopted the new religion of Buddhism, which appealed to Asoka because of its doctrine of *ahimsa*, or nonviolence. (That's the term Gandhi adopted in the course of driving the British from India, and Martin Luther King, Jr., used the English translation of it when he adopted a Gandhian strategy in the fight for civil rights in the United States.)

Buddhism also renounced Vedic ritual, caste, and the very idea of God or gods. Like Confucianism, it was more a philosophy than a religion in the Western, deistic sense, but unlike Confucianism it was strongly spiritual, concerned less with ordering social relations than with escaping from the sufferings of life. It's a difficult teaching, often corrupted into the simple worship of the Buddha. In Sri Lanka, at least, Buddhism also incorporates caste, even among Buddhist monks.

For a time, Buddhism flourished in India, and Asoka sent missionaries abroad. After his death, however, Asoka's empire disintegrated. Hinduism returned and by the 13th century had driven Buddhism almost entirely out of India. It survives today chiefly in India's neighbors, including Sri Lanka, Burma, Thailand, and Tibet, as well as more distant countries, such as Japan. That's why if you're interested in Buddhist monuments you will visit Anuradhapura in Sri Lanka, Pagan in Burma, Borobudur on now-Muslim Java, and Nara, near Kyoto. India still has some Buddhist relics—Sanchi is perhaps the most spectacular—but most have been eradicated. The greatest of them all, Amaravati, near the delta of the Krishna, is today only barren ground. A few magnificent sculptures survived on the site until the 19th century, but in 1880 they were excavated and carted off to the British Museum. There they are kept under far better conditions than Indian museums provide, but a strong case can be made that they belong back home, perhaps on legal grounds but also because works of art like these are weakened when uprooted—so it seems, at least, if you wander around the great collections of Indian art in the United States. A tycoon or tycoon's wife takes a fancy to the mystic East, and the result is as jarring as the pope gambling in Vegas.

Still, India is not finished with Buddhism. In 1956 a half million untouchables became Buddhists. Their inspiration was Dr. B. R. Ambedkar, himself an untouchable but also a government minister largely responsible for writing the Indian constitution. Conversions to Buddhism continue today. Late in 2001, for example, about 50,000 *dalits* went to Ambedkar House in Delhi and repeated a set of vows declaring themselves henceforth to be Buddhists. It's one way to escape the strictures of caste. *Dalits* of a more militant nature meanwhile have the option of joining the National Campaign on Dalit Human Rights. Still others have gone into politics, sometimes very successfully. Several times in recent years, the chief minister of U.P. has been Mayawati, a Dalit and a woman to boot.

A second threat to the caste system was Islam. It arrived in the 11th century with conquerors whose descendants held much of India for most of the next 800 years. High-caste Hindus seethed, much as many Muslims around the world today seethe at the thought of their humiliation by richer and more powerful nonbelievers. Low-caste Hindus, on the other hand, saw Islam as an escape. In large numbers they converted, so that the Muslim populations of India and Pakistan today are descended chiefly from former Hindus, not from the

Near Bhopal, Sanchi like other Buddhist constructions is a geometrical composition: it's circular, with profusely ornamented gateways at the compass points. The gates lead only to a processional path that is elaborately screened but merely circles the domical mound or *stupa*.

invaders who brought Islam to South Asia. For the strict Muslim, meanwhile, Hindu polytheism has always been an abomination, a denial of the unity of Allah; there are few graver offenses in Islam. There have been periods in India's history where the two groups got along amicably, but resentment and contempt are close to the surface, not far from violent eruption.

Fearful in the 1930s of being dominated by Hindus after the anticipated British departure from India, the Muslim elite pushed successfully for the creation of their own homeland. They got it with the creation of Pakistan, a country created in two parts—East and West—to reflect the two parts of India where Muslims were most concentrated. The Muslims of East Pakistan were Bengalis, however, not Punjabis or Sindhis, and they soon felt exploited and despised by their coreligionists in the West. In the 1970s they fought a civil war with West Pakistan and, with India's help, established an independent Bangladesh. West Pakistan became simply Pakistan, but not before blowing up East Pakistan's railway bridges—a parting shot, a farewell insult.

Some 130 million Muslims remain in India after these adjustments. Many Hindus continue to see them as a threat, and to defend Hinduism they often support a powerful fundamentalist organization called the *Rashtriya Swayamsevak Sangh* (RSS), or National Volunteers Association. Gandhi's assassin, Nathuram Godse, was an early member of the RSS, which was established in 1925. So is L. K. Advani, recently India's deputy prime minister. The RSS supports several sister organizations, including the BJP, the political party that led India from 1998 to 2004. (The initials stand for Bharatiya Janata Party, or Indian People's

Party.) Although the national leaders of the BJP, like Advani, speak the language of moderation and secular democracy, there are other RSS members who want to rid India of Muslims and who even believe that India will one day conquer Pakistan. Those are unlikely events, but RSS members have a major hand in writing India's history textbooks. Sample question from a booklet used in the RSS's own schools: "Who shed rivers of blood to spread Islam?" Answer: "the prophet Mohammed."

Textbooks like that help explain why India's Muslims lead unsettled lives. Thousands have died in recent years at the hands of Hindu fundamentalist mobs, often supported by state governments and local police forces sympathetic to the RSS. The violence is often connected to Hindu attacks on mosques. The most famous recent case was the destruction in 1992 of the Babri Mosque at Ayodhya, in U.P. Riots followed and killed 2,000 or 3,000 people. It's not a finished story, either. RSS activists say that the site was a temple before the Muslim invasion of India. They may well be right, because similar replacements of temples by mosques can be found in Delhi and elsewhere. A court-ordered archaeological excavation at Ayodhya found in 2003 the remains of a large building. Hindus claimed vindication, but Muslims retorted that the ruins were those of an older mosque.

Some 100,000 fundamentalist Hindus went to Agra in 2001 for a convention. They followed it up by going to the Taj Mahal, which of course is a Muslim tomb. They overwhelmed the security guards, poured into the grounds, and committed acts of petty vandalism. The damage could have been much worse than chalked slurs and trash, but the Taj occupies a special place in the minds of Indians, and its despoliation would be extremely provocative.

The next year, there were riots in the supposedly progressive state of Gujarat. Provoked by Muslim atrocities, Hindu mobs killed 2,000 Muslims. The murders were heinous: Women were raped then forced to swallow kerosene, which was set alight. Local government officers sat on their hands, yet the state government was reelected in the next year with an increased majority. To his credit India's prime minister, Atal Behari Vajpayee, was said to have been devastated by the riots and to have asked one of his ministers, "How can I show my face to the world after this?"

The 800,000 Muslims in Ahmadabad, Gujarat's capital, have taken refuge in ghettos. One of them told a reporter in 2003 that "it is better to live in Bihar." The reporter was surprised, because Bihar is much, much poorer than Gujarat. Bihar, however, is run by a low-caste Hindu—a cow herder by *jati*—who has won power by getting the votes not only of low-caste Hindus but of Bihar's Muslims, whom he promises to protect. Laloo Prasad Yadav tells his opponents: "You are privileged Brahmins and upper caste exploiters. You live in luxury and you have been treating us like animals for thousands of years. Now it is our turn to take control." Control in this case, unfortunately, means turning the tables. Laloo was convicted of corruption and forced from office in 1997. No matter, his wife was elected in his place. Such politics underlie the serious joke that in some countries you cast a vote; in India you vote your caste.

The government of India, in a rough equivalent to affirmative action in the United States, reserves university places and civil service jobs for low-caste applicants. The prime minister who introduced this policy in 1990 was quickly turfed for this offense, but the policy itself remains. Many upper-caste students have set themselves afire in protest. For a time, it's reported, salesmen made a good living selling fire extinguishers to worried parents.

The most dangerous flashpoint in South Asia, one threatening nuclear war, is Kashmir, another chapter in Hindu–Muslim relations. Inhabited chiefly by Muslims, India has retained Kashmir ever since the Maharajah of Kashmir, who was a Hindu, decided in 1947

Cobbled together near Delhi from masonry lifted from Hindu temples, the immense Kutb Minar, or Victory Tower, celebrates the advent in 1193 of Muslim power to India.

to join India instead of Pakistan. He could have argued that the Muslims of Kashmir were the victims of forced conversions to Islam in earlier centuries. Still, they were Muslims now, and given a choice would have joined Pakistan or formed an independent state. Decades have passed, but India remains determined not to relinquish the territory, especially because doing so would inspire other separatist groups. No Pakistani leader, on the other hand, can survive if he is seen in Pakistan as abandoning fellow Muslims.

Fifty years after they were divided, the peoples of Pakistan and India might as well occupy different continents. One sign of the rift is that transportation links between the two countries are extremely poor. Despite their immense populations, for example, there are fewer commercial flights between India and Pakistan than between Dallas and Denver. Sometimes there are none for months on end. At the end of 2003, when the two governments were making mutual gestures of accommodation, they agreed to permit a grand total of 10 flights weekly. The links would be between Karachi and Lahore on the Pakistan side and Mumbai (Bombay) and Delhi on the Indian. Analysts said that PIA, Pakistan's carrier, might make money on the routes, because it offers cheap onward connections to the Gulf and the jobs awaiting Indians there. Indian Airlines, they predicted, was going to fly boatloads of empty seats.

CHAPTER 9

A TECHNOLOGICAL CIVILIZATION

I want to reverse the flashlight 180 degrees now and look not at foreign civilizations but at our own. From here on out, in fact, we'll be looking mostly at ourselves. Bluntly, that's because our society, for good or ill, is the dominant one on the planet today. Do you think you can find a country that isn't striving for the kind of economic growth that was pioneered in Europe and the United States? Well, maybe Cuba, but that's momentary. Maybe Burma (Myanmar), but that too will pass. In a globalizing world, everyone plays the same game, win or lose.

As with my discussions of China and India, I'm going to focus on ideology. What's unusual in this case is that for the first time we'll encounter a civilization sustained by the promise of material wealth, obtainable in this life and this lifetime. We'll encounter a secular ideology sustained by a faith in technological progress.

CAPES AND BAYS

Once more the land! Apart from European Russia, which by convention extends east as far as the Urals and the Ural River, Europe is no more than a battered Asian peninsula. In the scattered mountains of northern Spain, Scotland, and Norway, it bears the scars of an ancient collision with North America. (The opening of the Atlantic Ocean separated the Old and New Worlds, which is why the Appalachians aren't attached to the Scottish Highlands.) The mountains rimming the Mediterranean, including those of southern Spain, Italy, and Turkey, are the bruises of a separate collision with Africa, and the Mediterranean is a relic of a once larger sea called Tethys.

The European peninsula's length is impressive—it's 2,500 miles from Gibraltar to Israel's beaches—but it's hard to find a spot in Europe west of Poland that's more than 300 miles from salt water. This means that European rivers tend to be short and divided by the Alps and Carpathians into rivers flowing north or south. The Rhine and Rhone, for example, start on opposite sides of the St. Gotthard Pass in Switzerland. The great exception is the Danube, originating in Germany but finding its way east and south to the Black Sea.

The peninsula's latitudinal orientation carries over to its agricultural character. There's a Mediterranean fringe of olives and citrus. Grapes extend north to a line just north of Paris, Frankfurt, and Budapest. Wheat survives to a line slightly north of Oslo, Stockholm, and St. Petersburg. Potatoes are possible still farther north. (A pioneering color photographer

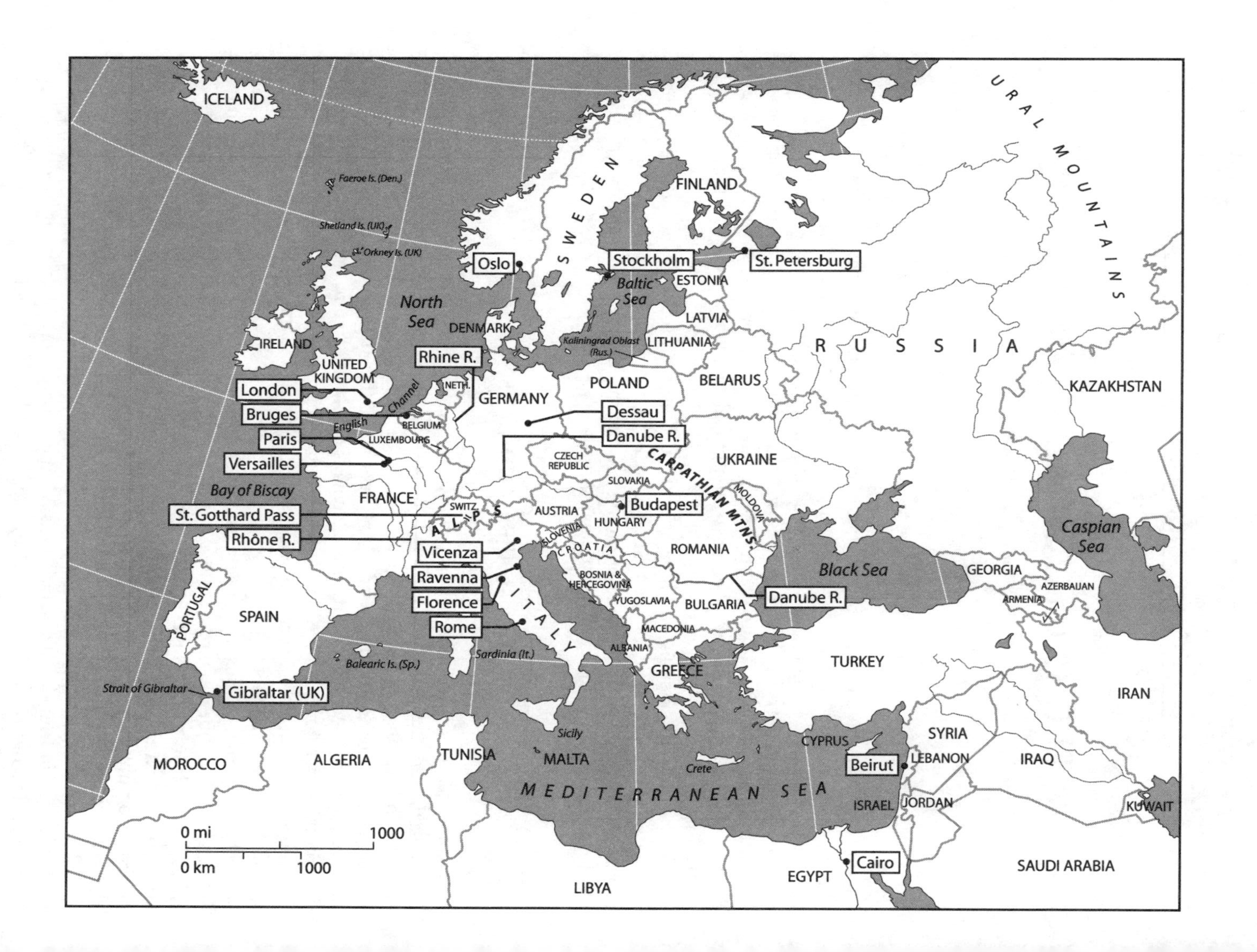
ICELAND
Faeroe Is. (Den.)
Shetland Is. (UK)
Orkney Is. (UK)
IRELAND
UNITED KINGDOM
London
Bruges
Paris
Versailles
English Channel
North Sea
BELGIUM
LUXEMBOURG
NETH.
Rhine R.
DENMARK
Oslo
SWEDEN
Stockholm
Baltic Sea
FINLAND
St. Petersburg
ESTONIA
LATVIA
LITHUANIA
Kaliningrad Oblast (Rus.)
POLAND
BELARUS
UKRAINE
MOLDOVA
GERMANY
Dessau
Danube R.
CZECH REPUBLIC
SLOVAKIA
AUSTRIA
HUNGARY
Budapest
CARPATHIAN MTNS.
ROMANIA
BULGARIA
Danube R.
FRANCE
Bay of Biscay
SWITZ.
ALPS
St. Gotthard Pass
Rhône R.
SPAIN
PORTUGAL
Gibraltar (UK)
Strait of Gibraltar
Balearic Is. (Sp.)
SLOVENIA
CROATIA
BOSNIA & HERCEGOVINA
YUGOSLAVIA
MACEDONIA
ALBANIA
GREECE
Crete
ITALY
Vicenza
Ravenna
Florence
Rome
Sardinia (It.)
Sicily
MALTA
MEDITERRANEAN SEA
MOROCCO
ALGERIA
TUNISIA
LIBYA
EGYPT
Cairo
RUSSIA
URAL MOUNTAINS
KAZAKHSTAN
Caspian Sea
Black Sea
GEORGIA
ARMENIA
AZERBAIJAN
TURKEY
CYPRUS
SYRIA
LEBANON
Beirut
ISRAEL
JORDAN
IRAQ
IRAN
KUWAIT
SAUDI ARABIA
0 mi 1000
0 km 1000

Continental watershed: Italy to the left, glaciered Austria to the right. These peaks belong to the Hohe Tauern, perhaps the most dramatic part of the Alps; the view is west.

named Sergei Prokudin-Gorskii traveled across prerevolutionary Russia and took amazing photographs along the way. One of them shows a line of black-robed monks digging potatoes in such a black-soiled forest clearing. See *Photographs for the Czar* [1980], edited by Robert Allshouse.)

DISCOVERING THE INDIVIDUAL

If Europe isn't quite a continent, as a civilization it's even more flawed. Unless you count the European Union, which has recently created a slight degree of pooled ego across 25 countries, nobody's been able to put forward even a plausible claim of sovereignty over the place since Martin Luther 500 years ago broke the Catholic Church's hold over Western Europe. European civilization, then, like India's, rests on an ideology sustaining a literate elite, not on a monolithic state of the sort that has ruled China for most of the last 3,000 years.

Rather than try a potted history of the fractured civilization that emerged on this ragged fringe of Asia, let me turn straightaway to ideology. I'll approach it through the revolutionary changes that occurred in European painting around the year 1450.

For centuries, painters had been portraying the Virgin, the Christ child, and angels. Gradually, these images became increasingly lifelike. A good example is Cimabue (1240–1302), whose paintings of the Virgin seemed to his contemporaries to capture the human form perfectly. A century later, Masaccio (1401–1428) painted Biblical characters whose expressions are intensely dramatic. His portrayal of Adam and Eve, which is in Florence's Brancacci Chapel, seems even now to capture and express their pain and shame as they are forced from Eden.

Painters soon shifted from tempera to oil. They continued to paint from the Bible—continued to do so for as long as religion remained a major force in European culture. In these late works they left a residue of often tedious sanctity on acres of palace and church walls, but painters also found a new subject. For the first time in European history, they began painting portraits of living people with photographic accuracy. For the first time since the Romans, individuals mattered enough to be artistically recreated.

Jan van Eyck (1390–1441) was the first or among the very first to master this art, there for all to see in "The Betrothal of the Arnolfini," now in London's national gallery. Giovanni Bellini (1430–1516) wasn't far behind, and van Eyck's influence is plain in Bellini's famous portrait of the Venetian doge Leonardo Loredan, also in London's national gallery. Self-portraits followed very quickly. Van Eyck may have created the first, if "Man in a Turban"—again in the London gallery—is him. Albrecht Dürer (1471–1528) certainly did self-portraits by about 1500, although they seem to be idealized, Christ-like figures. Portraits of women lagged, though the earliest extant one is far more famous than any of these male representations. It was painted about 1505 by Leonardo da Vinci and shows the famously enigmatic wife of a Florentine banker. Realistic representation came to sculpture, too. A good example might be Andrea del Verrochio's bust of the commanding Lorenzo de Medici, perched on a short column in the basement of the national gallery in Washington. Animated, given a body, and suitably dressed, Lorenzo would be instantly at home among Washington movers and shakers.

It was a remarkable achievement, linked to the emergence of a society in which people were proud to display their power—physical, economic, or political. Van Eyck worked in Bruges, now a small place favored by tourists but then a leading commercial center. His most famous subject, Giovanni Arnolfini, was the Bruges agent of Florence's Medici bank. Such people had worlds to conquer. Witness Hans Holbein the Younger (1497–1593) and his painting called "The Ambassadors," done about 1530 and yet again in London's national gallery. Two men, apparently French ambassadors to the court of Henry VIII, stand proudly next to a globe of the newly understood world. They are luxuriously dressed in fur-trimmed robes. Each has an elbow resting on a table covered with a Persian carpet, a globe, navigational instruments, and a lute. A silk cloth hangs in generous folds on the wall behind them; the floor is inlaid stone. The chief hint of the medieval past is the strangely distorted skull at the men's feet, a symbolic check on worldly pride.

Breaking away from photographically accurate images, painters began to paint what they felt while looking at their subjects. A good early example is Botticelli's "Birth of Venus," from the early 1480s. Botticelli was perfectly able to paint accurate representations of the human body, but here he instead shows the goddess in a position that no artist's model could hold without falling on her face. No matter. Botticelli was more interested in spirit than matter. Such departures from realism grew stronger with time, until realistic art almost vanished in the 20th century.

Nothing, however, was going to derail the growing importance of the individual. The Arnolfinis and Loredans and Medicis were powerful people, but today even the most socially insignificant people are routinely photographed. The rise of portraiture, it turns out, marked Europe's return to the ancient view of Protagoras, the Greek philosopher who had called man the measure of all things. Kenneth Clark makes the connection very clear in *Civilization: A Personal View* (1969), where he borrows that phrase for the title of his chapter on the renaissance. The medieval preoccupations with piety, salvation, sin, and damnation no longer dominated European culture.

If there's any doubt about this, look at how painters came to depict the human body. They had become conscious of three-dimensional space with Giotto (c. 1266-1337). What to fill it with? Giorgione (1478-1510) dared to paint nude and sensual women. His much longer-lived contemporary, Michelangelo (1475-1564), carved a defiant David completely unashamed of his nakedness, even arrogant in his self-display.

A precursor of this new outlook may lie in the Gothic style's love of light, where the material world possesses a beauty that can lead the mind to truth. Still, there's no definitive explanation for this emergence of the individual in Europe. It can be interpreted—my last major venture into irresolvable speculation—as an inevitable event in a fractured civilization too weak to suppress individuality. Renaissance Europe was such a place, with power divided between the church and state. A wonderful illustration of this tension is in a mosaic in Ravenna's church of San Vitale. There, the emperor Justinian stands next to the archbishop Maximianus. At first glance, Justinian is the focus of attention, because he's in the middle and wears purple instead of white. But Maximianus stands under his name, while Justinian does not. In addition, the body and head of the archbishop are shown in the same plane as Justinian, but from the hips down the archbishop is moved forward so that he clearly upstages the emperor. This was a strategic rivalry. Sons of heaven would have considered it a fundamental weakness.

Why, if this explanation is valid, did authoritarian Romans create very realistic portrait busts? Perhaps the answer lies in Rome's imitation of all things Athenian and in Athens, for all its glory, being a second-rate political entity on the political stage of its time. In any case, individuality insinuated itself like a weed in Europe's cracked stone. In *From Dawn to Decadence* (2000), Jacques Barzun writes that emancipation—breaking free from chains—is one of the great themes of Western Civilization. He points out, too, that the first literary expression of this new outlook is in the essays of Michel de Montaigne (1533–1592), which came to Shakespeare and influenced his own conception of complex, lifelike characters.

THE BIRTH OF A SECULAR IDEOLOGY

The emergence of the individual expressed itself in many ways besides portraiture. Some were heroic, like Faust striving to fly among the stars. Others were more modest. We say, for example, that a man's home is his castle. Even such a mundane thing as clothing suggests the importance of individuality in the modern world, not only because clothing is available in so much variety but because, at least in the world's richer countries, it has become very hard to distinguish rich from average people merely by looking at their clothes. Convergent taste may not seem to be an expression of individuality, but we look as we wish to look, not as authority dictates or poverty compels.

Of all these expressions, I want to emphasize only one. A society built around the importance of the individual can logically be expected to try to make life as comfortable as possible for everyone. And that's just what we've done, with amazing though as yet incomplete success. We call it progress, and we're generally persuaded that it can't and shouldn't be stopped.

The formula was spelled out by Francis Bacon in *The Advancement of Learning* (1605). Bacon urges his readers "to reject vain speculations" and seek instead whatever is "solid and fruitful." What will come of this? In *The New Atlantis* (1626), Bacon sets out to describe his utopia. He gets sidetracked and instead describes the institution—in modern language we'd call it a research institute that could create his utopia. He calls it the House of Salomon, an allusion to the reputed wisdom of King Solomon. The House is devoted to "the knowledge of causes and secret motions of things and enlarging the bounds of human empire, to the effecting of all things possible." Bacon means that the world is our oyster and that we should extract the pearl, learn nature's secrets, and control the world. Think again of Michelangelo's David, so scornful that he seems to despise doubt, hesitation, or fear. Bacon, by the way, was not the first to think these thoughts. He yields that position to Bernardino Telisio (1509–1588) and says that he himself is only the herald of the new outlook.

But is it moral to see nature as ours for the taking? Is it really here for us? Presumably the Gilbert Islanders who fell upon those dolphins would be shocked at the idea: They called the dolphins friends, not resources. The growing importance of the animal rights movement suggests that there are people today who would similarly answer these questions with a vehement "absolutely not." Still, the great majority of people today would probably answer both questions affirmatively. Certainly the modern world would not exist if everyone in the last few centuries had worried about the morality of "enlarging the bounds of human empire, to the effecting of all things possible."

In setting aside any doubts, they unconsciously follow not only the advice of Francis Bacon but the perhaps even more important teaching of René Descartes (1596–1650). It sounds very much like Bacon when, in the *Discourse on Method* (1637), Descartes attacks philosophy. He says that it "has been cultivated for many centuries by the best minds that have ever lived, and that nevertheless no single thing is to be found in it which is not subject to dispute. . . . " So much for the pursuit of vain speculation.

Descartes begins his *Principles of Philosophy* (1644) with another Baconian flourish, a metaphor in which knowledge is likened to a tree. Metaphysics is the root, physics is its trunk, and the various sciences are its branches. The point of the metaphor isn't just that knowledge is unified but that the fruits of knowledge grow in the branches, not at the metaphysical root.

Descartes goes beyond Bacon, however, by formally justifying human dominion over the earth. The word "dominion" reminds us that the idea of dominance isn't new with Descartes: It goes back at least as far as the first chapter of *Genesis*. But Descartes gives this position a scientific basis. He asserts that the world consists of two kinds of matter: *res extensa*, or solid matter including our own bodies, and *res cogitans*, or thinking matter, which only human beings possess. Animals, in his view, are only machines, as Descartes thinks we would be without *res cogitans*, or the power to think. By definition, then, animals lack mind or soul. That is why we need not question the ethics of dominating them, let alone dominating plants and the inanimate earth.

Descartes is giving a modern flavor to the ancient idea of philosophical dualism, or the separation of mind and body. Plato had made a comparable separation, but Descartes made it seem fresh. That's partly because he repudiated authority of every kind, from Aristotle to his own teachers. Insisting on trusting only his own judgment, Descartes's radical skepticism and rejection of inherited dogma have been an inspiration to scientists ever since. Descartes was additionally credible because he believed, as we do, that acquiring dominion over nature demands the development of the quantitative sciences, which he himself stimulated with his work in analytic geometry.

There's a third and more indirect way in which Descartes is strikingly modern, too. Very often, he presents his arguments not in the abstract but as part of a personal narrative describing his own search for truth. Imagine Michelangelo's David once again, this time as a thinker telling you how he came to the positions he holds. It's individuality in yet another guise. Perhaps it's hard to see David as a thinker, but you can certainly see Cartesian dualism in Michelangelo's sculptures for the Medici tombs in Florence. There, unfinished human bodies emerge from rock, like Descartes's souls from base matter.

Cartesian dualism is by now so deeply embedded in our culture that we no longer even realize we see the world through Descartes's eyes. Think, though, of the way we habitually reject the animism of our distant ancestors. They attributed spirit not only to plants and animals but to inanimate objects and even forces like thunder. For most Westerners, this is all rubbish and nonsense, along with the kaboodle of magic and mountebanks. It's amazing to see how far we've come without even realizing who gave us the ride.

We can defend Descartes by saying that it was science, not superstition, that unlocked nature's secrets, or at least a good many of them. It's true, apart from the anthropomorphic silliness of thinking that nature tries to do anything, let alone keep secrets. The familiar litany of science's triumphs begins with the astronomy of Copernicus (1473–1543), Tycho Brahe (1546–1601), Galileo (1564–1642), and Kepler (1571–1630). Then came physics with Newton (1642–1727), whose alchemical endeavors we'd prefer to forget. Medical science got started with William Harvey (1578–1657). Over on the botanical side there was Linnaeus (1707–1778). Explorers sought out new plants and brought them to splendid new botanical gardens established not only at Kew Gardens in London but in Algiers, Kolkata (Calcutta), Peradeniya (near Kandy on Ceylon, now Sri Lanka), and Buitenzorg (now Bogor, on Java).

Less comfortably for many, the dualistic eye turned to the human body with the study of human anatomy, pioneered by Andreas Vesalius (1514–1564). Rembrandt's "Anatomy Lesson of Dr. Tulp" (1632) shows a proud doctor and his colleagues grouped around a corpse, its arm opened to reveal muscles and tendons. It's no accident that in seeking to understand a person today, we say that we want to find out "what makes him tick." We test athletes for illegal drugs and declare that none were found in their "systems." It's all part of our Cartesian tendency to see the body as *res extensa*, with our consciousness neatly walled off as *res cogitans*. It's the same tendency that underlies the widespread fascination with body building, not to mention the even more common disgust with body fat. How strange to learn that "you've gained weight" is a compliment in Fiji; "skinny legs," a serious insult. China, by the way, offers a third ideal, with feminine beauty identified with light skin, small hands and feet, and a general air of weakness. Meanwhile, medical students tote textbooks with titles such as *The Machinery of the Body*.

THE IDEA OF PROGRESS EVOLVES

We're so accustomed to think of the 18th century as the Age of Enlightenment that it's easy to overlook the insult built into that name, the arrogance that dismisses a deeper past as the Dark Ages. History and economics were now approached from a recognizably modern—which is to say secular and progressive—viewpoint. The exemplars are Edward Gibbon (1737–1794) and his near contemporary, Adam Smith (1723–1790). Bacon and Descartes had at least claimed to be men of religious faith, but the tide of atheism was running long before Friedrich Nietzsche in 1882 wrote that God was dead. (Whether Nietzsche believed it is another matter. He seems only to have discarded the Christian God for one less devoted to the meek and poor.) Eighty years earlier, Napoleon had listened while Pierre Laplace (1749–1827) explained his theory of the origin of the solar system. Napoleon asked the scientist where God was in all this, and Laplace had famously replied that he had no need for that hypothesis.

In England this belief in the sufficiency of reason and the irrelevance of religion supported Utilitarianism, a philosophical school associated with Jeremy Bentham (1748–1832) and John Stuart Mill (1806–1873). The Utilitarians advocated the ordering of human affairs through a calculated balancing of pain and pleasure, but this was more than words. Both Bentham and Mill had close connections to the East India Company and were able to inject into the Company's policies a strong element of what we would today call economic development. The company set out, in the language of the time, to "improve" India by investing in telegraphs, railroads, agriculture, and irrigation. Even today you can look down from 37,000 feet over the Gangetic Plains and see the Ganges Canal, built in the 1840s by Proby Cautley, one of the East India Company's engineers. It runs straight as a die for miles and is as distinctive on the Indian landscape as center-pivot irrigation circles on the American Great Plains.

It's customary nowadays to say that the Company did these things for selfish reasons, but colonialism isn't necessarily a zero-sum game. The pursuit of profit could coincide with the pursuit of the public good, and the development of public works in Victorian India was in fact an unprecedented application of civilization's power to serve average people, not merely the elite. Wealth, for once, was not entirely diverted from the people who created it. The link to Francis Bacon was strong and explicit, too. One of the most famous names in Anglo-Indian history is Thomas Babington Macaulay (1800–1859). In India, Macaulay

Like all big canal systems, the Ganges Canal eventually splits into thousands of distributaries. Here's the bifurcation between the Kanpur and Etawah lines.

worked to establish English as the language of instruction in Indian schools, but in his spare time he wrote a long and rhetorically flamboyant essay on Francis Bacon. In it he writes that "the aim of Platonic philosophy was to raise us far above our vulgar wants. The aim of the Baconian philosophy was to supply our vulgar wants. The former aim was noble; but the latter was attainable. . . . " Macauley pushes his point: "the wise man of the Stoics would, no doubt, be a grander object than a steam-engine. But there are steam-engines. And the wise man of the Stoics is yet to be born."

When Macaulay wrote those words, the word "scientist" was 3 years old. The first recorded use is in 1834, with the justification that a word is needed to embrace the experimenters on all sides and with the further explanation that the word is consistent with others already in use, such as "artist," "economist," and "atheist." Much technological innovation of the time, however, had little or nothing to do with science. When Denis Diderot orchestrated the publication of the great *Encyclopedia* in the 1760s, for example, his subtitle was "a dictionary of science, art, and craft." He included scores of illustrations of industrial processes. Some, particularly in agriculture and fisheries, seem to employ no more than medieval tools. In other cases, however, there are elaborate machines. You will think of watermills and windmills. Fair enough, but Diderot also shows machines for papermaking. He shows machines to press grapes, apples, and sugarcane. Although the industrial revolution has yet to begin—let alone the modern linkage between science and technology—Diderot shows complex looms for weaving not only cloth but tapestries and carpets. (A few Jacquard looms, marvels of knotted string, still linger in Varanasi [Benares], where old weavers make fine brocades by hand and with threads of silk and gold-plated silver. The trade is dying even in India.) He shows machines that make pins and stamp coins. He also shows exceedingly complex, though nonmechanical, crafts such as shipbuilding; carriage manufacture; and the casting of guns, bells, and statues.

SYMBOLS AND EXPRESSIONS

From the egg of individuality and the attendant idea of technological progress came a new world.

The medieval garden was an enclosed space, a sanctuary. The renaissance garden, on the contrary, was open, with long axes to the horizon, befitting an owner who imagined himself in control of the world. No longer were plants allowed to grow untrimmed. Instead, they were forced into rectangular beds and trimmed into box-like shapes. The famous name here is André LeNôtre, who designed the gardens at Versailles for Louis XIV. That monarch had said that he and the state were synonymous. Why, then, not subject the world outside the palace to his will? The Versailles gardens eventually fell into disrepair but were restored in the 20th century with a donation from John D. Rockefeller, Jr.

Architecture changed, too. The leading renaissance writer on the subject, Leon Alberti (1404–1472), wrote that humans were "more graceful than other animals . . . [with] wit, reason, memory like an immortal god." Sound familiar? Maybe it's because 150 years later Shakespeare has Hamlet say much the same thing: "How noble in reason! how infinite in faculties . . . in apprehension how like a God!"

What kind of architecture suited such egos? Florence's renowned Pazzi Chapel, designed about 1430 by Brunellesco and adjoining the church of Santa Croce, is built to a human scale, neither overwhelming nor confining the viewer. Yet renaissance architecture isn't modest. Like many similar buildings, Andrea Palladio's Villa Rotunda, which was built near Vicenza about 1550, stands on a hill and has columned porticoes on four sides of a central hall, as though the owner's writ extends in all directions. It anticipates Francis Bacon's "affecting all things possible" by more than half a century.

Come closer to our own time. Until the late 19th century, the effective height limit on office buildings was about 10 stories, because going higher required masonry so thick that rooms shrank to closets. The steel-frame building, developed primarily in Chicago, changed all that. You might think that the most famous of the pioneering Chicago architects, Louis Sullivan (1856–1924), would have seen skyscrapers as engineering challenges or at least as a sensible way to produce plenty of rentable space. Instead, Sullivan saw the skyscraper as an expression of Baconian progress, of our Faustian yearning for power and control over the world. (Even the name "skyscraper" suggests the audacity of the undertaking.) In "The Tall Building Artistically Considered" (1896), Sullivan wrote that "the glory and pride of exaltation must be in it." Sullivan asks: "How shall we impart to this sterile pile, this cruel, harsh, brutal agglomeration, this stark, staring exclamation of eternal strife, the graciousness of the higher forms of sensibility and culture that rest on the lower and fiercer passions." Sullivan answered his own question by applying an extraordinarily rich, organically derived decoration, famously in Buffalo's terra-cotta-clad Guaranty Building (1896). Like his earlier Wainwright Building (1890) in St. Louis, it's treated today as a monument.

Sullivan's later career was rocky, however, and by the 1920s architects had grown hostile to ornament. The polemicist here was Le Corbusier, a Swiss architect who made a crusade of bringing methods of mass production to architecture. His thinking was hugely influ-

Florence's Pazzi Chapel, a Renaissance icon in its proportions, which seek neither to impress nor inspire, but to surround with comfortable spaciousness.

ential. *Towards a New Architecture,* for example, appeared in English in 1927 as a translation of the 13th French edition. In it, Corbusier called the house a machine for living. Here's how he explains it in the book: "If we eliminate from our hearts and minds all dead concepts in regard to the houses and look at the question from a critical and objective point of view, we shall arrive at the 'House-Machine,' the mass production house, healthy (and morally so too) and beautiful in the same way that the working tools and instruments which accompany our existence are beautiful."

In Germany, too, an architectural school in Dessau was pushing for a similarly purified architecture. The faculty of the Bauhaus rebelled against historic styles as an expression of social inequality. They wanted an architecture stripped of the trappings of aristocrats and the bourgeoisie. They developed, in the words of their leader, Walter Gropius, an architecture "avoiding all romantic embellishment and whimsy." Displaced in the 1930s by Hitler, whose taste swung from kitsch to imperial, the Bauhaus architects took up residence chiefly in the United States, where their style quickly became known as the international style, partly from an exhibition of that name and partly from the accompanying book, *The International Style: Architecture Since 1922* (1932) by Henry-Russell Hitchcock and Philip Johnson. Sullivan's "higher forms of sensibility" were gone now. Perhaps the finest practitioner of the sleek new style was Mies van der Rohe, who designed the Seagram (now Vivendi) Building in Manhattan, completed in 1958 and a symbol of elegance, with glass curtain-walls, bronze mullions top to bottom, and perfect proportions. He is famous for the aphorism "less is more," but he had the wit to add the tag: "and my architecture is almost nothing."

Mies's buildings were also very expensive. His followers sought ways to economize. Perhaps the firm that prospered most was Skidmore–Owings–Merrill (SOM), whose Gordon Bunshaft designed many office buildings in the 1950s. (As a prodigiously successful Mies imitator, SOM was destined to acquire the unkind nickname Three Blind Mies.) The new international style, stripped of ornament, appeared in a hundred cities. You can see it in the lines of the now destroyed World Trade Center. You can see it equally well and more modestly in the supermarkets, motels, and strip malls of the 1950s. The architects of these buildings wanted no higher praise than "efficient," the watchword of progress.

Between the 1940s and 1980s, another modernist architect designed what sometimes seems to be half of Southern California. Welton Becket had a significant international impact, too, because he designed Hilton hotels not only for Hawaii but for Havana, Cairo, and Beirut. For decades and for millions of people, those buildings almost defined the American dream—sleek, powerful, new. Now they seem naïve, anonymous in a world of warring architectural egos, each striving to capture the world's attention.

A Baconian city planning evolved alongside these architectural displays of technological power. Sixtus V (1585–1590) rebuilt Rome with great avenues lined by buildings built to a compulsory and uniform height, and the avenues converged in plazas with monuments symbolizing papal power. Peter the Great tried his hand, too. In 1710 he began to build a city to replace Moscow as Russia's capital. Fifteen years later, St. Petersburg had 25,000 people. Near the end of the century, the young United States invited a French officer, Pierre L'Enfant, to create a national city in Washington. L'Enfant alarmed his client by going straight to the book of Sixtus V and designing a city of boulevards and monuments for a city of 500,000 people—this, at a time when Boston had fewer than 20,000.

In the mid-19th century, Paris was famously redone at the none too tender hands of Georges-Eugene Haussman, Baron Haussmann, Prefect of the Seine. Haussmann cut boulevards through old neighborhoods, especially in the ancient Ile de la Cité. He located

Every American who has visited Washington, D.C., finds the monuments inspiring and the distances along the mall excessive. L'Enfant was interested in those great vistas and disregarded aching feet.

public buildings at visually commanding intersections, and he preserved Paris's old three-to-two ratio of building height to street width, even as he increased street width. Haussman insisted that the facades of new buildings be stylistically consistent with those of adjoining ones.

Over an equally long career from the 1930s to the 1980s, Robert Moses ripped and rebuilt New York City. His roads, parks, and public housing projects—recounted on a proportionately monumental scale in Robert Caro's *The Power Broker* (1974)—were the accomplishments of a titan who ignored objections. Like Haussmann, Moses lacked only Peter the Great's clean slate. If envy had been part of his emotional palette, Moses would have envied Walter Burley Griffin, the American who won the commission to design Australia's capital. Griffin explained that he laid out Canberra with "monumental government structures sharply defined rising tier on tier to the culminating highest internal forested hill of the Capital."

India at about the same time got a new capital. Edwin Lutyens structured New Delhi around a great processional way, Kingsway (now Rajpat), which descended from the viceroy's palace to a line of grand but unmistakeably smaller princely palaces. Inspirational new capitals became very popular after World War II. Le Corbusier in 1951 gave India's Punjab State a new capital in Chandigarh. In 1963 Constantinos Doxiadis gave Pakistan a capital in Islamabad. Over in the New World, Lucio Costa in 1957 laid out the vast plan for Brasília, whose monumental buildings were added by Oscar Niemeyer. As in so many other planned cities, the huge symbols of state were gradually surrounded by—symbolically sullied by—the very ordinary buildings built by or for the people working in these cities. In the case of

Brasília, the surrounding shantytown quickly became known as Cidade Livre, as though its residents had escaped the planners' clutches.

You can see the new, impatient, dynamic mentality in the countryside, too. It's there in the development of ever bigger, more powerful, more efficient machines for farmers, miners, and loggers.

Almost compulsively, it's there in the way that the public lands of the United States were surveyed. Plenty of people raised in the Midwest remember the shock of first driving the rural roads of the eastern states. Good-bye, grid; hello, crazy quilt. It's not the fault of the eastern road builders. Like Westerners, they were just following property lines. In the East, however, those lines followed or connected features like trees, hilltops, and streams. Thomas Jefferson would have none of this messy nonsense. He contrived instead a Cartesian grid system that was first applied in Ohio and later to the lands acquired by the Louisiana Purchase and farther west. Exceptions? Yes. Texas kept its public lands when it entered the United States, and the new survey system did not touch lands already privatized, such as the hundreds of ranchos in New Mexico and California.

Under Jefferson's system, first codified in the Land Ordinance of 1785, surveyors chose a basepoint through which they could run a prime meridian and a latitudinal baseline. They then struck grid lines at 6-mile intervals, creating a checkerboard of townships 6 miles on a side and easily identified as so many townships north or south of the basepoint and so many ranges east or west of it. Each township was subdivided into 36 sections, each of 1 square mile and each systematically numbered within the township. Each section, containing 640 acres, was subdivided into smaller pieces, including the iconic quarter-section, the 160-acre tract specified as the individual allotment in the 1862 homestead law signed by Abraham Lincoln.

For ease of property identification, the system is hard to beat. The University of Oklahoma's football stadium occupies the 20-acre plot precisely described as the east half of the southeast quarter of the southeast quarter of section 31, township 9 north, range 2 west, Indian Meridian.

As those words Indian Meridian imply, however, the public lands were never surveyed in one huge grid, because there was intense interest in the Far West long before anyone wanted to live on the Great Plains. And so a couple of dozen grids were created, each originating at its own arbitrarily selected basepoint. Most of Oklahoma is surveyed from the Indian Meridian, based in the Arbuckle Mountains, but the Oklahoma Panhandle has its own grid. Kansas and Nebraska are surveyed from a single basepoint, from which a grid was built that also covers most of Colorado and Wyoming. California has three basepoints, two of which have been extended (the technical term is "protracted") over large areas, while one covers only a bit of the north coast. Oregon is surveyed from a basepoint in the hills just west of Portland. The Willamette Meridian was then protracted north to cover Washington State. (Canada, by the way, beat the United States at its own game. The Dominion Lands Act of 1879 adopted the American system with only minor changes and then erected a giant grid over the prairies, from Manitoba west through Saskatchewan and Alberta. Only the Rockies defeated the continued spread of the grid to the Pacific coast.)

Both in Canada and the United States, roads mirror this Cartesian parcellation. You can wander to the horizon along east–west section lines, but on the north–south roads you have to jog at every fourth township line. Don't blame the surveyors: They were making necessary corrections to prevent the sections from getting narrower and narrower, as they inevitably would if the lines running north–south were true meridians. Such are the harsh realities that obtrude upon perfect rationality.

The Jeffersonian system has a wonderful simplicity for recordkeepers and people trying to find their way. Its disadvantage is that the survey lines don't respect topography. A quarter section can be bisected by a canyon that leaves its two sides on the equivalent of different planets. The same implacable logic accounts for the nearly impossible streets that were laid out in San Francisco. It would have been more practical to run many streets there on contours, but the progressive mind would not tolerate such irregularity. The roads were laid out as they should be, and if vehicles had trouble negotiating them, then someone would have to invent the cable car or, in the famous case of the Lombard Street hill, a street wide enough that it could be narrowed to a one-lane zigzag. Countless towns in the American Midwest and West are slaves to the grid. So are many in the East, from Manhattan and Philadelphia to Charleston and Miami.

Jefferson's debt to Francis Bacon echoes more broadly in the Declaration of Independence, with its Baconian "life, liberty, and the pursuit of happiness," and the subsequent economic development of North America can hardly be told without thinking of it as a story of "enlarging the bounds of human empire, to the effecting of all things possible." Even when Teddy Roosevelt thundered against the land skinners who were destroying the natural resources of the United States, he wasn't repudiating the Baconian philosophy. Far from it. He went straight to Jeremy Bentham in calling for an end to waste and for planning that would develop resources efficiently. He seems to be saying that with arithmetic we can create a perfect world. Not surprisingly, the 19th century saw the birth of psychology, which for many years sought to reduce the study of mind to the study of measurable behavior.

THE PAYOFF

The aristocratic Charles Maurice de Talleyrand (1754–1838) survived the French Revolution. In later years he said that nobody who had not enjoyed firsthand the aristocratic life of prerevolutionary France could understand how sweet it had been. A century later, that world of inherited privilege had been almost entirely replaced by one of bourgeois vulgarity, but John Singer Sargent's arresting portrait of Lady Agnew, painted in 1892, shows a woman whose relaxed elegance makes the images of luxury in our own time seem compulsively frenetic. As Talleyrand would be the first to say, something's been lost. But something's been gained, too: The number of people living comfortably today—vulgarly or not—is vastly greater than in Sargent's time, let alone Talleyrand's. We've gone a long way toward democratizing the comforts that were once the privilege of a hereditary elite. Even our language betrays our faith in the universality of progress. Statesmen and international civil servants, for example, would not dare to describe a country, however impoverished, as "poor." Instead, they seek a euphemism, and often as not they find it in the word "developing."

Four or five centuries after the discovery of the individual, we have a consumer society catering to tastes that our grandparents could not imagine, let alone imagine satisfying. New cars, new homes, new furnishings, new clothes, new medical technologies. . . . I almost forgot new food! Oklahoma doesn't have a Whole Foods store or anything like it, so newcomers are reduced to desperation runs to Dallas. "Don't forget the Sarawak peppercorns. And I want some of that mango nectar—the stuff made in France from fruit air-freighted from Douala. Oh, and some of that lovely flower-flavored sorbet." Well, what do you expect? These newcomers blow in from one coast or the other, and they're used to the amenities that diffuse—slowly, so slowly—from those cultural epicenters.

Think of the agribusinessman—he must have been in California, or at least of California—who about 1990 looked at all the small and misshapen lettuce heads he couldn't sell. He had an idea: Harvest them, clean them, chop them, then seal them in nitrogen-filled bags and put them on grocery shelves. Voilá, within a decade a $2 billion bagged-salad industry emerges in the United States. Almost the same story can be told about the formerly unmarketable baby carrots that now fetch top prices in every market in the land. Organic producers are getting into the act, too. Earthbound Farm, with sales topping $200 million in 2002, has machines in Yuma, Arizona, that bag 14,000 pounds of lettuce an hour. What's next? A good guess is cut and packaged fresh fruit. The entrepreneur who makes a fortune in it doesn't have to know beans about Bacon, but without Bacon and Descartes, without the discovery of the individual and the construction of a civilization dedicated to technological progress, the entrepreneur would be out of business. He never would have got *in* business.

It's hugely ironic that the technology of power plants, distribution centers, and kitchens filled with machinery—dishwashers, mixers, blenders, processors, toasters—should lead us back to peppercorns and mangoes, which were there in the first place. Still, mangoes once had a season and a location. Now we can get them almost anywhere, year-round. It's a feat that justifies our civilization in the same way that Confucian ideas justified imperial China. For better or for worse, we have almost no other belief system. Only 40% of Americans say that they attend church regularly. In Canada, only 20% do—a low enough proportion that no prime minister would think of concluding his speeches, American-style, with God Bless Canada. Not even a fifth of the Spaniards call themselves practicing Catholics. Then there's England, where only 5% attend. The number is lower still if you factor out parishioners from Africa and Asia, who form the majority at services in London these days.

The phrase "by the grace of God" still appears on British coins under an image of the queen, but the Archbishop of Canterbury said in 2001 that "a tacit atheism prevails. Death is assumed to be the end of life. Our concentration on the here-and-now renders a thought of eternity irrelevant." That's about right, and not just for the heathen English. The general secretary of the United Reformed Church in Britain, David Cornick, wrote in 2003 that "in Western Europe, we are hanging on by our fingernails. . . . The fact is that Europe is no longer Christian." It sounds extreme, but the 2003 treaty establishing a constitution for Europe refers vaguely to Europe's "cultural, religious, and humanist inheritance" and makes no mention of Christianity. This vagueness may please Europe's Muslims, but it's not there for their sake. It's there because Christianity is an embarrassment to the modern, educated Europeans who drafted the document. In religion's place they, along with most Americans, have Baconian progress. It's symbolized best, perhaps, by space exploration, which NASA enthusiasts never miss the chance to call our cosmic destiny. A more modest expression is the old General Electric motto: "progress is our most important product."

CHAPTER 10

GLOBALIZATION

Like the word "software," "globalization" is one of those words that nobody had heard of one day and everybody was using the next. It was apparently coined by Theodore Levitt, writing in the *Harvard Business Review* in 1983 to suggest that multinational companies were tending to market identical products around the world. Critics have pointed out that this hasn't happened, at least not yet. McDonald's serves congee, or rice porridge, in China, not Colorado; it serves Chicken Maharajah Macs in India, not Indiana.

But if globalization hasn't yet occurred in exactly the way Levitt meant, it certainly is occurring in other and perhaps more significant ways. Around the world, people are increasingly entertained by the same films, music, sporting events, and television programs. Around the world, students—the most avid consumers of that entertainment—are wearing jeans, or wishing they did. Their parents are no slouches, either. The 10 most trusted brands in India in 2003 were Colgate, Dettol, Pond's, Lux, Pepsodent, Tata Salt, Britannia, Rin, Surf, and Close-up—every one of which, except the salt, was the carefully protected property of an American or British company.

Globalization isn't just a matter of international tastes and brands, either. Around the world people are increasingly competing to provide goods and services in a single, global marketplace. Jobs that one day seemed permanent are gone the next, shipped overseas. It's happening in the United States and it's happening in Mexico; it's happening in Germany and it's happening in Korea. People themselves are moving to find work in foreign countries. It happens at the top of the labor scale and at the bottom.

How did the Baconian experiment—the creation of a society dedicated to technological progress—expand from Europe to the world? That's the question I want to explore here, beginning with the European exploration of the globe and proceeding to colonial empires, decolonization, and the integration of economies and cultures in a postcolonial era.

EUROPEAN EXPLORERS

We think from grade school of Christopher Columbus, of Vasco da Gama before him, of Ferdinand Magellan afterward. It's easy for all this to become rote, but think of the shock Magellan's crew felt when, 3 years after leaving Spain to circumnavigate the globe, they landed at the Cape Verde Islands and found that the date was July 10, 1522. Their log stated July 9. A mistake? Not at all: The log had been kept meticulously. (You should be able to

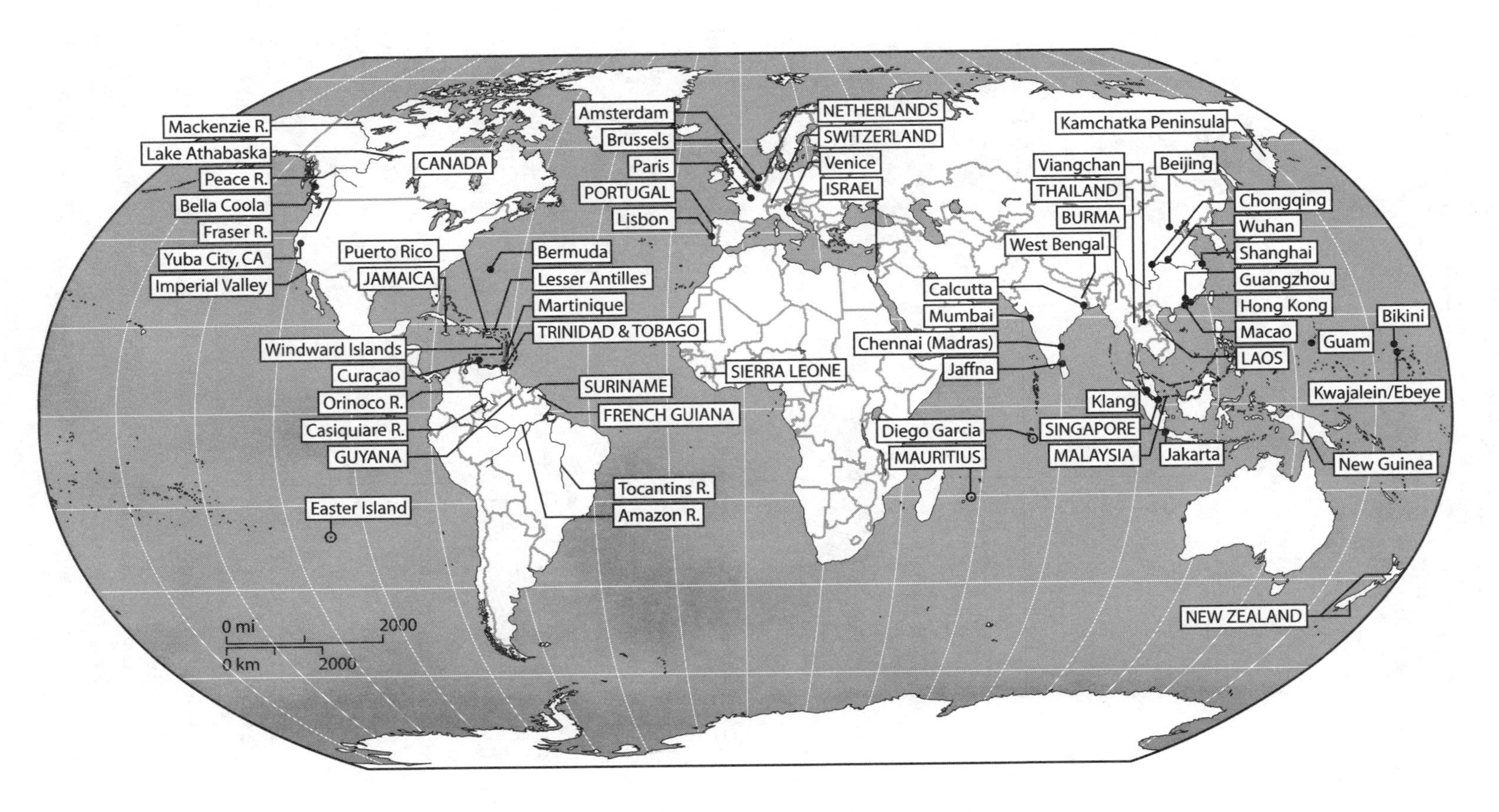
Mackenzie R.
Lake Athabaska
Peace R.
Bella Coola
Fraser R.
Yuba City, CA
Imperial Valley
CANADA
Puerto Rico
JAMAICA
Windward Islands
Curaçao
Orinoco R.
Casiquiare R.
GUYANA
Easter Island
0 mi
2000
0 km
2000
Amsterdam
Brussels
Paris
PORTUGAL
Lisbon
Bermuda
Lesser Antilles
Martinique
TRINIDAD & TOBAGO
SURINAME
FRENCH GUIANA
Tocantins R.
Amazon R.
NETHERLANDS
SWITZERLAND
Venice
ISRAEL
SIERRA LEONE
Calcutta
Mumbai
Chennai (Madras)
Jaffna
Diego Garcia
MAURITIUS
Kamchatka Peninsula
Viangchan
THAILAND
BURMA
West Bengal
Klang
SINGAPORE
MALAYSIA
Jakarta
Beijing
Chongqing
Wuhan
Shanghai
Guangzhou
Hong Kong
Macao
LAOS
Guam
Bikini
Kwajalein/Ebeye
New Guinea
NEW ZEALAND

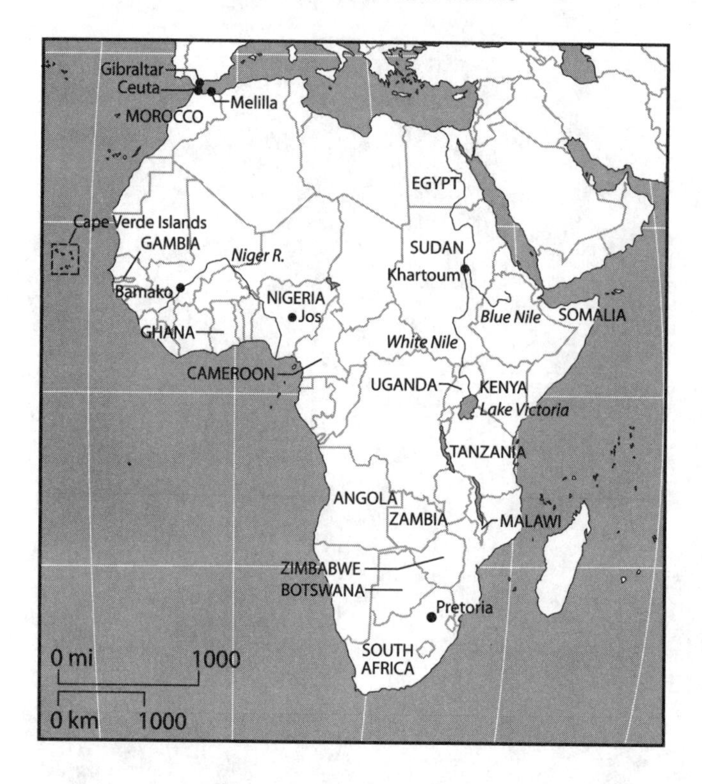

figure out just from this information whether their circumnavigation of the globe was accomplished by sailing east to west or west to east.) The solution to this enigma was the creation of the International Date Line, which zigs and zags more or less along the 180th degree of longitude and still confuses transpacific travelers. It's odd to welcome the new year in Hong Kong and, a long flight later, welcome it again in San Francisco.

Marine exploration continued to an 18th-century peak with Captain James Cook, who died in Hawaii. He wasn't the only explorer to die on his travels. Magellan, you may recall, was killed in the Philippines. Vitus Bering, a Dane in the employ of the czars, was fatally wrecked on Bering Island, near Kamchatka. Think about that someday when you're in 26A, somewhere over the cloud banks of the North Pacific.

Land explorers had it no easier. Francisco de Orellana said goodbye to Francisco Pizarro in 1538 and more or less accidentally floated down the Amazon to its mouth. He's the one who gave the river its name, perhaps because he saw long-haired tribesmen who he thought were female. A few years later he tried to repeat the journey the other way. He didn't make it. A century later, Jesuits would stake out missions in the deepest Amazon, but it would be another two centuries before an explorer in 1800 traced the course of the Orinoco, the great river of Venezuela. He was Alexander von Humboldt, and he discovered that the Orinoco is connected to the Amazon through an upper tributary called the Casiquiare. Still another century passed before Paul Ehrenreich in the 1880s traced the course of the Tocantins, a great southern tributary of the Amazon.

Over in Africa, Mungo Park (1771–1806) made a successful journey along part of the Niger River downstream of Bamako. Returning a few years later to complete his survey, he was attacked in his canoe and drowned. John Hanning Speke and Richard Burton got in a furious competition to find the source of the White Nile, and though neither man died in Africa, Speke shot himself fatally in the wake of a particularly acrimonious debate back in England. (But not before discovering an immense, shallow lake on the Equator and naming it after his sovereign, Queen Victoria.) Come to think of it, America's own contenders in this business, Lewis and Clark, didn't fare so well—especially Lewis, who later committed suicide. Up in Canada, there was a truly amazing expedition led by Alexander Mackenzie, who paddled down what is now the Mackenzie River from Lake Athabaska to the river's mouth. Since he had been looking for the Pacific, he wasn't amused to find himself on the shore of the Arctic Ocean, and a few years later, in 1793, he tried again. This time he struck west, following the Peace River, the Fraser River, and finally bushwhacking his way to Bella Coola. It was the first North American crossing north of Mexico, and Mackenzie was smart enough to know when to quit. He retired to Scotland and lived to a more or less ripe old age.

ESTABLISHING COLONIAL EMPIRES

There must be hundreds of tales like these, fascinating if you're in the right frame of mind. Even the word "fascinating" falls short of the grandeur that was once associated by Europeans with what they considered to be a heroic enterprise. To really grasp their emotions, you have to go to *The Lusiads*, the epic poem written by Luís de Camões in the 1550s. Celebrating the exploits of the Portuguese, especially Vasco da Gama, the poem's first stanza (in the recent translation by Landeg White), reads: "Arms are my theme, and those matchless heroes/ Who from Portugal's far western shores/ By oceans where none had ventured/ Voyaged to Taprobana [Sri Lanka] and beyond, Enduring hazards and assaults/ Such as drew on more than human prowess/ Among far distant peoples, to proclaim/ A New Age and win undying fame."

It's easy to mock these out-of-favor sentiments, especially if you recall the history of Vasco's voyages that was written by Diego Fernandez Correa, a near contemporary. On one occasion during his second voyage, Correa wrote, Vasco cut off the hands, ears, and noses of 800 Indian Muslims who fell into his hands off the coast of India. Not content with this mutilation, he ordered their legs tied together and—thinking that they might untie the ropes with their teeth—ordered their teeth to be knocked out with barrel staves. Then Vasco put his victims back on their ships, covered them with mats and leaves, set their ships afire, and ran them aground. Rescuers came, Correa writes, "to put out the fire, and draw out those whom they found alive, upon which they made great lamentations." It was such barbarity that inspired Jonathan Swift's famous excoriation in *Gulliver's Travels* (1726): pirates, Swift wrote, "see an harmless people, are entertained with kindness; they give the country a new name; they take formal possession of it for the King: they set up a rotten plank or a stone for a memorial; they murder two or three dozen of the natives, bring away a couple more by force for a sample, return home and get their pardon. Here commences a new dominion, acquired with a title by *Divine right*. . . . This execrable crew of butchers employed in so pious an expedition, is a *modern colony*, sent to convert and civilize an idolatrous and barbarous people." Nearly 200 years later, Joseph Conrad in the opening pages of *Heart of Darkness* says much the same: "The conquest of the earth, which mostly means the taking it away from those who have a different complexion or slightly flatter noses than ourselves, is not a pretty thing when you look into it too much."

Before you dismiss Camões outright, however, bear in mind that your own great-great-grandchildren are likely to mock or revile you, perhaps because you in your barbarity ate meat or supported capital punishment or even enjoyed athletic events in which people sometimes get seriously hurt. Camões's own fame, by the way, only arose posthumously, when Portugal fell under Spanish control. *The Lusiads* then served as a spur in the Portuguese fight for independence.

By then, Sebastian Münster had published the first modern geography. His *Universalis Cosmographia* (1544) pulled together the information that had been pouring into Europe about distant lands. Fifty years later, in 1595, Gerardus Mercater published the first atlas, so-called. Another 50 years passed before the publication of Joan Blaeu's *Atlas Major* (1662), with 600 very detailed maps in about 10 volumes. (Selections from Blaeu's atlas are available in a volume edited by John Goss. Don't bother rushing to the library for the real thing: A copy with 597 maps was recently on sale for $450,000, and abridgments with half the maps sell for over $200,000.) The only areas Blaeu didn't know much about were the west coast of Canada, the western South Pacific, and the east coast of Australia.

Readers wanted exploration narratives, too, and English readers got them in 1600 with the publication of Richard Hakluyt's *Principal Navigations . . . of the English Nation*. A Hakluyt Society was eventually founded, and it published (and continues to publish) a massive and still-expanding collection of travel narratives. It was the Hakluyt Society in 1869, for example, that published Correa's account of Vasco, which until then had remained only in dusty manuscript.

Flags followed in the explorers' wake. The old boast about the sun never setting on the British Empire was probably right in the 1920s, when Britain claimed sovereignty over 13 million square miles. That's 140 times larger than Great Britain itself, and it's larger by half than the Russian or Soviet Empire at its maximum extent. All one has to do is get a vintage atlas and look for the red bits. They stretch in Africa from Egypt down through the Sudan to Kenya and Uganda. South of the Equator they cover South Africa, Nyasaland (Malawi), and the Rhodesias (Zambia and Zimbabwe). In West Africa they include the Gambia, Sierra Leone, the Gold Coast (Ghana), and Nigeria. The red covers India in the prepartition sense of that word, plus Burma (Myanmar) and Malaya (Malaysia and Singapore). It covers the Dominions, too: more or less self-governing Canada, Australia, New Zealand, and South Africa. It covers Jamaica, Trinidad, and many smaller islands in the Lesser Antilles. It would have covered more, if the American colonists hadn't been so obstreperous.

One can similarly play with maps of the French Empire, or the Dutch, or the Portuguese. If not the match of Britain's, they were all many times bigger than the home country.

How does it happen that the Europeans were so successful in their conquests? In the Eastern Hemisphere, the answer is a combination of better technology and weak opponents. When Kitchener took the Sudan in 1898, for example, Sudanese armed only with spears and faith charged the British machine guns. The slaughter at the Battle of Omdurman may have been celebrated in London, but it prompted a mordant couplet in Hilaire Belloc's *The Modern Traveller*: "Whatever happens, we have got / The Maxim Gun, and they have not."

The British, by then, were old hands in India. They had arrived more than a century earlier not with vastly more powerful arms but with a unified force facing an India where power was shattered into hundreds of slivers, each held by a jealous sovereign. Robert Clive had to fight his way into Bengal, but few of the rajas put up any real resistance, and those who did were crushed. The British assembled the slivers into the grandly named British Indian Empire. Half of it was nominally still in the hands of Indian rulers. They continued to preen, but the British allowed them to keep their titles and wealth only if the princes acknowledged that they were finally just that—princes—not true sovereigns. Some Indian rulers refused to be humiliated in this way; some even joined the Mutiny of 1857. That's why the other half of the country was ruled directly by the British through governors: There wasn't a lot of forgiveness for princes who forgot their place.

Or consider Qing China, lucky in a way. It, too, was very nearly an open door, so defenseless that the Chinese were obliged in the 19th century to grant Europeans sovereignty over the locations that grew into Shanghai, Guangzhou (Canton), Chongqing (Chungking), and Wuhan. In the process, China came close to being dismembered by the beady-eyed European powers, including Russia.

The European takeover of the Western Hemisphere was different—a ghastly business of accidental genocide. Pizarro, with 168 men, was able to conquer the Inca in 1532 not because he had guns but because 7 years earlier smallpox had killed half the population and disorganized the rest. Thousands of Indians a few years later watched Hernando de Soto cross the Mississippi near Memphis; 150 years later, in 1682, La Salle came by and saw no villages and no Indians for hundreds of miles. They weren't on vacation. In 1792 Captain Vancouver explored the Northwest and found bodies "promiscuously scattered about the beach, in great numbers." One of his officers wrote that the surviving Indians were "most terribly pitted."

Smallpox wasn't the only European disease introduced to the Americas. The Europeans, or their animals, also introduced typhoid, plague, influenza, mumps, measles, and whooping cough. How many millions died of these diseases across the New World is unknown and much debated—current estimates of the precontact population of North America alone vary from 1.8 to 18 million. By 1600, deer, bison, and the passenger pigeon were hardly being hunted any longer, and their numbers skyrocketed. Charles C. Mann has suggested that what we think of as the virginal New World was really an ecosystem in disarray from the near extinction of its top predator.

Did the Europeans ever fail? It's debatable. Thailand and Japan never ceded sovereignty, but realizing that their survival depended on adopting Western ways they embraced many aspects of European culture. Rama IV, the Thai king patronized in *The King and I*, decided that to win recognition from Europeans his country better have a name—he chose Siam—and a flag, which he also adopted. Starting in the 1850s he signed treaties giving Europeans favorable trading rights in his kingdom. His successor went further, not only building grand palaces in the European style but collecting Thai antiquities to impress European visitors with the venerability and legitimacy of his rule.

ADMINISTERING COLONIAL EMPIRES

George Curzon, probably the most imperious Indian viceroy, once justified British rule in phrases of a *hauteur* almost inconceivable today: "to me the message is carved in granite," he said, "it is hewn out of the Rock of Doom—that our work is righteous and that it shall endure." His audience—this was London in July, 1904—presumably applauded lustily at this imperial equivalent of motherhood and apple pie. They were used to such sentiments. More than 50 years earlier, the London *Times* on February 4, 1847 had called editorially for better transportation in India, so cotton could be moved from the interior to the coast for export. Then the down-to-earth argument took wing: "The days are gone," the author wrote, "for the gathering of pearls and peacock thrones, of rubies and pagodas. There is little more barbaric gold to be gleaned from nabobs or rajahs. The treasures of India are now to be found in the people and the soil—by improving the condition of the one and employing the resources of the other—by transporting to the climate and capabilities of the East the science and steadiness of the North—by banishing famine and introducing plenty—by substituting intelligence and comfort for ignorance and want, and by turning the energies of an enlightened population and a beneficent government to the full development of the untold wealth of India."

It was pure Baconianism, but it wasn't just the British who thought this way. Francis Garnier, a French administrator in Indochina, wrote in the 1870s that Asian government "signifies stagnation," while European intervention signified "commercial freedom, progress, and wealth."

In a way, Curzon and Garnier were right. Local administrators in British territories would carry bland titles like commissioner or collector, referring to the taxes whose collection was part of their job. The Dutch in the East Indies were more honest: Their officers carried the Orwellian title Controller. There were nicknames, too. The British administrators isolated in the swamps of Sudan's Upper Nile, for example, were known only half in jest as bog barons.

Whatever their titles, district officers typically had great autonomy, especially in remote areas. Some put in their time and amused themselves with pigsticking and beer, but many went beyond the bedrock of tax collection and the maintenance of public order. In the mod-

In the cloisters of Westminster Abbey is this plaque, written by Sir George Cunningham, who may perhaps be forgiven the vanity of commemorating his own long career.

ern spirit, they found time to build roads and irrigation works, set up schools, and take measures to prevent famine. Although we reflexly sneer at anyone and everyone who was part of the imperial machinery, most of the places in which these men worked had never seen rulers who wanted to make life better for the people. The plaque at Westminster Abbey says "Let them not be forgotten, for they served India well." By and large, it's true.

The snag is that people prefer bad self-government to the most efficient and altruistic foreign rule. The Europeans would have been hated had they been saints, and saints they weren't. Racism was their most galling flaw. Leonard Woolf spent a few years in the colonial service in Ceylon (Sri Lanka) and in his autobiography recalls soldiers in Jaffna cursing Tamils with the word Americans think they invented in the plantation South. Officials at higher levels were too well-mannered to be vulgar, but they could be equally vile—perhaps more so, because they came from the upper ranks of their own class-ridden society.

Think of Edward Cecil, a younger son of Lord Salisbury, a British prime minister in the late Victorian period. Cecil was also a soldier, and for many years before World War I he was an important official in the Egyptian government. His posthumous *The Leisure of an Egyptian Official* (1921) was intended—and was taken by its readers—as light humor, but its humor comes at the expense of Egyptians. On the very first page Cecil jokes about his servant, who "weighs about as much as a big retriever dog." Such contempt explains why many colonial peoples in Southeast Asia in 1940 welcomed the Japanese as liberators. They were only substituting one foreign rule for another—and a harsher one at that—but they were at least partly immune from charges of racism.

The disease was most extreme in Africa. Outsiders in the middle of the 20th century were appalled by its pervasiveness there, even in the last days of empire. A well-known case is that of Ruth Williams, an Englishwoman who fell in love with Seretse Khama, the chief-in-waiting of the Bamangwato tribe of Bechanaland. This was 1948, and they married secretly in the face of opposition from both tribal and British officials. The South African prime minister at the time called the marriage "nauseating." The British asked Khama to renounce all claim to the chieftainship in return for lifelong payments; he refused. He was then banished, returning only in 1956, when he abandoned his claims to the chieftainship and instead entered politics and became president of newly independent Botswana in 1966. Khama died in 1980; his wife remained in Botswana, dying there in 2002.

DECOLONIZATION

Stand back far enough, and the European empires seem like a wave rushing forward and almost immediately slipping back. First, the colonies that became the United States broke away from England. Then nearly all of Latin America broke away from Spain and Portugal. After World II, Britain quit India, and the French and Dutch were pushed from Indochina and the East Indies. Independence swept Africa in the 1950s and 1960s. It's an image that comes from Harold Macmillan. As Britain's prime minister, he toured Africa in 1960 and in a famous speech in South Africa reported that "the wind of change was blowing over the continent." His meaning was that it could not be resisted. All the colonial powers departed except the dogged Portuguese, who held Mozambique and Angola until 1975.

Europe is still sifting the embers. A private British Empire and Commonwealth Museum is now open in Bristol. Belgium is meanwhile reviewing its record in the Congo with an exhibition scheduled to open in 2004 at the Royal Museum for Central Africa, which is at Tervuren, near Brussels. It will be interesting to compare the Belgian treatment of their empire with the lethal judgments in Joseph Conrad's *Heart of Darkness* (1902) and Adam Hochschild's *King Leopold's Ghost* (1999).

Fragments of the colonial world remain. In 1997 the British quit Hong Kong. The Portuguese in 1999 left nearby Macao. Still, the British retain Gibraltar, Bermuda and a few other bits of the Caribbean. The Spanish, resentful as they are of Britain's hold on Gibraltar, retain Ceuta and Melilla, two enclaves in Morocco. The French retain Martinique and French Guiana, and the Dutch still have the Netherlands Antilles, including Curaçao. All these French and Dutch possessions have been constitutionally transformed from colonies to overseas parts of the home country.

And let us not forget the supposedly anti-colonial Americans. Since 1898, they have controlled Puerto Rico. To this day, it is administered as a commonwealth, neither state nor independent nation. Most Puerto Ricans want it that way, but the United States also possesses some Pacific Islands—and not just Hawaii. After World War II, for example, it acquired the Marshalls. It ran them under a U.N. mandate until 1986 and now administers them under a free association, but American presidents don't talk about them much. Many of the residents were evacuated to make way for atomic tests on nearby Bikini, and there are problems with radiation poisoning. There are also problems with cholera and malnutrition, because some 15,000 Marshallese are crowded on the 3 square miles of Ebeye, where they live either on handouts under the free-association agreements or by commuting by boat to jobs serving the Americans on Kwajalein. The Marshallese can work there, but they can't shop in the subsidized stores. You can't, either: It takes special permission to get off the Continental flights that serve Kwajalein from Guam.

Have Americans come to regret this heavy-handed treatment of people they hardly know? Judging by behavior, the answer is no. They were quite willing, as recently as the 1970s, to pay the British to evict the 5,000 residents of the Indian Ocean island of Diego Garcia. A military base was needed, and those French-speakers were packed off to anglophone Mauritius. Against the odds, they're still trying to go home.

INTANGIBLE LEGACIES

The end of colonialism put an end to Europe's direct control over the colonial world, but it did little or nothing to reduce the influence of European civilization. Colonies, in other words, set in motion a process that continues today, probably more vigorously than ever.

Consider clothing. It's hard to find a soldier or police officer anywhere in the world who doesn't dress in a European-style uniform. Presumably it goes with the winning technology. So, too, the business suit, which leaves corporate executives around the world in clothes designed to erase physical idiosyncrasies and give the wearer a shape as uniform as a line of robots. Departures are not appreciated, as Dr. Mahathir Mohamad of Malaysia discovered when he dared to wear his own country's national dress to a meeting with U.S. Secretary of State James Baker.

Trivial? Want something more important? It's no accident that the best airline connections between Europe and Angola go through Lisbon. Want a flight from Europe to Suriname, formerly Dutch Guiana? Better figure on going through Amsterdam. How about Abidjan, capital of Ivory Coast? Figure Paris. The really big ex-colonies are exceptions, because they attract enough traffic to justify flights from all the major European countries. India's the obvious case. Also, privatization has forced European airlines to consider economics as well as hoary tradition. So has the threat of terrorism. It was the bottom line that caused British Air to quit flying a decade ago to Sudan. It's probably the risk of terrorist attack that more recently induced it to drop Pakistan, which joins the unhappy company of Guyana and Sierra Leone as formerly British-ruled places that the airline no longer serves.

Still not impressed? Consider the continued use of European languages outside Europe. Educated Indians from Chennai (Madras) can talk easily with counterparts from Kolkata (Calcutta), but they won't use their mother tongues, Tamil and Bengali. They might, but probably won't, use Hindi. Almost certainly, they will use English. An educated Moroccan will meanwhile have no trouble talking to an educated person from Senegal, the Ivory Coast, or Cameroon, thanks to their shared education in French. These survivals aren't the same as the preservation of French in Quebec, where French is spoken mostly by descendants of French settlers. The English spoken in India is spoken by Indians, just as the French spoken in West Africa is spoken almost entirely by Africans. No matter: Once a colony's elite speaks a European tongue, that tongue survives the colonial departure. It can fade, as Dutch has faded in Indonesia and Italian in Somalia, but it's likely to be replaced by English, on its way to becoming the world language. That's good for Anglophones on the road, but it just about kills their incentive to learn foreign languages. In Japan, on the contrary, ambitious employees of Matsushita, Toyota, NEC, and Hitachi are told that corporate advancement is impossible without English. It stands to reason. Japanese doesn't get you very far in Seattle or Seville.

If the globalization of English helps people around the world communicate with each other, it does so at a price. We're down from perhaps 20,000 languages 10,000 years ago to 6,000 today, of which half are no longer being taught to children, which means that they're dying. The Endangered Language Fund anticipates the loss of half the world's presently spoken languages over the next century and estimates that only 300 languages, each spoken by 1 million people or more, are safe from extinction in that period. To help retain at least a vestige of the world's dying languages, the Rosetta Project aims to put a thousand translations of the opening of *Genesis* onto disks designed to last 1,000 years. (The choice of text, ironically, is a symptom of the cultural homogenization that the project implicitly deplores.) Even in the United States you can see this trend at work with the standardization of English. This is not only a matter of pronunciation but of diction. Think of the colorful language once associated with the South. It's dying out fast—as fast, an old Southerner might say, as a rubber-nosed woodpecker in a petrified forest. How old a Southerner? Why, one so old that he doesn't even buy green bananas any more.

Is this coalescence good or bad? People often feel passionately about language preservation. One reason is that languages shape the people who use them, so that the loss of a language fundamentally changes the society that now speaks another. That's why some of the 2,000 residents of Easter Island, for example, want to preserve their language, Rapa Nui, against the invasion of Spanish. "Our culture, our cosmology, our way of being" is at stake, a teacher says. A Siberian Khanty-speaker, forced like her fellow tribespeople to speak Russian in school, says much the same thing, as she brings a Russian linguist from Tomsk to Lukashkin Yar to study this dying tongue. The Russians, she says, "destroyed our way of life." Both women are right, but it's tough competing against Spanish-language television or finding a job in Russia without Russian.

Even defenders of language have to acknowledge that it doesn't help New Guinea, as it seeks to modernize, that its people speak 800 languages. The complexities can be very challenging. Until recently, for example, the European Union, with 11 official languages, had to provide simultaneous translations for each language into all the others—a total of 110 combinations. With the entry of another 10 countries in 2004, the number rises to 420, unless one or more of the new entrants waives its right to use its own language in debate. That won't happen. In practice, many of the translations will be indirect. If there's no Finnish–Polish translator to be had, for example, the Finnish will be translated into an intermediate language known to both the Finnish and Polish translators, and eventually—with a delay of several seconds—the Polish translation will emerge. In some cases, two intermediate languages may be necessary, raising questions about the accuracy of the final translation of a speech delivered in Finnish, translated perhaps to German, then to Italian, and finally to Maltese.

The best solution is bilingualism or multilingualism. That's the position of Raymond Cohen, an Israeli who contrasts the "threadbare" English typically used by foreign speakers with the "luxuriant native language" used by native English speakers. Cohen argues that people whose native tongue is not English need to retain their native language precisely because their command of English is too weak to express subtle thoughts. He writes not as a linguist but as an expert in international relations, and his defense of multilingualism has political implications. Americans, for example, are in his view "locked into a monolingual view of the world" and are condemned to remain ignorant of it. (The reality is worse still. Americans aren't interested enough in the rest of the world even to read in translation what gets written or spoken there. Less than 3% of the literary titles published in the United States, for example, are translations. Foreign-film imports are about as slim.)

There are three other intangible legacies to consider, and they're probably more important still.

One is the division of the continents into precisely demarcated nation-states, or at least potential nation-states. This transformation came very quickly, mostly in the 19th century. The model was the European nation-state, which had emerged in the wake of the Protestant Reformation. Some kind of stability had been needed then, and it came in the form of absolute monarchs. France was one model, with French kings as absolute rulers. The English devised a different model, with absolute power residing not in a man but in a parliament. This was the form of sovereignty that gradually spread across Europe. Simultaneously, the leviathan of the state became a nation, which is to say a people unified by collective memory, typically of wars won or lost, triumphs and defeats shared. Sad to say, it seems to be true that you can't have a nation without national cemeteries. Americans have never been more united than in the days after September 11, 2001.

Venice may be seen as a precursor of the nation-state, with its proud citizenry on one hand and its business-oriented hierarchy on the other. Machiavelli (1469–1527) might be seen as the nation-state's intellectual guide, with his advice to princes that they can be successful or virtuous but not both. It's still good advice, though in public leaders repudiate it.

A political map of Africa as late as the 1830s would have shown Egypt on the Nile, kingdoms in West Africa, and colonies in South Africa. None of these places had demarcated boundaries, and in European eyes the rest of the continent was wide open, unclaimed. This changed very fast. By 1900, every inch of Africa was claimed by a European state or colonial dependency. Often the boundaries around these domains paid no attention to facts on the ground. A classic case is the boundary between Kenya and Tanzania, which runs as a straight line back and forth across the lazily meandering Umba River and in the process cleaves tribes in pieces.

No matter. Africa was seen as a game board. If not blank already, it could be wiped clean and begun anew. By the 1960s, the colonial powers were mostly gone from Africa, but their colonies were states, if not yet emotionally fused into nations. Despite the illusions created by maps, these postcolonial countries have proven to be neither sovereign nor immutable. Globally, the strongest nation-states are those that correspond to kingdoms or empires that antedated the Europeanization of the globe—countries like Thailand, Japan, and China. At the other extreme, the creation of nation-states has done little more than formalize the earth into organized domains on which everyone with an atlas and an ambition can hatch schemes and plot strategies, whether political or economic. With the passage of time, the results of these actions have become very tangible, as anyone with experience of the two Koreas can testify.

Perhaps even more important than the nation-state is the idea of progress itself. European flags may not fly over Asia and Africa, but many of the elites of those continents come to Europe or America for their education. They are not fully acculturated there. They arrive, for example, with the idea that progress is good and leave with that idea intact, which they would not do if they understood how ambivalent Americans and Europeans are about the rush of change. Some of these students become professors and impart the secular faith to their students. Others pick up an MBA and go back home to make money. There's nothing new in this diffusion of European culture—you can trace it back at least as far as Macauley's push to make English the medium of instruction in India. The forces behind that diffusion, however, are far more powerful now than in his time, even when they are in the hands of men and women who are bitterly hostile to the bad old days of colonialism.

Finally, consider hostility itself as a colonial legacy. The United States takes over the Philippines about 1900—builds roads and schools, allows a relatively free press, and in many ways changes the country for the better. What is the Filipino reaction? No doubt there are many conflicting answers, but one of them is encapsulated in the saying "I learned your language so I can damn you."

The most damning criticism, in English, probably comes from the Irish. Consider Maud Gonne, best known perhaps for her connection to William Butler Yeats but in her own right a lifelong participant in the struggle for a free Ireland. In 1900, Gonne wrote a collection of newspaper columns called "The Famine Queen." Who's she talking about? Why, none other than Queen Victoria. Here's an excerpt that gives a sense of Irish wrath:

> And in truth, for Victoria, in the decrepitude of her eighty-one years, to have decided after an absence of half-a-century to revisit the country she hates and whose inhabitants are the victims of the criminal policy of her reign, the survivors of sixty years of organized famine, the political necessity must have been terribly strong; for after all she is a woman, and however vile and selfish and pitiless her soul may be, she must sometimes tremble as death approaches when she things of the countless Irish mothers who, shelterless under the cloudy Irish sky, watching their starving little ones, have cursed her before they died.

The Ayatollah Khomeini understood this anger very well. Speaking on New Year's Day, 1980, about the need to purge Iran's universities of faculty who "look to the East or the West," Khomeini said that "most of the blows our society has sustained have been inflicted on it precisely by those university-educated intellectuals. . . . " He was right, as long as progress is seen as an attack on traditions that cannot be improved. It's useful in this connection to remember that many strict Muslims are proud to say that the precise way in which they pray has not changed in 1,200 years. In their view, it cannot be improved, nor can anything else that Islamic law or tradition regulates. Not a lot is excluded.

Westerners, on the contrary, see the idea of progress as a gift of great value. Even Asians and Africans who share that belief resent the patronizing package in which it came to them. Students in colonial-era schools were taught much more than Magna Carta and Dickens. They learned—came to believe—that in blood and culture they were inferior. Bharati Mukherjee, now an American professor but once a girl in a Calcutta convent school, has written that "the whole point was Calcutta (or wherever) did not exist. We did not have interesting lives. Our own cultures were vaguely shameful, and certainly not fit subjects for serious literature." Such experience perhaps explains the hostility toward the West of Malaysia's long-serving prime minister, Dr. Mahathir Mohamad. Speaking of a planned School of Occidental Studies, Mahathir said, "they study us, so let's study them." Fair enough, but Mahathir went on to describe Westerners this way: "Because they like to wage war and seize other people's territories, their main interest is in the development of weapons to kill people more efficiently." No need to look further to understand why Kuala Lumpur's port is now called Klang, instead of Port Swettenham.

Many Asians during the imperial period understood very well how their own culture was being undermined by the West. Consider Calcutta's Bankimchandra Chatterjee. A Bengali, he wrote in 1872 that "the stamp of the Anglo-Saxon foreigner is upon our houses, our furniture, our carriages, our food, our drink, our dress, our very familiar letters and conversation." This is from "The Confession of a Young Bengal," and I choose it because Chatterjee understood so well both the Baconian ethos and the threat it posed to his culture. He continues: "The very idea that external life is a worthy subject of the attention of a rational being, except in its connections with religion, is, amongst ourselves, unmistakeably of English origin. . . . Our ancestors . . . could never justify to their conscience any care bestowed upon food and raiment for their own sake. English civilization has pulled down the three hundred and thirty million deities of Hinduism, and set up, in the total space once occupied by them, its own tutelary deities, Comfort and his brother, Respectability."

Ironically, Chatterjee was a triple agent of the invading culture that he attacked, because he was a graduate of the University of Calcutta, wrote in English, and wrote the first Bengali novels, which were serialized, just like the novels of Dickens. As if to stack irony on irony, Chatterjee was also the author of "Vande Mataram." It's a poem that can be read as a song to a beloved Mother India, but it was understood by the British as an attack on their rule. People were regularly imprisoned for singing it, much as Israel for many years jailed Palestinians who dared to fly the Palestinian flag. Today, translated from the original Bengali to Hindi, "Vande Mataram" is India's national anthem.

TANGIBLE LEGACIES

And what about geography, the legacy visible in the cultural landscape?

Just as the Romans left a visible record of their empire, so did the Europeans, whether in recognizably French Vientiane (now Viangchan, Laos), Dutch Batavia (now Jakarta, Indonesia), or British Pretoria (South Africa). Earlier, the Spanish left a huge imprint on the cities of Latin America. When you see a church opening onto a square at the center of a street grid lined with arcaded public buildings, you're looking straight to the Law of the Indies, promulgated by Philip II in 1573.

Fly from Israel to South Africa, as El Al and South African Airways do, and in the northern Sudan you'll see an immense network of canals, laid out like the teeth of a comb. Pilots know it well but sometimes don't know what it is. It's the Gezira Scheme, a joint project of British capital and Sudan's colonial government. Built in the 1920s and enlarged later, it takes water from the Blue Nile and spreads it on land traditionally planted to patches of rainfed sorghum. The surveyors came in, are said in jest to have gone mad laying out canals on this almost dead-flat plain, and produced an agricultural machine devoted to the production of long-staple cotton. There was some resistance from the Sudanese. Their views were summed this way by one patronizing administrator of the 1920s: "We hate these straight lines. We would rather be hungry once every few years, with freedom to range with our cattle unconfined, than have full bellies and be fined if we stray outside these horrid little squares." Such complaints didn't stop the British, or a later generation of foreign experts, from judging the Gezira a triumph of African development. A few dozen miles away, in Khartoum, the city streets are laid out in the form of union jacks. The plan wasn't sentimental indulgence: The Sudanese were a turbulent people in British eyes, and the union-jack layout allowed machine guns to be placed at the center of each cross.

Too much ancient history? Then hurry out to the site of the 2008 Beijing Olympics: You can use a six-lane divided highway. On the way you'll pass subdivisions that look American: even their names are American. Translated from Chinese, for example, there's Orange County and Watermark–Long Beach. Both are products of SinoCEA, and both are based on research in California, topped up with on-the-spot help from an American architect. You'll see three- and four-bedroom single-family houses, trimmed in Tudor and fitted out with imported kitchens—so useless for Chinese cooking that homeowners take their woks outdoors. Still, the lure of the West is powerful. Watermark–Long Beach homes start at $1.5 million. There's competition, too, from Legend Garden Villas and Moon River.

Khartoum surrounds the confluence of the Blue and White Niles. That's the White on the left, the Blue at the middle right, and the Nile heading in the upper right toward Egypt. The British city hugged the Blue Nile, and a part of the original, flag-like street plan can be seen here.

SOHO China is a Chinese-owned company that has built the 2,000-unit SOHO New Town in Beijing, as well as the 50 consciously avant garde villas at what SOHO disarmingly calls Commune by the Great Wall.

Where do the railways of the world come from? The airports? The chemical fertilizer put on the fields? The chainsaws cutting the tropical hardwoods? Global corporations are often involved in these things, but even if the operators are strictly local and the capital is raised from banks on the spot, the master key—the "effecting of all things possible"—is a European import.

THE HUMAN COST

Has it been worth it? All in all, the answer is yes: We generally agree that Deng Xiaoping was right and that it's glorious to be rich. It's not just the rich who think so, or even the small rich. If anything, the poor are even more convinced. The good news is that we're getting there. The World Bank reports that the number of absolutely poor people in the world—those living on a dollar a day—declined from about 29% in 1990 to 23% in 1999. Critics—and there are souls brave enough to question the bank—argue with these numbers, but the evidence of growing prosperity is fairly overwhelming to anyone visiting Shanghai or Delhi, not to mention Singapore or Seoul. Yes, wealth is highly concentrated in the hands of a minority, but even in neighborhoods where people used to be lucky to have an electric fan,

Soho New Town, Beijing. Everything is very white, very blonde, very stylish and—in the height of style—very expensive.

many have air conditioners. Where you used to hear radios, now you see satellite dishes. The oligarchs don't have it all.

This doesn't mean that globalization comes free of charge. In 1977, the first missionary came to remote Babuyan Island, between Luzon and Taiwan. Twenty-odd years later, some of the islanders had been to school and earned money for the first time in the islanders' history. There was electricity on the island and even satellite television. Sounds good—sounds like progress—but one islander told a visiting reporter that he now felt "a clock ticking and time rushing by." Not so good: He had been bitten by the bug. The efficient organization of a technological society demands the precise measurement of time. It demands also that everyone not only be aware of time but be willing to structure their lives around it. Ditching the ticking is next to impossible.

You don't mind the time trap? Fine; I'll drop the psychological price of modernity and talk instead about money and poverty. Thanks to falling tariffs imposed by the World Trade Organization, pineapple growers in Mexico are being devastated by a flood of cheap canned fruit from Thailand. What to do? Let the pineapples rot in the fields, get on a bus, and head north. The other choices are Mexico City, drug trafficking, and political activism. And pineapples are only a part of the Mexican story. A half million Mexican smallholders, *campesinos*, have left their farms and moved to cities because Mexico, under the terms of the North American Free Trade Agreement (NAFTA), is dropping its duties against imported corn. It's now flooding in from the United States, partly because American farmers are efficient but partly also because the U.S. government, a tireless advocate of free trade and free markets, shamelessly subsidizes and protects American producers. The displaced *campesinos* struggle to find urban jobs. Some succeed and are better off than they were as farmers. Others do not and are not. Mexican price controls on tortillas and cornmeal have meanwhile been removed, and prices for those staples have tripled.

Or think bananas. In the late 1940s, three small Caribbean islands—St. Vincent, Grenada, and Dominica—were allowed to sell bananas in Britain without paying the steep tariff levied on other exporters. It was a policy designed to stabilize the economies of these small parts of the British Empire, but in an era of free trade the United States has successfully insisted that the islands no longer get a tariff break. The result promises to devastate the islands, because their production methods are much less efficient than those of the global producers—companies such as Dole, Chiquita, and Del Monte. Even while protected during the 1990s, the banana industry on the islands was in decline. Perhaps the best hope lies with Sainsbury's, a major British supermarket chain that is encouraging the islanders to take up the production of organic fruits and vegetables. That's less farfetched than it may sound, because the islands already use less pesticide than more important banana-growing areas.

You say that the cost of modernization is paid only by the distant poor? Not so. Around the world, tombstones are a reminder of how high the price of the Baconian experiment has been for the elite. Calcutta was the source of immense wealth for Britain, but visit the Park Street cemetery and you can see the tomb of Rose Aylmer, who died in 1800 at age 23. On it, there's a sentimental but not necessarily false poem written by Walter Savage Landor (1775–1864). The first of its two stanzas is: "Ah, what avails the sceptred race! / Ah, what the form divine! / What every virtue, every grace! / Rose Aylmer, all were thine!" (Landor himself is buried in Florence, where his tombstone carries some equally romantic lines by Swinburne.) On the coast of South Africa there's a town that used to be called Port Elizabeth. The name's been changed now to Nelson Mandela, but Elizabeth Donkin is still dead, the 27-year-old victim of a fever contracted in India. Her husband tried to rush her home, but she died at sea.

There's a North American version. It begins with the pioneer triumph. Well, it was a triumph for some. Perhaps it's a triumph still. A Belgian couple, tired of crowded Brussels, buys land in Oklahoma and raises quarterhorses. A house goes up, along with a big horse barn and fencing—lots of it. The couple work hard; so do their children. There's no end of things to do, but it's a great life: active, outdoors, changing all the time.

It's also a risky life. "Undercapitalized" is the usual explanation—the accountant's bloodless word for going broke. You can find plenty of British farmers who took their dreams to Kenya, lost everything, and packed up. (For an example, see *The Land that Never Was*, a bleak 1937 memoir by Alyse Simpson.) It was the same when Oklahoma was opened to settlement a century ago. Most of the homesteaders were quickly reduced to sharecropping, then driven off the land in the 1930s. From the viewpoint of the Indians, these white settlers were conquerors, but most of the settlers didn't have a lot to celebrate. Roxanne Dunbar-Ortiz has shown just how brutal their lives could be. Her memoir, *Red Dirt: Growing Up Okie* (1997), has the impact of James Agee's classic account of sharecroppers, *Let Us Now Praise Famous Men* (1941).

More ancient history? Then look back only to 1997. Ninety percent of Washington State's wheat was heading to Southeast Asia, but the economy there was in sudden crisis. Customers for the wheat went away, and the crop piled up in a heap big enough to fill a freight train stretching from Seattle to Chicago. The mountain is gone now, but you think it wasn't painful for the people who grew that wheat? We're in this globalization to our eyeballs, and we're in it together, regardless of our passports and bank books.

Keep the costs in mind, along with the benefits. Meanwhile, go out to Yuba City, California, and meet the man who, with 16,000 acres in cultivation, is the world's biggest peach producer. He's a Sikh, named Didar Singh Bains. Down in the Imperial Valley, you can meet the Okra King. He's Sikh, too: Harbhajan Singh Samra. These are unexpected but classic American stories, but they're no longer just American stories. The American Dream has been globalized. Want to buy gummy bears in Prague? No problem: In fact, they're made there. The company is Mexican owned, and in the year 2000 the plant was run by an ethnic Chinese who held a German passport and was raised in Indonesia. Not strange enough? Try going to church in Ireland. You could be in for a shock, though. The Irish economy is so hot that young men skip the priesthood and go into business. What's the Church to do? Answer: import priests from Nigeria, which in the St. Augustine seminary at Jos apparently has the world's biggest training institute for Catholic priests.

All these people migrating to foreign opportunity will retain the cultures of their childhood, substantially mixed with the cultures in which they now live. Judging from the American experience, their children will retain much less of their parents' childhood cultures. Possibly the children will mature in a world of democratically blended cultures. More probably, because they're living in a world where education and work are organized along Western lines, they'll mature in a world in which bits of this and that culture are blended into a world culture structured around the values of the West.

If that's true, globalization goes far beyond products, companies, and jobs. It goes to the way people think. It creates a global civilization with a single shared, sustaining ideology. Along with the psychological and economic costs of modernization, in other words, there's a cultural cost. It's easy to say that Confucianism and its respect for age will not yield to the youth-oriented West. It's easy to say that the cultural bedrock of caste will outlast us all. But the pace of globalization is accelerating, not abating, and culture may be able to change faster than we think. An unmarried woman is unthinkable in traditional India, but now there are 25-year-olds in Mumbai (Bombay) who say, "marriage and me will never mix." In an earlier generation, such women would have already spent years as the slaves of mothers-in-law who finally had someone to abuse in the way they themselves had been abused. But young Indian woman today can make $200 a month working at a call center, where they sell things to Americans or perhaps merely hound them nicely about an overdue credit card payment. An Indian mother says of her newly independent daughter: "Her values have changed and I blame it on this business." Mom's right. A young man at the call center has picked up the idea that American teens are on their own at 18. "I think that is very excellent," he says. His mother is baffled. Speaking about the changes that had come over her son, she says, "I felt a part of my life was gone."

PART III

LIVELIHOODS TODAY

CHAPTER 11

RESOURCE PRODUCTION

So far, we've operated mostly historically, bumping against largely unanswerable questions about the emergence of our species, the rise of agriculture, the establishment of urban societies, and the sources of a secular civilization. From now on, however, the focus will be on today, or at least on the very recent past. We'll be looking mostly at things we can see in the world now—things that are going concerns, not ruins or relics.

There are lots of ways of doing this, but I'm going to begin by looking at how people make a living. It's an odd approach for someone who stresses ideology, but we do live by bread, if not by bread alone. I'll start in this chapter with the production of resources. In the following two chapters I'll discuss manufacturing and the service sector. Then I'll indulge myself and devote a whole chapter to the transportation and communication systems that knit these things together.

MAKING AGRICULTURE A BUSINESS

The last time we looked at agriculture in any detail, we were considering its diffusion paths and the ways farmers adapted it to local conditions. We were talking about subsistence agriculture, in which farmers eat what they grow.

Plenty of peasant farmers still do that, but many other farmers would burst if they, their families, and even their neighbors had to eat even a tenth of what the farmers grow. Most of their production instead enters increasingly long and deep commercial pathways. Every year, more than 100 million tons of wheat are shipped internationally—often intercontinentally. (About 60% of it comes from the United States, Canada, and Australia. The importing countries are much more diversified, and no country imports more than one-tenth of the flow.) Despite the immensity of this traffic, less than one-fifth of the world's wheat moves internationally. There are other crops, however, that are grown primarily for export. About half the world's tea, for example, is exported, and about three-quarters of the world's coffee. Typically, the countries that produce these commodities are dependent on them and on commodity prices set in an international market over which they have almost no control.

This is a long way from medieval agriculture, that plow-based system of mixed farming on common fields. Seed was obtained by saving part of the previous crop. The only manure was fertilizer. There were no synthetic pesticides. No fieldwork was done unless a muscle—animal or human—made it happen. It's a world described almost a century ago for England

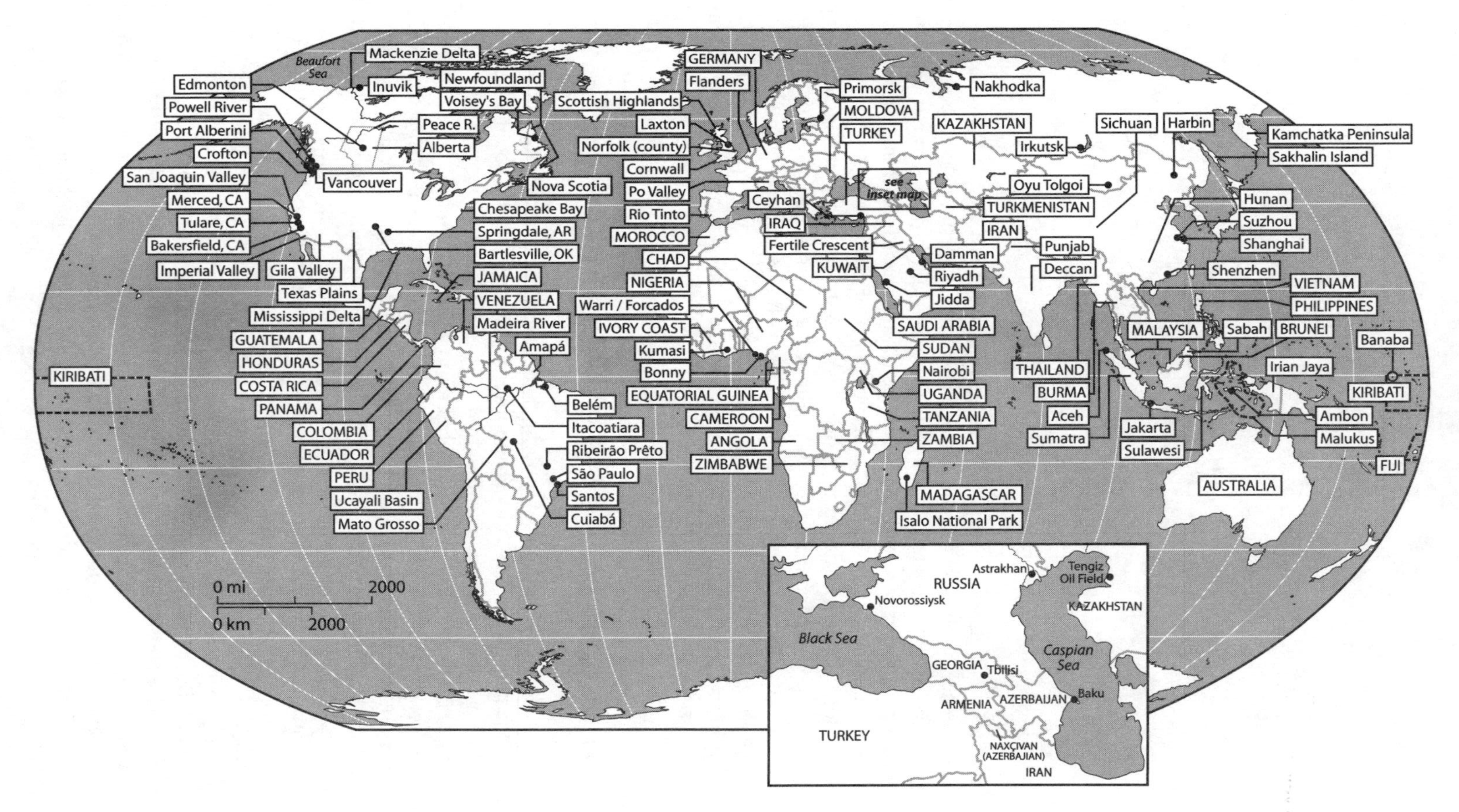
Beaufort Sea
Mackenzie Delta
Inuvik
Newfoundland
Voisey's Bay
Edmonton
Powell River
Port Alberini
Crofton
San Joaquin Valley
Merced, CA
Tulare, CA
Bakersfield, CA
Imperial Valley
Gila Valley
Texas Plains
Mississippi Delta
GUATEMALA
HONDURAS
COSTA RICA
PANAMA
COLOMBIA
ECUADOR
PERU
Ucayali Basin
Mato Grosso
Peace R.
Alberta
Vancouver
Nova Scotia
Chesapeake Bay
Springdale, AR
Bartlesville, OK
JAMAICA
VENEZUELA
Madeira River
Amapá
Belém
Itacoatiara
Ribeirão Prêto
São Paulo
Santos
Cuiabá
KIRIBATI
GERMANY
Flanders
Scottish Highlands
Laxton
Norfolk (county)
Cornwall
Po Valley
Rio Tinto
MOROCCO
CHAD
NIGERIA
Warri / Forcados
IVORY COAST
Kumasi
Bonny
EQUATORIAL GUINEA
CAMEROON
ANGOLA
ZIMBABWE
Primorsk
MOLDOVA
TURKEY
Ceyhan
IRAQ
Fertile Crescent
KUWAIT
see inset map
Damman
Riyadh
Jidda
SAUDI ARABIA
SUDAN
Nairobi
UGANDA
TANZANIA
ZAMBIA
MADAGASCAR
Isalo National Park
Nakhodka
KAZAKHSTAN
Irkutsk
Oyu Tolgoi
TURKMENISTAN
IRAN
Punjab
Deccan
THAILAND
BURMA
Aceh
Sumatra
Sichuan
Harbin
Kamchatka Peninsula
Sakhalin Island
Hunan
Suzhou
Shanghai
Shenzhen
VIETNAM
PHILIPPINES
MALAYSIA
Sabah
BRUNEI
Banaba
Irian Jaya
KIRIBATI
Ambon
Malukus
FIJI
Jakarta
Sulawesi
AUSTRALIA
0 mi
2000
0 km
2000
RUSSIA
Astrakhan
Tengiz Oil Field
Novorossiysk
KAZAKHSTAN
Black Sea
Caspian Sea
GEORGIA
Tbilisi
ARMENIA
AZERBAIJAN
Baku
TURKEY
NAXÇIVAN (AZERBAIJAN)
IRAN

in Lord Ernle's *English Farming: Past and Present* (1912) and Howard Gray's *English Field Systems* (1915). It's a world echoed today in the British village of Laxton, which has deliberately retained its old fields. (Willing to dig a bit? Shortly after William the Conqueror arrived in 1066, an inventory was undertaken of his new island. The scope and breadth of the survey seemed so apocalyptic in its finality that the work became known as the *Domesday Book*, the book of doomsday. According to it, Laxton was the property of Geoffrey Alselin, one of 30 landholders in Nottinghamshire. Geoffrey's man on the spot was one Walter. He had a plow. There were also 29 peasants in Laxton, and they shared five more plows. There were six slaves—five male, one female. The whole place was said to have a cash value of 6 pounds.)

There's another echo in the work of painters. Pieter Bruegel (d. 1569) is a good example. His "Haymaking" shows rakes and a large hay wagon. "The Harvest" shows scythes, pitchforks, and ladders. "Spring," a pen-and-ink drawing, shows long- and short-handled spades. "The Gloomy Day" adds hatchets to the list. "The Fall of Icarus" shows a plowman with a whip in his left hand, while the right guides a one-bottom moldboard plow, fitted with a colter and pulled by a horse wearing a horse collar. Brueghel's landscapes are fictional—a blend of things he had seen between Flanders and Italy—but his farm tools are instantly recognizable. Many of them are things you could buy at your local hardware store.

A third echo of that rustic world is heard in such common English surnames as Farmer, Fisher, and Shepherd. Very frequently, these names point to the range of occupations in Henry VIII's time, when surnames became a legal requirement in England. Some are connected to building trades: Sawyer, Barker, Carpenter, Mason, and Thatcher come to mind. Some relate to food, such as Miller, Baker, Brewer, and Cook. Others are connected to clothing, such as Taylor, Weaver, Dyer, and Fuller. Other trades are common, too, such as Smith, Carter, Cooper, Fowler, Hunter, Potter, Skinner, and Tanner. Another class of names indicates places where people lived and worked, such as Fields, Meadows, Woods, Waters, Lake, and Brooks. Many if not most of the names come straight from the rustic European world that Bruegel would paint a few decades later.

Even then, however, a profound change was under way. Subsistence farming was colliding with aristocrats who saw a way to make more money from their land. As early as the 13th century, Spain recognized the existence of a national cooperative of aristocratic woolgrowers. Known as the Mesta, the organization had the ruinous right to drive sheep across the countryside. Farmers were forbidden to protect their crops, even with fences. The Spanish crown was enriched with the proceeds from the exported wool, but long before the Mesta's abolition, in 1837, Spain's peasant farmers were almost ruined. (The classic treatment is Julius Klein's *The Mesta*, 1920.)

The British nobility meanwhile created sheep pastures from the old common lands. It took a long time. The enclosure movement started in the 13th century and rose to a peak in the late 18th and early 19th centuries. It was taken to an extreme in Scotland, where 350 people today own half of all the private land in the country. A mere 100 people own half the Highlands, while some 30,000 crofters rent the small plots assigned long ago to displaced clan members. In a delayed reaction, the Scottish Parliament in 2002 passed, by a vote of 101 to 19, a Land Reform Bill that gave crofters the right to buy their plots and gave the public the right of access to land in the great estates. Crofters thought it was high time—a century or more overdue—but the owner of Skibo Castle said that the law "borders on communism." His celebrity guests wouldn't come, he said, if he couldn't guarantee seclusion.

By the 18th century a new agricultural mentality had emerged, based on the experimental method and dedicated to greater profits. It wasn't just aimed at wool production, either.

Farmers experimented with larger varieties of cattle and posed proudly next to immense bulls that sometimes weighed 2 tons, much larger than animals today. A famous example is Robert Bakewell (1725–1795), though his English longhorn is scarcely known today.

Crop production changed, too. The medieval crop year commonly began with winter grain, followed the next year by a spring grain, and in the third year with a year's fallow. The 4-year Norfolk rotation got rid of the fallow and began with root crops like turnips, followed by barley, clover, and wheat. It wasn't long before soil chemistry began to be studied by Germany's Justus Liebig (1803–1873). Farmyard manure gradually became a nuisance, and hard times fell on the once profitable business of mining and exporting guano, which is mostly composed of seabird droppings. Some guano producers adapted successfully, like those of Micronesia's Ocean Island in the Gilberts. (Now it's called Banaba and is part of Kiribati.) Ocean Island shifted to mining its extensive reserves of phosphates, which were shipped to farmers in New Zealand and Australia.

European agriculture was so much a part of society, however, that innovation was impeded by social resistance. That's where empires came in handy. Even if there were people on the land, the slate there could much more easily be wiped clean to make way for a rational, scientific, capital-intensive agriculture. The Dutch wanted to monopolize the world market for cloves, for example, so they developed clove plantations on Ambon, one of the Spice Islands (today the Moluccas or Malukus). To reduce competition, they then uprooted every clove tree they could find on neighboring islands. Nobody could stop them.

The clove market was small compared to the market for sugar, for which Brazilian and then Caribbean plantations were established. Later, there was strong growth in bananas, pineapples, rubber, cacao, coffee, tea, and palm oil. For plantation owners, agriculture was a business, however much they loved it. When business was bad, their lives cratered. Think of Isak Dinesen, who as Karen Blixen wrote *Out of Africa* (1937). In the 1930s, falling world prices for coffee drove her from her farm just west of Nairobi. The house is still there. It's a museum.

The Great Depression devastated Brazil's coffee industry, too. In the 1920s you could go to Santos, Brazil's coffee-exporting port, stroll down Rua 15 De Novembro, and see lines of banks—American, British, Brazilian. They had come to serve an industry that was supplying 70% of the world's coffee. Brazil's rise had been quick, linked to the opening in 1867 of a railroad from Santos through São Paulo to the coffee-growing center of Ribeirão Prêto. Coffee exports through Santos rose from about 350,000 bags (each of 60 kilograms) to 10 million by the 1920s. By then, São Paulo State alone supplied half the world's coffee. It came from *fazendas,* plantations. The biggest in 1923 was the British-owned Fazenda Dumont, with 15,000 acres and almost 5 million trees. William Ukers, the editor of *The Tea And Coffee Trade Journal*, came to pay a visit. From his railway car, he wrote that he saw "towns with churches, shops, railway stations, main streets, to be sure, but they were always set among the coffee trees, which thrust their mossy-green branches loaded with golden-red clusters of ripening berries into the very windows and doorways. In some places the trees were so close to the tracks that one could almost touch them from the car windows, and everywhere, as far as the eye could reach, the rolling, coffee-clad hills."

The best known plantation company handled not coffee but bananas. This was the United Fruit Company, created in 1899 by a merger of the Boston Fruit Company and holdings of Minor Keith, a railway builder in Central America. Bananas had been a luxury item in the United States since the Civil War. Mostly they had come from Jamaica, to which they had been introduced. United Fruit, however, pushed the industry onto the mainland, and by 1913 the company was growing 150,000 acres of bananas—48,000 in Costa Rica, 35,000

Karen Blixen's Kenya home, now on the western outskirts of Nairobi.

in Panama, 27,000 in Guatemala, 23,000 in Colombia, and 9,000 in Honduras. The company still grew bananas on Jamaica, but its production there was smaller than from any of these other countries.

The United Fruit idea was to market huge volumes at a low per-unit profit. To implement that strategy, the company not only ran plantations but railroads to bring the fruit to shipping ports. It ran a fleet of ships—the "great white fleet," so-called because the ships were painted white to keep them cool in a time without refrigeration. The company ran ice-chilled railway cars in the United States, and in 1913 those cars carried 50,000 carloads of bananas. It was a tour de force: 25,000,000 bunches of bananas (a bunch consists of about 144 bananas clustered in nine hands), all consumed within 3 weeks of picking. Frederick Adams wrote an adulatory company history. As its title suggests, *Conquest of the Tropics* (1914) portrays the history of United Fruit as a military campaign. "An empire of agriculture," Adams writes, "was carved from the jealous and resentful jungle."

That's one side of the story. The other side begins with the United Fruit Company's involvement in Central American politics—so deep that these countries are still known derisively as banana republics. The company was even the subject of a vitriolic poem by Pablo Neruda, which ends with these lines: "The United Fruit Company Incorporated sailed off with a booty of coffee and fruits/ brimming its cargo boats, gliding/ like trays with the spoils of our drowning dominions./ And all the while, somewhere, in the sugary/ hells of our seaports/ smothered by gases, an Indian/ fell in the morning:/ a body spun off, an anonymous chattel, some numeral tumbling,/ a branch with its death running out of it/ in the vat of the carrion, fruit laden and foul."

Perhaps there's a degree of justice in the financial ruin of so many big plantation operators. Brazil's share of world coffee production fell from 70% in the 1920s to about 20% at century's end, and Chiquita Brands, the successor to United Fruit, saw its stock price crash from $50 in 1991 to 50 cents a decade later. The company had banked prematurely on deregulation of the European market and, analysts said, was not managed as well as its chief competitors, Dole and Del Monte.

Poor management wasn't the only problem facing plantations. Tropical crops often had to compete with temperate substitutes. Brazilian sugarcane growers, for example, can no longer count on supplying Europe, because Europe grows a very large crop of sugarbeets, which are refined into a product indistinguishable from cane sugar. Caribbean sugar producers are lucky enough to have European quotas, but those quotas are scheduled to expire in 2007: The growers will then either cut costs or go out of business.

Or consider the oil palm, best known to Americans from Palmolive soap. The oil palm is a major plantation crop in Malaysia and Indonesia, but it competes with oils from temperate plants such as soybeans, corn, cotton, and sunflowers. Natural rubber is a third example. Originally tapped in the Amazonian rain forest, it is now grown widely on plantations in Southeast Asia, chiefly Thailand. There it competes with synthetic rubber. It's losing: Two-thirds of all rubber is now synthetic.

Growing Chinese demand has raised many commodity prices in the last few years, but the historic trend is one of long-term decline. Tea prices, to take one amazing example, haven't risen in 30 years. It's unbelievable: A box of 100 tea bags still retails for under $2. That's terrible news for the world's tea exporters, led by Sri Lanka. Or consider coffee again. By value, it's the second most valuable commodity in world trade, after oil, but prices of arabica coffee, which is the superior type, fell in the summer of 2001 to a 9-year low of 50 cents a pound. The stronger robusta coffee, used in instant coffee and increasingly in arabica blends as well, fell even more steeply, to a 30-year low of $493 a ton. (Robusta is very bitter but acceptable if steam-cleaned or masked in those flavored coffees that have recently sprouted on grocery shelves.)

Why such low prices? Vietnam in 1975 grew 20,000 acres of coffee. By 1990, it grew 200,000. During the 1990s it cleared an additional million upland acres and planted them to coffee. In the process, it became the world's second largest coffee producer, after Brazil. The resulting oversupply, however, pushed prices down to 24 cents a pound. It's been ruinous for the world's 25 million coffee producers, most of whom are very small. It's been ruinous even for the Vietnamese, whose production crashed from 900,000 tons in 2000–2001 to 600,000 some 2 years later. Desperate Vietnamese began removing their coffee trees and scrambling into cocoa and cashews. The four major coffee buyers—Kraft, Sara Lee, Procter and Gamble, and Nestlé—meanwhile continued to buy half the world crop, which they then sold for an average of $3.60 a pound. The wide gap between producer and retail prices has not gone unnoticed. A director of the Uganda Coffee Development Authority says of these companies: "they are so powerful that they can determine the price. What makes us mad is that the retail price of coffee has never gone down." He continues: "I'm telling you it's us today. But it may be you tomorrow."

Both Vietnam and Brazil are major producers of black pepper, but they found it hard to sell their pepper even at cut-rate prices after India in 2001 had a huge crop. So Vietnam turned in a big way to the production of shrimp. Over 1 million acres are now in ponds, and the value of the country's shrimp exports in 2001 was $780 million. That's impressive, but shrimp prices are declining, too, and shrimp production is technically risky, especially when the producers are small and heavily indebted, as they typically are in Vietnam. (In a world as tightly knit as ours today, it's not just the Vietnamese shrimp producers who are in trouble.

American shrimpers on the Gulf Coast now supply only 15% of the American market, and the prices they receive at the dock are so low that many are going out of business. Perhaps their future lies in advertising their product as wild, instead of farmed. For the moment, they're hoping a new tariff on imported shrimp will save them. Restaurateurs are opposed to a tariff, of course, as are the Vietnamese.)

Sometimes commodity prices are high. Cocoa prices in 2000 were about $700 a ton; a year later, after civil war had broken out in Ivory Coast, the world's leading cocoa producer, prices rose to a 17-year high of $2,400 a ton. Producers in other countries, including not only nearby Cameroon but the distant Solomon Islands, have benefited greatly, but those in Ivory Coast cannot market their crop. Some cannot even pick it, because their laborers have fled. One bitter Ivorian producer says, "we are forced to live like rats."

Some governments protect their farmers from such cycles. It's not just the Americans who do it, or the Europeans, or the Japanese. The Mexican sugar industry has some 60 refineries, each supplied, on average, by 2,500 growers, each with 5 or 10 acres. Those growers expect to make a living growing cane, but the refineries are obsolete. For many years the government of Mexico solved this problem by subsidizing the cane crop. The absurd result was that a ton of harvested cane in Mexico was worth more than a ton of refined sugar. For a time, the Mexican government spoke about promoting a free-market economy, but it finally decided that it could not let farmers starve while it attended to what economists are fond of calling "structural problems." The Mexican secretary of agriculture said that farmers would simply have to find other jobs, but in September 2001 the government expropriated 27 of the nation's refineries and guaranteed payment of $500 million to the *campesinos*, or peasant producers. The plan was for the government to spend $300 million modernizing the mills, then privatize them, but production continued to exceed demand, especially when Mexico began importing cheap high-fructose corn syrup from the United States. The imports stopped when Mexico slapped a 20% tariff on high-fructose syrup imported for soft drinks.

Americans can hardly scoff at such market distortions. After all, Mexico would love to export sugar to the United States, but since 1934 the United States has fiercely guarded its own cane- and beet-sugar producers. Presently it limits sugar imports to about 100,000 of the country's 8-million-ton sugar market. As a result, Americans pay a premium of about $2 billion a year for sugar, which sells in the United States at roughly 20 cents a pound, twice the world-market price. Every now and then, a naïve foreign country tries to breach the wall. A proposed Central American Free Trade Agreement calls for a special 100,000-ton quota, but don't hold your breath. Late in 2003, Australia tried, too, but its prime minister gave up, saying that Australians were "the victims of a corrupted world trading system." Australia had sent troops to help the United States in Iraq, but that wasn't enough to counter the political influence of America's sugar producers, led by the Fanjul family of Florida. The sugar producers are the most generous political donors of any farm group, including the tobacco and dairy industries. They're up front about their concerns. The president of the American Sugarbeet Growers Association says bluntly that "if you go to free trade, Brazil wins and everybody else gets killed."

Speaking of Brazil, the sugarcane industry there has historically been organized in large companies that hire cutters. More than half the crop is now cut mechanically, however, and the growers have simply fired surplus workers. Half the country's cane cutters lost their jobs this way during the 1990s alone. Where shall they go? What shall they do? This is a recurrent question in all discussions of agricultural modernization. The usual answer is, "Go to the city. Write when you get work."

North America, like the other parts of the world that attracted settlement by Europeans, offered another kind of clean slate. The classic homestead in both Canada and the United States was a quarter section. That's 160 acres, a fourth of a square mile. Small by plantation standards, it's a principality from the perspective of peasant farmers. An acre, after all, is almost exactly the size of an American football field. Even 1 acre is a lot of ground to till by hand.

Although they might produce a variety of foods for their own use, homesteaders were primarily in the business of growing crops for sale. They always needed cash, both to buy the equipment that made them competitive and to buy the many things they wanted personally but could not make or grow. Under constant pressure to maximize income, they were always looking for more efficient production methods. Inventors obliged with a wave of innovative technologies that swept across these lands in the 19th and 20th centuries. Worker productivity kept rising. Fewer and fewer farm workers were needed. Farms grew larger.

Steam power was adapted from the locomotive to the tractor and thresher. Combines joined harvesters to threshers. The machines were expensive, but a custom industry emerged, with the machinery moving north in tandem with the ripening wheat. After a generation, steam yielded to the internal combustion engine. Pneumatic tires replaced steel wheels and most caterpillar tracks. Agricultural museums are full of the ingenious residue.

When did you last see a mule? In 1930 they constituted a quarter of the equine stock in the United States. Now, mules are outnumbered 25 to 1—and that's counting burros and donkeys along with the mules.

Tractors at planting are now often equipped with soil monitors that continually measure soil conditions and feed that information into a computer linked to a global positioning system. Later on, when the farmer comes through the field to apply fertilizer, the system delivers the precise amount of nutrient needed at every point in the field.

The hot topic today is genetic engineering. For many years, hybridization was based on experimental crossbreeding done at research stations belonging to the government or universities or private companies. Hybrid corn, developed about 1920, was the most famous of these new plants. Yields were so high that by 1950 almost no other kind of corn was planted in the United States, even though farmers had to buy seed every year, because hybrid corn doesn't reproduce faithfully.

Conventional crossbreeding is by no means an exhausted technology, but it's almost ignored nowadays by the media, preoccupied by gene-spliced transgenic plants and animals, or genetically modified organisms (GMOs). The big American players are Monsanto, Dupont, and Dow, but their agricultural operations are smaller than the European leaders, Syngenta and Crop Science. Merely by their corporate histories, the European giants illustrate the transformation of agriculture to an industry of giants. Syngenta was formed in 2000 by the merger of the agrochemicals businesses of Novartis and AstraZeneca. Novartis had been created by the merger of Ciba-Geigy and Sandoz in 1996. AstraZeneca had been formed by the merger of Swedish Astra and British Zeneca, which itself had briefly been a piece of Imperial Chemical Industries. CropScience, meanwhile, is a division of Aventis, a pharmaceutical company formed by Hoechst and Rhône-Poulenc. We're a long, long way from Laxton.

About 80 million acres of cropland in the United States are now seeded to genetically modified crops, mostly soybeans, cotton, and corn. Monsanto is particularly well-known for its Roundup Ready soybeans, which are unaffected by the company's powerful herbicide Roundup. Farmers plant the seed and hardly worry about weeds: they just spray the field, soybeans and all, because the soybeans are immune to the poison. The company's CEO says that "fifteen years ago, we were digging holes in the ground, extracting oil. . . . Now we're a seeds and biotechnology company." Elsewhere, he has said, "We've bet the farm on this."

The geography of American crop and animal production has changed over the last century as much as farm technology. Many old farms in New England have returned to trees, while the wheat and corn they once grew moved to the deep, level soils of the mid-continent. These two crops parted company, however. Soybeans replaced wheat as the rotation crop in the Midwest's Corn Belt. (Although the soybean is a Chinese domesticate, the United States now exports it to China, whose crop is huge but has lower levels of protein and oil.) Wheat moved west, to the drier Great Plains. The Cotton Belt of the Deep South, meanwhile, is gone, replaced by more trees. Cotton now comes from the Mississippi Delta below Cairo, Illinois, from the Texas South Plains around Lubbock, from Arizona's Gila Valley, and from the Imperial and San Joaquin valleys of California.

Animals have moved, too. Cattle used to be fattened on corn in Iowa, but they've moved west to be fattened in feedlots on the drier (and therefore healthier) Great Plains. Hogs are making the same move. Chickens, once centered on the eastern seaboard, particularly northern Georgia and the Delmarva Peninsula, east of Chesapeake Bay, have joined the westward movement. The biggest poultry producer in the world is now Tyson Foods of Springdale, Arkansas. The company disassembles more than 2 billion chickens annually, sales exceed $7 billion, and the company employs over 73,000 people. It owns Iowa Beef Packers, too, a major producer.

Wisconsin may call itself the dairy state, but California produces more than 50% again as much milk—and the gap is widening. The biggest milk processing plant in the United States today belongs to Land O'Lakes, Inc., but it's in Tulare, California, not Minnesota. The biggest cheese factory in the world is a few counties north, near Merced. There, the Hilmar Cheese Company takes the milk of 150,000 cows and makes 1 million pounds of cheese daily. So much for license-plate geography.

All these changes—mechanization, breeding, and industry mobility—exemplify the Baconian program. Food in America is abundant and cheap, a model for the world. And the family farmer? He's a living dinosaur. A model American producer today might be Paramount Farming of California. It's new, created in the 1980s when Stewart Resnick, a Los Angeles businessman who owned the Franklin Mint and Teleflora, bought some San Joaquin Valley properties belonging to Dole, Texaco, and Mobil. Now Paramount Farming is the biggest citrus producer in the United States. It's the country's biggest producer of almonds and pistachios. It's probably the nation's biggest producer of pomegranates, with 6,000 acres in cultivation. It falls short only in cotton. With 100,000 acres, Paramount is the nation's second largest producer.

Does the industry have problems? Sure. Remember Rachel Carson, who got royally hammered for attacking the pesticide industry in *Silent Spring* (1962). Still, she had the final word, not only with the near ban on DDT but with a booming market for organic products. The biggest American retailer of them, Whole Foods, now has over 140 stores. Each one has recently been increasing its sales by 10% every quarter. (The chain includes Bread and Circus, Bread of Life, Mrs. Gooch's, and Fresh Fields. The nearest competitor is Wild Oats, with just over 100 stores.) Monsanto got so badly burned over its aggressive marketing of bovine growth hormone for dairy cows that it had to cancel plans to spend $1.9 billion on the Delta and Pine Land Company, a cottonseed producer. Some say that it had no choice but to merge and be subordinated to another company, Pharmacia. (Pharmacia is now part of Pfizer, the world's biggest pharmaceutical company.) In Europe, the suspicion about pesticides runs deeper still. Even conventional British supermarkets have an amazing array of organic foods. The British Soils Association reported in 2003 that 58% of the cropland of Scotland was farmed organically and that three-quarters of the babies in the United Kingdom regularly ate organic food.

Gerber and Frito-Lay say they won't touch GMOs. In Europe, Nestlé and Unilever say they, too, won't touch genetically modified products. Tesco, Britain's biggest supermarket chain, says that it sells no meat from animals raised on genetically modified feeds, while Co-op, which runs 1,600 grocery stores and is Britain's biggest farming operator, announced in 2003 that it would no longer handle any GMOs. London's *Daily Mirror*, with its gift for language, in 1998 began referring to genetically modified foods as Frankenfoods. The more sober European Union announced in 2004 that all foods containing GMOs had to be labeled as such, even if the quantity of genetically modified ingredients in the product was as little as .9%. Monsanto was at that moment ready to begin selling Roundup Ready wheat to farmers in the northern Great Plains. It dropped those plans after farmers told the company they wouldn't be able to sell their crops in such a hostile international market.

Despite this resistance, separating transgenic foods from natural ones is next to impossible, partly because pollen from these plants mixes with plants in neighboring fields and partly because handlers can't isolate GMOs as they move through distribution chains. Consumers who want to avoid GMOs can try but will almost certainly fail. The GMO's acreage has meanwhile risen from 30 million in 1997 to 120 million in 2001. The United States that year led the pack with 80 million acres. It was followed by Argentina, with 20 million. Two years later, the global figure had risen to 160 million acres; 86% of the soybeans and more than 40% of the corn grown in the United States was genetically modified.

Then there's the social cost of converting American agriculture into an industry. In 1920 the United States was still a mostly rural country; now, of course, it's overwhelmingly urban. Perhaps that's a good thing, but it's been a painful transition, reminiscent of the old enclosure movement in Europe or the plight of Brazilian cane cutters today. Recall *The Grapes of Wrath*, John Steinbeck's 1939 portrait of Oklahomans tractored out in the 1930s—their farms taken over by neighbors eager to enlarge their own operations, especially when drought had forced down the price of land.

That was a long time ago, but potato growers on the irrigated plains of the western United States nowadays spend about $5 to grow 100 pounds of spuds; early in 2001, the crop was so big that farmers received $1 per 100 pounds. California milk producers were spending $1.05 to produce a gallon of milk for which they got 94 cents. California has only about 2,200 dairy farms, but one-tenth of them went out of business in 2002. An even bigger problem faces the producers of Thompson seedless grapes. There's been a doubling of grape production in California, and most of it has come in wine-grape and new table-grape varieties. That leaves little demand for Thompson seedless, either as a grape to blend into wine or as a table grape. The alternative is raisin production, but raisins face competition from Turkey. That's ironic, because the California industry was built and is still largely in the hands of the descendants of Armenians who fled Turkey early in the 20th century. They've seen the price they get for raisins fall in the last few years from about $1,500 a ton to about $500, far below their break-even point of about $800. Vineyards are being uprooted to save on taxes.

Cost-price squeezes such as these, coupled with the power of the farm lobby, have kept the federal government in the price support business. In 1996 it tried to phase out subsidies, but even under the Freedom to Farm Act of that year so-called emergency payments continued to be made. Supportive congressmen justified these payments by appeals to the plight of the family farm, but payments are proportional to production, so 10% of the nation's farmers collect 60% of the money. Only the organic farmers are increasing their acreage these days, but as public demand for their product rises, they, too, will be pushed to enlarge and consolidate. Carey McWilliams had it right six decades ago with the title of his book on California agriculture: he called it *Factories in the Fields* (1939). As for the farmers being squeezed? British farmers think that the public generally ignores them or, when it does on occasion notice them, resents them. Perhaps that's not true in the United States, where the Jeffersonian mythology of the hardworking but virtuous rural family remains powerful, but only 4% of the 460,000 members of the FFA (until 1988 the Future Farmers of America) say they plan on becoming farmers. School counselors in Iowa say that they haul out their thermometers when students tell them that they plan on a career in agriculture.

And what about overseas? You can find combines harvesting wheat in Morocco and rice in the Indian Punjab. Still, there is a huge world of peasant agriculture. In some ways or from some perspectives that world is immutable; in or from others, it's changing very fast.

The most brutal changes have accompanied Communist revolutions. By 1940, Russia's better-off peasants had been killed or starved to death. By 1950, huge collective farms operated across the broad swath of fertile land north and east of the Black Sea; gangs of machinery plowed enormous, unfenced fields. The collective farms were never very efficient, but the people of the Soviet Union were proud of them. Many Russians must be proud of them in retrospect, because after the breakup of the Soviet Union the collectives were reorganized as cooperatives. The results have been pitiful. Some 13 million Russians now own cooperative shares. There's enough money to pay them only a tiny fraction of their official salaries, and even that fraction is often paid not in cash but in kind. Many members survive by stealing from their cooperative. Others try to farm as smallholders—about a tenth of Russia's billion acres of farmland is now privately owned—but most of these new farms have failed for lack of capital. That's why in neighboring Moldova, when the workers on one state farm were each given three acres in a program called *Pamint*, or land, the workers joked that the program should have been called *Mormint*, or grave. Privatizing and adequately capitalizing Russia's farms, however, will be hugely difficult. There's no office charged with registering land ownership, and ownership shares are not tied to any particular piece of the collective's land. That's why fences are still, as in the days of collective farms, few and far between. There's emotional resistance, too, because many Russians see privatization as the path back to serfdom.

The Chinese repeated the Soviet Communist experience but emerged from it more quickly. China's landlords—"bloodsuckers" in Chinese literature of the period—were killed in the 1950s. Farmland was assigned to production teams grouped in production brigades, grouped into still larger cooperatives. Near Suzhou, for example, there was a subcounty township called Tungting, where some 45,000 people lived near the shore of Tai Lake. They were organized in 237 production teams clustered in 30 production brigades and 20 cooperatives. Production Team Number Eight of the Chenkuang Brigade had 161 people and 20 acres of irrigated land, which the team members could neither sell nor lease and whose cropping pattern had to be approved by the brigade management committee. The brigade meanwhile owned 4 tractors allotted between its 8 teams.

In 1958 the township's 30 brigades were grouped into the new Tungting Commune, which replaced the township government and in its own right owned a farm-machinery repair shop, a fruit-processing plant, a brick kiln, and a fish hatchery. Commune members were paid a salary calculated person by person, production team by production team, but at best those salaries provided members with enough money to buy grain and firewood. Most families relied heavily on their own private plots to grow vegetables and raise chickens and a pig.

It's easy to dismiss this revolutionary experiment as a disaster, but many Western experts at the time were filled with admiration. The New Zealand geographer Keith Buchanan, for example, wrote in 1970 that the idea that China would ever provide its people with a Western standard of living was "either downright dishonest or irresponsibly naive." The Chinese in his view—and in the view of many others—could only hope for "a meager and modest standard of living . . . no one rich, no one starving." He thought the Chinese were well on the way to providing this. Most impressively, they were motivated not by financial but by moral incentives. In short, they were building a socialist utopia.

The utopia's plug was pulled soon after the death of Mao Zedong. The communes reverted to townships, and individual farmers were allowed to lease the old production-team lands. Chinese farmers are ingenious at finding the most profitable crops to grow, and they now interplant their fields with an intensity that makes the fertile parts of China look like a horticultural garden, not merely green but filled with a complex assortment of crops grown in tiny patches. When possible, Chinese farmers adopt new technology, too, such as narrow combines sized for narrow fields. There's still a lot of Chinese wheat cut by sickle, but combines move north across the countryside in gangs, just as they do in North America. In fact, so many Chinese farmers have seen an opportunity to make money this way that too many have bought combines and tried to become entrepreneurs: They've managed to push the price of harvesting so low that the profit from custom harvesting has vanished.

Many Chinese fanners meanwhile say they cannot make a living because they are taxed

Thirty miles west of Xi'an, the farms still look like gardens, with an array of crops crowded on less than an acre.

so heavily by local officials. It's probably true, but even if the taxes were removed the farmers would face the underlying problem of uneconomically small holdings: They don't farm enough land to compete in a global market, at least with staple crops. The average holding in the fertile Red Basin of Sichuan, for example, is about half an acre. Holdings in the Northeast are bigger, but only one crop is possible annually in the cold climate, and water is in short supply.

The farms could be enlarged, but where would the displaced people go? The numbers involved are huge: China's agricultural labor force, about 400,000,000 people, could probably be cut in half without any decline in agricultural output. Four million Chinese farm laborers do leave the sector annually—they know that urban incomes are three times those in the countryside—but twice as many enter it, as children mature.

India has been spared Communism, but a typical Indian village consists of 300 acres divided into a thousand or more fields surrounding the villagers' homes. Land ownership is highly variable, but it's always minuscule. Whether a farmer owns a quarter-acre, one acre, or ten acres, it will almost always be scattered over several noncontiguous pieces. (The Punjab and Haryana are exceptions, because after the turmoil of partition in 1947 the government did a complete reparcellation, surveying the lands into millions of tiny squares, each about an acre.) How do you run a modern agricultural operation this way? You don't. Almost every family packs off most of its sons to find a life in one of India's towns or cities. The remaining son does the best he can with the plot he has and plots he can rent from others. Land-ownership records are generally held by a village clerk prepared to exploit that position. Moneylenders are happy to help farmers with high-interest loans; marketing is very difficult, with middlemen taking as much as they can; costs increase because crops are carted over barely passable roads. Like China, India is pushing hard to modernize its agriculture, but the problems outlined here—and others—won't go away easily or quickly.

Examples like those of Russia, China, and India highlight the immensity of the problems faced by countries around the world that seek to emulate the agricultural methods pioneered in the United States. On balance and with the exception of Africa, they have done amazingly well. Thirty years ago, Asia seemed destined for an immense famine, even a population die-off. Instead, it witnessed what by the 1960s was called the Green Revolution. One part of that revolution was the development of improved grain varieties. One famous newcomer was IR-8, a rice bred at the International Rice Research Institute in the Philippines. IRRI was a child of the Ford and Rockefeller foundations, which went on to incubate a half-dozen other agricultural research organizations that now operate across the tropical world under the umbrella of the Consultative Group on International Agricultural Research (CGIAR). Farms across the breadth of South and East Asia now grow high-yielding varieties of wheat and rice.

National research organizations, along with CGIAR institutions, have bred many superior crops adapted to many different growing conditions. To make them flourish, however, irrigation and fertilizer are needed. Huge sums have been invested to develop both, but comparatively little has been done for farmers in areas that aren't irrigated. Little, perhaps, *can* be done for them, short of capturing whatever water there is and using it with great efficiency. Still, famine has been staved off. A large fraction—some say one third, others say one fifth—of India's billion people are said to go to sleep hungry, but there's plenty of grain to feed them, if only they had the money to buy it or the government decided to release it from stockpiles. Call it a famine of jobs, not of food.

Africa remains the exception, chiefly because both its farmers and governments lack capital. Foreign aid might help, but donors are putting less, not more, money into African agriculture. Their inclination is to link Africa's economic future to the production of export crops and to nonagricultural activities such as tourism. They want Africans to import food. It's not bad advice—in Africa, too, smallholders growing staple crops cannot climb out of poverty—but converting subsistence producers to successful exporters or to service workers is a huge task. The U.N.'s Food and Agriculture Organization estimates that 718 million Africans are undernourished, and the number is going up, not down.

Is the world's food supply secure? With the exception of shifting cultivators, who tend to grow many crops together on the same piece of land, agriculture has always been a story of radical environmental simplification. It's been a story, in other words, of efficiency at the cost of ecological stability. A good example is cacao, the tree that is the source of chocolate. When grown by itself, as it is in most plantations, it's very susceptible to fungal diseases. The solution appears to be to mimic the habitat of cacao in the wild and grow it under the shade of other trees. This, however, calls for the root-and-branch reorganization of an industry, starting with a reduction in the number of cacao trees, and income, per acre.

Irrigation farmers have special environmental problems. African and Asian irrigation water is often contaminated by the waterborne disease schistosomiasis. Groundwater can rise so high from perennial irrigation that salts leach upwards and render the fields unusable: Pakistan is famous for old fields as white as snow. Reservoirs fill with silt, too, and though the process may take generations, it is inevitable.

Transgenic crops have meanwhile become as controversial in the world's poor countries as they are in the rich. Farmers have heard that Monsanto insists that anyone planting Roundup Ready soybeans must sign a form promising not to save a portion of the crop for planting next year. Monsanto means business: it has sued about 100 American soybean producers who violated the agreement. One grower in Mississippi saved $25,000 this way, was sued, fought it out, and lost. A federal court declared that he owed Monsanto $780,000.

Farmers overseas know, too, that there's a risk that transgenic plants will escape from the fields in which they were planted, will interbreed with local varieties, and even create Roundup-resistant superweeds. Mexico cautiously prohibits genetically modified corn, but it's there anyway, probably introduced by farmers who planted corn imported for cattle feed. Experts fear that this genetically modified corn will interbreed with the 60 or so native corns, leaving no uncontaminated or pure corn left in the country where corn was domesticated. Corn breeders in the future may have nothing to work with except genetically modified specimens.

Conventional plant breeders worry, too, that the giant companies developing GMOs may succeed in patenting not only transgenic but conventional crops. The announcement early in 2001, for example, that Syngenta had mapped the rice genome made many people uneasy, because the company said that it intended to patent certain genes that it would license to other breeders, whether they were gene splicers or traditional breeders who wanted to use genetic markers.

Such licensing can take a long time. Ask Dr. Ingo Potrykus, the German breeder of Golden Rice. Developed in 1999, Golden Rice contains beta carotene, a good source of Vitamin A. The grain has been promoted as a lifesaver for the million children who die every year for lack of that vitamin. For years, however, the seed was locked up in Zurich, while patent holders refused to waive their rights to parts of the new plant. The main patent holder, AstraZeneca, finally agreed to waive its rights to Golden Rice sold in poor countries, and the grain may yet complete needed certifications and be released. Meanwhile, in another controversy over the patenting of crop plants, RiceTec of Alvin, Texas, nearly succeeded in

1997 in patenting Basmati rice, a million tons of which are exported annually from Pakistan and India. In reaction to these developments, many poor nations now support the International Undertaking on Plant Genetic Resources. They hope in this way to profit from genetic research. By insisting on payment, however, they, too, may impede the work of modestly funded conventional breeders.

Despite all these problems, transgenic plants are coming to the world's poor countries. Malaysia is creating a research center it calls Biovalley. Indonesia is setting up Bioisland. The leaders, though, are China and India. The Chinese have invested heavily in bollworm-resistant transgenic cottonseed, and more than half their cotton crop is now genetically engineered for that trait. Farmers say they like genetically modified cotton because the price is right and because the alternative—pesticides—is dangerous. The Chinese are also working to develop a drought-resistant transgenic rice, as well as genetically modified corn, tobacco, and poplar trees. All told, 200 Chinese government laboratories have 20,000 employees working on genetically modified crops.

India meanwhile has the world's largest cotton crop, measured by acreage, but yields run about half the world average of 1,300 pounds of lint per acre. That's why India can have twice the cotton acreage of China or the United States but produce only one-third as much cotton as China and only about half as much as the United States. Part of the reason for such low yields is that the cotton bollworm destroys 15% of the Indian crop. The government has been working with Monsanto to test a genetically modified seed and has now approved use of one that may double Indian cotton production.

Other big countries are getting into the act. Brazil is America's chief competitor in the international soybean market. Ironically, it only got into soybeans when President Carter protested the Soviet invasion of Afghanistan in 1980 by banning soybean shipments to the Soviet Union. Now the great market is China, where an increasing number of people can afford tofu. Brazil has huge amounts of undeveloped land in the *cerrado* or savannas of Mato Grosso state, and that's where its crop is centered, with 8 million acres now in production. The trend is to plant farther north, in the Amazonian forest along the highway north from Cuiabá to Belém, and Brazil is likely to become very soon, if it is not already, the world's biggest producer.

Until 2003, Brazil outlawed transgenic crops, but that didn't stop smugglers of Roundup Ready soybeans. The cheating Brazilian growers immediately gained an advantage over their American counterparts, because Monsanto declined for the moment to sue them, although they saved part of their crop for planting the next year. In their defense, the Brazilians said they would go broke if they didn't cut costs. Now that Brazil is allowing genetically modified soybeans, the question is whether Brazilian growers will buy the seed each year.

Famine-struck Zambia has refused to accept donated grain if it is genetically modified, and Zimbabwe insists that genetically modified grain be milled so that it can't be planted. But South Africa is planting about 200,000 acres of genetically modified white corn, and Kenya is working to develop several genetically modified crops. They include corn resistant to maize-streak virus, manioc resistant to mosaic virus, and sweet potatoes.

Uganda, too, is trying to develop a GMO policy, especially for bananas. Perhaps bananas don't seem very significant, but they rank globally as number four in the list of agricultural export commodities, even though only a tenth of the crop is traded internationally. Big or small, banana producers have been hurt in recent years by a fungal disease that cuts production in half, not only for the dessert bananas that dominate international markets but also for the cooking bananas, or plantains, heavily used by subsistence farmers. Called Black Sigitoka, the disease was identified in Fiji's Sigitoka Valley in 1963; by 1972 it was in Honduras, and the next year it was in Zambia. The disease can be suppressed by fungicides, but treatment is expensive—about $1,000 an acre every year. That's bearable for the big commercial growers in the Philippines, Honduras, and Ecuador. They can afford the standard treatment of aerial spraying two or three dozen times a year. Subsistence farmers can't. That's why Uganda is pushing to find resistant varieties. Yet as one Ugandan official says, "say biotech here, all hell breaks loose." A colleague adds, "if we miss the GM revolution, then we're finished." Meanwhile, Sigitoka is becoming resistant to fungicides. Consumers are becoming more cautious, too, which means that even the commercial growers may soon need an alternative to spraying.

FORESTRY

How different is this story from what's happening in the world's forests? In broad terms, the answer is: not different at all.

In medieval Europe every village had its forest, from which its needs were met by the villagers themselves. Attitudes toward those forests were unbelievably conservative. Stewart Brand, in *How Buildings Learn* (1994), tells a good story in this connection about New College, which ironically has the oldest building at Oxford University. The roof is supported by massive oak beams. In the mid-19th century, the beams were about 400 years old and had to be replaced. Where were new beams to come from? Answer: When the trees were cut down for the original building, the College planted a new oak forest and then protected those trees for four centuries.

Until the 20th century, the logging business was incredibly labor intensive. The stories of Paul Bunyan hint at how hard men worked for wood. If in 1910 you had gone up to the Peace River country northwest of Edmonton, you'd have found pioneers building cabins with plank floors sawn by hand with pit saws. Each log was placed aloft in a wooden frame. One man straddled the log from above, while another stood in the pit below. Between them they pulled a long saw back and forth along lines chalked on the log. A pair of skilled sawyers could produce 100 board feet of lumber in a day. That's 10 planks, each 10 feet long, 1 foot wide, and 1 inch thick. Similar methods can still be observed elsewhere, for example, in China's Yunnan Province.

Things began to change in the 19th century, when German foresters began growing trees the way farmers grow cabbages: straight lines, single species, harvest at early maturity. That was the recipe, and you can see the results in the monotonous pine plantations that cover much of the Scottish Highlands. You can see it in the geometrically perfect rows of poplars in Italy's Po Valley. The new methods came to the United States later that century, as a wave of cutting spread across North America. This wasn't the old story of pioneers clearing land for farms; it was a story of logging companies that had no interest in the land, just the trees growing on it. The result, particularly west of the Great Lakes, was devastation, chiefly from fire. The loggers moved farther west, abandoning the Great Lake pineries for the spruce and fir and pine of the Cascades and coast ranges. After World War I, the same companies moved back east and began operations in the American South, where trees grow very fast in the moist warmth.

Today, 60% of the forest products made in the United States come from 13 southern states, where the natural forest has been almost entirely replaced by a secondary one of pine plantations. There are still fallers in the woods—in some parts of the country they're called choppers—who take chainsaws to trees, then de-limb the trunks. A partner skids the logs

Longcheng is a village near Lijiang, in China's Yunnan Province. There's a lot of construction, some modern with concrete, some traditional with wood. Tourists prefer the traditional wood, but it's laboriously sawn by hand, which helps make wood buildings more expensive than concrete.

out to a landing where the trees—or "stems," in the Bunyanesque language of the Maine woodsman—are usually loaded on trucks. In the old days, they were often floated down rivers. That's gone now, to the relief of the fish, but you still sometimes see logs bucked into 4-foot lengths for truck loading. It, too, is obsolescent. The future belongs to feller-bunchers, giant machines that roll through the woods, strip standing trees of their branches, cut the base with giant shears, and carry bunches of whole logs to waiting trucks. Those logs can be taken to a mill capable of producing 200,000 tons of wood chips annually and yet having a payroll with only seven workers.

The forest industry is now dominated by huge companies. Globally, the four biggest are International Paper, Georgia-Pacific, Weyerhaeuser, and the Finnish company Stora Enso. The American companies all have annual revenues of about $20 billion. Stora Enso is about half that size. All the companies claim to be engaged in sustainable forestry, but that's like a Kansas wheat farmer saying he's sustaining nature's grassland. It's not just a North American issue, either. In 2003, the Australian state of Tasmania clearcut 60 square miles of forest, chipped for Japanese pulp and paper mills. A diverse natural forest was then replaced, in the words of one environmental activist, with "plantations with as much diversity as a carpark."

Industry consolidation continues. In the year 2000, International Paper acquired Champion International, a major company in its own right. Meanwhile, the three biggest North American newsprint producers in 1997 accounted for one-third of newsprint production capacity. A few years later they accounted for 60%. One of the three leaders, Norske

Skog, took over the British Columbia Forest Products paper mills at towns including Port Alberni, Crofton, and Powell River. They were icons of the provincial economy, but that didn't stop them from being sold in 1987 to a New Zealand company which in turn sold them in 2000 to Norske Skog. The other two giants were Abitibi, based in Quebec and serving the big newspapers of the American Northeast, and Bowater, a South Carolina company tapping those fast-growing Southern pine forests.

Competition for these companies may come from Russia, which has immense timber resources but whose forest-products output fell 75% between 1989 and 1998. Now it's coming back, partly with new Russian companies and partly with outsiders like International Paper. There are huge problems with infrastructure and institutions, but the trees are there.

Forestry in the tropics is a different story, one where millions of acres are destroyed annually as land is cleared for agriculture, especially in Brazil, Indonesia, and Malaysia. Historically, these forests were not attractive to industrial operators, because merchantable timber species were few and scattered. (An exception is teak, whose cutting continues to enrich a few people in Thailand and Burma.) The forests weren't good for paper production, either, because the chemistry of paper mills wasn't suited to tropical species. These facts are changing, however, as consumers learn to accept furniture made from unfamiliar tropical hardwoods and as the chemistry of paper production changes so mills can use many kinds of feed. Tropical plantations are meanwhile raising introduced species. Brazil alone has over 5 million acres of eucalyptus, largely grown for charcoal used by steel mills. It's also used for pulp, however, and in the future will likely be used for sawlogs, even though eucalyptus is a very tricky wood to mill. Brazil also has over 3 million acres of pine plantations. The introduction of both eucalyptus and pine in South America has been bitterly and sometimes successfully resisted by indigenous peoples. Plans to establish eucalyptus plantations in Brazil's northeastern state of Amapá, for example, have been dropped.

METALS

I've ingeniously managed to avoid almost entirely the subject of metals, so here's a paragraph of deep background.

Malachite, or copper carbonate, was mined in the mountains behind the Fertile Crescent in the eighth millennium B.C., first for copper ornaments, then for cult objects, weapons, and—last of all—tools. Copper is soft, so it was mixed with arsenic to produce arsenic-bronze, the first alloy. Tin was then added to copper, producing a bronze that was harder though less ductile than arsenic-bronze. Low-grade ores were eventually used, including ones contaminated with iron. From this accidental origin, the use of iron itself originated in the second millennium B.C., perhaps at the hands of the Hittites in what is now Turkey. When their empire collapsed about 1200 B.C., the knowledge of iron working spread with the workmen themselves, who migrated in all directions except to the New World, which remained ignorant of iron until the Europeans introduced it with helmets and bloody swords.

Trade in these metals has been going on a long time. The Romans imported tin from Cornwall and copper from the Rio Tinto in Spain. (The river's name alludes to its ancient load of mineral waste; for anyone with environmentalist sympathies, it's ironic that one of the world's biggest mining companies today is RTZ, formerly Rio Tinto Zinc.) Fifteen hundred years later, the Spanish were shipping gold and silver across the Atlantic to the booming port of Seville. This may have been the first instance of the intercontinental shipment of minerals.

Jump to the heroic materialism of the 20th century. By 1913, the United States accounted for 36% of the world's iron-ore production. It was followed by Germany, France, the United Kingdom, and Russia. These five leaders together produced 80% of the world's iron ore and consumed 90% of it, with the 10% deficit coming from Sweden and Spain. Today, the world produces five times as much iron as it did in 1913, but the leading producers are Brazil, Australia, China, India, and Russia—still in last place. The top five producers produce two-thirds of the world's iron ore, but they export half of what they produce. The change is even more extreme with the two leading producers, Brazil and Australia. Together they produce about one-third of the world's iron ore but consume only 5% of it. The bulk of their output goes to Japan and Korea.

The mining industry has always been footloose, as miners chased ore. Americans remember California's gold rush, but 20,000 miners—many with their families to help with the digging—flocked in 1999 to an instant mining camp at Ilakaka, where sapphires had been found in the red clay of Madagascar's Isalo National Park. With no apparent irony, the U.S. Geological Survey describes the pandemonium there as artisanal mining. The global operators are tidier but equally mobile. Ivanhoe Mines, which was responsible a decade ago for a major nickel discovery at Voisey's Bay in Labrador, now has exploration rights to an Ohio-sized chunk of southern Mongolia, where it is busy assaying a site called Oyu Tolgoi ("Turquoise Hill"), which promises—forget turquoise—to have a gold and copper deposit worth $46 billion. Some 30,000 Mongolians aren't waiting for the mine to open. They're already digging shallow pits and tunneling between them.

Ivanhoe may have found the world's second biggest gold deposit, but there are many bigger mining companies. Until recently, the biggest of all was South Africa's Anglo-American. Now it's BHP Billiton, formed by the merger of Australia's BHP (Broken Hill Proprietary) and Britain's Billiton.

Companies such as these almost ignore national boundaries. Canada's Kinross, for example, is trying to mine gold in a national park in Kamchatka. Barrick Gold, also Canadian, is operating not only in the United States but in Tanzania.

The big mining companies must not only find and profitably sell minerals but work with governments that are often deeply corrupt and sometimes on the verge of collapse. Newmont Mining, a Colorado company, quit its gold operation on Indonesia's island of Sulawesi (Celebes), because the local authorities were unwilling to evict squatters from the company's site. Those same squatters used mercury to get at the gold and were creating a worse environmental problem than Newmont, which used cyanide with some care. In Irian Jaya (western New Guinea), the American company Freeport MacMoRan came under pressure to sell more than half its shares to Indonesians who wanted a piece of Grasberg, one of the world's biggest copper mines. As if that weren't enough, landslides in the open-pit mine slowed production from 834,000 tons of copper in 2002 to about 500,000 in 2004. Buyers around the world began scurrying for alternatives, which weren't easy to find for a mine that by itself had been producing more than 6% of the world's copper.

Whether they're working with corrupt governments or not, mining companies know that all major mineral developments are politicized. The gold mines of the East Rand, southeast of Johannesburg, were developed after 1917 by Ernest Oppenheimer's Anglo-American company. In 1928, the company began developing copper mines in what was then Northern Rhodesia. By 1969, Zambia was the world's third largest copper producer. The new nation optimistically nationalized the mines. Over the next decades, production collapsed by two-thirds. In 2000 the government returned control to Anglo-American, but after 2 years Anglo-American quit, explaining that copper prices were too low to justify investment in the development of new ore bodies and that keeping the old ones in operation had already eaten up $100 million. The jobs of 10,000 copper workers were suddenly in jeopardy. The company was busy elsewhere, however, producing copper in Manitoba, gold in Nevada, niobium in Brazil, sand and gravel in Spain, diamonds in Namibia—and even gold in South Africa, where it had made its original fortune.

Paradoxically, an Anglo-American subsidiary, AngloGold, merged in 2004 with Ghana's government-run Ashanti Goldfields. Ashanti was a company with a proud history going back to 1897. Its mine at Obuasi, south of Kumasi, had been nationalized in 1969, however. Thirty-odd years later, its managers and the government of Ghana decided that to survive in a world of giant mining corporations Ashanti needed a deep-pocketed international partner.

FOSSIL FUELS

It's time for a third and final story of resource development, this time one with huge economic and political implications. I want to glance at coal, which sustained the industrial revolution, then look more closely at petroleum, its successor as the world's premier energy source and now a resource over which governments and people both live and die.

In 1913 the five biggest coal-producing countries, in descending order, were the United States, Britain, Germany, the Austro–Hungarian Empire, and France. The first three were coal exporters. They dominated world trade, but the trade was local, with American coal, for example, going mostly to Canada. British and German coal stayed mostly within Europe.

Today, coal production rankings and the pattern of exports and imports have been as radically revised as iron-ore geography. Britain, Austria-Hungary, and France are gone from the list. China comes first now, followed by the United States, Germany, Russia, and India. The United States is still near the top and is, of all the top five producers, the only one able not only to meet its own needs but to export coal. The flow of American coal today, however, is more to Japan than Canada, and within the United States coal production has slowly shifted from the deep underground mines of the east toward the huge open-pit mines of the West, chiefly Wyoming.

Like every resource producer, the American coal companies want to cut the cost of production, and a surface mine with shallow coal beds is almost by definition cheaper to run than an underground one. A few men in Wyoming, with the right gargantuan equipment, can collect many more tons of coal in a day than the most skilled men following a subterranean coal face, drilling, blasting, propping, and excavating as they go. The decline of the eastern mines has been hard for the eastern miners and their families, but their lives were always as tough as any in the modern age. (In 1937, George Orwell saw the Welsh equivalent first-hand and wrote about it grimly in *The Road to Wigan Pier*.)

Coal produces half of America's electrical energy, and because it's cheaper than natural gas it's not likely to decline anytime soon; with coal gasification techniques that allow clean burning, it may become even more important. Still, two-thirds of America's total energy comes from oil and gas, a little more from oil than gas. Russia, Germany, Japan, and the United Kingdom are equally dependent on these fuels. The dependency is even greater in Latin America and Africa, which with the exception of South Africa hardly use coal. France is exceptional, because it relies on nuclear power for three-fourths of its electricity and almost half of its total energy consumption. So are India and China, which continue to rely very heavily on coal—too much so, considering the dismal quality of the air many Indians and Chinese breathe.

Oil, then, like the other resources we've looked at, has been radically reorganized during the 20th century. In 1918, hard as it may be to believe, the United States produced an amazing two-thirds of the world's oil. The runners-up were Mexico, Russia, the Dutch East Indies, and Romania. Export flows were minimal, and exploration had yet to reveal the bonanzas in Venezuela and the Middle East. Today the leaders, from the top, are Russia, Saudi Arabia, the United States (down, down, down to 12% of world production), Iran, and China. The United States and China are major importers; Russia, Saudi Arabia, and Iran are major exporters.

The industrial organization of the industry has been equally transformed. The early oil producers, whether at Baku in Russia or Bakersfield in California, were small, independent operators. They found oil and rushed to produce it before their neighbors took it. Oil, after all, is no respecter of property lines and happily migrates to the nearest well, regardless of who owns it. Producers often found that, for lack of pipelines and tank cars, they couldn't do anything with their oil except put it in a pond behind a hastily built dam. There, it often dried and formed thick tar mats.

Often, the people who controlled the pipelines took control of the fields; Standard Oil was famous for that tactic. Sometimes it was bankers who took control. That's what happened to Oklahoma's E. W. Marland, whose bankers not only took over Marland Oil but

Near Taft, at the southwestern corner of the San Joaquin Valley, two surviving redwood derricks face each other across a property line. The picture is from 1967; the derricks were obsolete even then and have since been removed.

renamed it Conoco, the Continental Oil Company. Marland vacated his Ponca City mansion and moved into its garage. He never got back into the house.

Over the years, oil production has become much more efficient, with bigger operating units, fewer wells per square mile, more oil per well, and far fewer dry holes. Not so long ago, 9 out of every 10 oil wells was a dry hole; nowadays, the average is 5. That's why the cost of finding a barrel of oil has fallen from $4 in 1980 to about 66 cents. Old-timers tried to read geology from surface features; their successors not only map deep stratigraphy from seismic data and well logs but visualize structures in computer-simulated 3-D caves that give them a worm's-eye-view of conditions underground.

Production methods, too, have grown very sophisticated. There are platforms operating off the coast of Brazil in 8,500 feet of water. Brunei and Malaysia are arguing over control of the Kikeh oil field, where production is soon to start of some 700 million barrels of oil lying under 4,500 feet of water 100 miles off Sabah.

Many, though not all, of the biggest oil companies are American. The biggest publicly traded one in recent years has been ExxonMobil, formed by the merger of two fragments of John D. Rockefeller's old Standard Oil Trust. That company, ironically, had been broken up in Teddy Roosevelt's day by a famous antitrust action. One fragment became the Standard Oil Company of New Jersey, which after many years took the name Exxon. Mobil was another fragment, the Standard Oil Company of New York. Merged once again, ExxonMobil now in most years ranks by sales as the biggest publicly traded corporation in the world. It produces about 4 million barrels of oil daily from about 25,000 wells. More impressive, in recent years it has found more oil each year than it has produced.

Exxon's traditional competitor is Royal Dutch Shell, a dual-national company with roots in the Royal Dutch and British Shell Trading companies. Trailing both but on a rapidly rising curve is BP (formerly British Petroleum), a company that lost 70% of its reserves when Iran's fields were nationalized in the 1970s. To make up the loss, BP bought into Alaska and the North Sea. Now it's the biggest producer of oil and gas in the United States and is finally approaching its production levels of 30 years ago. The company is banking heavily on the Gulf of Mexico, where perhaps 50 billion barrels of oil wait to be discovered in areas under more than 1,000 feet of water. In 2003 the company invested about $7 billion in a merger of its Russian resources with those of Tyumen and Sidanco, Russian companies whose assets will increase BP's reserves by a third. All these measures, and more, paid off when the company in 2004 moved to first place in the industry, measured by revenues, though not profits. Other giants are merging to survive. Chevron, for example, merged with Gulf, then added Texaco. (Chevron, by the way, was another of the Standard Oil fragments, Standard of California.) More recently, Conoco and Phillips have merged.

Despite their huge size, none of these public companies ranks among the four biggest oil-producing companies. Big as they are, they produce less oil than Saudi Aramco, National Iranian Oil, Pemex (Mexico), and PDV (Venezuela). Exxon comes fifth in this ranking, followed by impatient BP, eager to overtake it. Establishing and maintaining good relationships with these giant state-owned producers is a key task for the public companies that consumers know so well.

Stand back now from corporate sagas and consider the globalization of this critical resource.

The towering fact of energy geography is that two-thirds of the world's proven oil resources of about 1 trillion barrels lie in the Middle East, far from the consuming centers. The United States is less directly vulnerable to interruptions in the flow of Middle East oil than Europe and Japan, because it obtains only about 2 or 3 million barrels a day—one-tenth of its consumption—from the countries around the Persian Gulf, chiefly Saudi Arabia but also Kuwait and Iraq. (The United States imports more oil from Latin America, chiefly Mexico and Venezuela.) Europe, however, relies on the Persian Gulf countries for about 20% of its oil, and Japan does so for a startling 70%. Americans may take comfort in thinking that interruptions in the supply of Middle Eastern oil will hurt them less than other countries, but oil prices are set in a world market dominated by Saudi Arabia, the largest exporter. There will be pain all around. The preeminence of the Middle East is meanwhile likely to grow in the new few decades, because consumption in the Far East by 2020 will exceed consumption in the United States and Europe combined. The Persian Gulf, which now supplies about 30% of the world's oil, will then supply 40–60%.

So far, the supply system is holding together amazingly well. The world consumes 80 million barrels of oil a day, and with rare exceptions customers can buy the fuel they want, when they want it. That figure is projected to rise to 115 million barrels in 2020. Where will it come from? Or, to narrow the question, what is the United States going to do? In its oil-production peak year, 1970, the United States produced 9.6 million barrels of oil daily; now it produces less than 6 million, excluding oil produced in association with natural gas. That's not very much for a country that consumes 20 million barrels a day. In 1975, after an Arab oil embargo left Americans determined to reduce their dependence on foreign oil, the United States imported 37% of its oil. Since 1996, it has imported over half its oil, primarily from Mexico, Saudi Arabia, Canada, and Venezuela. True, the active rig count has risen from less than 500 in early 1999 to over 1,100 in early 2004, but almost all the rigs are looking for gas, not oil, and even 1,000 rigs aren't so impressive compared to the 4,000 that operated early in the 1980s. On the gas side, things aren't much better. American consumption is projected to rise 50% by 2010, but production is likely to be flat.

The answer of course is to think globally.

Alberta is booming with exports of natural gas to the United States. The ebullient premier likes to say that New York City is lit by Alberta gas. A soberer statistic is that Canada provides about one-eighth of America's natural gas—twice its contribution in 1990. High prices are pushing Canadian producers into the high Arctic, where huge gas reserves underlie the Beaufort Sea. Remote Inuvik is likely to thrive if the Mackenzie Delta pipeline is built from the Canadian Arctic south to Chicago. Even bigger things may lie ahead if Alaska's Arctic gas reserves are shipped through the Canadian Arctic, rather than south through Alaska and then via the Yukon to Alberta.

Canada is set to become America's largest supplier of oil, too. Partly this increase will come from offshore developments in Nova Scotia and Newfoundland, but mostly it will be from the Athabaska tar sands. These sands lie in northeastern Alberta, near Fort McMurray, and they may have more oil than exists in all of Saudi Arabia. (Official estimates are more modest: 180 billion barrels, compared to Saudi Arabia's 262 million.) Refining cost has been an impediment for decades, because you have to process 2 tons of sand to get a barrel of oil, but the current producers say that they can now extract the oil for about $9 a barrel. That's a green light. More than $11 billion were invested in the tar sands between 1996 and 2001, and production is now approaching 1 million barrels of oil daily.

There's plenty of potential elsewhere, too.

Venezuela has a supply of near-solid heavy crude that is about equivalent to the entire reserves of Saudi Arabia. The major impediment is politics, because the government in the name of social justice is marching into a Cuban-style quagmire that will repel investors for years to come.

Russia is a bit more promising. Daily production fell in the post-Soviet years from 11.5

million to 6 million barrels in 1996. Now it's rising again. In 2002 Russia inched ahead of Saudi Arabia, and production is approaching 8 million barrels daily, restoring Russia to its position as the world's biggest producer. (We think of Saudi Arabia as the leader because Russia exports only about 5 million barrels daily, a bit less than Saudi Arabia.) Russia's reserves are also ramping up, and the country seems likely to overtake Iraq, Iran, and Kuwait and rank second in reserves only to Saudi Arabia.

Foreign companies have been sensibly wary of investing in the countries of the former Soviet Union. An executive with RTZ, the mining company, summarizes foreign investment in Russian resource-development projects this way: "we have seen a lot of money incinerated there." Ask officials of the China National Petroleum Company about the team they sent to Moscow in 2002 to bid on a Russian producer called Slavneft. One of the team members was kidnapped at the airport but freed as soon as CNPC dropped out of the auction. Slavneft was then sold to Russian bidders at a price more than $1 billion less than the $3 billion the Chinese were prepared to bid.

Still, Russia has a lot of oil, and international companies find its allure irresistible. ExxonMobil, Shell, and BP are all investing heavily in fields off the coast of Sakhalin Island, where 13 billion barrels of oil are thought to lie. (That's about two-thirds of the 22 billion barrels still in the United States.) That's one strategy; the other is to follow in BP's footsteps and buy into an existing Russian producer. ExxonMobil tried to buy a big piece—Lee Raymond, its chief executive, hoped for a controlling interest—in Yukos, Russia's biggest private oil company, but he could only watch as its chief executive, Mikhail Khodorkovsky, was arrested and imprisoned on tax charges. Looking for safer investments, Western companies in the 1990s invested more money in Azerbaijan than in Russia—and twice as much in Kazakhstan. Despite strong efforts to build cordial relations with Kazakhstan's dictatorial president, Nursultan Nazarbayev, investments there are only comparatively safe. ChevronTexaco, for example, has a 50% interest in Kazakhstan's big Tengiz field but halted expansion when the government, Russian-style, decided unilaterally to revise the company's 1993 production contract. The safest players are those like Halliburton, which owns nothing but provides oil-field services. Schlumberger does the same and has 2,000 employees in Russia.

Special attention is being paid to the Caspian Sea, which Russia shares with Iran, Azerbaijan, Kazakhstan, and Turkmenistan. The Kashagan field alone, under the North Caspian, has 7 to 9 billion barrels of oil and is the world's biggest discovery since Alaska's North Slope. Things were simple before the breakup of the Soviet Union, because the Soviet Union and Iran divided the resource 50:50. Now Iran is willing to accept 20% of the resource, but its neighbors are pushing for 13%, proportional to Iran's share of the Caspian's coastline. Efforts to agree on a joint regime have failed, but Russia has at least agreed on a joint-development program with Kazakhstan.

There's been a major contest, too, over how the Caspian's oil should be transported to market. Russia is already collecting transit fees for oil shipped across its southern territories in a pipeline built by the Caspian Pipeline Consortium. The line, 40 inches in diameter and almost 1,000 miles long, starts in Kazakhstan's Tengiz field and winds around the north shore of the Caspian to Astrakhan, where it beelines west to Novorossiysk, on the Black Sea. More is going to be sent this way, despite Turkish objections. (Within a few years, three tankers a week will be threading their way through the Bosporus.) The Russians are getting ready with an alternative, an expanded terminal at Primorsk, at the head of the Gulf of Finland. It will ship Siberian oil to European markets but could also handle oil from the Caspian.

A third line is now being built. This is the BTC (Baku–Tbilisi–Ceyhan) pipeline long favored by the U.S. government, which doesn't like the idea of Caspian oil flowing through Russian pipelines. The Turks were happy with this alternative, and in the summer of 2001 BP announced its support for the route. Work began the next year with a completion date of 2005. Most of the oil sent through the BTC line will be from Kazakhstan, even though the Kazakhs prefer a pipeline south through Iran, either to the Persian Gulf or possibly to Pakistan and India. The United States has been adamantly opposed to such a route.

To the east, China became a net importer of oil in 1996 and is now the second biggest importer in the world, after the United States. Production at its best known field, Daqing, near Harbin, peaked in 1997. There are other Chinese fields, but they can't keep up with the country's roaring demand, which has already made China the world's second largest consumer of oil, after the United States. China imports 2 million of the 5.4 million barrels it consumes daily, and the International Energy Agency estimates that by 2030 China will import 10 million, as much as the United States does today. That's why in 2003 China signed a contract to buy $150 billion worth of oil from Russia. Yukos planned to send 400,000 barrels a day through a 1,500-mile pipeline to be built from Angarsk, near Irkutsk, to Daqing. This plan was thrown into doubt late in 2003, when the head of Yukos was jailed on those tax charges at the same time as the Russian government began promoting an alternative pipeline from nearby Taishet to the ice-free Pacific port of Nakhodka. The change quadrupled the cost of the pipeline to about $10 billion, but Russia could then sell to other customers besides China—Japan, for example.

China's seeking ways to boost its use of natural gas, too, a much cleaner fuel than the coal on which it relies so heavily. Right now, gas provides only 2% of China's energy, but within a decade the government would like to see it supply 10%. To do this, the Chinese plan to develop the gas fields of the Tarim Basin and to pipeline that gas to Shanghai. This requires a 2,500-mile pipeline likely to cost about $18 billion, enough for China to reach beyond its own PetroChina and seek help from Shell and Russia's Gasprom. Is there enough gas in the Tarim Basin to justify the pipeline? Will Shanghai consumers buy it at a price that's profitable for the companies? Will those consumers be able to buy power more cheaply from the huge Three Gorges dam on the Yangtze? What will be the price of the imported liquified natural gas that China will soon import from Australia to Shenzhen? (There's a 25-year, $12 billion contract there, involving Shell, Chevron, and other companies.) There have been hopes that such imports would mean less pollution from China's coal-burning power plants and perhaps fewer deaths among its coal miners (5,000 Chinese miners die in mine accidents every year; the figure for the United States, which produces a bit more coal than China, is 30). More likely, coal production will stay constant, while imports help satisfy rising demand.

Africa is a major source of oil, too, but this is a wretched story. Nigeria's last ruler, Sani Abacha, stole at least $2 billion of oil revenue that could have been used to help his country. Under his successor, the skim is distributed to more hands but still not to the people of Nigeria. Since 1990, Nigeria has exported $200 billion worth of oil, but 70% of its 130 million people still live on less than $1 a day. That's why the village of Ugborodo, 40 miles west of ChevronTexaco's Escravos Terminal at Warri in the Niger delta, has no plumbing, no phones, and not even a gas station. That's why villagers say, "Americans are like terrorists to us. They come, take and leave without putting back." Two high, parallel, barbed-wire fences separate the terminal—with its paved roads, air-conditioned quarters, and imported food—from the village. Hundreds of Ugborodo women invaded the terminal on July 8, 2002, and occupied it until the company agreed about a week later to provide the village

One of China's many thousands of coal mines: The boy with the loaded pannier had just emerged from a dark tunnel entering the ground on a steep angle.

with water and electricity, to build a school, and to help village women produce fish and poultry for the terminal's 1,800 employees. No doubt the government should have done these things—it collects 60% of the value of the oil, after all—but that wasn't about to happen, and the terminal managers had to meet their production schedule of 350,000 barrels a day. The following year things got worse, and the government killed dozens of Ijaw villagers who were seeking a larger share of oil revenues. The Ijaw, who number 8 million, threatened to blow up the terminals. ChevronTexaco closed its Escravos Terminal, Shell closed its terminals at Bonny and Forcados, and the French company TotalFinaElf shut its operations at Opumami. All are in the Niger delta, described by a frustrated government official as "one of the most richly endowed and yet one of the least developed deltas in the world."

There's a variation on the theme in Angola and Sudan, where oil revenues have mostly funded civil wars. Global Witness says that $1.4 billion of Angola's oil revenues disappeared in 2002. Human Rights Watch puts the figure for 1997–2002 at $4.2 billion. Angola has 11 million people, 80% of whom are poor; assessing its quality of life, the United Nations ranks Angola 164 out of 175. Two-thirds of the country's oil is produced by ChevronTexaco; where do its royalties go? It's hard to say: the International Monetary Fund has written that the country's oil revenues pass through "a web of opaque offshore accounts."

Even tiny Equatorial Guinea has become modestly notorious. Its oil production has risen from 17,000 barrels daily in 1996 to 220,000. Where has the money gone? The answer is a state secret, but the country's president, General Teodoro Obiang, has sole control over an account with more than $300 million in it. It's at the Riggs Bank in Washington, D.C., not far from the Maryland suburb where in 1999 Obiang paid $2.6 million cash for a house. Sound fishy? Riggs came under such intense scrutiny of its handling of accounts for Equatorial Guinea and Saudi Arabia that its long-time head resigned in 2004, while the bank paid a penalty of $25 million for "willful, systemic" violations of federal money-laundering law.

You could argue that the oil wealth of the United States, too, has gone to a few hands. Why should Africa be different? But America's policy arose in the aftermath of gold rushes, when finders–keepers seemed a fair exchange for the hazards of the chase and when each finder competed with others to stake a small and limited mining claim. With the development of modern industrial organization, the principle of finders–keepers amounts to a plutocratic theory of mineral development. The United States itself corrected its policy in 1920, when it began collecting royalties from oil found on public lands and, after 1947, from oil found offshore and beyond the narrow fringe of state-owned land.

As Nigeria demonstrates, even the collection of royalties doesn't guarantee a dime for roads, schools, or clinics, and the rot of corruption isn't limited to Nigeria, or Africa. When Azerbaijan was listed by Transparency International as one of the most corrupt countries in the world, a joke made the rounds. Azerbaijan was actually the most corrupt, but it had bribed Transparency International to get a lower ranking. When Nigeria didn't come first in the listing, cynics inverted the joke and said that somebody should have bribed Transparency so Nigeria could come first in *something*. Brunei has meanwhile seen some $15 billion blown for the amusement of the sultan's favorite brother, who's now reduced to a monthly allowance of $300,000. The sultan's fortune seems to have collapsed from $40 billion to $10 billion.

Saudi Arabia is yet another country whose immense wealth, concentrated primarily in the Ghawar Field between Riyadh and Qatar, has been largely wasted. Seventy billion barrels of oil sit under Ghawar, and another 50 are at Safaniya and Shaybah. The kingdom extracts oil from these reserves far more slowly than it could and has been called the central bank of oil because it uses its surplus capacity to minimize disruptions in the world supply. That's good for consuming nations, but the relatively slow, steady income has been of little benefit to the Saudi people.

The country's oil reserves were first developed in the 1940s by a consortium of international oil companies. The Arabian-American Company, or Aramco, was nationalized in the 1970s, and it was easy to believe that a fabulously prosperous future lay ahead. After all, per capita income in Saudi Arabia in 1980 was equal to that in the United States. Since then, however, per capita income has risen in the United States to about $34,000, while in Saudi Arabia it has sunk below $7,000. Oil revenue has fallen per capita from $24,000 in 1980 to less than $3,000 today. One reason: The average Saudi woman has 6.2 children. So much for the rule that population growth rates decline as income rises.

What will these young Saudis do when they grow up? A generation ago, Saudi men with college degrees were guaranteed government jobs where they had very little, if anything, to do. No longer: The unemployment rate for men ages 20–24 is about 35%. (For women, the rate is much higher, because Saudi law forbids men and women working in physical proximity to one another, where the men would face apparently irresistible temptation.) Gone, too, are the days when any Saudi could get an $80,000, interest-free mortgage for the asking; now there's a 10-year wait. The country's infrastructure, only a few decades old, is deteriorating, with blackouts and water rationing.

Eighty percent of the country's people now live in Riyadh, Jidda, or Damman, and there is raw sewage in the streets between the mud-brick slums in these places. Meanwhile, 90% of all private-sector jobs in the country are done by foreigners—Pakistanis, Indians, Filipinos—who repatriate $16 billion every year, equivalent to $1,000 for each Saudi citizen. The government intends to substitute Saudis for 3 of the 7 million foreigners working in the country, but getting back to your grandparents' toughness isn't easy. For the Saudis it's especially tough, because their role models are 7,000 mostly useless princes, who typically see everyone around them as servants. After all, Saudi Arabia is the only country in the world named for a family, and those princes naturally believe that they own the place. One Saudi who prudently remains anonymous says, "We have one party, one ruler, corrupt judges, and all we're supposed to do is praise the government." Another says that the government expects Saudis "to drink camel's milk, ride dune buggies and sit by the fire. After a time you begin to go mad."

Late in the 1990s, the big oil companies were asked to return to Saudi Arabia. The country aimed to attract some $25 billion in investment capital to develop natural gas for the country's own use and industrial development. Plans called for production to rise from 2.5 billion cubic feet per day to 14 billion in 2025. ExxonMobil took the lead, but Shell and others were heavily involved, too. The odd thing is that these projects promised little oil or gas for the companies, which were investing almost as service contractors. Why would they do such a thing? The answer seems to be that they wanted to position themselves for the day when the country decided it needed private companies to develop more of its oil fields. There were plenty of well-positioned Saudis who thought that their own national oil company, Saudi Aramco, could do the job without help, however, and in 2003 the Saudis put an end to the negotiations and said they would proceed with a number of much smaller projects.

You could be forgiven for concluding that the lucky countries are the ones without resources: resource-poor countries often have higher economic growth rates, less corruption, and greater political stability than the ones cursed with oil. Think of Aceh, the rebellious province at the northern tip of Sumatra. Since 1978 it has loaded 3,600 tankers with oil worth a total of $47 billion, but the revenues have gone almost entirely to the national government in Jakarta. There has been talk of peace and even of a share of the oil revenue, but peace has not come and Chevron officials still travel in armored vehicles. You might think of Peru, too, where protests grew strident enough that Shell in 1998 pulled out from plans to develop a major gas field in the Ucayali Basin. Indigenous groups celebrated, but the government looked for new partners and found them in Hunt and Halliburton of the United States. Work is proceeding once again on the Camisea Project, from which liquified natural gas is supposed to be shipped by tanker to the United States via Baja California by 2006. "If we'd only known," says the director of the Center for Development of Amazon Indigenous Peoples. "We would, one thousand times over, have preferred Shell."

Can poor countries with oil ever do things right? The economic boom in the United Arab Emirates and Oman suggests that countries can use oil money to develop a diversified and self-sustaining economy. Perhaps another success story is beginning in Chad, where the $3.5 billion Chad-Cameroon Petroleum Development and Pipeline Project will soon bring Chadian oil 660 miles to a coastal terminal in Cameroon. The project is funded by the World Bank, though oil production is in the hands of Exxon and partners. It's a unique arrangement, but Exxon was reluctant to move into this politically unstable environment without World Bank participation. For its part, the bank would not normally partner with multinational oil companies but was persuaded to do so after Chad agreed to put its oil revenues—$125 million annually for about 20 years—in a special account devoted to development projects. Chad has 8 million people, of whom 80% live on less than $1 a day. Will the revenues find their way to schools, clinics, and infrastructure? The bank has developed elaborate plans to sequester the oil revenue and apportion it for social programs, but early signs are that the government covets the money for arms. It has one of the worst records of corruption of any country in Africa.

If you can overlook these social issues, the mineral production system is a great one. So, for that matter, is the whole resource production system. Despite concerns that we'll run out of food, trees, and oil—and we've had these concerns for a century now—commodity prices are low and generally falling. And it's not just supply and price that are to the consumer's advantage. Production methods were once famously wasteful. You don't see much of that nowadays, although when accidents occur they can be huge. So what's the point? I've already mentioned on many occasions our ambivalence toward uncivilized peoples and, by implication, toward ourselves. And so it is: We can tell the story of modern resource production in a way that's almost terrifying, whether viewed socially or environmentally. Or, we can tell it as a triumph both of technology and of improving living conditions. Odd that from so many hard facts comes such uncertainty.

CHAPTER 12

MANUFACTURING

Windmills, town clocks, and church organs are handy reminders that there were machines long before the industrial revolution. Europe by the 17th century had a tradition not only of making machines but of importing goods made abroad with sophisticated mechanical technologies. Loomed textiles are probably the best example, and English itself tells the story. The word satin comes from the name of an unknown city in China that Arab traders called Zaitun. *Khaki* is the Hindi word for dusty. The word "calico" comes from India's southwestern coastal city of Calicut; chintz, from the Hindi name for a printed calico; cashmere, from Kashmir. Percale comes from the Farsi word *pargalah*. Another Farsi derivative is seersucker, whose bands of alternating smooth and puckered fabric prompted a Farsi name that literally means milk and sugar. Still another Farsi borrowing is taffeta, which comes from the Farsi for "spun." The coarse cloth we call muslin is named for Mosul—the town in Iraq—while damask is a short form of Damascus. Cotton takes its name from *qutun*, the Arabic name of the fiber. Even jeans are part of this geography, because their name comes from Genoa. The cloth from which jeans are made is historically the cloth of Nîmes, or denim.

THE INDUSTRIAL REVOLUTION

Despite the machines creating this textile geography, manufacturing in the 19th century changed so profoundly that we justifiably speak of an industrial revolution in that century.

First, the energy driving machinery changed. Muscles had already been supplemented by wind and waterpower, but now those technologies became more sophisticated. A fine example is at Lowell, Massachusetts, where a textile mill opened in 1814. There, waterwheels on the canalized Merrimack River drove a complex set of leather belts that ran over drums of different diameters and impelled rotating shafts to much higher velocities than the waterwheels below. The spinning shafts drove carding machines, spinning frames, and looms. Such systems operated in Lowell until the 1920s, although by that time they were obsolete.

Steam then replaced waterpower. In the United States in 1820, the ratio of water-driven mills to steam-driven ones was 100 to 1. By 1900 the ratio was one part water to 4 parts steam. Even then, however, electricity was beginning to replace steam. It held on longest—until the 1950s—in railway locomotives. Diesel-electric engines finally replaced steam even

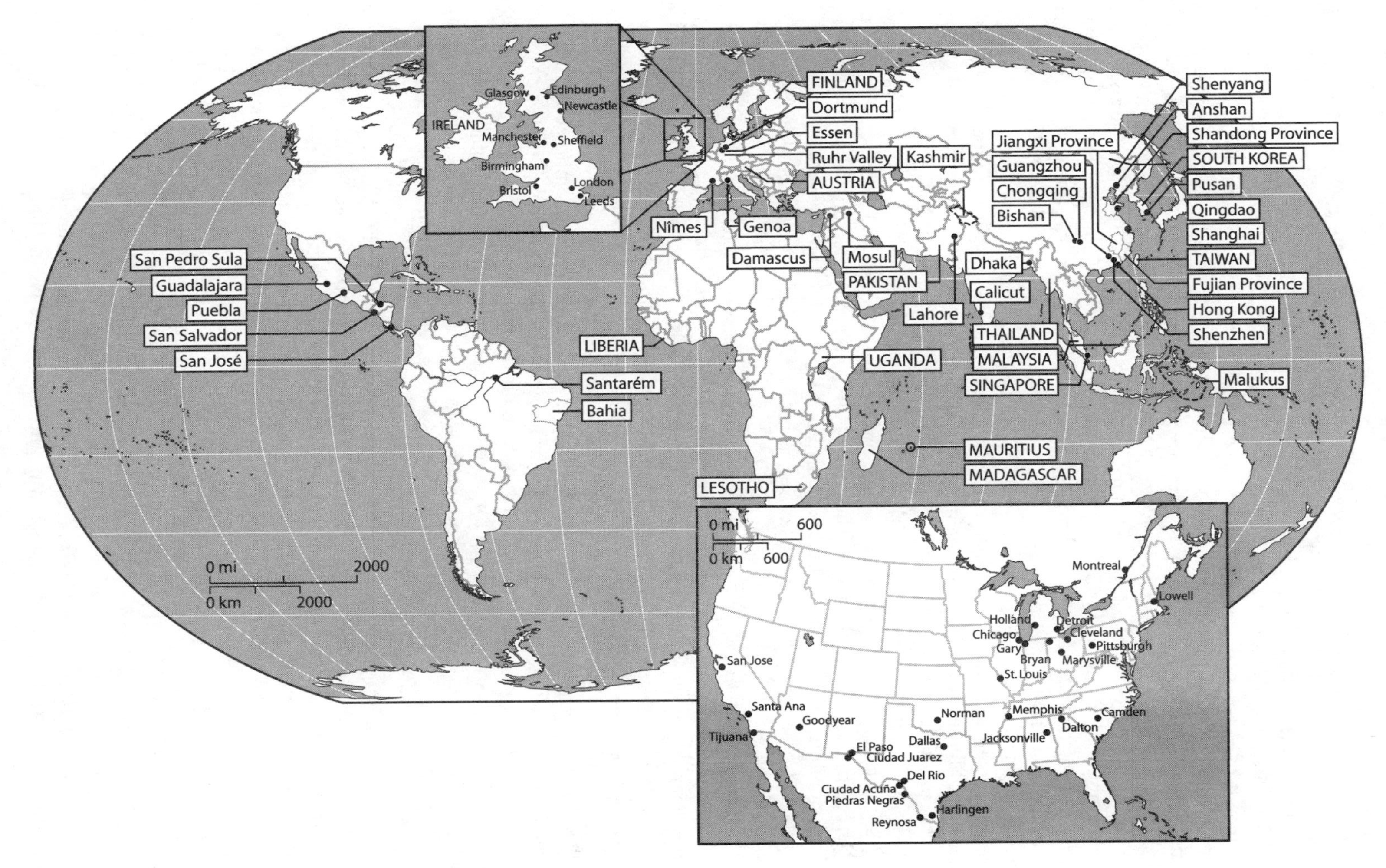

Glasgow
Edinburgh
Newcastle
IRELAND
Manchester
Sheffield
Birmingham
London
Bristol
Leeds
FINLAND
Dortmund
Essen
Ruhr Valley
Kashmir
AUSTRIA
Nîmes
Genoa
Damascus
Mosul
PAKISTAN
Lahore
Shenyang
Anshan
Shandong Province
SOUTH KOREA
Pusan
Qingdao
Shanghai
TAIWAN
Fujian Province
Hong Kong
Shenzhen
Jiangxi Province
Guangzhou
Chongqing
Bishan
Dhaka
Calicut
THAILAND
MALAYSIA
SINGAPORE
Malukus
San Pedro Sula
Guadalajara
Puebla
San Salvador
San José
LIBERIA
Santarém
Bahia
UGANDA
MAURITIUS
MADAGASCAR
LESOTHO
0 mi
2000
0 km
2000
0 mi
600
0 km
600
Montreal
Lowell
Holland
Detroit
Chicago
Cleveland
Gary
Pittsburgh
Bryan
Marysville
St. Louis
San Jose
Santa Ana
Goodyear
Norman
Memphis
Camden
Dalton
Tijuana
Dallas
Jacksonville
El Paso
Ciudad Juarez
Del Rio
Ciudad Acuña
Piedras Negras
Harlingen
Reynosa

there, partly because at low speeds a diesel engine is much more powerful than a steam engine of the same rated horsepower and partly because several diesel engines can be operated by a single engineer in the lead engine.

Second, an expanded range of materials became available. Steel, to take the most important of them, had been produced since medieval times, but controlling carbon content—not too much, not too little—was so difficult that a fine sword seemed magical, like King Arthur's mythical Excalibur. By the 18th century, steel was produced by the slow and extremely arduous process called puddling. Then came the Bessemer converter, developed in the 1850s and producing large batches of steel at unprecedentedly low cost. Carbon content could now be precisely controlled by injections of air blown into huge vessels of molten metal. Again the technology was short-lived, and the converter was replaced by the open-hearth furnaces that sustained the first billion-dollar corporation, U.S. Steel. That company, which derived from Carnegie Steel, combined Minnesota ore and Pennsylvania coal to make steel at Pittsburgh and Gary, Indiana. (Gary takes its name from the founding U.S. Steel chairman, Elbert Gary.) The open-hearth furnace in turn became obsolete, replaced by basic oxygen and electric furnaces.

Third, assembly lines cut the cost of production. Butchering is an early example. By mid-19th century, hogs were killed in huge numbers after they had walked up a long ramp to the top floor of a large building. They were killed there, shackled on rails, and allowed to roll downhill until they left the building as so much pork, lard, and by-product. (For more on this and other early assembly lines, see Siegfried Giedion's *Mechanization Takes Command,* 1948.)

Mass production was applied in less grisly ways, too, such as the manufacture of pocket watches, sewing machines, and typewriters. (These small but intricate machines are a good reminder of the importance of the machines tools making them and of the critical importance of screw-cutting machines to produce the screws whose adjustment was essential in making precision parts.) Early in the 20th century, Henry Ford applied it to a bigger machine. Cars had been for the rich, but Ford changed that. His secret was simple: "The way to make automobiles," he wrote, "is to make one automobile like another automobile, to make them all alike, to make them come through the factory just alike; just as one pin is like another pin." Between 1912 and 1914, he built three assembly lines. The one at Highland Park was four stories high. Up top, engine blocks were milled and engines assembled, fuel tanks were tested, and fenders and hoods were painted. The third floor had stockrooms for lamps, as well as a tire-and-wheel assembly shop. On the second floor, there was an assembly line for the car body, including upholstery. Down on the ground floor, the frame was fitted with its engine, wheels, and other components. It was then driven off to pick up the body, which was ramped down from the second floor and only temporarily fastened to the frame. Final assembly was done after the vehicle was unloaded from a railroad boxcar. (The Highland Park building was used for tractor assembly into the 1970s. Since then, it's been used for storage. The city of Highland Park is now a very poor, heavily black enclave of Detroit.)

Ford continually sought ways to build the Model T more efficiently, with the amazing result that the price of his car fell from $825 in 1908 to $290 in 1924. By 1914, Ford had a 48% share of the American market. By 1920 he was producing half the cars made worldwide. An industry established in Europe had become quintessentially American and would remain so for another 50 years.

The pace of Ford's assembly lines and the repetitiveness of the work provoked resistance and high worker turnover. Ford was implacable. He wrote that every job had been broken down to such a few components that "the most stupid man" could learn every job in the plant "within two days." How to make workers accept such jobs? Ford's solution was simple: pay $5 a day, twice the going rate. If Ford had been interested in history—which he notoriously wasn't—he might have enjoyed reading the description of Manchester, England, written by Alexis de Tocqueville in 1835. "The footsteps of a *busy* crowd, the crunching wheels of machinery, the shriek of steam from boilers, the regular beat of the looms, the heavy rumble of carts, those are the noises from which you can never escape in the sombre half-light of these streets. . . . [H]ere civilization works its miracles, and civilized man is turned back almost into a savage."

The industrial revolution redistributed national populations. By 1900, for example, the densely populated areas of Britain were the parts of Britain densely populated today. A relatively even distribution of population has been replaced with the clottings of an industrial nation, with concentrations in London and Bristol, Birmingham (the Midlands), Liverpool (Merseyside), Manchester, Newcastle, Glasgow, Edinburgh, and the coal-producing region of South Wales. Some of these concentrations were older than the industrial revolution. Some had more to do with mining or transport or commerce than with industry. Some are in steep decline today. Still, it was the industrial revolution that helped them rise as a group to their modern pattern of dominance. With the exception of London, all were near or on actively mined coal fields.

The contrast with continental Europe was sharp. In 1920, 40% of all Frenchmen farmed. Seventy percent of all Russians farmed. Six percent of the English farmed.

Industrialization created the basic pattern of population distribution in the United States, too. By 1900, Americans clustered in a block stretching east from St. Louis and Chicago to the country's old coastal cities. Glance at a population map from that time. Unless your eye drifts to Miami, Atlanta, or Phoenix, you can easily mistake the map for one showing population distribution today.

DOMINANT CORPORATIONS

The capital requirements of the new enterprises were immense, which is why giant companies like U.S. Steel and Ford evolved. Consider the tires that carried Ford's cars. Despite the efforts of the Brazilian government to prevent the export of rubber tree seeds, an Englishman named Wickham succeeded in getting some to Kew Gardens, near London. Malayan rubber plantations—orchards, really—soon undercut the Brazilian rubber-collecting industry, which depended on *seringueiros,* collectors wandering through the forest and tapping its naturally scattered rubber trees. Goodyear, Dunlop, and Michelin obtained their rubber from the Malayan plantations, while Harvey Firestone imported rubber from his own plantations in Liberia. Goodyear tried vertical organization, too, and at a time when the cord used in tires was cotton, Goodyear ran cotton farms west of Phoenix, in a town still called Goodyear.

The larger these corporations grew, the more cheaply they could make things. They could also crush budding competitors by pricing goods below cost until the competition went away. They could wield political influence, often unscrupulously. Last but not least, they had money to spend on advertising. Procter and Gamble (P & G), originally a soap and candle maker, became a powerhouse of consumer products partly because it began advertising on the radio and sponsoring what were soon called soap operas. (In 1944, there were over 40 nationally broadcast soaps, or serial dramas. Perhaps the most indestructible was

"The Guiding Light," which began in 1937, switched to television in 1952, and continues today.) By 2002, Procter and Gamble was spending almost $4.5 billion annually on advertising, more than any other company in the world.

P & G is a handy reminder of the huge role such corporations play in our daily lives. A piece of cheese likely comes from Kraft, but Kraft brands include Breakstone, Cheese Whiz, Cool Whip, Country Time, Cream of Wheat, DiGiorno, Grey Poupon, Jell-O, Kool-Aid, Knudsen, Life Savers, Louis Rich, Miracle Whip, Oscar Meyer, Nabisco, Oreo, Planters, Post cereals, Ritz, Sanka, Shake 'n Bake, Tang, Triscuit, and Yuban. Huge as Kraft is, it was acquired in 1988 by an even bigger company, Philip Morris. In 2002 the giant changed its name to Altria, perhaps because more than half its $80 billion in sales that year still came from tobacco, and company executives wanted some linguistic camouflage.

Competition for Kraft comes from General Mills, whose revenues exceed $10 billion, not only from cereals like Cheerios and Wheaties but from Yoplait and Colombo yogurt, Betty Crocker cake mixes, Pillsbury products, Hamburger Helper, Progresso Soups, Old El Paso Mexican foods, Green Giant, Häagen-Dazs, Jeno's pizza, and health foods sold under the labels Muir Glen, Nature Valley, and Cascadian Farm.

A soft drink in the afternoon probably comes from Pepsi or Coke, whose annual revenues are about $20 billion each. A delightful Hostess Twinkie or Dolly Madison cake comes from Interstate Bakeries, which puffs Wonder Bread and other baked goods into annual sales exceeding $3 billion. Some companies straddle categories. Sara Lee is a good example, with revenues of $18 billion not only from the frozen cakes found in every supermarket in America but from the not-so-tasty combination of Earthgrains bread, Jimmy Dean sausage, Ballpark franks, Kiwi shoe polish, Hanes underwear, Wonder bras, and socks. (It sounds odd, but the company is one of the two largest manufacturers of socks in the United States.)

COMPETITION

Although the size of such corporations has always generated fears about monopoly, most of the giants are locked in competitive combat. In 2001, for example, P & G went dumpster-snooping to learn more about Unilever's strategy for hair care products. There's nothing illegal about rummaging through garbage, but Unilever sued for $20 million and got a P & G apology. Two years later, Schick-Wilkinson announced its four-bladed Quattro razor. The product had been under wraps, but Gillette somehow knew enough about it to sue for patent infringement on the day the product went on sale.

Or consider aircraft. Worldwide, there are only two suppliers of large airliners. Boeing is way ahead of Airbus if you measure by the number of aircraft in service: 14,000 for Boeing to 2,000 for Airbus. Until recently, Boeing was also ahead if you measured by aircraft deliveries: in 2001, for example, it delivered about 500 planes, while Airbus delivered about 350. But Airbus in 1999, 2000, and 2001 had about as many new orders as Boeing, and in 2003 it delivered more planes. By then, Airbus had 1,500 planes on order, while Boeing had only 1,200. Airbus had announced a giant, completely double-decked plane, the A380. Boeing had toyed with several projects but was committed to none. At the end of the year, its boss was sacked.

Many airlines that for decades have flown only Boeing aircraft know that their training and maintenance costs will rise if they start operating a second family of aircraft. Airbus replies that it has only two cockpit designs, so pilots can fly several aircraft models. Airbus has another advantage, too, because airline executives don't like depending on a single supplier. China's a good example. It's the second biggest aircraft market in the world today, after the United States, and both Boeing and Airbus are scrambling to win as many orders there as possible. China Southern, the biggest domestic carrier, has historically relied on Boeing, but its president says, "If I have some of this plane and some of that plane, it gives me negotiating room." It's a recipe for continuing and intense competition, so intense that some analysts think that Boeing will retreat from civil aircraft production and concentrate on its military aircraft division at St. Louis. At the least, Boeing will subcontract more and more of its parts production to foreign suppliers. Even that may be optimistic. One British supplier to Airbus concludes that Boeing is "in a terrible dilemma," because it can't catch Airbus without spending more money than investors will tolerate. Boeing "left it too long," he concludes. This epitaph may be premature: Early in 2004 Boeing announced its firm commitment to build the 7E7 "Dreamliner," scheduled to enter service for All Nippon Airways in 2008.

Or consider the automotive industry. Under the leadership of Alfred P. Sloane, in 1931 General Motors displaced Ford as the leading car manufacturer. Sloane divided the corporation into several largely autonomous divisions, including Chevrolet, Oldsmobile, Buick, and Cadillac. (Pontiac came a few years later.) Each had its own price range. Each was made of parts largely purchased from suppliers. (This was a major change from Ford, whose obsession with vertical integration led him to operate his own steel mill at the River Rouge plant in Dearborn. That wasn't enough: Ford also set up rubber plantations in the Amazon—Fordlandia was one, near Santarém—so he could make his own tires with his own rubber.) Sloane also had the stroke of marketing genius to introduce new models each year, even if there was nothing new except the shape of a fender. This was a huge change from Ford, who was oblivious to style and status. By the time Ford realized how much these things mattered, it was too late. By the 1960s, GM's profits were three times Ford's. Young Americans began with Chevies and worked up the GM ladder to Cadillac.

By the 1990s, however, Cadillac sales were sliding, down from 350,000 cars in 1978 to 172,000 in 2001. By then, the average Cadillac buyer was 66 years old and a poor candidate for future purchases. GM products were perceived as terminally dull, and the company announced that in 2004 it would stop making Oldsmobile, a brand that went back to 1897 and was older than GM itself. Ford hoped it might regain the lead, but it was hit with a scandal over defective Firestone tires, and in 2001 it lost over $5 billion—over $1,500 for every car it sold. GM made about $300 on each of its cars that year. It also set out to resuscitate Cadillac and make it competitive with Mercedes, Lexus, and BMW; its Escalade quickly proved a hit.

Both Ford and GM are now scrambling to centralize their engineering and purchasing—scrambling, in effect, to copy Toyota. A now-retired GM chief executive officer says: "we were run by divisions and countries. Our major competitors were run with one engineering center, one selling effort. Boy, we needed to get that kind of organization." GM may well keep its global lead, but Ford is on shakier ground. Its Kansas City plant is the biggest automotive factory in the United States and assembles one vehicle a minute, year round, yet some analysts think Ford may still slip into the ranks of smaller carmakers. Chrysler is already gone, merged into Germany's Daimler, which also controls the biggest heavy-truck manufacturer in the United States, Freightliner. For the moment, Ford is concentrating on profits, not market share. The one is up, but the other is down.

Over in Europe, similar tales can be told of automotive companies waxing and waning. Daimler's flagship Mercedes is selling into a flat market everywhere except China, and the company's disastrous investment in Mitsubishi destroyed Daimler's hopes to build a global company. Italy's Fiat has seen its share of the European car market decline from 14% in 1990 to 8% in 2002. That may not seem like much, but Fiat was once Europe's biggest car maker, and now there are industry analysts who think its days in the business are numbered. In 2002, the company had losses of $4.5 billion and faced very tough Japanese and Korean competitors.

INNOVATION

Manufacturers with a will to survive innovate.

The Japanese are famous for their just-in-time manufacturing system, pioneered by Toyota when it began making cars in the 1930s. Japan had no established parts industry and no tradition of ordering parts from a warehouse, so Toyota adopted Ford's assembly line and combined it with the idea that parts should arrive from the supplier at the moment they were needed on the assembly line. The idea is now commonplace, used by all kinds of manufacturers worldwide, including American ones. In *Cities in Civilization* (1998), by the way, Peter Hall argues that Henry Ford was the real inventor of just-in-time manufacturing but could not implement it because American railroads could not or would not operate on the precise schedules it demands.

Lean manufacturing is another example of innovation. It can probably be traced to the time-and-motion studies pioneered a century ago by Frederick Taylor (1856–1915), an insomniac industrial engineer who famously studied, among other things, the best way for steel workers to handle a shovel. In its modern guise, however, lean manufacturing is another Japanese innovation. The object is to set up the assembly line to use minimal space and allow workers to do their job with as little movement as possible. The Boeing 777, for example, used to take 71 days to assemble. With lean manufacturing it comes together in 37. The Boeing 737 model is pushing this approach even further. Good-bye to planes parked in a hangar while workers swarm over them. Now, as if the planes were automobiles, unfinished aircraft are towed through the Renton, Washington, plant at 2 miles per hour.

Another area of innovation is quality control. *Kaizen* is the term used by Toyota to signify continuous product improvement. Good-bye, good enough; hello, make it better still. American manufacturers have adopted the principle. GM, for example, used to build cars, then repair the defects. GM now wants 90% of the cars coming off its 81 assembly lines in the United States, Canada, Mexico, and Brazil to be defect-free when they arrive at inspection stations. A similar story can be told for Ford, which strove in 2001 to make its reintroduced Thunderbird flawless—to the point of holding hundreds of plastic-wrapped cars in storage while awaiting delivery of replacement parts for a possibly defective cooling system. Not that it helped: Sales were poor, and Ford announced in 2003 that it would stop Thunderbird production in 2005. By then, Toyota was poised to take over Ford's position as the second largest car maker in the world.

Another innovation is contract manufacturing. The public scarcely has heard of Ingram Micro, which builds computers for IBM, Hewlett-Packard, and Apple on the same assembly line in Memphis. Those companies are meanwhile free to invest their capital in product development and advertising.

Within the next few years, nearly half of all electronics manufacturing worldwide will likely be handled by contract manufacturers. Ingram is going to be fighting to survive against competitors like Solectron and Jabil Circuit. They'll all be fighting Singapore-based

Flextronics, the biggest electronics manufacturing service. Its main factories are in Guadalajara, Shanghai, and Hungary, but it has a total of 87 factories in 27 countries. Annual Flextronics sales approach $15 billion, and its products include cell phones for Sony-Ericsson, printers for Hewlett-Packard, and copiers for Xerox. On one of its shop floors there's this sign: "It's not the big companies that eat small ones but the fast that eat the slow."

Many contract manufacturers, including Flextronics, have in recent years located in Mexico, particularly in Guadalajara, where 60,000 people were working in the electronics industry in 2000. What was Mexico's edge over Asia? Location: the factories shipped via UPS and FedEx and on both time and cost beat more distant suppliers. Still, it's very hard to compete against Chinese wages, and by 2003 these factories were running at 60% of capacity. Low-cost products had moved to China, and the only way for the Mexican plants to survive was to make higher-value products.

Jabil Circuit, for example, saw its work force in Guadalajara cut in half, from 3,500 to 1,700. Then it started making computer routers for Nokia, internet firewalls, and electronic controls for washing machines. Its work force rose to 3,900. The losers are the American workers who used to assemble these complex devices. Jabil is an American company, and in the late 1990s three-quarters of its products were made domestically. Less than a fifth are today. The Nokia routers had been made at a Boise, Idaho, plant that was 2 years old when it closed.

Subcontractors have taken over laptop production, too. Sixty percent of world production comes from Taiwan. The leading manufacturer is a subcontractor, Quanta Computer, which makes half of Dell's laptops but also makes machines for Gateway, H-P, Apple, and Compaq. Making 5 million machines annually, Quanta is now the world's biggest manufacturer of laptops, followed by Toshiba. The company takes 48 hours to build a computer after a purchaser orders it. Soon that number will be cut in half. Despite these improvements, China's nearby Jiangsu Province hopes to overtake Taiwanese manufacturers. To cut costs, Quanta is likely to move at least some of its factories to China and help Jiangsu do so.

Contract manufacturing is likely to become important in many sectors. The Finnish company Valmet builds the Porsche Boxster, and a Canadian company named Magna builds Chrysler Voyager vans in Austria. Not every manufacturer is completely sold on subcontracting, however. The Japanese think they're the best manufacturers in the world and subcontract only about 4% of their electronics, compared to 30% for American companies. Still, Japanese wages are high, and subcontractors in China are likely to play a larger role in the manufacture of Japanese products.

There are American doubters, too. IBM makes its own Thinkpads at Shenzhen, near Hong Kong, though it has now subcontracted its more specialized laptops. Dell, too, uses subcontractors but also has its own superefficient plants. Despite its use of subcontractors, it continues to make computers at five plants in the United States and at other company plants in Brazil, China, Ireland, and Malaysia. These plants are pitted against each other. They operate in 2-hour production runs in which all the parts needed in that run are assembled at the start. If the company is short on a particular component, it will offer an upgrade or even a free upgrade for the part in short supply. Dell is the pioneer in another innovation, too: the sale of virtual products that do not exist until they've been ordered and paid for. It's a great way to reduce inventories, and it works especially well for a company where every product is built to customer specifications.

Innovations are sometimes ill-conceived, of course. The historic leader of the jeans industry in the United States tried a decade ago to replace individual piecework with team-based production to reduce worker boredom. It was another innovation from Levi Strauss, a company renowned for socially conscious management. Back in the 1950s, for example, Levi Strauss's Virginia plant had been a pioneer in racial integration. The new production teams backfired because fast workers resented slow ones. Other well-intentioned Levi's strategies failed too. Long after its competitors had shifted production overseas, for example, Levi's continued making jeans in the United States. It saved American jobs but kept Levi's prices high. To make matters worse, Levi's styles were no longer considered stylish. Only after company sales fell from $7 billion in 1996 to about $4 billion did it face the inevitable. In 2002, it shut six U.S. factories and said that it hoped for annual savings of $100 million. The company announced the next year that it would start selling a budget-priced Signature line in Wal-Mart and announced the closure of its last American plants. From now on, Levi's would be made only by contract manufacturers. Americans might still buy 575 million pairs of jeans annually, but they would all, or nearly all, be imported.

Campbell's Soup is another good example of corporate failure. It is about as familiar as any brand in the world and has been around forever—since 1869. But the core business—less than appealingly, stock analysts call it wet soup—is dying. "It's not what the consumer wants," one analyst says. Campbell's has peripheral businesses that are doing well, such as Godiva, Pepperidge Farm, Prego, and Pace, but company revenues in 2002 were less than those of a decade earlier. It's a common problem with big companies. Lou Gerstner, who oversaw a turnaround of IBM, has compared the challenge to teaching elephants to dance.

SECTOR REALIGNMENTS

It's not just companies that come and go: Industries as a whole can flourish one generation and languish the next.

As late as 1958, the steel produced in the United States was worth more than all the automobiles it made. Today, however, the positions are drastically reversed, with the nation's new cars worth about $100 billion, compared to $20 billion for steel. In 1955, U.S. Steel ranked third on the Fortune 500 list of the country's industrial giants, and runner-up Bethlehem ranked eighth. By 2003, "the Corporation," as U.S. Steel was known in the industry, had fallen to number 209, and Bethlehem, "the Steel" in its Pennsylvania hometown, was in bankruptcy, after having fallen in 2002 to number 440. (One of the chief causes of Bethlehem's demise was the unbearable cost of pensions and medical benefits. In 1957, 167,000 people had been on the payroll, but by 2002 that number had plummeted to 12,000. Those workers were supporting 60,000 retired workers—double that, counting dependents. A federal program saved most of the pensions, but the workers' hard-won medical benefits were lost.) Bethlehem's properties were bought for $1.5 billion in 2003 by International Steel, a new company formed from the wreckage of another bankrupt producer, LTV Steel of Cleveland, which itself had been formed chiefly from Republic Steel.

Steel is still important to the American economy, but other things have become more important. Meanwhile, the world needs only about three-quarters of the roughly 800 million tons of steel made each year, and so imported steel is cheap. The United States would see more of it if the federal government didn't erect trade barriers. As it is, the United States imports only about 15 million tons of steel annually, much less than the 100 million tons it makes. Ironically, although the United States is no longer a big steel exporter, it does export old and abandoned steel mills. One such mill that used to be in Cleveland is now in Shenyang, China.

Worldwide, the United States ranks a distant third in the production of steel, after China and Japan, and no American company ranks any longer among the world's 10 largest steel makers. The big producers in recent years have been from Europe, South Korea, and Japan, where the leading companies have been Nippon Steel and Pohang (POSCO) of Korea. Nippon is being replaced as Japan's biggest producer by a merger of Kawasaki and NKK, however, and both Kawasaki/NKK and Pohang are smaller than Arcelor, another new company formed by the merger of Usinor of France with Belgian and Spanish partners. As things stand, Arcelor will be the world's biggest producer, followed by an international newcomer, LNM, which specializes in makeovers of bankrupt mills across Asia, Europe, and Africa. Even Arcelor will only produce about 6% of the world's steel, but it will be far ahead of such old titans as Thyssen and Krupp of Germany, which merged to form ThyssenKrupp in 1999, and British Steel, which is now part of a joint British–Dutch company called Corus.

It's quite a ballet. Meanwhile, if you see a heavy bridge girder being trucked into place somewhere in the United States, you can be sure that it's imported: No American mill makes that stuff any more. Travel around the country, and you'll also get used to the sight of closed mills from Utah to Georgia to Ohio. Even shut down, they're impressive. The economic impact of closure is equally impressive. When the LTV mill in Cleveland closed, 18,000 jobs were lost.

Or, consider the personal computer. In 1990 the value of electronics and electrical equipment produced in the United States was about $105 billion. The value of primary metals, including steel, was about $43 billion. By 2000, the value of electronics and electrical equipment had risen about 80%, to $181 billion. The value of primary metals, however, had risen only about 20%, to $52 billion. Personal computers were a key part of the booming sector, and they were made mostly by new companies. The 10 largest computer manufacturers in the United States included only three with roots deeper than World War II: They were IBM (International Business Machines), NCR (formerly National Cash Register), and Pitney Bowes, originally a postage-metering company. The seven newcomers were Hewlett-Packard, Compaq, Dell, Xerox, Sun, Gateway, and Apple.

Today, almost all those companies are hurting. Falling prices are pushing the business overseas, just as they did a generation ago with television manufacturing. Gateway has closed its 18 computer stores in Australia and New Zealand, as well as the factory in Melaka, Malaysia, that supplied them. Then, in 2004, it closed its remaining 188 stores in the United States and said that it would sell through third parties and direct by phone or internet. That leaves the field of combat largely to Dell, which dominates in the United States, and to the merged HP/Compaq, which is strong in Europe. Dell is trying to export its lean model to Europe, while HP is sticking with its more traditional, staff-rich organization. Both are targeting Asia, where they face entrenched local manufacturers such as China's Legend and Japan's NEC and Fujitsu. Legend is planning a counterattack in American and European markets.

Seagate is a related case. It's the world's biggest maker of hard-disk drives, but the cost of one megabyte of storage has fallen from about $12 in 1988 to 10 cents. Miniaturization is largely responsible for this, and although it yields better computers, the parts in those machines are now so small that people can't build them. Only machines can. The consequence is that Seagate's work force in Malaysia, Thailand, and Singapore has been slashed, along with local hopes of transitioning to a high-tech economy. For Malaysia, manufacturing for the computer industry is on the verge of turning out to be a dead end, like the plantation economy it largely replaced. Talk to investment gurus today, and they're likely to say that fortunes in the next decade will come from biotech. Don't go overboard, though. Smart planners, like those working for the government of Singapore, will invest in biotech but work to diversify their economies, too.

GLOBALIZATION

Like resource production, manufacturing has been spatially rearranged in the last century.

An early step was the move of the American textile industry after World War I from New England to the American South, where labor was cheap. During World War II, defense industries and petrochemicals moved to the West Coast. Electronics began moving there, too. The Santa Clara Valley, centered on San Jose, had been famous for the production of fresh and dried fruit. In 1939, however, Bill Hewlett and David Packard, both of Stanford's school of electrical engineering, set up a small shop. In 1952 IBM joined the party and set up its first plant dedicated to making disk memories. Intel came in 1968. The Santa Clara Valley was becoming Silicon Valley.

In 1930, California's value added by manufacturing ranked behind Pennsylvania and New York and was only slightly ahead of Ohio and Illinois. Today, California ranks number one in the nation and adds more than twice as much value to manufactured goods as any of the neck-and-neck runners-up, New York, Ohio, and Texas. Texas, too, has come a long way since 1930, when its manufacturing output was half that of New York.

It didn't take long for manufacturers to realize that they could find even cheaper labor farther afield. In the 1970s, several clothing companies set up cut-and-sew shops in or near Harlingen, Texas, where wages were much cheaper than in the Carolinas. By 1991, Levi Strauss, Carter's, Dickies, Vanity Fair, Haggar, and Fruit of the Loom employed 9,000 people around Harlingen. By 2003, however, all of them had been laid off. Some of the most energetic considered themselves lucky to get new jobs paying less than half what they had made before.

The factories had all gone to Mexico. The export of American jobs there began in 1965, when Mexico permitted American companies to set up assembly plants without paying duty on the manufacturing equipment or materials, as long as the finished goods were exported. Such plants, known as maquilas or, more mellifluously, maquiladoras, now employ about 1 million workers in 3,500 plants. There are some 350 maquilas in Juárez alone, while on the other side of the border, El Paso, like Harlingen, no longer has the factories that once made it a center for jeans and leisure wear.

There's no mystery about it. The American clothing companies at Harlingen were paying $9.45 an hour in 1999. Wages across the river were about $2. For a company like General Motors, which pays its American assembly-line workers $25 an hour, the savings were irresistible. That's why GM makes about $20 billion worth of cars annually in Mexico, one-tenth of its North American production. Volkswagen is in Mexico, too, originally drawn there by a Mexican law requiring assembly in Mexico of cars sold in Mexico. Now, low wages have induced the company to build all its new Beetles, worldwide, in Puebla. About 450 Beetles are made daily in a plant there with 15,000 employees. Nearby, another 10,000 workers make parts. The Jetta, too, is made only in Puebla. So is Daimler-Chrysler's PT Cruiser. Toyota is now joining the party with a pickup-truck plant in Tijuana.

It sounds like a huge success story, and it's true that maquilas shipped $77 billion worth of goods in 2001. But since 2000, about 350 have closed. Juárez, after devastating El Paso, has itself now lost 60,000 jobs since 2001. Between 1999 and 2002, almost

5,000 textile jobs vanished from Piedras Niegras, where wages subsequently fell from several dollars to 80 cents an hour. You don't have to look far for the explanation: China has replaced Mexico as the second biggest exporter to the United States. (Canada comes first.) One sign of the change: Hasbro has closed its Tijuana toy factory and moved to China.

Low Chinese wages aren't the only problem for Mexico, because the tax breaks that favored establishment of the maquilas are due to expire under the terms of the North American Free Trade Agreement. Laid-off workers are already moving back to southern Mexico. Maquilas likely to endure and even expand are those, like Flextronics, that make increasingly complex products or those that make products with low value but high shipping costs. Maytag, for example, pays its American workers $15 an hour and has only recently begun moving some of its refrigerator manufacturing to Reynosa, just across the Rio Grande from McAllen, Texas. Maytag's American workers are livid, but the company won't be alone in Reynosa. Whirlpool is already there.

Manufacturers are looking south of Mexico, too. Intel invested half a billion dollars in a computer-chip factory outside San José, Costa Rica. It was part of an effort by the government there to attract high-tech industries to what was being called San José South, a hopeful allusion to Silicon Valley. Microsoft, Acer from Taiwan, and Lucent, the former manufacturing arm of AT&T, had already set up shop. Meanwhile, money was pouring into Brazil, where GM and Ford built plants serving the Brazilian market. The Ford operation was at Bahia, where Ford Escort modules are assembled by some 20 suppliers working under the Ford roof.

Impressive as these developments may be, they pale alongside China. Foreign investment in the automotive industry there goes back to a Volkswagen joint venture that opened in Shanghai in 1985. By 2003, China was making 3.8 million cars and light trucks annually, enough to put it in fourth place after the United States, Japan, and Germany. Volkswagen retains about 40% of the market, but GM now has about 10%. Its Shanghai plant began with Buicks—common across China—but now also makes a budget car called the Sail. Soon the Sail will be made farther north, in Shandong. Honda produces Accords in Guangzhou and is planning to serve not only the domestic market but the international one, with exports of 50,000 cars annually. Nissan and Toyota are coming and, with Honda, will invest $3 billion in this market. All will operate with Chinese partners, just as GM does with Shanghai Automotive Industry Corporation. Luxury cars are next. BMW is investing 450 million euros in a plant at Shenyang; by 2005, it should be making 50,000 cars a year. Mercedes is close to building an assembly plant at Nansha, a new town midway between Guangzhou and Hong Kong. The gleam in the eyes of all these companies is that China still has only 8 cars for every 1,000 people of driving age.

It isn't just cars. Whirlpool is one of GM's Shanghai neighbors. It's invested $60 million in China, and its Shanghai factory is making washing machines for sale not only in China's big cities but in much smaller places. It's slow going, because there are big Chinese manufacturers like Haier and also because, in a country with few chain stores, markets have to be built one store at a time.

Coca-Cola is already experimenting with ways to increase its presence in China's small towns. Rather than continue marketing through big wholesalers who distribute through more middlemen, Coke has teamed up with 365 partners in Fujian Province. Some are small shopkeepers themselves. Each is assigned a territory, each delivers directly to shopkeepers, and each, instead of having to buy the product and resell it, is paid for how much Coke he sells to retailers. The other part of the formula is selling Coke in small, returnable bottles, so prices are kept down to about 12 cents. European brewers are in the same school of hard knocks. Several have entered the Chinese market, which is now the world's biggest, ahead of the United States, but many have had a hard time. Not many Chinese are willing to spend 60 cents for a can of beer when for half the price, or less, they can get a local beer in a returnable bottle that's twice the size. "There seemed to be no end to the red ink," says an executive with Australia's Foster's. The market remains very fragmented, with China's 25 biggest brewers sharing less than half of it. A Singaporean now working in China for Belgium's Interbrew explains that "China is a nation, but not a national market."

Regardless of who owns the factory and whether production is exported or consumed at home, China is becoming the world's factory. Remember Etch A Sketch? Popular since it was introduced in 1960, the toy was made by the Ohio Art Company in Bryan, Ohio. Facing competition from other toys, the company concluded that Etch A Sketch had to retail for no more than $10. With its Ohio workers making $9 an hour, this was impossible. Workers at the Kin Ki Industrial plant in Shenzhen, China, however, were available at 24 cents an hour. "It tore our hearts out," says the company's chief executive. Some of the workers at Bryan were kept on to open the crates that arrive from China.

How about the clothes on American backs? In 2002 about 96% of the clothes that Americans bought were imported. Surviving American plants keep closing. The Union Yarn Mill of Jacksonville, Alabama, for example, was for almost a century the town's largest employer, but now Fruit of the Loom, the underwear manufacturer that operated the mill, is in bankruptcy. If you look for exceptions, you'll probably settle on socks. They require little hand labor and so are the only standard clothing item that the United States still makes mostly at home. *Made* mostly at home, more accurately: The percentage of socks made domestically is declining quickly, from 76% in 1999 to 40% in 2004.

By 2005, China is expected to make half of all the clothing exported worldwide. Whatever the label says, those clothes will be made by companies that consumers have never heard of. How about the Yue Yuen shoe company of Hong Kong? It makes shoes for Nike, Adidas, New Balance, and Reebok. The company is controlled by a parent company in Taiwan. Its factories are in Guangdong. Workers typically make $30 a month and live in factory compounds which they scarcely leave, especially as 7-day workweeks are common.

Furniture is going to China, too. The business is already much more consolidated than American consumers realize. They go to a furniture store, see furniture from Drexel, Broyhill, Thomasville, Lane, and Henredon, and never realize that these brands all belong to a single company, unimaginatively named Furniture Brands. With the exception of upholstered furniture, which continues to be made domestically, Furniture Brands is following its competitors and sourcing in China. In one Shanghai plant alone, Lacquer Craft International plans to employ 5,000 workers. Not surprisingly, the number of Americans working in furniture plants has fallen from 150,000 in 1980 to about 90,000.

Will America retain any manufacturing? Think of Life Savers, a candy for many years made mostly in Holland, Michigan. In 2000, Kraft bought the company and decided to move production to Montreal. It wanted to escape a unionized work force. It also wanted to escape the cost of health insurance for its workers (Canada has socialized medicine and is proud of it). It wanted, finally, to escape America's artificially high sugar price: This main ingredient in Life Savers was costing 6 cents a pound in Canada but 21 cents in the United States. From the corporate viewpoint, it was a no-brainer.

Oddly enough, one likely survivor is automobiles. It sounds bizarre, with the percentage of foreign cars sold in America rising from 7% in the 1970s to about 40%, but half of the foreign-branded cars sold in the United States are made in it. It's a remarkable change since 1980, when Honda opened its Marysville, Ohio, plant—the first Japanese production line in North America. The Japanese plants take about 18 man-hours to assemble a vehicle, while Ford and GM take about 26. The Japanese companies have another huge advantage too, because they're nonunion. This saves a little money on salaries, but it saves a pot of money on lower benefits. General Motors, for example, pays its workers an average of $24 an hour in pension and medical benefits. That's twice what the Japanese pay. In other words, every GM car carries a benefits penalty of $1,350—this in an industry that tries to earn $800 per vehicle. The disparity won't decline until GM's huge pool of 460,000 retirees begins to die off. Those retirees outnumber current workers 2.5 to 1, and they know how lucky they are. Speaking of future retirees, one current pensioner says, "No one will have what we have. Those days are gone."

Carpeting is another good example of a surviving domestic industry. Huge investments have been made in existing factories, mostly in Georgia, because carpet is so heavy that transportation costs become a major deterrent to foreign manufacture. But look closely. The carpet capital is Dalton, a town of 28,000 people in the northwestern part of the state. Who's working in Dalton's factories? Mostly Mexicans, who now comprise 11,000 of the town's 28,000 people.

WORKING CONDITIONS

By the standards of American factories, working conditions in overseas factories are terrible. Nike is the whipping child here, although it has been pushed to sponsor a Global Alliance for Workers and Communities, which shouldn't be dismissed entirely as window dressing. Nike actually looks pretty good alongside Calyx and Corolla, a Florida company whose motto is "the flower lover's flower company." That company's chief marketing officer doesn't comment on worker conditions and says that the environment "is not an issue we have any business being in." It's a common-enough mentality. The workers in Shenzhen who make Etch A Sketches work 12 hours a day, 7 days a week. They're told that if an American inspector comes around and asks to see employment documents they should "intentionally waste time and then say they can't find them."

Then there's Mandarin International. Despite its name, chosen by its Taiwanese owners, the mill is in San Salvador. Until the mid-1990s, it produced clothing for Gap, Eddie Bauer, Target, and J. C. Penney. A strike closed the mill, however, and three of those companies left. Gap could have left, too. After all, it has contracts with about 4,000 suppliers around the world. Instead, it was persuaded to stay and work to make conditions better. An American reporter came for a visit and found the mill back in operation. There was a functioning clinic, and workers who were sick were no longer refused permission to see a doctor. The restroom was unlocked, and there was a functioning cafeteria. Still, salaries were 60 cents an hour, almost unchanged from the days before the strike. The company explained that competition simply didn't allow it to pay its workers more. "I'll take my business elsewhere" is a powerful argument, even when used by consumers against a company as large as Gap. It's also powerful when used by Gap against the government of El Salvador, which knows that if it insists on higher wages Gap will move. (Gap continues to be unusually involved with worker conditions. In 2004 it put on its website a "social responsibility report" on the factories that make its clothes. The report detailed contractor violations of company policy and revealed that in the previous year the company had stopped doing business with 136 suppliers with repeated or very serious code violations.)

Next door in Honduras, a woman at the Cosmos clothing factory in San Pedro Sula sews the sleeves on 1,200 shirts daily. That's one sleeve every 15 seconds for 10 hours. "There is always an acceleration," she says. "The goals are always increasing, but the pay stays the same." That's $35 a week. A nearby factory is owned by Oxford Industries of Atlanta, which over the last 15 years has closed all of its 44 American plants. An American manager says he expects the Mexican ones to close as the company seeks still cheaper Asian labor.

This race to the bottom is inevitable, unless you think that consumers can be persuaded to pay a fair-price premium or the governments of rich nations erect legal barriers against products made in foreign sweatshops. The odds aren't good. Wal-Mart used to promote American products. Sam Walton even had a slogan: "Bring It Home to the USA." That slogan's gone. A Wal-Mart executive explains that customers simply won't pay a premium for American products. That's why between 1995 and 2005 the percentage of imported Wal-Mart merchandise skyrocketed from 6% to over 50%.

There's nothing peculiar to the apparel industry in all this. In 1982 the Aluminum Company of America, or Alcoa, established an assembly line for automotive-wiring systems in Ciudad Acuna, across the Rio Grande from Del Rio, Texas. Alcoa said it had no choice but leave the United States. Its American plants, even though they were in low-wage Mississippi, were unable to compete with Taiwanese manufacturers. Ciudad Acuna, which was booming—its population quadrupling from 40,000 people in 1970 to about 160,000 today—welcomed Alcoa. The company's wages were at the top of the local scale, about $80 a week for a 6-day week. The cost of living in the town is high, however. A typical family, recruited from southern Mexico and arriving with dreams of prosperity, will spend $11 a week on drinking water and $20 a week on electricity. Why so high? Because taxes are low and city services—schools, hospitals, garbage collection, sewage, water, and power—almost nonexistent. How pay for food? Answer: Have a working spouse. At stockholder meetings, Alcoa executives have been embarrassed by such facts, but, like Gap in El Salvador, they say that they are powerless to make things right.

Bad as these things are, working conditions in export-oriented plants, especially American-owned ones, are usually better than in factories serving domestic markets. Twenty-five miles west of Chongqing, for example, Bishan County is a center for the Chinese domestic shoe industry. It's also a center for often fatal diseases caused by benzene in the glues used by the factories. The big foreign companies don't use benzene glues any more. The local ones do, because they're cheaper. It's not unusual: More than half the locks, razors, eyeglasses, and electrical transformers sold in China are made in Wenzhen, with close to 8 million people in southern Zhejiang. The city is dangerously polluted by factory waste.

Workers in export-oriented factories are, on the other hand, uniquely vulnerable to international politics. Pakistan's 3.5 million textile and apparel workers have recently learned this the hard way. In 2000, they made $3.5 billion worth of clothes for the American market. Highnoon Textiles of Lahore, to take one case, was making clothes for Gap, Levi Strauss, and Eddie Bauer. Its workers could make $88 a month, good money by local standards. After September 11, however, American companies got nervous. Perry Ellis and Tommy Hilfiger canceled orders, and Highnoon laid off 500 workers. By the end of the year, about 70,000 textile workers had been laid off nationwide. Some of them were asking if this was the friendship that America promised in exchange for Pakistan's support of America's war in Afghanistan.

COUNTRIES AGAINST COUNTRIES

Not so long ago, Americans perceived Japan as a juggernaut that was going to destroy their manufacturing economy. Now, despite a boost from China's appetite for Japanese imports, Japan is in danger of having its manufacturing base decimated by China. Toshiba already makes all its televisions there. Olympus makes its digital cameras there. Pioneer has moved its manufacture of DVD recorders there. Sharp operates a research and development laboratory in Shanghai. One of its executives says that Japanese engineers are still better than Chinese ones, but "comparing Chinese and Japanese engineers on a cost-performance basis, Chinese are superior. They are hungrier. Most Japanese are no longer hungry." The Japanese even have a word for this gutting of their manufacturing economy: *kudoka.*

Korea has the same problem. In 1990 it exported $4.3 billion worth of shoes. Now the figure's down to less than $700 million. The shoes are still being made, but they're being made in China. Pusan, the center of the Korean industry, is full of deserted factories.

Or look at Madagascar, one of the countries now entering the global textile market. Wages are about 37 cents an hour, low enough to snag contracts from Victoria's Secret and Gap. Workers are keen to get out of the paddy fields, which pay less than half that: These are people who want to buy shoes and ride bicycles. Over the next 25 to 30 years, they will become more skilled. They'll shift from making cheap clothes to more expensive ones. Their wages will rise, and buyers will start looking for another low-wage country.

That's already happened next door, on the island of Mauritius. Through the 1960s, sugar was Mauritius's only significant export. Then, in 1972, the country set up an export-processing zone. Wages on the island have risen to $1.47 an hour, and the annual per capita income is 15 times the $250 it is in Madagascar. It's a success story, but Mauritius now looks at Madagascar the way Japan looks at China. It must find new industries requiring highly-skilled workers.

For the moment, Africa is benefiting from America's African Growth and Opportunity Act of 2000, which waived tariffs and quotas on African exports to the United States. Investors have taken advantage of this law, which is why factories in Uganda send clothes to Target stores. Oddly, the country that has taken the most advantage of the opportunity is Lesotho, an enclave in South Africa that in 2003 sent about $300 million worth of clothing to the United States. Unless amended, however, the law will expire in 2008. It hardly matters: As of 2005, quotas on textiles and clothing will be prohibited among members of the World Trade Organization. Africa's window may be closing. Ironically, most Africans are too poor to buy the clothing they make. There's a huge trade in "bend-down boutiques." Used American clothes are baled up, sent back to Africa, and resold in open-air markets, where they are heaped on the ground.

What happens when a country fails to climb this ladder? The value of Indonesia's shoe exports has fallen from $2.2 billion in 1996 to about $1.5 billion. Manufacturers say Indonesia's salaries are too high. Besides, the atmosphere on the streets and even in the shoe factories is charged with violence. Nike is still responsible for 170,000 Indonesian jobs, mostly through contractors, but it's scaling back. Reebok has shut down. Investors are looking elsewhere, especially to Vietnam and China. Indonesian workers are likely to follow in the footsteps of the Filipinos who years ago learned to leave home and find work abroad.

Does that sound good? Filipinos moving to Taiwan or Malaysia or South Korea can quintuple their salaries. These workers generate a sliver of the world's $90 billion of annually repatriated earnings. Some of it comes from people at the high end of the wage scale, including Americans who have moved to China to supervise factories making the things—furniture, for example—that the Americans used to make back home. Most of it comes from people at the low end. Their labor is certainly a good deal for employers. A Taiwanese company that would pay a Taiwanese $700 a month will pay a Filipino doing the same work $460. The workers have a problem, though. To get these jobs they must pay labor brokers both at home and in the host country. Typically, the fees run $1,500 to $2,000 at each end. The workers don't have this kind of money, so they borrow it from the brokers at high rates of interest. The result is that a Filipino making $460 a month in Taipei may pay a Taiwanese broker $215 a month for 18 months, more than is deducted for Taiwanese income tax and room-and-board combined. And this isn't counting the broker back home, whose loan continues to accrue interest.

The Philippines and Taiwan impose caps on the amount of loans made by brokers, but the laws aren't enforced. Maids are particularly abused. Singapore has a bad habit of losing Indonesian maids off high-rise apartment balconies. Some have probably fallen while cleaning windows or hanging clothes, but many almost surely jumped to end their misery. An Indonesian official says that his country can't protect its citizens overseas. "We have no people, we have no budget," he says. "We just have a big problem."

It's a cruel business. Textile factories are clustered along Dhaka's Malibagh Chowdury Para Road. That's because since 1974 Bangladesh has developed a clothing-export business that now employs almost 2 million people—overwhelmingly women. They work at rock-bottom wages: about $35 a month, less than half the Chinese average. These women produce about $2.8 billion worth of exports, or four-fifths of the nation's hard-currency earnings. It's an achievement, because Bangladesh ranks as the most corrupt country in the world, and its transportation and communication systems are in terrible shape. It's also an achievement in jeopardy, because the 1974 Multifiber Arrangement, which guarantees Bangladesh a fraction of the world's clothing trade, will expire at the end of 2004. What happens then? Without a quota guarantee, and despite low wages, many owners are likely to move to better-organized China. There are very, very few jobs waiting for a million laid-off Bangladeshis.

For a long time to come, China will be the great exception to the pattern of national work forces becoming more expensive. China's labor costs stay flat year after year, because there's a endless supply of people happy to work for 60 cents an hour. Millions of peasants have already migrated toward the coast, where manufacturing is concentrated. There's no end in sight, with 200 million Chinese expected to leave the countryside between 2000 and 2010. If they find factory jobs, they will typically work from 8 in the morning until midnight, with a 90-minute lunch break and a 30-minute dinner break. At Christmas, workers rush to sew eyes and ears on an endless line of stuffed animals. "Overwork death" has become a recognized term for workers—often teenagers—who simply collapse from the stress of these jobs, which can run for weeks without a day off. Consider Daxu Cosmetics, a Korean company that hires Chinese women to make artificial eyebrows in Anshan, in Northeast China. The job may sound funny, but the work is terrible, with women expected to use tweezers to make a neat pattern of 464 bits of human hair on a glue strip. It takes skilled workers an hour to make a pair, but the eyebrows retail for only 50 cents. Some of the women in the plant in 2003 broke out of their locked seventh-floor dormitory, shinnied down a rope of bedsheets that was too short, and fell to the ground with serious leg and spine injuries.

China, in other words, will for years to come set the basement price against which other countries compete. And China isn't limited to textiles. At the expense of Japan, it has grabbed 70% of the motorcycle market in both Vietnam and Indonesia. China is now setting up assembly plants in the United States. A good example is Haier, which in its present form started out in 1986 with a single factory in Qingdao. It's grown so fast that now it's the world's second-largest maker of refrigerators, after Whirlpool. A few years ago, Haier started making small refrigerators for Wal-Mart and Home Depot. U.S. sales are now about $500 million annually, and Haier is moving into freezers, air conditioners, microwave ovens, and other small appliances. It has a plant in Camden, South Carolina, and it wants by 2005 to sell a tenth of all standard refrigerators in the United States. If it succeeds, it will be at the expense of Whirlpool, General Electric, Maytag, and Frigidaire. Can there be more dramatic examples of Chinese market penetration? How about cheap straw hats sold by Mexican street merchants? They're made—you guessed it—in China.

Don't blame the Chinese, though, for exploiting their people. Chinese manufacturers are pressed by buyers always looking for a cheaper source. A Wal-Mart vice-president for global procurement says of the fierce competition among Chinese manufacturers, "Yeah, we try to take advantage of it." Wal-Mart does have wage, hour, and hygiene standards for its suppliers, but enforcement is poor. Besides, if China stopped exporting nearly 5% of the world's industrial products, it wouldn't be able to import nearly five percent of them. That's what it does, generally exporting cheap things and importing expensive ones, largely from the United States. And don't forget that many of the factories in China are at least partly owned by American companies.

All in all, the future's dim for small or even medium-size producers. Consider the career of one merchant in Norman, Oklahoma. Harold Powell opened an apparel shop at campus corner after World War II and built a company with 1,500 employees, 52 full-line stores, a catalog operation, and annual sales of about $125 million. He created a vertically integrated operation, with buyers getting textile art from Italy, woven cloth from Japan, and finished clothes back from China. The clothes were air-freighted to Dallas, trucked to Norman for bar-coding, then distributed within 24 hours through the retail system. But Powell by the 1990s was worried about the competition. Talbot's, he knew, had 100 stores, supplied through a 45-acre distribution center. How would he weather a recession? Sales at Harold's fell from $127 million in 2000 to $90 million in 2002. The company shrank the following year from 50 full-line stores to 39. There weren't many choices, and Powell did the necessary. Early the next year, two investors bought control of his company. One was the vice chairman of Saks and the chairman of a 350-store European chain, We International. The other was W. Howard Lester, who had built Williams-Sonoma. Wells Fargo was now pleased to extend a $22 million line of credit to the company called Harold's.

CHAPTER 13

SERVICES

The promise of modern technology has always been that it would free us from the drudgery of hard physical labor. To a very great extent it has done that. It has taken the American farmer from behind a mule and put him in an air-conditioned tractor cab. More often than not, it has taken him out of farming altogether and put him in a city. For a time, it seemed as though he would work there in a factory, but that too is changing, as manufacturing requires fewer and fewer hands. The result is the overwhelming importance in modern economies of jobs in which people neither grow nor extract nor make anything.

THE CHANGING LABOR FORCE

In 1960 about 8% of the American labor force worked in agriculture. Americans thought of themselves as far ahead of Europe, where 21% of all workers were on farms. They felt more progressive still when they thought of China and India, where more than 70% of the work force was in agriculture. Today, when full-time farmers account for 2% of the American work force, 8% sounds archaic. Other nations are tracing the same curve. A quarter of Brazil's workers are in agriculture, for example, down from 54% in 1960. The countries that *aren't* making this transition are the poorest countries on earth: Nepal, Burkina Faso, Rwanda, Niger, and Guinea all have more then 80% of their labor still on the land.

Then there's the second path of decline. The United States had 15 million manufacturing jobs in 1954, and it has almost exactly that many today, even though its population during the 50-year period has risen from 162 to 286 million. During the same period, the percentage of the American work force engaged in manufacturing declined from 30% to 15%.

If most Americans aren't farmers, miners, or factory hands anymore, what are they? The work force has about 130 million people. About 25 million are in manufacturing, construction, or resource production. About 20 million work for governments—federal, state, or local. That leaves about 80 million doing other things.

Like what? The three biggest categories, each with about 15 million workers, are retail trade, health services, and business services. The biggest subcategories in retail trade are general-merchandise and grocery-store jobs. The biggest subcategories in health are jobs in hospitals and doctor's offices. The biggest subcategories in business services are clerical and custodial jobs. Of jobs in other categories, the largest are the 8 million in places serving food

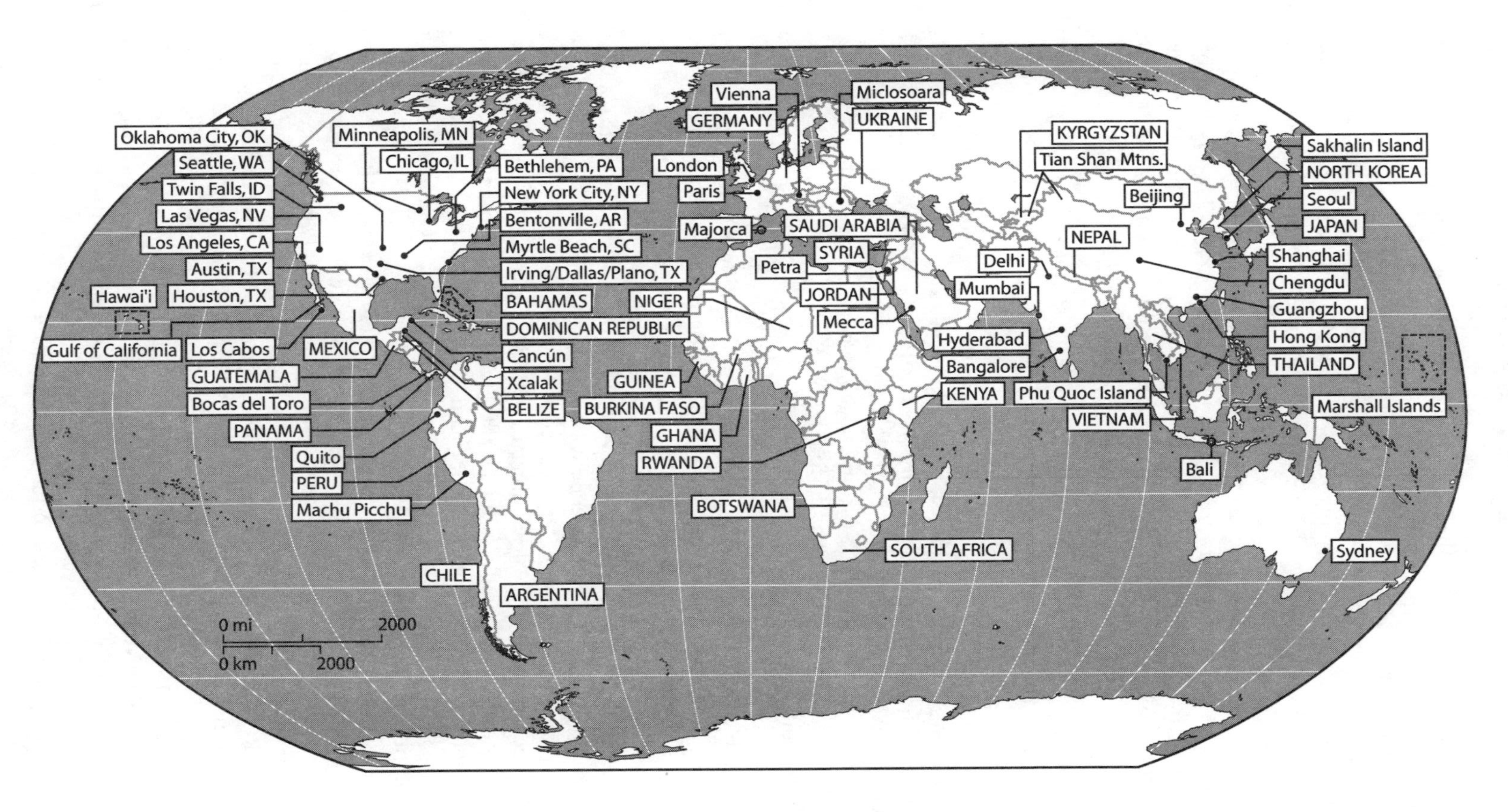

Oklahoma City, OK
Seattle, WA
Twin Falls, ID
Las Vegas, NV
Los Angeles, CA
Austin, TX
Hawai'i
Houston, TX
Gulf of California
Los Cabos
MEXICO
GUATEMALA
Bocas del Toro
PANAMA
Quito
PERU
Machu Picchu
Minneapolis, MN
Chicago, IL
Bethlehem, PA
New York City, NY
Bentonville, AR
Myrtle Beach, SC
Irving/Dallas/Plano, TX
BAHAMAS
DOMINICAN REPUBLIC
Cancún
Xcalak
BELIZE
CHILE
ARGENTINA
0 mi
2000
0 km
2000
Vienna
GERMANY
London
Paris
Majorca
NIGER
GUINEA
BURKINA FASO
GHANA
RWANDA
BOTSWANA
Miclosoara
UKRAINE
SAUDI ARABIA
SYRIA
Petra
JORDAN
Mecca
SOUTH AFRICA
KYRGYZSTAN
Tian Shan Mtns.
NEPAL
Delhi
Mumbai
Hyderabad
Bangalore
KENYA
Beijing
Phu Quoc Island
VIETNAM
Sakhalin Island
NORTH KOREA
Seoul
JAPAN
Shanghai
Chengdu
Guangzhou
Hong Kong
THAILAND
Marshall Islands
Bali
Sydney

and drink, the 6 million in finance and insurance, and the 4 million in transportation and warehousing.

Some kinds of service jobs are declining. The number of people employed in depository institutions such as banks, for example, fell during the 1990s by about 200,000. Thank the rise of the ATM and online banking. Also, the federal government has shed 400,000 jobs since a 1990 peak. (Don't worry: New state and local government jobs more than made up for the decline.)

Overall, however, the service sector continues to grow. One particularly fast-growing area is child care, where jobs have risen from fewer than 150,000 in 1972 to about 750,000 today. Measured by dollars, however, the fastest-growing major category is health. It accounts for about 13% of the United States' $10 trillion economy, up from about 4% in 1950 and heading to 16% by 2010 and perhaps 20 or 30% by 2030.

COMPETITION

As these numbers suggest, most Americans have been liberated from hard physical labor. The Brave New World of services isn't much kinder, though.

Consider Burger King. The company was created in 1954 and sold to Pillsbury in 1967. Pillsbury was acquired in 1988 by a British Company named Grand Met, and Grand Met merged with Guinness in 1997 to form Diageo, whose brand portfolio included not only Burger King but Guinness, Smirnoff, Johnny Walker, Bailey's, and Tanqueray. In 1997, Burger King sought to establish itself as having the best french fries in the fast-food world. (French fries, with 80% profit margins, are the most profitable part of the fast-food business.) A starch-coated fry was introduced with a $70 million marketing campaign. The new fry was too hard to make, however. Quality suffered, sales fell, and in 2002 Diageo announced that it was selling Burger King to an American company, Texas Pacific. Over the course of 15 years, the company had nine CEOs.

Despite its 30,000 restaurants worldwide, McDonald's announced that it lost money in the last quarter of 2002—the first time it had done so since going public in 1965. Month after month, sales kept declining. Determined to keep its market share, which hovers slightly over 40%—compared to 20% for Burger King and a bit over 10% for Wendy's—McDonald's spent some $400 million to build 13,500 "made for you" kitchens in the United States. Half that money came from reluctant franchisees, who had other uses for the money, but McDonald's burgers had traditionally been precooked and preassembled, then put in warmers until they were ordered.

The new kitchens weren't a success: customers were waiting 3 minutes or more for food that they complained still tasted like cardboard. Concerned about very public linkages between fast food and obesity, the company in 2003 announced a reformulation of its cooking oil to a new, secret blend of corn and soy oils with less saturated and transfats. The company was looking hard for alternatives to burgers, too. It experimented with a group of fast-casual restaurants including Donato's Pizza, Boston Market, and Chipotle, but they kept losing money, and in 2003 Donato's was sold back to the same businessman who had sold it to McDonald's in 1999. The company tried more diverse menus, too, but new items often failed. Both the 1991 McLean sandwich and the 1996 Arch Deluxe, as one journalist wrote, went "to their respective McGraves." Persisting, the company introduced sit-down McCafes both domestically and internationally.

Wendy's may look small alongside McDonald's, but it's the American leader in minimizing the drive-through wait. Wendy's is fast because the company-owned restaurants (not necessarily the franchise ones) have separate kitchens for the drive-through. The result is that the wait at Wendy's is 150 seconds, versus 167 at McDonald's. Everyone wants to cut the wait, ideally to 90 seconds, because drive-though business is growing three times faster than on-premises eating. That's why you see experiments with windshield transponders, so customers can prepay a lump sum and shave 15 seconds from the wait. Seconds are a lifetime in this business, where a six-second reduction translates to a one percent increase in sales.

That's also why you see drive-through experiments at Dunkin' Donuts and Starbucks. Even 7-Eleven is getting into the act, at least at its experimental store near company headquarters on Preston Road in Plano, just north of Dallas. Customers seem to like the convenience, and there are 22,000 7-Elevens waiting. The quickest drive-throughs, by the way, are those at Tim Horton's, a Canadian chain that's now part of Wendy's. Some of its stores are operating in the northern states. At the higher end of the scale, to-go food now accounts for one-tenth of the business done at Chili's. That's the most important brand of Brinker International, which also runs On the Border and Corner Bakery. Chasing Darden Restaurants, which runs Red Lobster and Olive Garden, Brinker's chief executive is thrilled by the growth in to-go food. He says: "not only are people not cooking, they're also substituting to-go and bringing that back to the kitchen without shame or guilt—and of course, we applaud that."

Until very recently, the franchise phenomenon in the last few years has been Krispy Kreme. It's been around since 1937, but it took off in the mid-1990s. There are only about 225 of these doughnut shops, compared with 5,500 Dunkin' Donuts, but each one costs its owner about $2 million. That's a lot, but Krispy Kreme revenues are extremely high—somewhere between $2 million and $4 million annually—and 20% of the revenue is profit. Store openings have been a phenomenon, with people camping out to be the first in the door, where they can watch doughnuts on a very visible production line. In 2003 the company opened its first overseas store, near Sydney, Australia. Mexico followed, as well as Canada. Other countries are in line, creating a franchise drama almost as interesting as the doughnut theater in the stores. Dunkin' Donuts is fighting back, partly by moving into the western United States, where it's sparse, and partly by introducing new products, including fancy coffee. Meanwhile, both companies have been hit hard by the low-carb diet craze and can only hope that it's brief.

For every success, the retailing landscape is littered with failures. Thirty years ago, the great merchandising chains in the United States were Sears, Montgomery Ward, and Penney's. They were American institutions. Now sales at Sears are only a third those at Wal-Mart. Montgomery Ward is gone. Penney's survives but is still closing stores, especially in the small towns where it once did so well. People drive longer distances than they once did. They go to bigger towns with bigger stores. Penney's has centralized its purchasing and distribution systems, and that has helped. It also has relied heavily on its subsidiary Eckerd, which it acquired in 1993 and which, with 2,600 stores, is the fourth largest drugstore chain in the country, after Walgreens, CVS, and Rite-Aid. The pharmacy business will probably grow faster than department stores, and Eckerd already accounts for almost half of all Penney's revenues. Still, there's so much competition from the other pharmacy chains—both Eckerd and Walgreens have almost 600 stores in Florida alone—that Penney's in 2004 sold Eckerd to CVS and a Canadian firm—and lost money on the deal. That will leave Penney's struggling in the difficult niche between upscale retailers and discounters.

Think of glamorous Hollywood and the fortunes made by actors there. But think again:

there are about 100,000 members of the Screen Actors Guild. Seventy percent of them make less than $7,500 a year acting. Ninety-five percent make less than $70,000 a year. That leaves 5% making more than $70,000—and fewer still making the big bucks. That's why from time to time you read about actors going on strike.

Things are hardball even at the top. During the 1990s, the biggest selling recording artists were Garth Brooks, the Beatles, and Mariah Carey—and the Beatles hadn't recorded anything since 1969. Carey's album *Glitter* had the wretched luck to be released on September 11, 2001. Sales were miserable—and she was let go from a contract worth almost $90 million. (Her consolation prize was a $28 million buyout.) Recording artists in general seem to have increasingly brief shelf lives. A good example is Alanis Morrissette, whose first album sold 28 million copies but whose second sold 5 million. Stars such as Rod Stewart, David Bowie, and Sinead O'Connor have been dropped. Executives bite the dust, too. The bosses at Sony, for example, weren't happy that their own salaries were one-tenth what they were paying the boss of Sony Music Entertainment, a money-losing subsidiary. Thomas Mottola decided in 2003 that it would be smart to resign before he was told to leave.

INNOVATION

Competition has produced entirely new kinds of service businesses. Consider the department store, which appeared simultaneously in the United States and France in 1852, when Marshall Field opened in Chicago and the Bon Marché opened in Paris. For decades, these stores offered the height of both style and convenience. They also offered fixed prices, something new in 1852. Department stores still strive for the upscale image, which is why they do not provide shopping carts. (Speaking of innovations, those carts were invented in 1937 by Sylvan Goldman, who ran a chain of grocery stores in Oklahoma City. Compare the fate of companies that use them, like Wal-Mart, with those that don't, like Sears.) Now, however, there's plenty of style in specialty stores. Department stores seem positively inconvenient, not only because you have to carry everything but because shoppers have grown impatient with stores that deliberately confuse you in the hope that you will get lost, wander around, and buy something on the spur of the moment. (The only commercial environment that is intentionally more confusing than the department store is the casino, whose owners want gamblers to forget the outside world.)

Seeking to cut costs, department stores have made things worse by cutting staff. They've also done a lot of corporate merging, which has stripped them of individuality. The story is in the numbers. Department stores accounted for 5.5% of all American retail sales in 1990; by 2002, the number had slipped to 3.3. In 2004, Target decided to sell Marshall Field.

True, department stores are trying to reinvent themselves. A well-known case is England's Selfridges, established in 1909 and a long-time model of British stodginess. Now it rents half of its main London store to designers who decide how to stock their space and who pay Selfridges a percentage of their sales.

American chains are watching and following. Between 2002 and 2004, Federated Department Stores spent $300 million renovating its 120 Macy's stores. Marshall Field in Chicago has introduced 30 vendor-operated boutiques in its flagship State Street store. Similar overhauls are happening at Saks Fifth Avenue, Bergdorf's, and Bloomingdale's. Will it pay? Nobody knows, but there are critics who think that the restoration of walnut balustrades makes more sense for a retailing museum than a successful store. The innovations that have proven most successful in a laboratory store operated by Federated in Columbus, Ohio, have been simple things such as directional signs, fitting rooms with a place for spouses to sit—and shopping carts.

In the 1980s, it seemed as though the future belonged to category killers. These are the big-box stores like Toys 'Я' Us that specialize in one class of item. Lately, though, many of the category killers have gotten into trouble, mostly at the hands of the Beast of Bentonville. Wal-Mart's share of the toy market, for example, rose from 13% in 1994 to twice that much a decade later. Toys 'Я' Us is surviving, but other big sellers such as FAO Schwarz and KB toys have gone into bankruptcy. Just For Feet went under in 1999, though now it's back in business as a unit of Footstar. Office Max and Office Depot have been hurting, too, though Staples has done well in the same niche, perhaps because its stores stand alone, while the other two tend to locate across the street from each other.

Home Depot is one category killer that has done phenomenally well. Started in 1988, it now sells about 30% of all home-improvement supplies in the United States. It has about 1,500 stores and plans on doubling that number. On the assumption that 50,000 people can support a store, Home Depot has built over 40 stores in Atlanta alone. Still, the company has grown so large so fast that problems have arisen. A new chief executive, appointed in 2000, said that "it wasn't like you had 1,500 stores; you had 1,500 businesses." He imposed tighter financial controls and replaced nine regional buying centers with one, in Atlanta. There's no time to rest. Lowe's has 780 stores and more appeal to women. Its sales in 2003 were projected to grow at an annual rate of 18%, compared to 8% for Home Depot.

Nobody threatens Wal-Mart. Established in remoter Arkansas in 1962, the leviathan grew slowly at first. In 1970 there were 38 stores and 1,500 employees. Today there are 1,600 Wal-Marts, 1,100 Supercenters, 525 Sam's—and that's not counting 1,100 stores outside the United States. The company tries to open one new store in the United States every day and about one every 3 days overseas. Along the way, it pushed K-mart, with 2,100 stores, into bankruptcy. The companies both began in 1962; by 1990 Wal-Mart was the bigger of the two, and its revenues are now almost nine times those of K-mart. For the moment, Target is doing well, but it feels the hot breath on its neck and knows that it can't survive if it's kinder and gentler. It tells its own suppliers to wait 60 days for payment—that, or sell to somebody else.

Sam Walton's company now employs 1.3 million people and plans on adding another 800,000 by 2007. Each week, Wal-Mart's 4,750 stores have 100 million customers. Sales in 2001 hit $226 billion; the next year they rose to $245 billion. Daily sales often exceed $1 billion; peak days have hit $1.5 billion. A quarter or more of all the toothpaste, dogfood, and disposable diapers sold in the United States are sold by Wal-Mart. Every hour, the company sells 20,000 pairs of shoes.

Private labels like Great Value, Sam's Choice, and Special Kitty are still a small part of Wal-Mart's business—less than one-tenth—but they're growing. Some are strategically important. No Boundaries cosmetics and the Mary-Kate and Ashley clothing lines are aimed at teenagers, for example, a crucial age group for any company with long-term plans. Wal-Mart is experimenting with lots of other things, too. For a time it even sold used cars at fixed prices at some Houston stores. One consultant says, "The world has never known a company with such ambition, capability, and momentum."

Wal-Mart has prospered by buying and selling cheap. About 200 of its suppliers have offices right next to the company's headquarters in Bentonville, Arkansas. For those that don't, Bentonville has the only airport in Arkansas with nonstop service to New York City.

Either way, the suppliers enter a spartan negotiating room—blue, with a portrait of Sam Walton—and meet with a buyer. One survivor says: "We were grapes, but now we are raisins. They suck you dry." The owner of Lakewood Engineering, a manufacturer of box fans, says that Wal-Mart's prices forced him to stop making fans in Chicago. Instead, he makes parts in China and assembles them in Chicago. He's not happy. "My father was dead set against it," he says: "I have the same respect for American workers, but I'm going to do what I have to do to survive."

Suppliers rue the day they tried resisting. Rubbermaid, long an American fixture in the provision of household articles, raised the price of its products to reflect higher resin prices; Wal-Mart responded by turning to a supplier who did without resin. To avoid bankruptcy, Rubbermaid was forced into a merger and became Newell-Rubbermaid. Its products were meanwhile shoved to lower shelves at Wal-Mart. A new CEO at Newell-Rubbermaid set out to repair the damage. He explained, "if we don't win at Wal-Mart, we can't win."

Suppliers say that Wal-Mart negotiates one price and sticks to it without add-ons for displays, damages, handling, and rebates. An impressed senior employee at the new Rubbermaid puts it this way: "It's very pure. . . . They'll negotiate hard to get the extra penny, but they'll pass it along to the customer." In fact, they'll pass it on to the customer before they've paid Rubbermaid: 70% of Wal-Mart merchandise is sold before Wal-Mart pays the supplier.

So many merchants have closed their doors after Wal-Mart's arrival that the company has picked up a range of nicknames including not only the Beast of Bentonville but the Merchant of Death. Still, consumers benefit, even the ones who refuse to shop at Wal-Mart. That's because the surviving stores in the community cut their prices by an average of 13%. A dramatic example is jeans in Britain, where the brand Wal-Mart sells—George—now sells for an average price of $8 a pair, down from $27. Multiply this across the span of merchandise and you see why Wal-Mart has become a significant factor both in keeping inflation low and in raising manufacturer productivity.

Perhaps Wal-Mart's greatest triumph is that after only 13 years selling groceries, it is the nation's biggest grocer. Shoppers can cut their food bills 30% at Wal-Mart, partly because Wal-Mart accepts thinner margins than supermarkets and counts on customers walking to the other side of the store and buying higher-margin, nonfood items. It's a winning formula and explains why the major supermarket chain Winn-Dixie announced in 2002 that it was abandoning the Texas and Oklahoma market at the periphery of its Southern core. Nationally, the rule of thumb is that for every Wal-Mart superstore that opens, two conventional supermarkets go out of business. In states where supermarket employees are unionized, 200 union jobs are lost with every new Wal-Mart superstore.

Perhaps the biggest threat to Wal-Mart is local governments fighting with the support of labor unions to save jobs: Sprawl-Busters, a Massachusetts consulting firm, calculates that 164 towns have barred the company. The company has yet to enter Chicago, for example, though it has a site—a former steel mill—and support from the mayor. The president of the Chicago Federation of Labor, however, bluntly calls Wal-Mart "Public Enemy No. 1 in the eyes of labor."

Wal-Mart isn't the least chastened. In 2003 the company was planning on building 40 supercenters by 2007 in California. It opened one in La Quinta and had others coming along in Stockton and Hemet. Inglewood City, near Los Angeles, however, passed a law barring stores that exceeded 155,000 square feet and sold more than 20,000 food items. The law didn't mention Wal-Mart but applied to nobody else. Wal-Mart sponsored a measure in Inglewood City that would have bypassed the city's rules, and it spent $1 million pushing for passage of the measure, which failed 7,000 to 4,500. A union spokesman replied that the voters had put "Wal-Mart on notice that L.A. County is not Arkansas." A Wal-Mart spokesman responded by saying, "we are not going to get pushed around or bullied by unions. We are here to state our case, and we are not going to go away quietly."

California's labor unions know that two-thirds of Wal-Mart's employees get no health benefits, work at the minimum wage, and counting wages and benefits earn $10 an hour less than they do. Their own employers—companies such as Albertsons, Kroger, and Vons—are getting ready for Wal-Mart by cutting benefits. That's why 70,000 California workers went on strike in 2003. The strike was settled the next year, mostly because workers couldn't hold out any longer. Existing workers would be protected, but new ones would have lower salaries and poorer benefits. One of the strikers said: "The middle class is just becoming the lower class."

A long-time employee of Ralphs said of the new workers, "they won't be able to have a career in this kind of job any more." She herself had had "a good life with the income, the health benefits and the retirement." A Safeway executive defended the cuts by saying, "if we don't change, you bet we'll lose jobs—and it will be in the thousands." Wal-Mart meanwhile replied that it is a good neighbor and that shoppers want low prices. The company became a sponsor of National Public Radio, and its spokespersons ceased making belligerent comments, such as this one from 2002: "The union is a business, and it's a declining business." Safeway employees in Washington, D.C. could see what was happening as they came into contract negotiations in 2004. A cashier with seniority could make $45,000 a year at $17.66 an hour—double that on Sunday. But Safeway lost $828 million in 2002 and $170 million in 2003; the future almost certainly held a two-tier wage scale here, too, with an end to medical and pension benefits fully funded by the company.

Not surprisingly, IGA stores are in trouble. During the 1990s, the number of these independently owned supermarkets fell from 2,700 to 2,000. Ace Hardware, with about 5,000 independent outlets, seems to have settled on service as a survival strategy. Conventional supermarkets will have to make comparable adjustments if they hope to survive the arrival of Wal-Mart. They can't beat Wal-Mart on price and will have to offer either better service or better selection.

They face huge risks, because innovation in the service sector is as risky as in any other sector of the economy. Remember the tale of Gap. It aims at customers in their twenties and had almost $12 billion in 1999 sales. Then sales slowed and the company responded by stocking trendy merchandise. Customers walked. Sales tanked. The company's problems weren't helped by its budget stores, operating under the name Old Navy. Gap customers started thinking that Gap quality had deteriorated to the budget-brand level. Some went to the company's upscale unit, Banana Republic. Others went down to Old Navy, then drifted outside Gap's group of stores and started buying from still cheaper Target or Kohl's. The company lost money in 2001 and its CEO departed, later to reappear as boss of J. Crew, a smaller company. His replacement, fresh from running Disney's theme parks, saw a small profit in 2002 but was selling into a poor economy. Looking back, the company's chairman said that "it took us 30 years to get to $1 billion in profits and two years to get to nothing." The store was running 3,000 stores worldwide and having to cope with different tastes in different markets—different color preferences in Germany, for example, and a different body shape in Japan. This time, the company coped well; by 2003 sales were up to almost $16 billion.

GLOBALIZATION

The overseas push of these retailers is now almost notorious. It's not just fast food, either. Starbucks has about 1,700 overseas stores, compared to 4,300 domestically. (The numbers change fast with a company opening 1,300 stores every year.) There are Starbucks shops in Vienna, where the coffee house is an institution. There is a Starbucks in Beijing's Forbidden City, where the ghosts of the Manchus occasionally sip a latte. With annual sales of $4 billion, Starbucks sees itself, according to its chief executive, not as a coffee company serving people but as a people company serving coffee. Its atmosphere is said to appeal to the aspirational side of Americans, always looking for something better. Will it work overseas? As of 2003, the foreign stores weren't profitable. Still, Starbucks has little choice, because investors expect continued growth, and the U.S. market is almost saturated. In Manhattan alone, Starbucks operates 168 coffee houses.

As for fast food, KFC operates in about 80 countries, including Saudi Arabia, where it has a restaurant in Mecca. Along with Pizza Hut and Taco Bell, KFC is part of the world's largest restaurant chain. Sold by PepsiCo in 1997, it was called Tricon for a while; now it goes by the odd name of Yum Brands. Call it what you like, it's huge: 30,000 restaurants, if you include the recent acquisitions of Long John Silver's and A&W. Yum plans to open 1,000 restaurants annually, indefinitely. It already has 300 KFCs in Thailand, where during 2001 it also opened almost 90 Pizza Huts. Since the late 1980s, Yum has opened about 850 KFCs in 120 Chinese cities, and it's aiming at 5,000 within the next 20 years. To get things right when it entered China, the company insisted on company owned stores. The first franchisees were allowed to set up shop only 4 years later, and the company insisted not only on a million-dollar investment but also on the franchisees taking over company-owned restaurants and staying clear of the big cities. KFC is now the biggest fast-food operator in China, and though its prices are high by Chinese standards, the restaurants are flourishing. It's a common sight in China today to see long lines of customers at McDonald's and Pizza Hut, while Chinese restaurants next door are almost empty.

Supermarkets and discount stores are following the same international path. Italy used to be the kind of place where you went one place for bread, another for vegetables, a third for eggs. You can still do it, but the law requiring product separation has been repealed, and the country now has 6,500 supermarkets. They sell as much as the country's 193,000 small shops. The number of supermarkets is up from 3,700 in 1996, and the number of small shops has declined from a quarter million in 1990.

Wal-Mart now gets about 16% of its sales from overseas stores. Ten years after entering Mexico, it has over 600 stores there, accounts for half of all Mexican supermarket sales, and is the country's biggest private employer. Wal-Mart's Mexican stores are supplied by 10 very efficient distribution centers, and its competitors, chiefly Controladora Comercial Mexicana ("Comerci"), Grupo Gigante, and Soriana, have been trying to get government permission to pool their purchasing.

Wal-Mart operates 22 stores in China, too, including 5 in Beijing. It isn't alone there. Carrefour, a French company that operates hypermarkets in many countries, including 31 in China, has 3 in Beijing. Tesco, a British chain, is planning on entering the Beijing market with Wal-Mart supercenter equivalents. There are local chains, too. The largest is Lianhua Supermarket Company, which has 1,600 outlets, though many are convenience stores. China's modern retailers, including hypermarkets, supermarkets, and convenience stores, now handle more than half the consumer goods sold in Beijing, Shanghai, Guangzhou, and

Chengdu. The goods are overwhelmingly domestic, but the stores have sections with imported products.

Globally, the retail landscape is evolving from one dominated by domestic companies to one dominated by American ones. An advanced case is Canada, which is a relatively easy market for American companies. Famous Canadian retailers such as Eaton's have gone under, and many national enterprises quickly sell out as soon as American competition arrives. Cases in point? Best Buy said that it was going to open a Canadian division, and Canada's Future Shop promptly sold itself to the American company. The same thing happened when Home Depot arrived: Aikenhead's lowered the blinds and sold itself. In both cases, the American companies were so big that they could buy at lower prices than the Canadian firms.

India remains the one great Asian market that continues to exclude the global retailers. Domestic chains are beginning to take shape, however, and they will put increasing pressure on India's tens of thousands of very small retail shops. India's major cities in the last decade have also opened shopping centers, and they're extremely popular, stocking plenty of international brands. India has grocery chains, too. One is called Food World. Though the stores are brightly lit and carry a range of products never seen before under one roof in India, they're small, perhaps twice the size of an American convenience store. The goods are almost all made in India but many carry American brand names such as Kellogg's, Tropicana, Coke, and Gillette. The multinational fast-food companies have also arrived in India. Their penetration is greatest in Delhi, which has 23 McDonald's, 17 Pizza Huts, and

The local shopping center, the most popular place in Noida, suburban Delhi.

15 Dominos. There's only a single Subway in town so far, but it reports 500 customers daily.

There are exceptions to the pattern of American dominance in global retailing. Hilton is British, except for the American hotels of that name. On the food side, the Japanese own 7-Eleven. The Thai government a few years ago announced plans to open 1,000 Thai restaurants in the United States, including a fast food chain called Elephant Jump. A Dutch Company, Royal Ahold, is now the fourth-largest supermarket owner in the United States, where on the East Coast it operates Giant Foods and Stop and Shop. A Mexican bakery, Bimbo, owns Entenmann's cakes, Mrs. Baird's bread, Oroweat, and Boboli pizza crusts. And there's Pollo Campero, an exceedingly popular Central American fast-food operator specializing in adobo-flavored fried chicken. Aiming at Central Americans living abroad, it now has takeaway operations in Spain, Poland, and of course the United States, where the company operates in states with large numbers of Central Americans, such as California, Texas, and New York.

Some international commercial empires operate entirely outside the United States. Since 1987, for example, 2,000 "100-Yen" stores have opened in Japan. They're dollar stores, with Chinese slippers, Korean toothbrushes, Vietnamese floor mats, Turkish glassware, Portuguese bowls—and plenty of Japanese products. The privately held company that owns the stores, Daiso Industries, is looking to expand to Taiwan, South Korea, and Singapore. The company owner says, "My philosophy is that merchandising is a combat sport."

Or consider Tesco, the major British supermarket entering China: It already operates in Eastern Europe, Taiwan, Thailand, and Korea. It recently opened its first Seoul store, in the southwestern district called Youngdeungpo. The store is a seven-story monster with an in-house dentist and ballet school, but the company also pays close attention to Korean tastes, and it stocks such a variety of goods that it makes Tesco's U.K. stores seem provincial. Small shops account for three-fourths of all retail sales in South Korea, and there are big changes ahead for Korea if chains like Tesco and Wal-Mart can push that number down to where it is in the United Kingdom or United States, at about a quarter of all sales.

Not all international ventures succeed, of course. Sears flopped a few years ago in Chile. So did Home Depot, which has pulled out of both Chile and Argentina and now operates outside the United States only in Canada and Mexico. Even Wal-Mart has had trouble overseas, for example in Germany and Indonesia. Of all these markets, perhaps Japan has been the toughest. A few years ago, Merrill Lynch took over the failed Yamaichi securities firm. The securities business has not been a happy one of late in Japan, and Merrill has since closed its branch offices. That's on the consumer side. It's a different story with investment banking, where both Merrill Lynch and Goldman Sachs have done very well in Japan. When it comes to advising on mergers and arranging equity offerings, they now surpass the Japanese giant, Nomura Securities. They must see a strong future here, because Goldman in 2003 announced its intention of investing over $1 billion for a 7% stake in Sumitomo Mitsui Financial Group.

On the retail side, Sephora, a French cosmetics and perfume chain, pulled out of Japan after entering in 2000 with plans for 40 stores. Office Max and REI have also quit Japan. Japan is Starbucks' biggest foreign market—the company entered in 1996 and now has about 450 stores—but sales are slow, plans for 1,000 stores have been delayed, and some stores are being closed. Even Wal-Mart tried and failed a few years ago, but Japan is such an important market that Wal-Mart is back again, this time saying that it will adapt to local demands for quality and freshness rather than trying to impose American standards on the Japanese consumer. It will be a challenge, because Japanese consumers think that low prices indicate poor quality. If successful, however, it will have a huge impact on the organization of Japan retail businesses, 58% of which are family-owned. Starbucks meanwhile announced in 2003 that it was quitting Israel, where it had set up in 2001. Sales had been disappointing, partly because the company doesn't advertise much and partly because potential customers worried about security in an American icon.

Meanwhile, there's been a huge surge in the export of European and American service jobs. At the top end, investment banks are hiring Indian workers to help New York do financial analysis. Citigroup, for example, has 8,000 employees in India, including a small group in investment banking. J.P. Morgan and Morgan Stanley are both going to Mumbai and hiring junior analysts—compilers of financial data, financial modelers, balance sheet analysts—at a tenth the cost they'd pay back home. Besides, first-class office space rents for about $100 a square foot, per year, in the City of London. It's $10 in Bangalore.

A bit further down the salary scale, there's a $35 billion global business in outsourced software development, and over 150 of the Fortune 500 companies now outsource software development to India. Wages are low, and English is widespread. There is a growing number (though still a shortage) of trained information technology workers in India, and by 2001 they had replaced about 20,000 American programmers. Three years later, the figure hit 80,000. That's still only a small fraction of the 3 million software programmers in the United States, but the average pay of the survivors is declining, and fewer American students are choosing to study programming at university. Both India and China now have more computer science graduates than the United States.

Simpler jobs are being exported, too. Facilitated by consultants such as Houston's Technology Partners International, India's Wipro, Infosys, and Satyam each are paid hundreds of millions of dollars annually for data management for companies like American Express and British Airways. Call centers are moving overseas, too, especially to India, the Philippines, and Ireland. Dell, for example, has call centers in Austin, Nashville, Twin Falls—and Bangalore. Microsoft has contracted with Wipro for a call center there, too. The people who answer the phones are often trained to adopt an American accent and even a phony American identity, just in case a customer gets personal. (Do they fool anyone? Probably not many. Dell pulled some of this business back to the United States in part, it seems, because the phony accent annoyed callers.)

Still, in 2003 there were over 50,000 of these jobs in India alone. In Scotland, which has 40,000 people working in call centers, there are fears that a quarter will be lost between 2003 and 2008. How can it be otherwise, when an employer can cut labor costs 40% by moving to India? The stakes are high in the United States, too, where over 3 million jobs may be exported in this way by 2015. The Communications Workers of America are trying to resist the tide, but the lure of capable and steady employees happy to work for $200 a month is irresistible. In India, of course, the trend is great news for an ambitious and hopeful country. Indians who have already made the move to the United States, ironically, learned that in 2003 the U.S. unemployment rate for computer engineers was higher then the national average, and they counselled their own children to plan on a different career.

The work's not all heading to India. Metropolitan Life has its data entry work on dental insurance claims sent to Affiliated Computer Services of Dallas, which then sends the work to operations in Mexico, Guatemala, or Ghana. Affiliated Computer had revenues exceeding $3 billion in 2002. Its workers in Guatemala sit at long tables loaded with PCs. Paid by how fast they work, the clerks work furiously with their own personal keyboards while listening to music through headphones. Salaries average $250 a month, well above the national average. The jobs are popular, partly because they're safe in a country that's not.

Some of the clerks used to be truckers and remember when armed guards accompanied them on soda pop deliveries.

ADVERTISING

Much of this economic growth depends on advertising and its world of legally and morally tolerated lying. That's a strong word for which the customary euphemism is "puffery." We see it all the time, with claims for the world's best pizza or most comfortable shoes. Whatever you call it, this kind of exaggeration can be defended by saying that most consumers don't take it seriously. It can be attacked, however, not only because some consumers do take it seriously but because everyone grows cynical when surrounded by endless acts of polished deception. That cynicism isn't confined to advertising, either. It has spread to politics, partly because commercial advertising has inured us to expect lies and partly because the techniques of advertising are being applied by politicians. Consumers mute the tube; as voters, they stay home.

Before Albert Lasker transformed the business about a century ago, advertising had been little more than helping retailers place their copy in publications. Lasker began writing copy for those retailers. Going far beyond the announcement of factual matters, he sought to link products to our deepest desires. "Salesmanship in print," he called it. The goal is to bond consumers to products—to go beyond trademarks, as the advertising executive Kevin Roberts suggests, and create "lovemarks." Devotees of BMW, Coca Cola, or Sony should understand what he means.

Lasker worked for and eventually controlled an agency called Lord and Thomas, but a rival firm, the eponymous J. Walter Thompson, grew to dominate the industry. Starting in 1912, Thompson published periodic editions of a reference book called *Population and Its Distribution*. The purpose of the book was to help advertisers understand what we would now call market demographics. The techniques were primitive in comparison to the tools advertisers have today. In the 1926 edition, for example, Thompson attempted to map consumer purchasing power. At the time, income tax returns had to be filed by everyone making more than $1,000 in taxable income. So Thompson counted income tax returns and compared them to the census. Advertisers found it helpful to know that in New York City one family in eight was rich enough to file a return. It was the same in California, but in Texas the figure was one family in 26, and in Mississippi it was one family in 67. The book also organized the country into shopping areas—619 of them in 1926. An advertiser could look at the San Francisco area, for example, and see that it had 10 grocery chains with a total of 122 stores. This applied research was useful to Cream of Wheat, Libby's, and Swift, all of which were Thompson clients.

Thompson was also working on the emotional side of advertising. The firm had been sold in 1916, and Helen Resor, the wife of one of the purchasers, was perhaps the first person to link a product explicitly to sex appeal. "The skin you love to touch" was her slogan for Woodbury soap. Thompson in 1924 also became the first agency to use celebrity testimonials, with Queen Marie of Romania plugging Pond's cold cream. You couldn't get much tonier than that: The queen was a granddaughter of Queen Victoria, and she was as beautiful as she was vain.

J. Walter Thompson held its lead until about 1960 but is now part of the London-based WPP Group, which also owns both Young & Rubicam and Ogilvy & Mather. WPP is one of four conglomerates that dominate the industry. Another is New-York-based Interpublic, which owns a stable of agencies including McCann-Erickson and Campbell-Ewald. The third giant is Omnicom, also based in New York. The last is Paris-based Publicis. Each of the four contains competing units. MindShare, for example, is a part of WPP and works for Burger King. Mediaedge:CIA, another part, works for Yum Brands.

The advertising budget in the United States in 2000 exceeded $220 billion. About $60 billion of that went to television, $50 billion to newspapers, $45 billion to mailings, $20 billion to radio, and $13 billion to telephone directories. Run down the list of the hundred companies that advertise most heavily in the United States, and you'll recognize them all, from the leader, Chevrolet, which spent over $800 million in the United States in 2000, to number 100, poor Clorox, which spent only $115 million.

Advertising builds name recognition globally, too. The world list starts off with Procter & Gamble and its $4.5 billion advertising budget. That's almost twice as much as it spends at home. Coca-Cola doesn't spend so much, but it works hard overseas, with an ad budget of $1 billion there, more than three times the home budget. Mars, the candy manufacturer, spends three-quarters of its billion-dollar ad budget overseas. McDonald's also spends more abroad than at home. So do Gillette and Kellogg. There are some companies that seem less global in their ambitions. General Motors, for example, spends twice as much at home as abroad. Other stay-at-homes include Altria (Philip Morris), Walt Disney, and Time-Warner. Foreign companies have to make the same strategic choice. Daimler-Chrysler spends more in the United States than it does in all the other countries in which it operates, including Germany. The United States is such a huge market that of the world's hundred largest advertisers, only five don't advertise in the United States. They are Peugeot-Citroën, Renault, Carrefour, France Telecom, and Vodaphone.

Think you're immune to the hucksters? Not likely. The chemistry is subtle, and the big agencies have given it a lot of concentrated thought. It's probably no exaggeration to say that Coca-Cola is loved at least as much as any celebrity. A world without the name Chevrolet would seem oddly silent. It's all part of the advertising goal of making consumers like the product or its maker. Consider the experience of a teenager walking into an Abercrombie and Fitch store. The loud music interposes itself between teen and parent. Having broken down that line of communication, the music also suppresses private thoughts. Meanwhile, the walls are covered with huge photographs of seminude older teens and a profusion of canoes and fake moose heads, as if to say that the teen is now with a group of rich friends with lots of cool stuff. The boy or girl is invited to become part of this consumer theater, whose doors are wide open and welcoming. What's for sale isn't shirts and pants; it's beautiful company. This is what it means to make the customer like you. The labels saying that the clothes were made in the Dominican Republic or Marshall Islands might as well be written in Sanskrit.

It's almost true that nothing sells unless it's sold—in the sense of marketed. That's why the push to understand the consumer is relentless. Omnicom runs Once Famous, a Minneapolis retail store that serves also as a merchandising laboratory. Customers are filmed so Omnicom clients can see how consumers react to new products. People in the business call it retail ethnography, and it's a pretty accurate label. Customers are told what's going on by a sign at the entrance, but most seem not to mind. Perhaps it's an opportunity to become a budget celebrity.

One of the biggest challenges for advertisers now is to expand megabrands without debasing them. L'Oréal, for example, is the world's biggest cosmetics producer, with brands including Lancôme, Helena Rubinstein, and Maybelline. It uses celebrity models like Catherine Deneuve and Andie MacDowell to make its products seem glamorous. It's locked in combat, however, with Procter & Gamble, which markets Noxzema, Max Factor, Clairol, and Pantene. To compete, L'Oréal has to get on Wal-Mart shelves. Can it do that without losing its glamor image? P&G is also fighting Unilever. The brands in play are P&G's Olay versus Unilever's Dove. Unilever was hurt a few years ago when it started selling a cheap dishwashing liquid under the name Dove; now it's experimenting with Dove antiperspirant.

LEISURE

The ultimate purpose of all these services is to make us happy, so it's not surprising that travel and leisure is the biggest industry of all, with annual sales of $450 billion just in the United States. Americans took almost 1 billion person-trips in 2000. (Definition: a person-trip is one person traveling 50 miles from home and/or staying overnight.) Only 14% of those trips were for business. The overwhelming majority were for shopping, outdoor activities, and visiting museums and historic sites.

There's no better place to start than with Disney, whose annual revenues exceed $25 billion. A part of that comes from the company's amusement parks, which go back to the 1955 opening of Disneyland. It's a crowded market now. There are parks like Disneyland and Disneyworld, drawing people from around the world, but there are also parks aimed at a local or regional audience. A good example is Premier Parks, which operates some 35 parks around the country, including Six Flags. There are more admissions to these parks each year than there are Americans. That's why Disney has headed overseas. In Europe, there are three times as many people as tickets sold each year, so there's plenty of room to grow, at least according to the demographics. It may not be so easy. In fiscal 2003, Euro Disney lost $66 million. In Asia, Disney is investing $300 million in a park on Hong Kong's Lantau Island. The Hong Kong government is spending a whopping $2.8 billion in park and island infrastructure. Maybe it will be more successful.

Speaking of fantasy brings us to sports. Most of this is for spectators, like the 16 million who watch the average NFL game on television. That number explains why the NFL's 8-year television contract with ABC/ESPN, CBS, and Fox is worth $18 billion. Our devotion to watching these games, and the opportunity it creates for advertisers, explains why the Boston Red Sox were sold in 2001 for $700 million. (The team of purchasers was led by a commodities trader and a television producer.) It explains why Tiger Woods and Michael Schumacher each made more than $50 million in 2001. It explains Alex Rodriguez's 10-year contract worth $252 million and the New York Yankee payroll running at $200 million annually. Numbers like these aren't an American monopoly, either. British soccer clubs now take in more than a billion pounds annually through ticket sales and TV coverage. The team with the most revenue in the world is Manchester United, whose revenues of $340 million in 2003–2004 exceeded by $60 million the revenues of the runner-up New York Yankees. It's a long way from Turrialba, Costa Rica. That's where official baseballs are made in a plant owned by Rawlings Sporting Goods. One ball retails for $14.99. The workers who make them, by hand, get 30 cents.

Extracting dollars from leisure pockets can be as tough as anything in the business world. The United States has 25 million golfers engaged in the world's most relaxed competitive sport. That number is level, but the number of golf courses keeps going up. An extreme case is Myrtle Beach, South Carolina, in whose neighborhood there are 118 courses. Each of them annually needs, on average, 40,000 paid rounds to make a profit. The total number of paid rounds nationally, however, is declining.

Maybe foreign visitors can help. The United States had 50 million in 2000. Half were from Canada and Mexico, and only four other countries sent more than 1 million people. From the top down, they were Japan, Britain, Germany, and France. Where did these people go, once they were in the United States? The chief destinations, attracting over 5 million visitors each, were California, Florida, and New York. Hawaii and Nevada came next but had fewer than half as many visitors as the three leaders. Even the numbers at those secondary destinations were impressive, however, and Las Vegas is a phenomenon. In 2000 it had 35 million visitors—domestic and foreign. A quarter million are in town at any moment. Since 1990, the city has added over 60,000 hotel rooms. Luxor and the MGM Grand opened in 1993, and Bellagio followed 2 years later. Bellagio, costing $1.8 billion, represented the apogee of megahotel superluxury, but the Aladdin opened in 2000, and Bellagio's developer began working on a new project, La Reve, scheduled to open in 2005. Steve Wynn plans to spend $2 billion on this 2,700-room hotel.

Overseas, the numbers aren't so large, but the economic impact of travel and tourism is even greater than it is in the United States. Malaysia, for example, has 10 million tourist arrivals annually. Proportional to its population of 25 million, that's equivalent to well over 100 million visitors to the United States.

Mexico tapped the international market when in the 1970s it pushed development at Cancún and its 220-mile coral reef. Now Cancún has 46,000 hotel rooms and generates about 40% of Mexico's tourist revenues. The government plans a repeat performance around the Gulf of California. There's already a string of hotels at Los Cabos, the strip of land between the towns of Cabo and San José del Cabo. They sprang up after the completion in 1973 of the 1,000-mile transpeninsular highway and, 4 years later, an international airport. Now there are over 8,000 hotel rooms at Los Cabos, 800,000 visitors are expected annually, and the permanent population has grown to 100,000. Enter the planners of Escalera Nautica. They intend to spend $1.7 billion by 2014 on a chain of yacht stops providing some 10,000 more hotel rooms targeted at Americans. The future of Baja California Sur is at stake, if it is not already a foregone conclusion. The state has a population of 400,000, of whom about a quarter are American. Some are retirees; others, in places close to the border, like Rosarito, are commuters looking for relatively cheap real estate.

Eighty thousand jobs are supposed to be created by Escalera Nautica, but locals doubt that they will benefit. One shrimper commented to a reporter, "What are we supposed to do, sell gum to tourists?" Does the fisherman know what happened in Cancún? The former village of Cancún now has 700,000 people, most of whom have one of the unskilled jobs the resorts offer. They live in fenced-off squalor and have no access to the beach. A leader of Yuxcuxtal, a local organization that means Green Life, says "our entire ecology has been 'concessionized.' " Just to the south are the towns of Xcaret and Solidaridad, which have grown from 15,000 people in 1990 to 150,000. One big employer in Xcaret is Puerta Cancún-Xcaret, which owns an aquatic theme park that attracts 800,000 visitors annually, despite entrance fees approaching $50. Now, in partnership with Carnival Cruise Lines, the park's owners want to build a pier so an additional 800,000 visitors can come ashore.

Want unspoiled Yucatán? Try Xcalak—"sh-ka-lak"—near the Belize border. It's got a road now, and it's got Banco Chinchorro, a biologically rich reef. The government's given this stretch of the coast a name—the Costa Maya—but they're aren't any big hotels yet. You can go a bit farther, too, and try Bocas del Toro, off the northwest coast of Panama. It's a quiet place, calmed down from the days when it was a shipping port for United Fruit.

Closer to the East Coast markets, a South African firm, Sun International, recently spent $1 billion building a 2,300-room hotel in the Bahamas. It's called Atlantis; rooms average $250 a night. The hotel employs 6,000 people, about 5% of the work force of the Bahamas. Atlantis is the world's biggest hotel outside Las Vegas, and it's supposed to generate about an eighth of the nation's gross national product. There are other big hotels on the islands, too. Our Lucaya, for example, is owned by Hutchison Whampoa of Hong Kong and has 1,350 rooms. With over 4 million tourists arriving annually, the Bahamas now rely on tourism for four out of every five jobs. Bahamians can scarcely get to the beach, however, because almost all of it is now private.

Too far to go? During 2002 and 2003 about two dozen new hotels opened along the beaches of California's Orange and San Diego counties. Between them, they offered 2,000 rooms with rack rates averaging over $250 a night.

Europeans, too, head for the sun. That's why there are 100 summer flights daily just between Germany and Majorca. Those with a taste for more distant places can visit Vietnam's Phu Quoc Island, off the Cambodian coast. Fly in and catch a ride on the back of a motorcycle to a simple hotel fronting virgin beaches. You can go seriously upscale, too. AmanResorts offers very handsome accommodations at $500-a-night-plus. After starting in Indonesia, it operates widely now, including in the United States.

Or you can take a cruise. The venerable P&O (the initials stand for Peninsular and Oriental Steam Navigation Company) operates the *Grand Princess,* a cruise ship that cost $450 million and carries 2,600 passengers. In 2001 the company announced plans to merge with its bigger rival, Royal Caribbean Cruises. Such a merger would have pushed Carnival Corporation to number two in the industry. Carnival fought back. Not for nothing is it known in the business as Carnivore, and ultimately it pushed Royal Caribbean aside and, like a fish in the sea, swallowed P&O. Carnival now operates the *Conquest*, a half-billion-dollar vessel with room for 2,974 passengers. Carnival also owns the biggest European cruise operator, Costa Crociere, which is the owner of the *Costa Fortuna*, with a capacity of 3,470 passengers. It's a long way from the 1840s, when P&O ships carried the Royal Mail to Lisbon, Alexandria, and Calcutta.

You can try ecotourism. Central America is famous for it, but there are more exotic contenders. How about the Tian Shan, the heavenly mountains of Kyrgyzstan? Lodges await. Want to do more than look? Tourists in Thailand can work at elephant camps. Tourists in Bali can get involved in programs for endangered sea turtles. A Friends of the Reef program on the island's north coast helps protect the coral from fishermen who work with dynamite and cyanide. If you want to vacation really hard, see World Leader, a Japanese travel agency. It offers the opportunity to go to Bali and slog bags of brine up to drying ponds.

We're verging now on cultural tourism. Taybeh, an almost abandoned Arab village just south of Petra, in Jordan, has been gutted and redone for tourists. Or how about a visit to Machu Picchu, the lost city of the Incas? Sounds exotic, but you'll be one of 300,000 visitors every year. It's tempting to try going in the wet season (October to April), just to avoid the 2,000 people who visit every day in the dry. That's more than twice the population Machu Picchu had in its 15th-century heyday, when it was Inca Pachacuti's country retreat. If you want something in Europe, try Miclosoara, Romania, where you can visit the recently abandoned villages of Germans who have returned to the country their ancestors abandoned.

You can combine eco- and cultural tourism. Fly to Kopawi, 120 miles southeast of

This way to your room! An old house in Taybeh, Jordan, is now part of a hotel that sprawls across the village grounds.

Quito, and you can stay in simple but expensive accommodations offered by an Indian tribe seeking an alternative to oil money. You can try Posada Amazonas in Peru, too, or, over in Africa, Ill'Ngwegi in Kenya or Gudigwa in Botswana.

There's adventure tourism, like riding the *pororoca*, the Amazon's tidal bore. At the March equinox, the wave comes every 12 hours, is often 12 feet high at the river's mouth, and can be ridden upstream—though this is the record set in 2003—for an exhausting 34 minutes, 10 seconds.

There's industrial tourism, with visits—courtesy of the government of Ukraine—to the site of the Chernobyl nuclear power plant. A healthier variant may be tours of the old Bethlehem steelworks in Bethlehem, Pennsylvania. The five giant blast furnaces there were shut down on November 18, 1995. They're rusting and are likely to be dismantled unless somebody spends millions stabilizing them.

There's agri-tourism—"agritainment" in one chipper formulation. It's most common in Europe, but Ag Venture Tours takes groups by van around California's vegetable-growing coastal valleys.

You can even order up a tour to a rural corner of North Korea, arranged by Hyundai and including a daily fee of $100 per tourist, payable to the government of North Korea. (If you're a South Korean, you can book a tour to Pyongyang through a travel agency controlled by Rev. Moon's Unification Church. One condition of the tour is that you're not permitted to complain about the food.) Farther north, you can visit Sakhalin, most readily by flights from Sapporo. You can stay at the Dr. Zhivago Inn, which serves bear dumplings and cutlets. Syria meanwhile is encouraging tourists, though few come. Even reclusive Saudi Arabia a few years ago encouraged decorous and high-paying tourist groups.

We pick from this leisure buffet whatever appeals and we can afford. Limited as we may be, our plates are much fuller that those of our parents or grandparents. That's what you'd expect from a society dedicated to progress. For some, the trend is depressing. They look at Las Vegas and recall that phrase of H. L. Mencken, the curmudgeonly Baltimore journalist who once wrote that Americans have a libido for the ugly. Or they look at tourism abroad and think that it's ruinous to local cultures. Judging from history, however, we're not going to change course in any fundamental way. Civilizations don't do that. They're too fragile to risk swapping an ideology that works for another that might not.

CHAPTER 14

TRANSPORTATION AND COMMUNICATION

I mentioned at the beginning of Chapter 11 that I'd indulge myself with a whole chapter on transportation and communications. I could justify this by arguing that a country's economic geography is mirrored in its freight flows: If you know what moves where and when you know a lot about how people make a living. Really, though, I choose to discuss this subject at length because I spent many hours as a child watching trains. I still can't understand how people landing at London Heathrow can ignore the geography on show.

I'll take it up mode by mode, roughly from oldest to newest.

INLAND AND MARINE SHIPPING

Between the 17th and early 19th centuries, huge investments were made in canals, both in Europe and the United States. The key innovation was the development of locks. The Erie Canal, for example, opened in 1825. Forty feet wide and 4 feet deep, it had 83 locks that raised barges from tidewater at Troy on the east to about 565 feet above sea level at the Niagara River. The canal cut the time for freight shipments between Albany and Buffalo from 2 weeks to 6 days. (The course of the original Erie Canal is now a challenge to archaeologists, because it was heavily modified in the 1840s and again early in the 20th century, when it was rebuilt with fewer locks as the New York State Barge Canal.) The Pennsylvania Mainline Canal opened in 1834 as a competitor to the Erie Canal. Locks couldn't get across the steep face of the Allegheny Front, however. Everything had to be put onto the cars of the Allegheny Portage Railroad, which used stationary engines to raise and lower cars on a set of 10 inclined planes. It was so unwieldy that the canal closed 20 years after opening.

There have been plenty of other American canal projects. In 1862, as the American rail network was developing, the Illinois River Canal linked Lake Michigan to the Mississippi. A century later, and despite opposition from railroads and environmentalists, the 10-lock Tennessee–Tombigbee Canal opened in 1985. At a cost of $2 billion, it links the Tennessee River to Mobile Bay, shortens the barge distance between Knoxville and the Gulf of Mexico by hundreds of miles, and carries some 2 million tons of freight annually—mostly forest products and coal.

That may sound like a lot, but 1,000 times more freight—2 billion tons—is handled

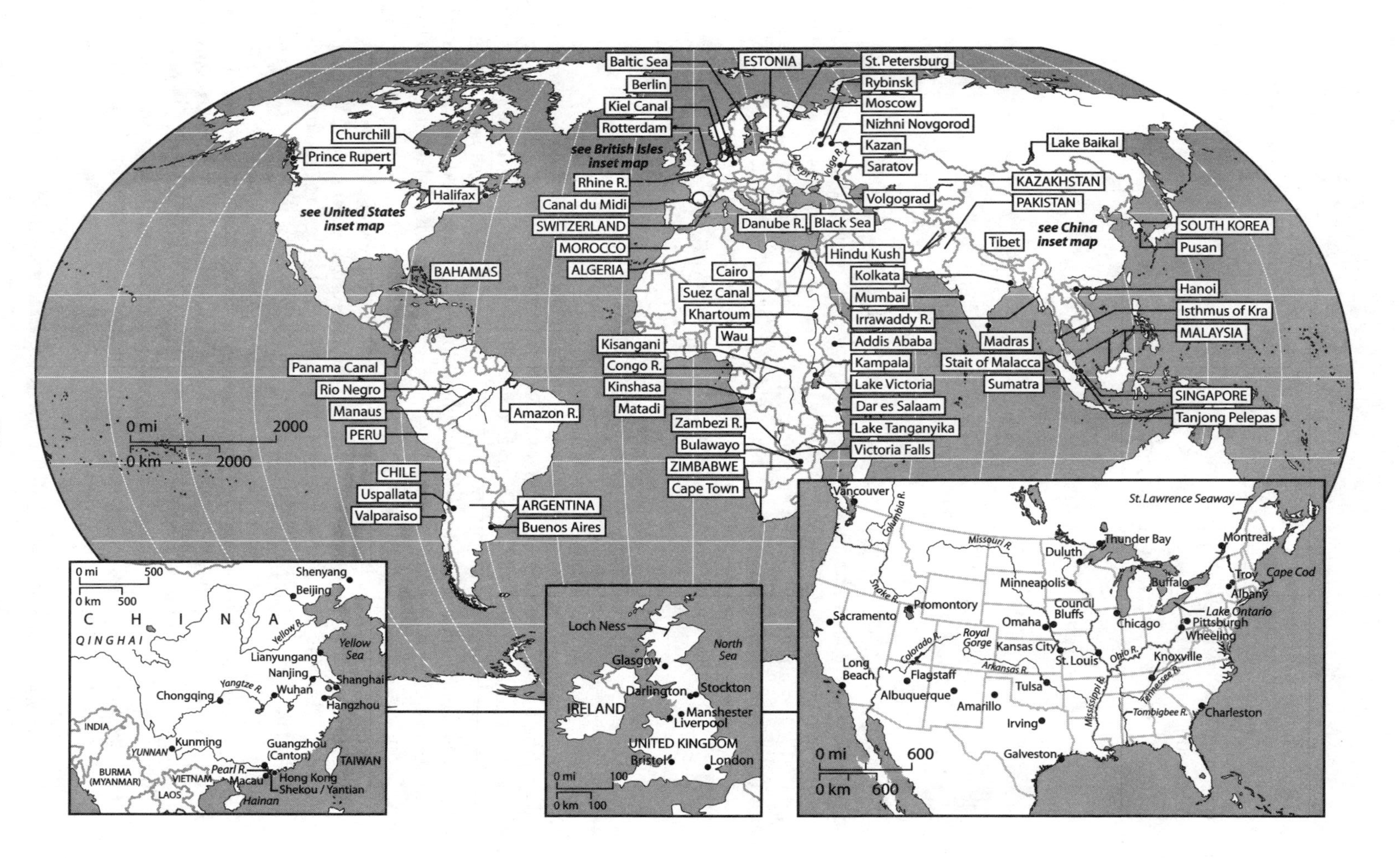

Churchill
Prince Rupert
Halifax
see United States inset map
BAHAMAS
Baltic Sea
Berlin
Kiel Canal
Rotterdam
see British Isles inset map
Rhine R.
Canal du Midi
SWITZERLAND
MOROCCO
ALGERIA
ESTONIA
Danube R.
Black Sea
St. Petersburg
Rybinsk
Moscow
Nizhni Novgorod
Kazan
Saratov
Volgograd
Volga R.
Dnepr R.
Lake Baikal
KAZAKHSTAN
PAKISTAN
Tibet
see China inset map
SOUTH KOREA
Pusan
Cairo
Suez Canal
Khartoum
Wau
Hindu Kush
Kolkata
Mumbai
Irrawaddy R.
Addis Ababa
Madras
Kampala
Stait of Malacca
Lake Victoria
Sumatra
Dar es Salaam
Lake Tanganyika
Victoria Falls
Hanoi
Isthmus of Kra
MALAYSIA
SINGAPORE
Tanjong Pelepas
Kisangani
Congo R.
Kinshasa
Matadi
Zambezi R.
Bulawayo
ZIMBABWE
Cape Town
Panama Canal
Rio Negro
Manaus
Amazon R.
PERU
CHILE
Uspallata
Valparaiso
ARGENTINA
Buenos Aires
0 mi 2000
0 km 2000
0 mi 500
0 km 500
Shenyang
Beijing
C H I N A
QINGHAI
Yellow R.
Yellow Sea
Lianyungang
Nanjing
Shanghai
Wuhan
Yangtze R.
Chongqing
Hangzhou
INDIA
YUNNAN
Kunming
Guangzhou (Canton)
TAIWAN
BURMA (MYANMAR)
VIETNAM
Pearl R.
Macau
Hong Kong
Shekou / Yantian
LAOS
Hainan
Loch Ness
North Sea
Glasgow
Darlington
Stockton
IRELAND
Manshester
Liverpool
UNITED KINGDOM
Bristol
London
0 mi 100
0 km 100
Vancouver
Columbia R.
St. Lawrence Seaway
Missouri R.
Thunder Bay
Duluth
Montreal
Troy
Cape Cod
Albany
Snake R.
Minneapolis
Buffalo
Lake Ontario
Promontory
Council Bluffs
Sacramento
Omaha
Chicago
Pittsburgh
Wheeling
Royal Gorge
Colorado R.
Kansas City
St. Louis
Ohio R.
Knoxville
Long Beach
Flagstaff
Arkansas R.
Tulsa
Tennessee R.
Albuquerque
Amarillo
Mississippi R.
Tombigbee R.
Charleston
Irving
Galveston
0 mi 600
0 km 600

annually by the 19-lock St. Lawrence Seaway, which allows vessels 700 feet long to sail from the Atlantic Ocean up the St. Lawrence and through the Great Lakes to Duluth. About 2,000 ships, loaded mostly with iron ore, coal, or grain, use the system annually, although only some of them traverse its complete length.

If you want an even busier canal, try the Kiel Canal, which cuts through Schleswig-Holstein and reduces the distance between the North Sea and the Baltic by 250 miles. A hundred ships use it daily—38,000 annually. It's a relatively new canal, delayed for many years by resistance from the German Navy, but it finally opened in 1895 as the Kaiser-Wilhelm-Kanal. After 1945, it officially became the Nord-Ostsee Kanal.

By then, histories of European canals had been written. One of the milestones in that history was the completion in 1681 of the Royal Canal de Deux Mers, which saw its name changed during the French Revolution to the Canal du Midi. Like the Kiel Canal, it provided a shortcut, allowing French vessels to pass from the Bay of Biscay straight to the Mediterranean. It was a masterpiece, not only with some of the earliest locks but with reservoirs to feed them, and it took England a century to catch up. It began to do so when the Duke of Bridgewater in 1761 opened a canal bringing coal from his mines to Manchester. Within 30 years, the Grand Union Canal linked the Midlands to London with a canal of 160 locks. The British tried their hand at shortcut canals, too. Thomas Telford in 1872, for example, finished the Caledonian Canal, which cut through Scotland at Loch Ness and bypassed the long and rough journey north around Cape Wrath.

Russia developed a very extensive network of canals linking the rivers flowing north into the Arctic Ocean to those flowing south to the Black and Caspian seas. The network continued to develop in the Soviet era, when great reservoirs were built along many of these rivers. The Dnepr and Volga became strings of ponds, across which tugs dragged barges from the Black Sea to the Baltic. The pattern of cities in European Russia reflects the course of these rivers to a very high degree: they're like beads on a string, especially along the Volga, which passes Rybinsk, Nizhni Novgorod, Kazan, Saratov, and Volgograd.

Canals continue to be dug, including one that recently linked the Rhine through the Black Forest to the upper Danube and Black Sea. In 1979, however, the last commercial barge passed through the Canal du Midi, and in 1996 that canal was designated a UNESCO World Heritage Site. It's symptomatic of the change that has come to many canals, where most of the traffic nowadays is recreational. In both England and France, tourists rent boats and slowly traverse the countryside. The Chesapeake and Ohio Canal was restored in the 1970s as one of Washington, D.C.'s urban amenities. Montreal's Lachine Canal, which closed in 1970, reopened in 2002 after the expenditure of $60 million to develop it for recreation. The upgraded Erie Canal is now used mostly by pleasure craft. It's sufficiently popular that a private developer in 2003 bought from the state of New York the right to build link canals into the state-owned one. His plan, to develop the property fronting these link canals, was put on hold after a scandal broke out over what appeared to be his sweetheart deal.

Don't count inland waterways out, though. If you measure by ton-miles instead of freight revenues, and if you include freight hauled not only on canals but on navigable rivers, then about 13% of America's freight moves by water. Most of the traffic is heavy, low-value stuff, especially coal, grain, and sand and gravel. Barges pass along the Mississippi as far upstream as Minneapolis, along the Ohio to and even beyond its origin at Pittsburgh; along the Missouri to Omaha; and along the Tennessee to Knoxville. Out West, the Columbia is navigable to the lower Snake. Among major American rivers only the Colorado is not used for navigation. There's not enough water in it.

The most important European inland waterway is the Rhine, which carries freight from Rotterdam upstream to Basel. Ships go up the Amazon as far as Peru, though traffic is minor above Manaus, at the mouth of the Rio Negro. The waterfall at Matadi, near the Congo's mouth, impedes transport, and the country's political mess makes things even worse. It's hard to believe that a century ago the Belgians were running steamboats up and down not only the Congo above Leopoldville (Kinshasa) but along a dozen of the river's tributaries, including the Bangui and Kasai. Nowadays, it's a small miracle when a barge appears at Kisangani (formerly Stanleyville); the captain explains that with engine problems it took him 3 months to make a 1,000-mile journey that should take 2 weeks. The old railway that circumvented the rapids near Kisangani is inoperable; freight is carried on bicycles.

Once, the British ran steamers through Sudan on the White Nile. These services are long gone, and presumably the vessels are rotting somewhere. In Asia, both the Irrawaddy (in Burma transliterated as Ayeyarwaddy) and Yangtze remain heavily used. The Yangtze in particular is very important. In 2004 China announced a $2 billion Golden Waterway Project to improve navigation on the river so that ocean-going ships could go upstream as far as Nanjing, fleets of 5,000-ton barges as far as Wuhan, fleets of 3,500-ton barges as far as Chongqing (above the Three Gorges Dam), and fleets of 1,000-ton barges across the Red Basin and as far as Shuifu County, in Yunnan's Zhaotong Prefecture. Quadruple-deck ferries are loaded at Chongqing with shiny new minivans and sent downstream to Wuhan.

Meanwhile, there are ships at sea, some 40,000 of them. The first steam-powered crossing of the Atlantic was in 1819, and steam soon cut transit time westward from 34 to 17

On the Yangtze below Chongqing: an hour earlier, the vans were parked on the riverbank; loaded quickly, they then set off to downstream markets.

days. Sails survived for a while: The famous China clipper Cutty Sark, for example, was launched in 1869 and once sailed 363 miles in a single day. But the Suez Canal opened in 1869, too, and it annihilated sailing ships, because winds in the canal are so calm that vessels literally cannot sail through it. The word "steamship," of course, is atavistic. By the 1980s steam had been replaced in the world's fleets by diesel power.

Oceangoing freighters today are of overwhelming importance in the handling of bulk commodities, especially oil and grain, but they are hardly less important in the long-distance shipment of manufactured goods. Very few of these ships, by the way, are American: Only 4% of the cargo entering and leaving American ports, for example, is carried in an American vessel. The figure would be zero if it weren't for the 1920 Jones Act, which requires that domestic freight—for example, between the West Coast and Hawaii or Alaska—is carried in American ships. The decline of the American fleet has a simple explanation. An American seaman makes $70,000 a year, while seamen from the Philippines, Indonesia, Bangladesh, and China are lucky to make $7,000.

Just 50 years ago, crates used to be piled in the holds of freighters, carried to their destination, hoisted out in rope slings, shoved by stevedores onto pallets, then forklifted into dockside warehouses. It was a show to watch, but goods were frequently damaged or stolen.

The show's over now, and so are most of the stevedores. In 1957 an American company, Sea-Land, introduced the revolutionary *Gateway City*, a ship built specifically to handle standardized steel boxes that could go directly from a ship's hold onto a railroad flat car. A decade later, in 1966, Sea-Land began offering container service to Europe, and by 1969 it offered service to Japan and Hong Kong. The great ports of the world now are all recognizable from a distance by the huge cranes that lift and stack containers, "cans" in trade lingo. Often, the ports have moved to new locations. Freighters bound for New York, for example, used to dock in Brooklyn, especially in and near Brooklyn's Red Hook. Now they head to New Jersey. It's the same in San Francisco, whose obsolete docks have been nearly abandoned as container ships unload more sensibly in Oakland.

Sea-Land is now part of CSX, the corporate descendant of the Chesapeake and Ohio railroad, but the trend it started is still evolving. The average container ship today holds 4,000 20-foot containers—or, more properly, 4,000 TEUs, 20-foot equivalent units. Some are able to carry more than 6,000. Shipbuilders in Korea and Japan had orders in 2003 for 100 ships that would each carry 8,000. These giant ships will cost about 25% less to operate, per container. That's a big enough savings to justify investment in the special deep-water docks that these ships will need. They are much too big for the Panama Canal, which means that Chinese exports to the United States will generally land on the West Coast and be carried east by truck or rail.

Americans like to imagine that they're in the forefront of technology. It's true that some 34,000 trucks pass daily in or out of Long Beach, the biggest container port in the United States. Long Beach is growing, too, with new megaterminals to handle the giant ships. (One of those terminals is being built on land formerly occupied by a giant Navy drydock.) To accelerate the movement of freight through the port, the Alameda Corridor was built at a cost of $2.4 billion. It eliminated 200 grade crossings and doubled train speeds across metropolitan Los Angeles.

Even so, Long Beach is less sophisticated than the world leaders. A crane in Singapore's billion-dollar Pasir Panjang terminal is operated by a crew of one, compared with a crew of four at Long Beach. The Asian facility will unload and load a ship in 40 hours, while an American crew needs almost twice that much time. The American crane operator will make $100,000 a year. His Singapore counterpart makes a fifth that much. Oddly enough, the superefficient port at Hong Kong, which stacks containers an amazing 14 high, is owned by CSX. Clearly, it's not American ignorance that's to blame. The CSX facility in fact includes the world's biggest multilevel industrial building, one that handles 10,000 trucks daily, each arriving with goods ready for shipment overseas. What's holding American ports back? Lots of things, including union rules, old facilities incapable of stacking containers very high, and terminals that only handle ships from one shipping company.

America isn't the only country behind the curve. Japan operates its ports about 18 hours a day, versus 24 in Singapore, Hong Kong, and Pusan, Korea. With inefficiencies like that, Japan's ports are getting left behind. Pusan, for example, handled 7.5 million containers in 2000, while Tokyo handled only 3 million.

Even the leaders can't rest easy. Singapore may handle 15 million containers annually, but it is so expensive that in 2000 Maersk Sealand International shifted its business—2 million containers a year—to the new, nearby, and cheaper port of Tanjung Pelepas, in neighboring Malaysia. The Taiwanese shipper Evergreen followed, moving its million-container-a-year business to the new port.

Hong Kong is facing similar pressures. Its port at Kwai Chung handles 18 million containers annually, more than any other in the world and almost a tenth of all container traffic in the world. Turnaround time is only about 10 hours, yet two serious rivals, Yantian and Shekou, are taking shape farther up the Pearl River delta. These two new ports (especially Yantian, which is on deep water) already handle more than half as many containers as Hong Kong, and they are projected to overtake it within the next 20 years, mainly because they charge shippers less money. A manufacturer who sends a container from Guangzhou to Japan through Kwai Chung, for example, will pay $1,800 in trucking and terminal fees. The same container going through Yantian will cost about $1,160.

Who owns these new ports? None other than Hutchinson Whampoa, controlled by Li Ka-shing. He also controls Hong Kong International Terminals, which handles more than half the containers moving through Hong Kong.

Li is looking ahead not only to the continued boom in Guangzhou but to the Yangtze, where Shanghai has risen from being the 20th most active container port in 1995 to the third most active in 2003, when it handled 11 million TEUs. It's likely to become the world's busiest port with the development of a giant new port at Yangshan, an island near the mouth of Hangzhou Bay. The Chinese are hard at work there, with plans to raise capacity by 3 million containers in 2005 and another 3 million in 2006. The target is 22 million TEUs by 2020. And farther still: Hutchinson has developed a container port in the Bahamas. The idea is to use it as a transshipment point for traffic flowing between China, Europe, and North America. The Freeport Container Port already has a capacity of 1 million containers a year, and it's growing; Maersk is already using it. Li Ka-shing has a competitor, though. Gordon Wu of Hopewell Holdings wants to build a bridge from Hong Kong's Lantau Island to Macau, where he plans a deepwater port. Not to be outdone, the city of Guangzhou has its own plans to dredge the Pearl River to a depth of 48 feet so ships can come to nearby Nansha.

Other major changes in marine transport are linked to climate change, particularly the melting of Arctic ice. The Hudson Bay port of Churchill is on an almost straight line from the Canadian prairies to Europe, but ever since it opened in 1931 it has been open only 3 months each year, starting in mid-July. In 1995 it was slated for permanent closure. Then a Denver-based company, Omnitrax, bought the whole port. Perhaps the company anticipated milder Arctic winters. In any case, there is certainly less ice in Hudson Bay now than there was in 1931, and Churchill is on its way to handling 2 million tons of freight annually.

This would match Thunder Bay, its competitor on Lake Superior. Thunder Bay still has a longer season than Churchill, but grain routed through Thunder Bay must follow a longer route to Europe.

Thanks to climate change, there are prospects now of both the Northwest and Northeast Passages opening to marine shipments. The Northwest Passage, running north of Canada, is the more difficult of the two because there are more islands en route, but both routes are shorter by 5,000 miles than the existing routes from Europe to Japan through the Panama or Suez canals. Who's unhappy? Canada for one, because it fears accidents whose damage will be hard to contain or ameliorate. Too bad: Canada claims sovereignty over these waters, but other countries consider them international.

Panama's unhappy, too. Like the Suez, the Panama Canal was an epochal achievement in the time of George Goethals, who pushed it to completion. Now, it's being enlarged at the Gaillard Cut so it can handle more than the 14,000 ships that already use it annually. The limiting factor is water from Gatun Lake, because every passing vessel requires the release and loss of 52 million gallons from the lake, whose watershed is suffering the effects of deforestation.

What about the Suez? It's a lock-free canal, and although it's too small for supertankers heading from the Persian Gulf to Europe, it's still used by some 16,000 vessels each year, mostly container ships. About 12 million containers pass through the canal annually, and some of that business will be lost if the Northeast Passage emerges as a practical route.

Beyond containerization and climate change, there's a third issue facing the shipping industry. It's piracy, particularly in Indonesia but more generally throughout Southeast Asia and around the east and west coasts of Africa. This is a sometimes deadly business, aimed at the theft not only of cargoes but occasionally of whole vessels. The Piracy Reporting Center counted 48 acts of piracy in 1989, 285 in 1999, and 469 in 2000. The number in 2001 fell to 335, but the number of whole-ship thefts doubled, to 16. The numbers rose in 2003, when the Center reported 445 incidents, including 19 ship hijackings. Indonesia came first, with 121 acts of piracy; it was followed by Bangladesh, with 58, and Nigeria, with 39. The danger of pirates in the Strait of Malacca is so great that Thailand, with support from China, began in 2004 to develop a land bridge—two ports and a pipeline—that would allow tankers from the Persian Gulf to unload on the west side of the Isthmus of Kra. Oil would be picked up on the other side by tankers continuing to East Asia, shaving 600 miles from the journey and bypassing the pirates.

RAILROADS

The heyday of American canals was brief, a fact neatly suggested by the coincidence that Europe's first railroad opened in the same year as the Erie Canal. The railroad was the Stockton and Darlington, in northeastern England. On its first trip, it hauled 36 cars 9 miles in 2 hours. Five years later, and over the stoutest objections of canal owners, the 31-mile Liverpool and Manchester opened. A year later, it carried almost half a million passengers, and the railroad boom was on. The London and Birmingham railroad connected those cities in 1837, and the Great Western—with its expansive track gauge, exceeding 7 feet—connected London to Bristol in 1841. European countries quickly adopted the technology and built rail networks, usually in a star-like pattern centered on the capital city and reinforcing its national importance. Moscow by 1910, for example, had 10 railroads converging on it from all directions.

In the United States, the Baltimore and Ohio was conceived as an alternative to the Erie Canal. It did not get to the Ohio River, at Wheeling, West Virginia, until 1852, 17 years after it began providing service between Baltimore and Washington. Things accelerated very fast in the 1850s. In 1853 the New York Central was formed by an amalgamation of ten railroads linking Albany to Buffalo, and a year later the Pennsylvania Railroad connected Philadelphia and Pittsburgh. By the time of the Civil War, a dense railroad network was operating east of the Mississippi.

In 1869 an American railroad reached the West Coast. It was built by the Union Pacific, or U.P., from Council Bluffs, Iowa, west to Promontory, Utah, where it met the Central Pacific building east from Sacramento. This was a Civil War undertaking, highly political, an instrument of union, but by 1883 a Southern Pacific route ran from Los Angeles to New Orleans. It was controlled by the same quartet who built the Central Pacific: Collis Huntington, Charles Crocker, Mark Hopkins, and Leland Stanford. They soon folded the Central Pacific into the longer railroad, and the name Central Pacific vanished.

A Northern Pacific route opened across the northern tier of western states, but the Big Four had more to fear from the Atchison, Topeka, and Santa Fe, which also opened in the 1880s and which for many years operated the only through route from Los Angeles to Chicago. The Santa Fe main line today still connects those cities with a line that runs east from Los Angeles through Flagstaff and Albuquerque, then turns northeasterly through Amarillo, Emporia, and Kansas City.

Throughout the 20th century, the history of American railroading has been a story of financial disaster and corporate consolidation. By the year 2000, all the nation's major railroads had been consolidated into four systems, functionally paired into two nationwide ones. In the East, the New York Central and Pennsylvania railroads had failed and been consolidated into a system called CONRAIL, which was then divided between the two surviving southeastern companies. One of them, the Norfolk and Southern, itself the product of many mergers, took over the Pennsylvania's lines. The other, the Chesapeake Southern, or CSX, also a product of mergers, took over the New York Central's. Between them, the Norfolk Southern and Chesapeake Southern cover the East. Out West, the U.P. swallowed the Southern Pacific, while the Santa Fe joined the Burlington Northern to form the BNSF. The Burlington itself had been formed by a merger of the Great Northern, Northern Pacific, and Chicago, Burlington, and Quincy.

By 2001, transcontinental freight services were provided by the two eastern giants working with the two western ones. The Union Pacific tended to coordinate shipments with CSX, while the BNSF paired with the NS. Freight cars had always been sent interline, but now it's almost as common to see engines from one railway on the distant tracks of another. (For more on the recent evolution of these systems, see Richard Saunders, Jr., *Main Lines,* 2003.)

Americans are inclined to think of railroads as an industry in decline. That's probably because they don't travel much by train anymore, but it's true also that track mileage in the United States has declined from a 1916 peak of 254,000 miles to about 170,000 miles. Entire lines have been abandoned, including the Milwaukee Road east from Seattle and the spectacular Denver and Rio Grande route through the Royal Gorge of the Arkansas River in Colorado. (A tourist train continues to operate on the 12-mile section through the Gorge proper.) More significantly, American railroads carry a smaller proportion of the nation's freight than they used to. At the 1916 peak and measured by ton-miles, railroads handled 77% of all intercity freight in the United States. Now the figure is down to about 40%. If you measure by revenues, the picture is even gloomier, with the railways handling only about 20% of America's freight. Out West, for example, in the area served by the U.P. and BNSF, trucks handle almost 80% of the business and leave the two railways to divide most of the rest, which they do about equally.

American railways remain very important in several freight categories, however. One is containerized shipments. This business took off in 1986, when the Southern Pacific launched double-stacked container trains running east from Long Beach. Soon the BNSF and U.P. were doing the same, with the BNSF in particular having a reputation for hotshot service to Chicago. Another category in which railways are doing well is bulk commodities like coal, ore, sand, chemicals, and forest products. Coal is the most important of these, with dozens of unit trains—their cars never uncoupled—going back and forth between Wyoming mines and power plants around the country. A third is automobiles, loaded on autoracks shielded against bored teenagers with rocks.

These categories dominate rail freight today. In the typical week ending June 12, 2004, 563,000 freight cars were loaded in the United States. Of those, 218,000 were trailers or containers. Coal came next, at 135,000 cars. About 61,000 carried ores and other minerals. About 40,000 carried agricultural products. About 35,000 carried chemicals. About 25,000 were autoracks. About 18,000 carried forest products, such as lumber and paper. That left only 16,000 other loads, which is a good indicator of the huge decline in the general freight business that railways have lost to trucks. Count the number of old-fashioned boxcars on the next freight train you see. The total will be surprisingly low.

How well railroads do in the future depends largely on how dependable they are. Historically, they have had a bad reputation among shippers, and there have been true fiascos in recent years. A good example occurred in 1997–1998, when the U.P. swallowed the Southern Pacific and almost ground to a halt as a traffic jam in Texas backed trains up across the system. BNSF's largest customer these days, however, is United Parcel Service (UPS), which pays BNSF about $350 million annually to carry its trailers on flatcars. That's a good indicator of the railroad's reliability, because UPS won't tolerate delays. UPS also does business with U.P. Every Tuesday a special train leaves Los Angeles loaded with UPS trailers full of packages to be delivered in New York City on Friday. To make that happen, other freight trains in the U.P. and CSX systems have to be sidelined, sometimes for hours, so the UPS train can stay on schedule.

The railroads also see potential in transloading, which amounts to containerized freight without the container. The BNSF, for example, plans to bring to Dallas each year about 1,500 flatcars loaded with Pacific Northwest lumber. The cars will be unloaded there and the lumber transferred to trucks that will distribute it to local builders. If the yard operates efficiently, costs will come down enough to steal business from long-haul truckers.

It's hard to imagine railroads not continuing to play a major role in moving America's freight, but horror stories keep bubbling up. The U.P. in April 2004, for example, was so short of train crews that it asked UPS to temporarily stop using its services for the Los Angeles–New York run. Fearful of losing this prime business permanently, the U.P. agreed to reimburse UPS for the added cost of using trucks to make the cross-country journey.

What about railroads elsewhere in the world? Canada is traversed by both the Canadian Pacific and, farther north, the Canadian National (CN). Eager to enter the American market, the CN bought the Illinois Central in 1999 and instantly had a track to the Gulf of Mexico. It tried a merger with the BNSF, too, but this was opposed by shippers and put on hold when the U.S. Surface Transportation Board announced a moratorium on railway mergers. Sooner or later, however, such a combination is likely to take place, particularly as NAFTA is increasing freight shipments between the two countries. Don't worry about

Canadians running America's railroads, however. If anything, it's the other way around, because most CN stock is already owned by Americans.

The Canadian railroads are meanwhile trying to take a share of America's containerized freight. The CN's port at Prince Rupert in northern British Columbia is considerably closer to Asia than California's ports. The CN also provides service to Halifax, Nova Scotia, which is closer to Europe than any American port. In theory at least, the CN could offer faster connections to both continents than any American railroad.

In South America, 19th-century railroads traversed Argentina's grain-producing pampas and Brazil's coffee country. In 1910 a transcontinental line connected Valparaiso, Chile, to Buenos Aires, 900 miles to the east. The transandine section was a spectacular achievement, with a 2-mile-long summit tunnel at an elevation of 2 miles, but passengers had to change trains at either end of the mountain section, which was narrow gauge. In 1982 the mountain section was abandoned, and traffic between Chile and Argentina now goes by highway over approximately the same route. The arid summit town of Uspallata has faded away, though it was briefly revived in 1996 when it became the movie set for *Seven Days in Tibet.*

In Africa Cecil Rhodes dreamed, he wrote in 1900, of a Cape-to-Cairo railroad "to cut Africa through the centre." He continued optimistically: "the railway will pick up trade all along the route." Track even then reached north from Capetown to Bulawayo, Rhodesia (now Zimbabwe), and Rhodes foresaw correctly that it would continue north and cross the Zambesi near Victoria Falls. "I should like to have the spray of the water over the carriages," he wrote. The line did ultimately continue north as far as Lake Tanganyika and the coast at Dar es Salaam. Meanwhile, the British built another railway south from Cairo as far as Wau (pronounced wow) in Sudan. That left a gap of almost 1,000 miles, and it has never been bridged. Apart from South Africa and the French-settled parts of Morocco and Algeria, most of Africa remains barely served by railroads. Short lines reach inland from ports to capital cities like Khartoum, Addis Ababa and Kampala, but they don't form a network.

In Asia, railroad development was more thorough, especially in India. The British government offered capitalists a guaranteed 5% return on approved Indian lines. On this basis Bombay, Calcutta, and Madras all got railways in the 1850s, and by 1869 almost all of India's big cities were on the network. In 1879 the government bought the East India Railway, the largest of the private companies, and as contracts expired it bought the others. After 1925 it took over their operation as well as their ownership. By 1910, India had more track than the rest of Asia put together, including Siberia and China; only the United States, Russia, and Germany had more. This was prepartition, of course, so the Indian network extended into what is now Pakistan and, to a lesser extent, what is now Bangladesh, where huge rivers were an impediment.

The Indian system was never linked to Europe, even though many surveys were run. The cheapest route would probably have been through Russia and Afghanistan and would have crossed the Hindu Kush by a 13-mile tunnel, but the British feared that this could be a pathway for invasion. A more expensive route ran through Turkey and Iraq, but in British eyes it might have helped Berlin as much as London. Plans foundered on such strategic considerations, even though by World War I both Russian and Indian railroads approached the Afghan border.

East of India, rail networks have always been very sparse, with separate networks in Burma, Thailand, and Vietnam. Railway development was especially slow in China, where a weak government and competing foreign investors delayed the development of the network for 50 years after the establishment of the Indian one. The important Pinghan railway, financed by British and Belgian investors, finally opened in 1905 and connected Beijing to the Yangtze, 750 miles to the south at Wuhan. In 1907, the British and Chinese Corporation opened the Peining railway, which ran from Beijing to Mukden (now Shenyang), in the Northeast. It also built a railway from Shanghai northwest to the then capital, Nanjing, and a railway in 1909 south from Shanghai to Hangzhow. Meanwhile the French in 1910 opened the Yunnan railway, which connected Kunming to Hanoi. Still, there was no north–south arterial from either Beijing or Shanghai to Canton (Guangzhou). Those crucial links weren't completed until 1936, when the Yuehhan railway opened from Wuhan to Canton. It was funded indirectly by the British, who in 1922 forgave Chinese debts arising from the Boxer Rebellion and recommended using the funds on the railway.

Today, railways carry about half of China's freight over a 47,000-mile system. As it is in so many other things, China is investing aggressively in railroads; it's spending about $8 billion annually on them, not only with new lines such as an ambitious one from Qinghai to Tibet but in upgrading old routes to handle both freight and passenger trains at higher speeds.

Far to the north, there's the famous trans-Siberian. Czar Alexander had been shocked by the completion of the Canadian Pacific. Thinking of Siberia, he wrote in 1886 that he had to admit "with grief and shame that until now the Government has done scarcely anything towards satisfying the needs of this rich but neglected country. It is time, high time."

A handbook published in 1875 gives some idea of the conditions of travel across Siberia before railroads. The first part of the journey east from Moscow was an easy 6-day journey by stream-powered riverboat from Moscow 880 miles to Perm. "From Perm," however, "the only mode of travelling is by post." Travelers must "be provided with everything they may require on the journey in the shape of tea, coffee, sugar, wine, spirits, preserved meats, milk, etc." Three kinds of carts are available: open without springs, covered without springs, or covered with wooden springs. The handbook says that "post service across the Ural into Siberia is excellent, but the state of the roads varies a good deal with the season." Early summer is best. Along the way, the travelers will see that "caravans of merchandise follow one after another, and consist generally of 50 to 60 carts or sledges, and frequently of 500 and 600." At Irkutsk, the major city of eastern Siberia, the guidebook warns bluntly that hotels are "very bad and dear."

In 1905, almost 20 years after Czar Alexander's confession, the trans-Siberian railway was completed from Chelyabinsk in the Urals to the Pacific at Vladivostok. For a decade, it included a ferry trip along the length of Lake Baikal in summer or, in winter, a trip on temporary tracks on lake ice. (A huge icebreaker kept the lake open until the ice was 4 feet thick.) The railway also followed a shortcut through Manchuria over the Chinese Eastern Railroad, built by Russian capital with the understanding the ownership would pass in 80 years to China.

In 1916, the Russians bypassed both Baikal and Manchuria. The distance on the all-Russian route between Chelyabinsk and Vladivostok was 4,776 miles. Including the European section from St. Petersburg, it was almost 6,700 miles, more than twice the length of the Canadian Pacific from Montreal to Vancouver.

Like other transcontinental undertakings, the trans-Siberian initially made little or no economic sense. There was very little freight that could be economically shipped along the line's huge length, and there was little local traffic. Still, a Northern Pacific railway engineer named Benjamin Johnson went to Siberia in 1917 as an American expert, and he wrote home 2 years later that "there is not one single American transcontinental line in the splendid physical condition of the Trans-Siberian." While Americans abandoned or severely reduced service on several transcontinental lines, the Russians after World War II expanded the trans-Siberian with a loop north of Lake Baikal to create the Baikal Amur Mainline, or BAM.

These lines remain the only transasian railroads in operation, with the exception of a line from Kazakhstan to Shanghai. Like Russia, however, Kazakhstan uses a broader track gauge than China and Europe, so passengers on the one weekly train each way between Kazakhstan and China wait 8 hours while each car is hoisted and different wheels put on. Farther south, there are plans to build other transasian railroads, but gaps remain in the network, particularly around the old Iranian and Afghan bottlenecks. Conditions are likely to improve, however. The government of Kazakhstan in 2004, for example, announced its intention of changing its rail gauge to help trains cross a Eurasia Land Bridge running from Europe through Russia and Kazakhstan to China and its Pacific port at Lianyungang, just south of the Shandong Peninsula.

I've neglected passenger service because in the United States it's been a financially losing proposition for a century or more. Compelled by government to carry people, railroad companies after World War II worked long, hard, and successfully to escape the burden. Nowadays most passenger service in the United States is provided by federally subsidized Amtrak, a separate corporation that runs on railway-company tracks.

Elsewhere in the world, particularly in Europe and India, travelers rely much more heavily on railways. Service in continental Europe is often fast and efficient. The same thing can't be said for India—or England, where trains heading to the Continent roll slowly through the British countryside before speeding up dramatically in the Channel tunnel. The British have been trying to upgrade some of their arterial lines. A few years ago they planned to spend about $18 billion modernizing the track from London to Glasgow. Plans originally called for trains on that route running at 140 miles an hour, and Virgin (the same company that runs the airline of that name) spent $3 billion buying tilting trains to make the run. When the estimated cost of the project had risen to over $70 billion, however, the government decided that Virgin's new trains would have to run far below their design speed.

Even with the heavy use of European railways for passenger travel, the European rail-freight business is in decline. Trains in 1970, for example, carried 20% of the freight of the 15 members of the European Union. By 2000 the figure had dropped to 8%. Motorists on European highways know where it has gone. There is one area where Europe is likely to see major new rail investment, however. Rail Baltica proposes to realign the railways of Estonia, Latvia, and Lithuania so that they are no longer tied primarily to Moscow but to Berlin. At present, there's no direct connection to the west, but the project is so expensive that even with funding from Brussels it won't be finished before 2016.

ROADS AND TRUCKING

What about roads? Stone ones were built by the Romans as well as the Chinese and Japanese, but they were expensive. The 18th-century British engineer Robert McAdam recommended spreading crushed rock to create a well-drained, flat surface. Still, many roads in Europe and North America were dirt until well into the 20th century, which meant that they were often wretched mud. Even when dry, the movement of goods by road was costly compared to water or rail.

Under optimal conditions, the journey by road from New York City to Charleston in 1800 took more than 2 weeks. Even so, a network of roads covered the coastal plain and

Appalachian piedmont by the time roads began crossing the Appalachians, first along the Mohawk River and then with Daniel Boone's Wilderness Road to Kentucky. By 1830, you could take a stagecoach from New York City to southern Illinois in about 2 weeks. Famous roads began pushing farther west, like the National Road from Cumberland to Wheeling and St. Louis.

The growth of railroads impeded road construction for the next 50 years. A road map of Oklahoma in 1925, for example, shows no paved road between Oklahoma City and Tulsa. From both those cities, pavement reached out for no more than 10 to 20 miles in any direction. If one headed south from Oklahoma City to Dallas, there was pavement to Norman, pavement authorized from there to Lexington, then dirt as far as Ardmore. Streets there were paved, but the pavement gave out on the far side of town, and it was more dirt from there to the Red River.

All this changed very fast with the introduction of the federal highway system in 1926. This was the start of the U.S.-numbered highway network, a 96,000-mile system of roads selected by the states in coordination with the Department of Agriculture. Even-numbered roads ran east to west and odd-numbered roads ran north to south. The lowest numbers were on the eastern and northern margins of the country. That's why U.S. 1 runs from Fort Kent, Maine, to Key West, Florida, and why U.S. 2 runs from Houlton, Maine, to Everett, Washington.

Pennsylvania in 1940 opened America's first freeway, or superhighway as it was then called. Modeled on the German autobahns, the Pennsylvania Turnpike became the model for later toll roads in the northeast and then for the national system of interstate highways, which were authorized in 1956 as a 42,000-mile system. Numbers on the Interstate system

Old U.S. 66 in western Oklahoma; now supplanted by Interstate 40, the old road evokes memories of the Great Depression, when people walked this road or drove jalopies on one-way trips to California.

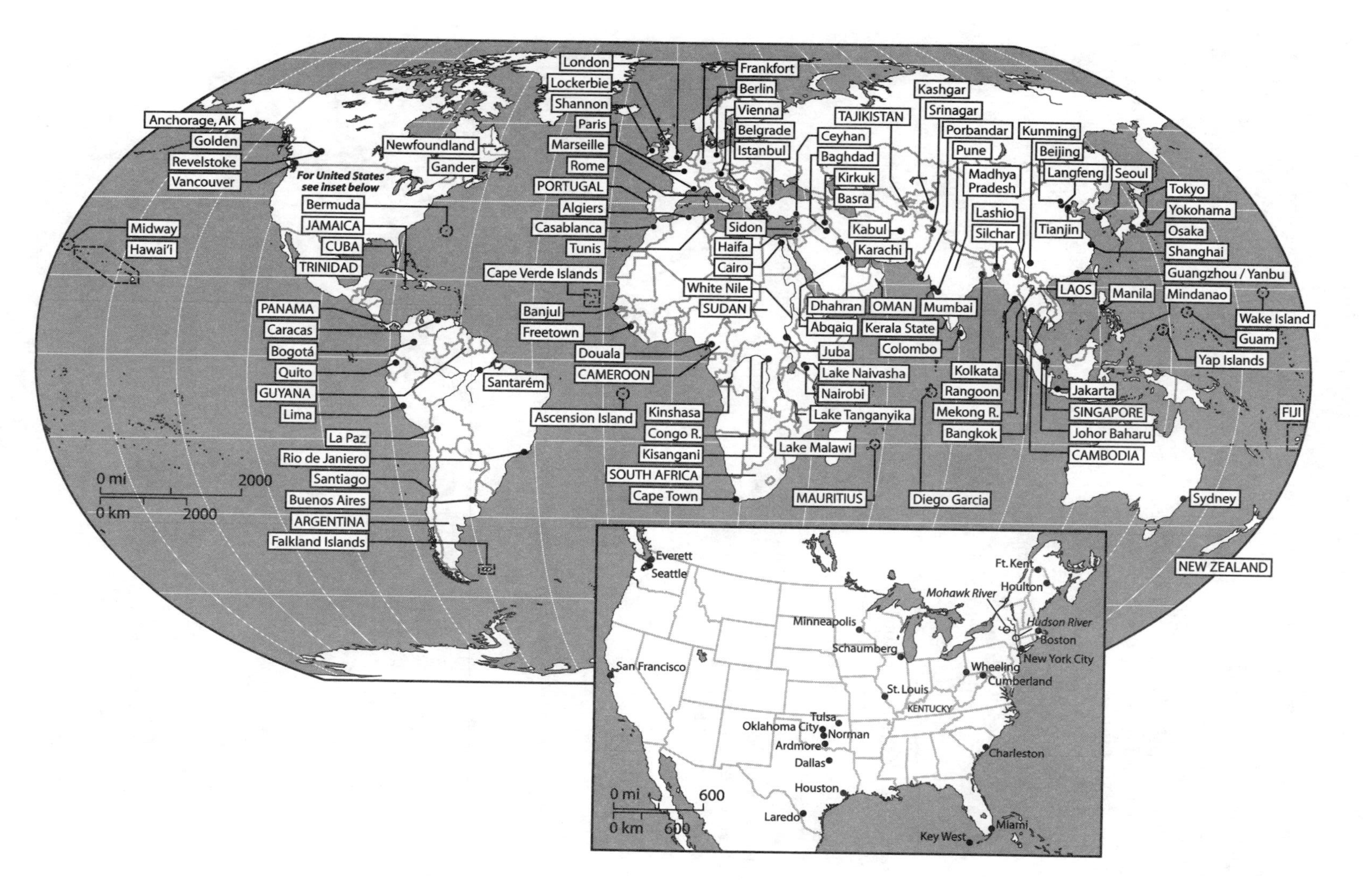

Anchorage, AK
Golden
Revelstoke
Vancouver
Midway
Hawai'i
For United States see inset below
Newfoundland
Gander
Bermuda
JAMAICA
CUBA
TRINIDAD
PANAMA
Caracas
Bogotá
Quito
GUYANA
Lima
La Paz
Rio de Janiero
Santiago
Buenos Aires
ARGENTINA
Falkland Islands
0 mi 2000
0 km 2000
London
Lockerbie
Shannon
Paris
Marseille
Rome
PORTUGAL
Algiers
Casablanca
Tunis
Cape Verde Islands
Banjul
Freetown
Douala
CAMEROON
Santarém
Ascension Island
Frankfort
Berlin
Vienna
Belgrade
Istanbul
Sidon
Haifa
Cairo
White Nile
SUDAN
Kinshasa
Congo R.
Kisangani
SOUTH AFRICA
Cape Town
TAJIKISTAN
Ceyhan
Baghdad
Kirkuk
Basra
Kabul
Karachi
Dhahran
Abqaiq
OMAN
Kerala State
Juba
Lake Naivasha
Nairobi
Lake Tanganyika
Lake Malawi
MAURITIUS
Kashgar
Srinagar
Porbandar
Pune
Madhya Pradesh
Mumbai
Colombo
Kunming
Lashio
Silchar
Kolkata
Rangoon
Mekong R.
Bangkok
Diego Garcia
Beijing
Langfeng
Tianjin
Seoul
Tokyo
Yokohama
Osaka
Shanghai
Guangzhou / Yanbu
LAOS
Manila
Mindanao
Jakarta
SINGAPORE
Johor Baharu
CAMBODIA
Wake Island
Guam
Yap Islands
FIJI
Sydney
NEW ZEALAND
Everett
Seattle
San Francisco
Minneapolis
Schaumberg
Mohawk River
Ft. Kent
Houlton
Hudson River
Boston
New York City
Wheeling
Cumberland
St. Louis
KENTUCKY
Tulsa
Oklahoma City
Norman
Ardmore
Dallas
Houston
Laredo
Charleston
Miami
Key West
0 mi 600
0 km 600

are arranged so that east–west routes have even numbers, while north–south ones have odd, but the numbering is inverted from the U.S. numbered system, so Interstates 5 and 10 cross in California, not Maine. The system also provides for loops and branches in and around cities. They're given a third digit. If the first digit is odd, the road's a branch; even, it's a loop.

Roads overseas have histories, too. The modern Chinese road network began very slowly with roads built in the 1920s and 1930s by the American Red Cross and the China International Famine Relief Commission. Between 1921 and 1929, China's highway mileage rose from a negligible 730 to a still very modest 21,000. By 1943, highways connected China to the Soviet Union, both through Mongolia to the trans-Siberian railroad near Lake Baikal and west through Kashgar (Kashi) to Tajikistan.

A similarly germinal network evolved in South America, where the Pan-American Highway was conceived to run from Laredo, Texas, to Buenos Aires. The Central American section was delayed, but by 1941 motorists could drive in August and September from Caracas all the way to Buenos Aires via Bogotá, Quito, Lima, and either Santiago or La Paz.

In 1949, the Canadian government undertook to build a trans-Canada highway. The road was supposed to be finished in 1959, but the last paved bit, between Revelstoke and Golden in the Canadian Rockies, was not opened until 1962. A few years later, the highway was extended through Newfoundland.

The famous Burma Road—700 miles from Lashio, Burma, to Kunming, China—was built, unbelievably, in all of 14 months, but travel was very slow. Now it's being improved by the Chinese, who are eager to have a land route to the Indian Ocean. Courtesy of Japanese foreign aid, Cambodia meanwhile has got its first bridge across the Mekong, which slices the country almost in two. Another long link of paved road has been announced in Brazil, where BR163 is apparently going to be paved from Mato Grosso north to the Amazon at Santarém.

Oman in the last 20 years has built an incredible system of superb highways, as smooth or smoother than any American ones, and often both lit and lavishly landscaped in populated areas.

In a few countries, paved roads remain a rarity. There have been reports, for example, of North Korean peasants approaching the nuclear reactors under construction at Kumho, on the east coast. This is the huge facility that has raised concern about North Korea's development of nuclear weapons, but the peasants aren't interested in that: They come to see the plant's approach roads. They've never seen asphalt before, and they kneel to touch it.

African roads have been in decline for many years. Starting in the early 1930s, bus services began operating on a road running from Douala, Cameroon, all the way east to Juba, on the White Nile in southern Sudan. A couple of years later, the French and British colonial governments proudly opened a shorter road from Juba southwesterly to Stanleyville (Kisangani) on the Congo River. Combined with rail and river transport, it provided a link from Cairo to South Africa. Both these roads, however, are next to impassable today. Work does continue, however, on a paved road across North Africa. The gaps are in Mauritania, but they are being closed with Arab donations. It should soon be possible to drive on pavement from Nigeria to Gibraltar.

Controlled-access highways are a good sign of a country with a modern economy or every intention of having one. India has a road network that is still dominated by one- or two-lane British-era roads, but it's building a modern system. An eight-lane highway now connects Pune (Poona) to Mumbai, and it's just a small part of the 4,000-mile Golden Quadrilateral, a system of highways with a minimum of four lanes connecting Mumbai,

Chennai (Madras), Kolkata (Calcutta), and Delhi. More such roads are planned to bisect the country from north to south (Srinagar to Kanyakumari) and east to west (Silchar to Porbandar).

China is even more active. It presently has over 10,000 miles of controlled-access freeways, which makes it number two, behind the United States, and China's opening more at the rate of over 1,000 miles annually. The Jinghu Expressway, for example, runs 780 miles between Beijing and Shanghai. Another expressway links Beijing to its port, Tianjin. What used to be a 6-hour trip now takes 2, and the town of Langfeng, at the midway point, is booming. The roads are built to American freeway standards, but they aren't free. The Chinese hope that high tolls will discourage private use, but rapid growth in car ownership means that cars are already about as numerous on these roads as trucks and buses.

Meanwhile there's freight to move. You can sit along Interstate 40 in Oklahoma City and count 300 trucks heading west every hour of every weekday. Trucking in the United States isn't a very profitable business, however. In 2002, for example, the nation's biggest carrier, 72-year-old Consolidated Freightways (CF), failed. Its business was picked up by others carriers, including Yellow and Roadway. So were some of CF's instantly unemployed 15,000 workers. Yellow then took control of Roadway to form the nation's biggest less-than-carload, or LTL, freight handler. Together the two firms operate 19,000 tractors. Still, 80% of the business is in smaller hands, which continue to look for a competitive edge. U.S. Xpress, for example, specializes in coast-to-coast hauls. It has over 5,000 trucks, and they roll 24 hours a day when they're on the road, yet they waste a quarter of their time waiting at destinations. There's another industrywide problem, too, analogous to piracy at sea. More than $600 million worth of freight is stolen annually from trucks in Southern California alone.

Entering the tollroad at Suzhou.

AIRLINES

Air travel in Europe began in a very small way in the later days of World War I, but by 1922 flights connected London, Paris, Berlin, Vienna, Belgrade, and Constantinople (not yet known abroad as Istanbul).

Long-distance routes served imperial requirements. Britain's Imperial Airways (now British Air) began service through the Middle East to Karachi in 1929. It began service to Capetown in 1932 and Australia in 1935. All of it was slow by today's standards: London to Sydney, for example, took 11 days. As late as World War II, the flights back to London stopped at Batavia (Jakarta), Singapore, Rangoon (Yangon), Calcutta (Kolkata), Karachi, Baghdad, Rome, and Marseilles. Many of the stops were overnight ones. Still, flying was faster than the alternatives, and Holland was quick to follow with KLM (for Koninklijke Luchtvaart Maatschappij, or Royal Air Traffic Company), which offered service to Batavia in 1930.

Air France was created in 1933 through a merger of Air Orient, which flew from France to Indochina, and Aéropostale, which linked France to South America. (Antoine de Saint-Exupéry, the author of *Night Flight* [1932] worked for Aéropostale, first in Casablanca, then in Argentina.) Sabena, the Belgian carrier that declared bankruptcy in 2001, was created in 1923 and began offering regular service to the Belgian Congo in 1935.

The largest European system in the mid-1930s was none of these, however. It was the German carrier, Deutsche Lufthansa, which began running international service to Rio de Janeiro in 1931 via Bathurst (Banjul), on the west coast of Africa. A few years later the airline was serving Kabul, while a subsidiary flew domestically in China.

Service in the United States meanwhile began primarily for airmail. In 1924, letters were first sent by air from San Francisco to New York. In 1930, a particularly energetic postmaster general, Walter Brown, forced a great many small companies to merge into Eastern, Western, United, and TWA. (That acronym referred to Transcontinental and Western Airlines until after World War II, when it was ambitiously changed to Trans World.) United carried the mail between New York, Seattle, and San Francisco. TWA carried it from New York to Los Angeles via Kansas City. American had New York to Los Angeles via Oklahoma City and Dallas, and Eastern had New York to Miami and Houston. It didn't take long, however, before these systems began to overlap and for other carriers, like Delta, to come along. TWA in 1930 offered the first coast-to-coast passenger service: It took 36 hours, with 11 stops including an overnight in Kansas City. Passenger service grew quickly after the introduction in 1936 of the 21-passenger DC-3. Eleven thousand DC-3s were built before the model was discontinued about a decade later. By 1941, American Airlines alone was carrying over a million passengers annually.

International air travel on American carriers was a much smaller though more glamorous business, almost synonymous until World War II with Pan American Airways. Pan Am began in 1928, with mail service from Cuba to Trinidad and Panama. Texas to Mexico followed, then service from Panama to Chile and Argentina. The growth of the system was as much a political as a technical story, because many people in Latin America feared what they called the Octopus of the North. The company's president, Juan Trippe (no Spanish speaker, despite his name), quietly but tenaciously pursued his plans to develop a global network.

The refusal of the British government to grant landing rights in London kept Pan Am out of Europe until 1939, when it began flying to Lisbon. Surprisingly, because of the greater distance, Pan Am had already begun transpacific service in 1936, with what (in a reference to the era of sailing ships) it called clipper service to Hong Kong. These clippers were huge seaplanes, and they stopped en route at Hawaii, Midway, Wake, and Guam. (Pan Am established airfields and hotels on Midway and Wake, which were uninhabited.) The trip took several days, but sailing on the Canadian Pacific steamers from Vancouver took 3 weeks. (A ship left Vancouver on May 1, 1936, and arrived at Yokohama on May 12. It also stopped at Nagasaki, Shanghai, and Manila before getting to Hong Kong on May 22.) By 1941, Pan Am had a large network, although prices were so high that the airline was carrying only about 1,000 passengers daily, or 400,000 annually.

The transatlantic link had been delayed by the British refusal to allow Pan Am to offer service until a British carrier could compete with it, but there were also technical difficulties of flying west against severe headwinds. These obstacles were overcome during World War II, when several American airlines offered temporary service to London via Gander (Newfoundland) and Shannon (Ireland). After the war, these flights continued in the hands of Pan Am and Trans World, which began flying to Paris and Rome in 1946 and to London and Frankfurt in 1950. The TWA network grew rapidly, and for a while the airline almost lived up to its name, with regular flights not only to Europe but to Algiers, Tunis, Cairo, Tel Aviv, Basra, Dhahran, Bombay (Mumbai), Colombo, Calcutta, Rangoon (Yangon), Mandalay, Bangkok, Hanoi, Canton (Guangzhou), Shanghai, Manila, and Tokyo. If you lament the passing of the American Century, you'll note right away that most of those cities haven't seen an American civilian airliner for decades.

After World War II, many new carriers were created. Air India and Saudia, for example, were organized in 1946. Thai Airways began flying in 1947; Garuda (Indonesia), in 1949; and Japan Air Lines, in 1951. Foreign carriers began service to the United States, too. KLM offered it in 1945, and Scandinavian Airlines System (SAS) followed a year later. By 1957, more people were crossing the Atlantic by air than by sea, and by 1965 six passengers were flying for every one that sailed. The *S.S. United States* had been launched in 1952 with 3-day transatlantic service. It was extremely fast by historic standards but hopelessly slow, and the ship was retired in 1969. A new chapter in transatlantic sailing opened in 2004 with the launch of the Queen Mary II. Carnival's Cunard line hopes to attract customers who have the time—6 days—for a luxurious crossing.

Northwest began flying a polar transpacific route in 1947, undercutting the longer route used by Pan Am's clippers. In 1955, JAL was allowed to offer competing service. For a time, Anchorage airport was busy with stopovers not only from these airlines but from European carriers such as Lufthansa, Air France, British Air, and SAS, all of which used Anchorage on polar flights between Europe and Japan.

The introduction in 1958 of the long-range, jet-powered Boeing 707, which flew at over 600 miles an hour, took Anchorage off the map. By the mid-1960s, Pan Am was flying 250 jet flights weekly. Remarkable as it now seems, four airlines were offering round-the-world service: Pan Am, TWA, BOAC (the successor to Imperial Airways and predecessor to British Air), and Australia's Qantas (the name is an acronym for Queensland and Northern Territories Aerial Service).

In 1970 Pan Am introduced the wide-body Boeing 747, a plane originally designed for a competition Boeing lost to Lockheed for a military freighter. Boeing made more than 1,300 of them, and fares continue to fall. It's easy to get a round-trip ticket across the Pacific for $800. The same trip on the Pan Am clippers of the 1930s cost $1,300 in 1930 dollars.

An even more important introduction for the expansion of air travel was the Boeing 727, which entered mostly domestic service in 1964. This three-engine plane was very economical for its time, and it could operate on short runways. With it, airlines could serve many new airports and bring air travel to millions of new passengers. Boeing sold 1,832 of the planes, and during its first decade in service passenger numbers rose by 500%. Eventually the 727 was replaced by twin-engined planes like the 737, introduced in 1967, and the 757, introduced in 1984.

Today, about 11,000 jetliners operate worldwide. About 40% of the traffic, measured in seat-miles, is handled within the United States. About 20% each is in Europe and East Asia. Near the other end of the spectrum, only 2% of Brazilians have ever flown.

The business is divided among hundreds of carriers, but 17 of them have three-quarters of the business. The biggest, measured by revenues, are American, United, and Delta, followed by British Air, Japan Airlines, and Lufthansa.

National carriers such as Egyptair, Varig (from Brazil), and Garuda are the pride of almost every country that can afford an airline—and quite a few that can't. The key to the problem is the 1944 Chicago Convention. Under its provisions, countries allow flights from other countries only on a reciprocal basis. There is a negotiated and fixed number of flights between France and Turkey, for example, and the only carriers allowed to provide them are French or Turkish. In theory, the convention allows a "fifth freedom," in which an airline from one country is allowed to provide service between two others. Such services exist—Northwest flies between Manila and Tokyo, for example, and Cathay Pacific between Mumbai and Bangkok—but they are uncommon. In short, the convention encourages an uneconomic proliferation of airlines.

For decades, governments subsidized their carriers, but the losses have become unbearable. Australia's Ansett folded in 2001, West Africa's Air Afrique collapsed a year later, and Nigerian Airways—"Air Waste"—folded in 2004. Brazil's Varig, established about 1920, is on the edge of financial ruin. Bankruptcies came to Europe in 2001 with the collapse of Sabena and Swissair. There is no future for SAS, the Scandinavian carrier, which has only a quarter of Lufthansa's traffic but more than half as many staff. It will likely merge with Lufthansa. In 2004, KLM and Air France merged, and Alitalia was expected to join them. Long-haul traffic will probably be left to British Air, the enlarged Air France, Lufthansa, and their non-European competitors.

Will an industry with fewer airlines be more profitable? Nobody can make money, it seems, on the fiercely competitive transatlantic corridor, so everyone looks for niches or, in industry lingo, segments. Air France thought it had found one on its Paris–Kinshasa route, because diamond merchants will go anywhere. Kinshasa, in the Democratic Republic of the Congo (formerly Zaire), is in such terrible shape, however, that the airline couldn't find local accommodation for its cabin crews. On arrival, they were taken instead by charter flight to Brazzaville, in neighboring Congo, which has safe hotels. Air France has dropped the route, but SN Brussels, the successor to Sabena, now flies diamond merchants to Freetown, Sierra Leone. The cheapest return ticket costs $2,000.

Owned by investors instead of national governments, American airlines have been free to fail for many decades. The list of departed carriers is long, including Eastern, Western, and Trans World—three of the four companies formed during the 1930 consolidation. The fourth, United, was in bankruptcy in 2004. Terrorist attacks at Karachi and over Lockerbie, Scotland, drove Pan Am into bankruptcy, and the company was liquidated, with its valuable routes sold to other carriers. United bought the Pacific ones; Delta, the Atlantic ones. United tried reestablishing a round-the-world link from Hong Kong through Delhi to Frankfurt but could not make it pay: The company dropped the Hong Kong–Delhi link. Delta offers many flights across the Atlantic, and American is strong to Latin America, where it took over the routes of Braniff, another failed carrier. No American carrier has any presence in Africa, however, except for Delta to Cairo via Paris. There are many big Asian countries, including Pakistan and Indonesia, to which no American carrier flies these days.

After the destruction of the World Trade Center, world air traffic fell by more than 10% and about 2,000 planes were mothballed. Unprofitable routes can cost an airline a lot of money fast, and many segments were dropped in the slump. United announced that it would drop a new nonstop between New York and Hong Kong. Delta dropped its flight from New York to Lyon. American dropped its Los Angeles–Paris flight. Northwest dropped its Seattle–Osaka flight. In 2001 Dallas/Ft. Worth lost its Sabena flight to Brussels and at least temporarily its Japan Airlines flight to Tokyo, as well as its Korean Air link to Seoul. The only bright note was that Singapore Airlines began freight service from Dallas to Singapore, via Brussels and Mumbai. In 2004, Singapore announced the world's first nonstops from Singapore to both Los Angeles and New York.

Within the United States, the surviving carriers have operated for 30 years with hubs, a system developed first by American. With a given number of planes, you can serve far more points if you route passengers through a hub than if you offer only direct flights. That's why American has hubs in Dallas and Chicago and St. Louis. Delta has them in Atlanta and Miami; United, in Chicago and Denver; Northwest, in Minneapolis and Memphis; and Continental, in Houston. Air travel is far commoner and much cheaper today than it was 30 years ago. Still, the nine biggest U.S. airlines managed to lose $11.2 billion in 2002. Between 2001 and 2003 it lost $23 billion, more than it had earned since 1947. By 2004, Delta was skirting bankruptcy.

Budget carriers like Southwest have meanwhile enlarged their market share. Southwest began in 1971 but didn't fly outside Texas until 1979. It expanded rapidly in California in 1990 and on the East Coast in 1993. Today Southwest has about 360 planes serving 58 cities. There are half a dozen American carriers with more seat-miles, but Southwest hasn't lost money since the early 1970s. It thrives by focusing on flights of less than 750 miles, keeping turnaround time to a minimum, and using a fleet composed only of Boeing 737s. Its planes are in the air 11 hours a day, compared to 9 for United and Delta, and its labor costs are about 30% of revenue, compared to over 40% for United and Delta. Such differences explain why a seat-mile costs Southwest 7.6 cents, 4 cents less than it costs the big carriers. They also explain why Southwest, though its flights are generally short, usually carries more passengers domestically than other carriers. In December 2003, for example, only Delta had more passenger boardings. Some observers conclude that the traditional or legacy carriers are doomed; one former executive describes them as "icebergs drifting south."

Budget carriers are appearing overseas, too. They have about 10% of the market in Europe, far less than their American counterparts, but you wouldn't know it from the media-savvy CEO of Dublin's Ryanair. Speaking of British Airways, a cocky Michael O'Leary says, "They are such a mess, most people just feel sorry for them." (The problem with Ryanair, however, is that it flies to peripheral airports, often an hour's drive from the nominal destination.) There are Asian wannabes, too. One is Malaysia's Air Asia, which would like to build a hub at Johor Baharu and take business from Singapore Airlines. China, too, has a Southwest clone: It's Hainan Airlines. Tiny in comparison with Air China and the other state-controlled carriers—China Southern in Guangzhou and China Eastern in Shanghai—Hainan is efficient, aggressive, and is building a wide, mostly domestic network. India has Air Deccan, which hopes to serve the smaller markets that are missed by the bigger domestic carriers, particularly Indian Airlines and a private newcomer, Jet Airways.

Air freight remains a major part of the airline business, as it was in the pioneer days when carriers depended on mail contracts. In ton-miles, air freight is trivial, but measured by revenue, air freight is huge. There are many industries that couldn't function without it. Fifteen thousand acres of roses grow in greenhouses near Bogotá, for example, and at the peak season, before Valentine's Day, they fill 35 jet freighters daily. Another example is tuna, caught by thousands of Filipino fishermen. They set out from General Santos, on the south coast of Mindanao. They return, their outrigger boats loaded. The fish are quickly frozen and loaded in a 747 for the flight to Tokyo and its huge wholesale fish market at Tsukiji. British supermarkets meanwhile sell string beans in winter. They're grown in Kenya by a company called Homegrown, which packages them in shrink-wrapped plastic trays. The plastic trays have already been flown empty from London to Nairobi. Now they're filled and loaded into containers known as coffins for the trip back to London. At the end of the day, the women working at the Homegrown facility will each have topped and tailed 400 pounds of string beans and been paid $3. One woman says: "I would do another job now if you would give me one. Why don't you take me to England?" Kenya has other companies dependent on air freight, too. Oserian Farm, for example, is a Dutch company with 40,000 acres near Lake Naivasha; not only does it have about 400 acres in greenhouses devoted to roses, but it operates its own airline from Nairobi to the Netherlands.

Then there are package delivery services, which are rapidly expanding into remote areas. Between 2002 and 2007, for example, FedEx plans to begin service to about 100 new cities in China. It's way behind DHL, which is controlled by Deutsche Post and already serves over 300 Chinese cities through a partnership with a Chinese freight forwarder, Sinotrans. (It's a different story in the American domestic market, where UPS and FedEx dwarf DHL.)

Like international passenger routes, air freight is tightly regulated. Singapore Airlines, for example, in 2003 became the first carrier to win fifth-freedom opportunites in China, but the Civil Aviation Administration of China gave it the right to pick up shipments only at Xiamen and Nanjing—not Guangzhou and Shanghai. That's slim pickings.

PIPELINES

Pipelines are perhaps as far removed from airline glamor as any transport mode. Used primarily for oil, both crude and refined, they carry about one-fifth of all freight ton-miles in the United States. The routes are complicated, which is why Houston, at the center of the American system, is sometimes called the Spaghetti Bowl. Pipelines have their own histories, too, including that of the Big Inch, which was a World War II pipeline from the Gulf Coast to the Northeast. Later came the Alaska Pipeline, from Prudhoe Bay south to the tanker terminal at Valdez, on Prince William Sound. Overseas, perhaps the most famous pipeline is the now-closed Tapline or Trans Arabian Pipeline, which was opened in 1950 from Abqaiq, at the center of the Saudi Arabian fields, to the Lebanese port of Sidon. Even earlier, in 1934, there were pipelines to the Mediterranean from Kirkuk, Iraq. One of them terminated in Haifa and was shut down in 1948 with the creation of Israel. In 2003, while the regime of Saddam Hussein was crumbling, Israel proposed reopening that pipeline, but the proposal ran almost at once into heavy opposition from Israel's neighbors.

Nowadays, there are two important pipelines bringing oil from the Persian Gulf west. One crosses Saudi Arabia from east to west and terminates at the Red Sea port of Yanbu. The other runs from Iraq to Ceyhan, the Mediterranean port in southeastern Turkey. Both pipelines, however, carry only about 1 million barrels daily, far less than the 15 million that are exported daily by tanker through Persian Gulf ports.

COMMUNICATIONS

Let's look finally at the impalpable but vital world of communications.

The story of electronic communication begins in 1844, when Samuel F. B. Morse sent that famous first telegraph message—"What Hath God Wrought!"—from Washington to Baltimore. Within 50 years, a dense network of telegraph lines ran across the inhabited parts of the Western Hemisphere, through Europe and European Russia, through the Orient, and through the settled parts of Australia. The lines ran under the oceans, too, with the first marine cables laid in the 1860s off the stern of the huge steamship *Great Eastern*. The largest cable network belonged to the imperial British. For better or worse, Indian viceroys no longer had to wait months for answers to questions that had always been sent by sea mail. By 1900, the British had five cables crossing the North Atlantic from Ireland to Newfoundland. They had cables running from England and Portugal to the Cape Verde Islands. From there, British cables ran to Brazil and Uruguay and also to South Africa, Mauritius, Australia, and New Zealand. The British also had lines through the Mediterranean, Red Sea, and Arabian Sea to Bombay (Mumbai). They had lines running from Madras (Chennai) to Malaya and Australia. They even had a cable across the Pacific, from Australia through Fiji to Victoria, on the west coast of Canada. The United States denied the British the right to bring the cable to Hawaii, so the cable went instead through Yap, in the Carolines. Decades later, the United States would deny Imperial Airways landing rights in Hawaii.

The impact of the telegraph probably ranks with that of the internet. A young Cambridge graduate took a trip across Africa just before 1900 and met up with a telegraph crew at work near Karonga, at the northern tip of what is now Lake Malawi. "Behind them," he wrote, "lay many hundreds of miles of perfected work which brought the far interior of Africa within a minute of Cape Town; before them stretched an arrow-like clearing to Tanganyika (two hundred miles long) waiting for the transport service to bring poles and wire." A surviving echo of those imperial days is the fact that telephone service today in Jamaica, Bermuda, the Falklands, and even such remote places as Ascension Island and Diego Garcia is provided by Cable and Wireless, the now private but venerable descendant of the old imperial telegraph network.

The telegram has very nearly passed into history; not so the telephone. Alexander Graham Bell steadfastly refused to have one in his own house, but by 1890, the Bell system provided long-distance service between Boston, New York, Buffalo, and Pittsburgh. By 1898 the system reached as far west as Minneapolis and Houston. A separate system began growing on the West Coast about 1900, and by 1917 the network covered almost the entire country. Long-distance was expensive. A 3-minute coast-to-coast call in 1920, for example, cost the equivalent of more than $100 today. Thanks to fiber-optic lines, the cost of long-distance calls has collapsed since then to almost nothing.

One of the big changes in recent years has been the proliferation of about a billion cell phones globally; in 2003 alone, more than 500 million were produced. The United States has about 140 million of them, along with 192 million conventional phones. That's a lot of cell phones—four for every 10 Americans—but the figure in Europe is seven in 10. Perhaps the most spectacular advances have been in poor countries with few fixed lines. In Latin America, for example, the number of cell phones jumped from 3.6 million in 1995 to 61 million in 2000. By 2005, it's expected to be 138 million. China now has 200 million cell phones, and in places like Beijing, Shanghai, and Guangzhou there's one in every two households. China Mobile alone has more than twice as many subscribers as Cingular/AT&T Wireless. In India, Reliance Infocomm was promoting in 2003 a 3-year contract that cost

$63 up front and $14 a month in exchange for a phone, 400 minutes a month, and unlimited long-distance within India to other phones on the Reliance network. India in mid-2004 had 40 million subscribers and was adding more at 2 million every month.

With prices so low, taxi drivers in India have cell phones. So do many autoricksha drivers. Aid organizations sometimes lend a villager the money to buy a cell phone, which then becomes a portable pay phone and the foundation for a small business. Other villagers don't need anybody's help to see an opportunity. In a remote village on Sri Lanka, a villager buys a cell phone from Mobitel, Sri Lanka's wireless provider. He sets up a phone booth, pays the company 8 rupees a minute for calls anywhere on the island—that's about 8 cents—and charges his customers 12. He says that he has 12 or 15 customers daily and that it's a profitable business. The phone's not just for chitchat or personal emergencies. It's mostly for business, as farmers seek technical advice or want to comparison-shop.

A shrimp fisherman in Kerala, on the southwest coast of India, heads back to land, picks up a cell phone, makes a few calls, and heads to the port whose buyers offer the best price. Multiply that by 1,000 equivalent stories, and you can see why a 1% increase in a nation's phone calls correlates roughly to a 1% increase in gross domestic product.

Phone lines are also heavily used for data transmission. This began with the telex system, now obsolete. Faxes are still important, though heartily despised by the computer *cognescenti*. The real growth, of course, has been with the internet. Linked computers go back to ARPAnet, developed by the Pentagon in 1969. By 1990, about 160,000 computers around the world were linked by what was already called the internet. In that year, a British software engineer named Tim Berners-Lee developed the universal record locator (URL), the hypertext transfer protocol (HTTP), and hypertext markup language (HTML). Since the addition of the first efficient browser, introduced by Netscape in 1994, the internet's growth has been explosive, with traffic increasing 1,000% annually in 1995 and 1996 and then doubling annually.

Between 1996 and 2001, almost half a trillion dollars were spent building data networks around the world. Fiber-optic lines now run under both the Atlantic and Pacific, but the highest-capacity line is in the Indian Ocean and connects Singapore to Chennai (Madras). Laid by Alcatel of France and Fujitsu of Japan, it belongs to Singapore Telecommunications and Bharti Enterprises, India's largest private telecommunications company.

The promise of such systems reaches far beyond the American consumer looking for the best hotel or airline deal. Cambodian villages with internet access can market handmade silk scarves directly to consumers. Far away in southwestern Guyana, Wapishiana women in the Rupununi savannas have set up a cooperative to market their handwoven hammocks over the internet. In Laos, the Jhai Foundation is setting up computers for farmers in the Hin Heup district. The machines have no delicate hard drives—only flash-memory chips. They have LCD screens to cut power requirements and bicycle-crank generators to produce that power. They're equipped with wireless internet cards linked to solar-powered relay stations through which the farmers, far from electricity and phone lines, can get access to market information.

American Assistance for Cambodia has provided internet access a different way. The solar-powered computers are connected to nothing, but once a day a man comes by on a motorcycle fitted with a computer and wireless internet antenna. Cruising slowly past the village computer, he collects all the e-mail of the previous day and delivers e-mail to the villagers.

In the Dhar District of Madhya Pradesh, a program called Gyandoot provides 39 computer kiosks to farmers who want to check produce prices. The farmers say they can now skip middlemen and ship to whichever market offers the best price. The government of Kerala has chosen the educationally backward and heavily Muslim district of Malappuram to be the test case for Akshaya, a program that will set up 615 computer centers in the district, each with six computers and run mostly by women as profit-seeking ventures funded mostly by advertisers. Probably the biggest Indian venture is sponsored by the Indian Tobacco Corporation (ITC). It has set up 2,100 e-choupals, or electronic meeting places, in village houses. The solar-powered setups cost about $3,000 each, but through them the corporation is already annually buying $50 million worth of crops—wheat, coffee, shrimp—sold by farmers eager to escape middlemen. The system works the other way, too, and ITC annually sells through the choupals $200 million worth of farm supplies like seed and fertilizer. It's begun selling household goods, too, adds five e-choupals every day, and plans on ultimately having 100,000 of them. A sign of the times is that Microsoft in 2002 announced plans for 10 Information Technology Academy Centers in India.

The world's telecommunications companies are presently buried under about $650 billion in debt. (France Telecom comes first, owing $60 billion.) A good part of the debt—about $100 billion—was incurred by European companies bidding on license rights for G3 wireless phones capable of high-speed data transmission. Now the companies can't afford to build the systems, which continue to have software problems anyway.

To cut costs, British Telecom and Deutsche Telecom have announced plans to work together. The same pressure seems likely to push AT&T toward a merger with one of the Baby Bells carved from it by an antitrust action in 1984. Sprint's future as an independent company is in doubt. Newcomers that performed sensationally a few years ago are gone or in desperate straits. Global Crossing, which raised $750 million in 1997 to lay cable between the United States and Europe, is bankrupt. Investors in WorldCom have been devastated by that company's implosion. People look back at Level 3, which raised $6.5 billion before it had a single lit, or operating, fiber, and wonder at the craze. Companies that make network hardware have been equally hard hit. Industrywide, 500,000 American jobs were lost between 2000 and 2002. In that brief period, Lucent went from 106,000 workers to 61,000, Motorola went from 150,000 to 57,000, Nortel went from 94,000 to 52,000, and WorldCom went from 80,000 to 30,000. Only 5% of the fiber-optic cables laid since the 1980s are even lit.

Like atmospheric clouds, however, this one will pass. Broadband usage has grown more slowly than people anticipated, but eventually it will become commonplace. In Korea, almost 75% of all internet households have broadband connections—more than three times the percentage in the United States. DSL is not only cheaper in Korea than in the United States but, for users willing to pay $50 a month, much faster: 20 megabits per second, compared to three for DSL, American-style. The explanation, at least in part, is that Korean providers got in the business late, after equipment costs had fallen 80% from what American providers had paid a few years earlier. Li Ka-shing of Hong Kong is meanwhile once again putting billions of dollars into G3 phones in Europe. India's Reliance and Singapore investors are separately laying new lines from East Asia to Europe via the Middle East: the Falcon and Sea-Me-We 4 lines depend on building internet use in the comparatively starved Middle East. In a sign of the emerging communications world, the BBC announced in mid-2001 that it would stop its shortwave transmissions to North America and Australia. It explained that the technology was obsolete and that most of its audience there was already listening over the internet. It's a long way from the 1930s, when Philco sold state-of-the-art shortwave radios to Americans who wanted to hear London, Berlin, and Tokyo. Hearing those places was a challenge and an adventure.

PART IV

SOCIAL AND ENVIRONMENTAL CONSEQUENCES

CHAPTER 15

ROMANTIC RESPONSES

It's easy to conclude about now that the driving force behind the development of today's cultural landscapes is the Baconian pursuit of ever-more-powerful technologies. What else can one think after this parade of fields and factories, fast-food chains and superhighways? But to leave the human geography of our technological civilization as merely an account of its technology would be like those biographies of famous people that tell you whom they knew and what they did but nothing about how they felt. We, too, are more complex than that. We, too, have an interior life, and it doesn't simply mirror our exterior one. Paradoxically, more often than not it's at odds with it. That's because the pursuit of efficiency doesn't simply give us the freedom to do as we like. It makes us pay a high price for that freedom. For hours we're confined to tedious offices, classrooms, or meetings; later, we're hurrying down a crowded airport concourse or driving along a highway walled with signs of prodigal ugliness; soon enough, we're standing at a line at a supermarket in a gauntlet of idiot tabloids. That's why the cultural landscape reveals a society simultaneously pursuing progress and trying to escape from it. That escape, too, shapes the cultural landscape.

ROMANTICISM

We're back to ambivalence. In Chapter 3, I contrasted Hobbes's state of nature with Dryden's noble savage. We can descend into historical subbasements, too. Think of dour Tacitus, the Roman whose *Germania*, written in the first century, portrays the barbaric Germans as morally superior to the soft Romans. It was a favorite theme of the Romans: the savage but virtuous Gaul who, rather than lose his freedom, kills his wife and then plunges his short sword downward into his own neck, just above the left collarbone. (The Gaul is always shown with body-builder muscle definition, while his lovely but dead wife slips gracefully from his powerful grasp.) Call it classical primitivism. You could also call it humanism, a term familiar from the Renaissance and suggesting that society should serve the individual—rather than the individual, society. A good early modern example is the Spanish friar Bartolomé de Las Casas (1474–1566), who spent his life fighting, mostly in vain, to help the American Indians suffering under Spanish rule.

The label I prefer when discussing this secondary landscape signature, however, is neither primitivism nor humanism but romanticism. I use it to suggest a revolt against reason when reason is carried to unreasonable lengths. As I'm using the term, "romanticism" rests

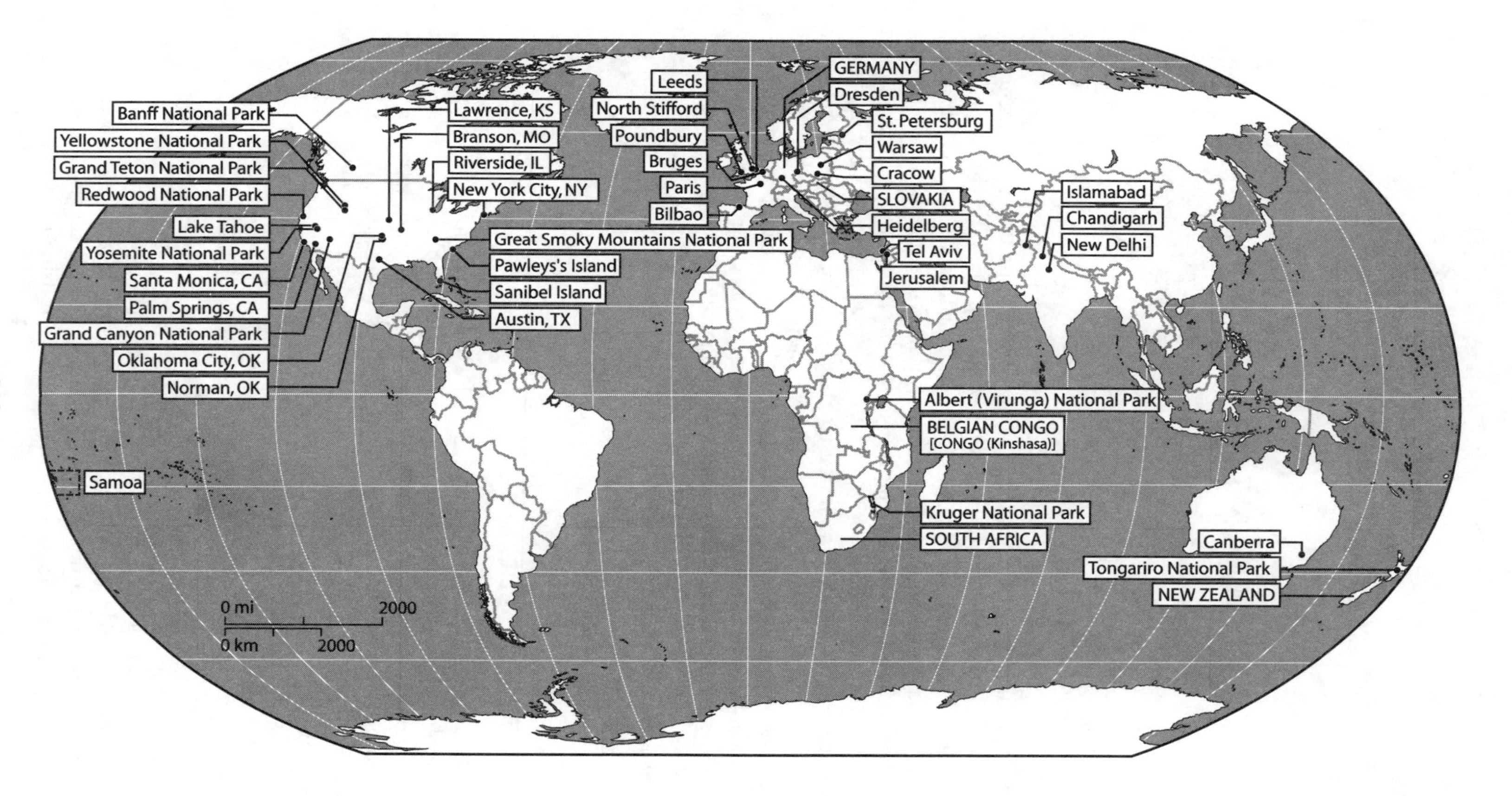

Banff National Park
Yellowstone National Park
Grand Teton National Park
Redwood National Park
Lake Tahoe
Yosemite National Park
Santa Monica, CA
Palm Springs, CA
Grand Canyon National Park
Oklahoma City, OK
Norman, OK
Lawrence, KS
Branson, MO
Riverside, IL
New York City, NY
Great Smoky Mountains National Park
Pawleys's Island
Sanibel Island
Austin, TX
Leeds
North Stifford
Poundbury
Bruges
Paris
Bilbao
GERMANY
Dresden
St. Petersburg
Warsaw
Cracow
SLOVAKIA
Heidelberg
Tel Aviv
Jerusalem
Islamabad
Chandigarh
New Delhi
Albert (Virunga) National Park
BELGIAN CONGO
[CONGO (Kinshasa)]
Kruger National Park
SOUTH AFRICA
Samoa
0 mi
2000
0 km
2000
Canberra
Tongariro National Park
NEW ZEALAND

on the belief, or merely the intuition, that the Baconian and Cartesian view of reality is false. It rests on the belief that living in a fractured reality is painful, even if the pain is so continuous that we become numb to it.

These are perilous waters, but recall Rembrandt's "Anatomy Lesson of Dr. Tulp." When I mentioned the proud doctor and his colleagues in Chapter 9, I neglected a central figure staring at the viewer in disbelief. What is the hesitation implicit in his gaze? It's very different from the attentive expressions of the other observers in the painting. I think it reveals shock at how the people around the table have made the Cartesian disconnect and now see the dead man in front of them as only so much inert stuff. It's the shock felt during an autopsy, when the cutter is horrified to realize that the body is that of a close friend. It's the difference between a nurse who deftly gives a thousand injections but faints when it's her turn.

A high school biology teacher chloroforms a rabbit so its innards may be inspected. A few students balk, and the teacher scolds them for being sentimental. If they're cheeky, they might say that the teacher is following in the footsteps of Descartes. They might explain that distancing ourselves from the world is emotionally alienating and that the only real happiness comes when we put things together, not when we take them apart. Might they dare to say that the name for this fusion is love? They'd want backup. I'd suggest a quote from the *Lyrical Ballads* of William Wordsworth (1770–1850): "Sweet is the lore which nature brings; / Our meddling intellect / Mishapes the beauteous forms of things;—We murder to dissect."

Words, words, words: A minute later, the rabbit would kick pointlessly and the students would get to work with their delicate knives. Such is the power of science. The protesting students, however, were tapping a vein of protest that runs all the way back to an acquaintance of Descartes named Blaise Pascal (1623–1662). A founder of probability theory, Pascal was also a deeply religious man who rejected Descartes's belief that human beings can understand themselves and their lives with logic alone. At one famous point in his *Pensées,* Pascal writes: "The heart follows its own path, one that reason cannot understand. We have learned this from a thousand cases."

Pascal wasn't the first with such doubts. Long ago I mentioned the pioneering anatomist Vesalius. I didn't mention that he eventually decided that dissection was sinful. In penance, he went on what proved to be a fatal pilgrimage to Jerusalem. Perhaps the students might mention it to their teacher.

You can see the same tension in Shakespeare. In Chapter 9, I quoted Hamlet talking about God-like man, but I omitted the line that follows: "And yet, to me, what is this quintessence of dust? man delights not me." There you have it, point and counterpoint: Renaissance arrogance and an early expression of the alienation that accompanies the loneliness of dualism. You encounter the same bleakness when Macbeth calls life "a tale told by an idiot; full of sound and fury, signifying nothing."

Such nihilism scarcely exists, if it exists at all, in other cultures, and it arises from the fact that the Baconian or Cartesian world is fit for machines, not people. People can be comfortable in such a world—many millions are to the point of obese lassitude—but they don't look happy. Look at the Dürer engraving "Melancholia I" (1514). It shows a figure surrounded by the scientific tools of the time. The figure is winged, as if to suggest the power of technology, but it positively broods, as if something's deeply wrong.

You can trace this cultural counterpoint to progress all the way to the present. In 1719, Daniel Defoe ends *Robinson Crusoe* with the arrival of European rescuers, who, it turns out, are worse than savages. That's the romantic impulse. A decade later, Jonathan Swift's

"A Modest Proposal" suggests that we raise and slaughter Irish children as food. Swift lays out the benefits of this proposal with the insane logic of the wholly rational man, who operates in a mental universe devoid of everything except the pursuit of efficiency. In another three decades, Voltaire, that beacon of the Enlightenment, paradoxically ends *Candide* (1758) with the recommendation that we work in our gardens—about as far from the Baconian program as can be imagined. Voltaire, it seems, understood that we must be guided by what feels right, not by systems of thought resting on inherited dogmas, even modern ones.

Early in the 19th century, Wordsworth warns of the barrenness of lives spent "getting and spending." Near century's end, Robert Louis Stevenson packs up and leaves Europe for Samoa. Most Americans have never heard of Germany's best-selling novelist: He's Karl May (pronounced "my"), and he was a 19th-century thief and jailbird who wrote dozens of novels about the American West. He never saw the West, but just like the movie *Dances with Wolves*, the Indians in May's novels are noble, while the white men are brutes. That's the Romantic spirit. By the end of the century, H. G. Wells was writing very popular science fiction in which the earth is nearly destroyed by Martians who are as merciless as any anti-imperialist might say Englishmen were toward the peoples the English ruled. We're now bumping into science fiction. Plenty of it eagerly awaits the future, but as much or even more dreads it. Samuel Butler's *Erewhon* (1872) portrays a utopia where machines are destroyed before they destroy humanity. Fritz Lang's *Metropolis* (1926) is a film with a similarly dark theme, and there's only a small step from it to the Terminator.

You can see the same trend among painters. In the midst of industrializing England, John Constable (1776–1837) paints pictures of rustic landscapes untouched by metal and steam. Two centuries later, his pictures remain hugely influential. Walk around an art festival on some Saturday afternoon and you'll search in vain for paintings of shopping malls, hospital corridors, or multilevel parking structures. Instead you'll find meadows and mountains, horses and dogs, the occasional Indian, and just maybe a gas station from the 1950s. Higher up the price ladder, there's a thriving business in very expensive cowboy art. Critics laugh at these paintings, but they can't afford to buy them. Plenty of people who can, do. America's landscape painters of the 19th century's Hudson River School—Thomas Cole, Frederic Church, Albert Bierstadt, Thomas Moran, and a raft of others—are back in style, after a long period when they were scorned to oblivion by modernists. Then there's the phenomenon of Thomas Kinkade, a painter whose golden-hued cottage landscapes have become an industry in themselves, printed by the thousands and sold coast-to-coast.

Why do so many art buyers want evocations of a world that's gone, a mostly natural world in which people live without machines? Why should they not happily decorate their homes with paintings of assembly lines? If we were wholly content with a high-tech world, we would be happy to gaze upon these things. Apparently we're not, and so we fill our homes with gadgets but choose to decorate with art that indulges our desire to escape.

It's a fantasy, because the people buying those cloyingly sentimental paintings of cozy cottages in a friendly forest don't have any intention of doing without the benefits of modern technology. Leo Tolstoy insisted on the moral value of physical labor, but his primitivism was tolerated only because of his towering stature as a novelist. For most of the rest of us, life is juggling. We can't imagine doing without our technology—medical technology is just the most obvious, but we're almost as wed to air conditioning and central heat—and we shouldn't doubt that the vast majority of the world's poor people—the ones who still live without that technology—would grab it if they could.

As our taste in paintings and books shows, however, we also cling to values that are diametrically opposed to progress. That's why bookstores devote miles of shelving not only to evangelical Christianity but to other religions and philosophies. That's why airlines and hotels run endless campaigns evoking tropical beaches and exotic cultures.

On the one hand, the mechanistic view is so embedded in our society that we want to be machines ourselves: The diet and exercise industries thrive off our doomed desire to have flesh as hard as steel. Yet we admire the trumpeter who becomes one with his horn, the potter who becomes part of the clay on his wheel, the fly fisherman whose arm is fused to the rod.

COMMODIFYING ROMANTICISM

There are plenty of people smart enough to make money from the romantic side of our divided cultural personalities. Advertisers know that people want the sense of power that comes from a 300-horsepower engine, and they know that people also want at some emotional level to escape their modern lives, so the ads boast of horsepower in a sport-utility vehicle helicoptered to some crag in southern Utah.

You don't have to be a high-powered ad agency to play us this way. The restaurant in the Oklahoma City bus station is called The Chuckwagon. There's not a wagon in sight, or a cow, but the owner knows that people getting out of a metal box that's been vibrating for hours will be drawn to the solitude and quiet they associate the old West. The bus station's newsstand, meanwhile, is called Ye Olde Gift Shoppe. Don't bother looking for a snuff box. The name is supposed to entice exhausted customers to buy chewing gum and Kleenex, and the owner draws them in by making them feel good. He does that by evoking a past in which peasants froze in the mud but were part of the natural world.

More examples? Many American towns each year have a medieval fair, full of craft objects that appeal to people bored with factory-made things. Every town, too, seems to have an antiques mall, where people hunt for objects that have not come from a product-development chain. Restaurants advertise entrées as home cooked, even if the food is cooked in a distant factory. Frozen-food manufacturers label their boxes of factory-made products with pictures not of the assembly line where the food was made but of a grandmotherly figure purporting to be the inspiration for whatever's inside the package, which usually included plenty of ingredients that nobody's grandmother ever heard of. It's no accident that commercial cafeterias, which do not disguise their efficiency, have fallen on hard times. It's no accident that the Formica Corporation, whose shiny countertop plastic was ubiquitous in the 1950s, has declared bankruptcy. Home buyers today want a slab of granite, the thicker the better.

HOUSE DESIGN AS ROMANTIC EXPRESSION

Step back now from art, literature, and commodities; contemplate romanticism in the cultural landscape. We could begin with the external appearance of houses, very few of which look modern—certainly not those built by the million in suburban tracts. Internally, they're full of machines, as well as high-tech moisture barriers, insulation, and wood substitutes. Externally, however, they are decorated so realtors can describe them as Tudor, French Provincial, or Florentine—anything rooted in the preindustrial past. It's almost a law that a builder who builds truly modern houses—the kind that the Bauhaus architects would have admired—is either working for a rich client or about to go broke.

Every now and then, an architect breaks the law. A good example is the Gregory Ain tract in Mar Vista, near Santa Monica, California. Ain was a militant modernist, and in the 1940s he designed and built about 50 severely functional houses in a three-block neighborhood. They were small—about 1,000 square feet—and they sold in 1948 for about $11,000. Today, realtors say they're worth $600,000 and carry a premium of at least 15% over conventional homes. They're also the first modernist neighborhood in Los Angeles to be designated a historic-preservation zone. There's some irony in this—antitraditional architecture acquiring traditional status—but true modernity is so rare in tract housing that when produced on this scale it becomes culturally noteworthy. Back in 1948, however, the homes were a flop. Both Ain and the builder, Advance Development, lost money. A second phase was canceled. Has the law ever been broken with impunity? Perhaps in Palm Springs, California, where some 2,500 modernist homes were built in the 1950s and 60s by George and Bob Alexander. More are being built today with glass walls to bring the desert indoors. High land values, however, compel much greater densities than 50 years ago.

Meanwhile, builders around the world—not just the United States—prosper while indulging in archaic fantasies. England's Chafford Hundred consists of some 2,000 houses built in mock Tudor and Georgian styles near North Stifford, Essex. It's a long way from Le Corbusier (1887–1965) and his machine for living. Poundbury, an extension of Dorchester, consists of a growing community of 750 people—in 2020, 5,000 people—living in a spatially integrated community of houses and businesses in buildings with double-paned windows but very traditional lines and materials, especially brick and slate. Poundbury homeowners are forbidden from putting satellite dishes on their roofs, and traffic is deliberately

A house in the Gregory Ain Tract, Mar Vista, Los Angeles County.

impeded with bumps and curves. (Poundbury is on a sliver of the 150,000-acre Duchy of Cornwall, which is owned by the Prince of Wales, famously outspoken in his opposition to modern architecture.) Farther east, in Slovakia, decrepit palaces are now for sale by the government for as little as a penny, provided the purchaser invests $75,000 or more in rehabilitation. If Slovakia doesn't grab you, try Santa Fe. Tell a realtor that price is no object, and she'll take you to the most expensive neighborhood in town. Amazingly, it has gravel roads. Plenty of Santa Fe neighborhoods are properly paved, but if you have money to burn, you'll get gravel. You'll also be able to stroll down to Canyon Road, which has a single lane lined with galleries and restaurants with $50 lunches.

ESCAPING THE CITY

Which brings us to the American infatuation with the suburb, which is the usual name for the antiurban city.

Many Americans in the late 1920s saw the dramatic paintings of Hugh Ferris, full of skyscrapers and highways drawn to astonish the viewer with the promise of the future. (Ferris' work, collected in *Metropolis of Tomorrow* [1929], is still impressive.) Or they read Le Corbusier's *La Ville Radieuse* (1933), which amounted to the theoretical underpinnings of Ferris's paintings. They couldn't actually see a Radiant City, but we can come pretty close with Crystal City, that chilling line of high-rises that front the Potomac River south of the Pentagon.

You might think that a nation infatuated with fast cars and faster computers, not to mention warp-drive spaceships in distant galaxies, would welcome Ferris's and Le Corbusier's future, but just as Americans want traditional-looking houses so, too, when it comes time to buy a house, they want to live in some proximal imitation of nature. That's why a subdivider in California will market his tract homes under an absurd name like The Wilderness. Like most developers, he'll trim his houses with barebones landscaping, and for decades to come people will strive to make the grass around these homes into a biological desert, with one plant endlessly cloned in a world without a single insect. It's absurd to call such an environment natural, but in choosing it Americans choose, as best they can, to live in an environment that is reminiscent of nature, not the city.

It's a dream that's been reinforced for a long time, notably by Ebenezer Howard and his *Garden Cities of To-morrow* (1902). Howard himself was inspired by *Looking Backward*, an 1888 utopian novel by the American Edward Bellamy, and Howard in turn influenced Frank Lloyd Wright's 1935 proposal for an extraordinarily sprawling settlement to be called Broadacre City.

Broadacre City was never built, but the Metropolitan Life Insurance Company picked up the garden-city idea and after World War II built several housing projects based on it. One was Park LaBrea, in Los Angeles; another was Parkmerced, in San Francisco. The houses in both were attached to one another but grouped to enclose courtyards of shared lawn and trees. Streets were curved, as they almost universally are in modern suburbs. "It slows traffic," people say. "It's good for kids." But the breakaway from the grid began in the 1860s, long before the automobile. The originator was Frederick Law Olmsted (1822–1903), who designed the Chicago suburb of Riverside.

Riverside had predecessors, including Llewellyn Park in New Jersey and Lake Forest near Chicago, but Riverside was the first subdivision that brought nature and natural shapes to the middle class, instead of the wealthy. It should come as no surprise, because a few years earlier Olmsted had designed New York's Central Park. Fortunately for the middle class and their suburban dream, developers can squeeze as many houses onto a curved layout as a gridded one, and the footage—and cost—of roads and infrastructure can be much less. By 1950, all but the most hidebound developers had abandoned the grid.

There's an exurban option, too. The second-home industry in the United States is worth $50 billion annually. In the year 2000, 400,000 second homes were sold in the United States. The next year, the average 1,500-square-foot condo or house on the Nevada side of Lake Tahoe cost $400,000. Sanibel, on Florida's west coast, was a bit more expensive. Pawleys Island, near Myrtle Beach in South Carolina, was still higher, with prices starting at about $650,000. Poor Branson, Missouri, was almost cheap at a measly $200,000.

Who was buying? Mostly baby boomers, the same people who send their children to colleges whose campuses—the word comes from the Latin for fields—are rarely in big cities. It's no accident that American state universities were set up in places like Berkeley and Austin and Urbana, rather than wicked San Francisco, Houston, or Chicago. Even the most hard-boiled parents are romantic enough to think that lawns and flowers are good for their children, or at least safe.

POSTMODERN ARCHITECTURE

What about downtown, the belly of the beast? Miesian high-rises have become historical monuments, a testament to the lost faith of the modernist architects who believed that buildings could revolutionize society, could bring about an era of social justice merely by putting everyone into buildings from which status symbols had been stripped. Le Corbusier, in the words of Fil Hearn, had wanted "an austere, rational lifestyle that rejected luxuriant excesses imputed to the bourgeoisie. . . . " But people, it turned out, much preferred the pursuit and enjoyment of status. No wonder: Hearn says that Le Corbusier's "conception implied a world view that was more socialist than democratic, for the implementation of which he always imagined an unspecified but controlling central authority."

What replaced the unremitting efficiency of the International style? If you base your answer on the buildings discussed by architectural critics, you'll say that architects today seek novel and eye-catching forms. An early case was the Pompidou Center, a Paris museum opened in 1977 and designed by Richard Rogers and Renzo Piano. Brightly colored plumbing and ventilating systems were exposed on the outside of the museum to flaunt its mechanical nature without regard for traditional notions of architectural elegance. Twenty years later, the talked-about building was Frank Gehry's Guggenheim Museum in Bilbao, Spain. It opened in 1997 and immediately began attracting 100,000 visitors a month, less for the collection than the building housing it.

Celebrity architects don't build much, however. When the governor of New York State insisted on hiring Daniel Libeskind to plan the new World Trade Center, the main lessee of the site insisted on partnering him with David Childs, from Skidmore-Owings-Merrill. "Danny," he told Libeskind bluntly but honestly, "you've never designed a tall building." If, then, you answer the question by considering what actually gets built, you'll conclude that architects today are as traditional as home buyers in their fondness for historical allusion.

The pioneer here was the 1988 AT&T (now Sony) building in New York, in which Philip Johnson, a one-time assistant of Mies, topped that structure so that it looks like a Chippendale cabinet. Its influence has been particularly strong on college and university campuses, which are a gushing fount of fraudulent tradition. Fifty years ago, for example, the University of Oklahoma built several severely modern buildings: In the case of the building that now houses the School of Architecture, hallways were broader at their ends and narrower in the middle, because fewer people would pass there. The campus planners were openly contemptuous of the pseudo-Gothic buildings dominating the campus at the time. Yet the buildings they were so proud of are no longer popular, and when architects today add a new wing, say, to the student union, they call for the cast-stone gang. They build a lounge that pretends to be the nave of a church, and they install wooden trusses that hold up nothing but themselves. It's all intended to make students feel as though they're in a community as devoted to learning as they or someone else imagines old Oxford was. Call it the delusional phase of the romantic, or perhaps just history repeating itself as farce.

I say farce because the Victorians had their own Gothic Revival, associated with architects like Augustus Pugin (1812–1852) and George Gilbert Scott (1811–1878). Passionately defended by John Ruskin (1819–1900), the Gothic Revival rested on the idea—as romantic as they come—that the Renaissance worship of reason and classical restraint was a sign of cultural decay from the creative passion of earlier, or Gothic, times.

For a century, the classical and Gothic styles did international battle, and European and American cities today are full of examples of both. You can hardly find a better contrast than the United States Capitol, rebuilt on classical lines after the War of 1812 chiefly by Benjamin Latrobe and Charles Bulfinch, and the British houses of parliament, built to a

The tapered corridor of the University of Oklahoma's building housing the School of Architecture. It was designed by Walter and Robert Vahlberg, who were determined to build rationally, regardless of how the building felt to users or whether any money was saved by what they thought was efficiency.

Non-weight-bearing trusses over a lounge in the University of Oklahoma Student Union: instant heritage.

Gothic design of 1840 by Pugin and Charles Barry. The British building is younger than the American one, but it looks older because in this war of stylistic revivals the neoclassical outlasted the Gothic and so seems newer. Ironically, Britain's Gothic Revival reversed the outcome of a much earlier battle, when Renaissance architects turned away from the then dominant Gothic and revived the architectural forms of Greece and Rome.

Meanwhile, it's hard to find a European country where architectural heritage is not a bankable resource. Perhaps an exception is Moscow, which in the flush of new wealth is busily pulling down Soviet-era buildings like the old Moskva Hotel on Red Square. Everywhere else, there are signs proclaiming neighborhoods as part of the patrimony of humanity. The phrase is pretentious and a good sign that you'll be hit up hard at the ticket office. (Sad that nations make visitors pay to see the places of which they're proudest. It cheapens the experience—all that lining up, fishing for coins, handing over a ticket. Apparently these treasures just aren't prized quite enough to be supported by the nations they own them.)

Preservation and restoration are now bankable. An early case was Bruges, the first Belgian city to capitalize on its medieval heritage. Planners deliberately choked traffic by making vehicle access to the historic core very difficult. They prohibited parking there, paved the historic-core streets with noisy stone blocks, and widened the sidewalks to confine what little historic-core traffic there was to single-lane, one-way streets. New buildings had to have the height, mass, and color of existing ones, and shop owners were forbidden to hang signs projecting from their facades. Blinking and flashing signs of course are out of the question.

If there's a downside, it's the tourists who overrun Bruges in summer and come close to dominating it even in the damp winters.

In some European cities, such as Warsaw and Dresden, historic urban centers destroyed in the war have been rebuilt exactly as they were. (The peripheries of those cities did not fare so well: Pastel Old Warsaw is encircled by gray, concrete apartment blocks, once the pride of Communist rulers who took their cues from Stalin's Moscow. Ranks of monumental apartment blocks stand as proud symbols of Communist humanity.) Elsewhere, decayed industrial districts have been recycled. West of Leeds, the immense factories of Dean Clough, in Halifax, once housed the Crossley mills, the pioneer of power-loomed carpets. Crossley closed in the 1970s, but the buildings reopened in 1983 and now house over 100 companies whose 3,500 workers are almost as numerous as the 5,000 people who once worked in the same buildings for John Crossley & Sons.

Real skill is needed to combine the modern and the traditional. McDonald's, to take a surprising example, has restaurants in Heidelberg and Cracow that have no electric signs and which highlight the buildings they're in, rather than mask them. Somebody someday should take a look at how often the colossus of Oak Brook has departed overseas from its toy-box ways back home.

PARKS AND GARDENS

Then there are gardens. They're always a barometer of how a civilization understands its relationship to nature. The romantic story here is how Europe turned away from the geo-

Steenstraat, Bruges: an early example of a city capitalizing on its history by appealing to consumer romanticism.

metric, controlling style of Andre Le Nôtre, who I mentioned in Chapter 9 in connection with gardens at Versailles. In its place is the British landscape park, derived from Chinese gardens and pioneered by Lancelot "Capability" Brown (1715–1783). Great lawns were now dotted naturalistically with trees. Rather than erect walls at the perimeter, fences were sunk in trenches called ha-has so they couldn't be seen from the house. The garden had now become proof of the owner's absorption in the natural world, rather than a sign of the owner's control of it.

It's a style that came to the United States when Frederick Law Olmsted laid out New York City's Central Park after 1858. The state legislature in 1853 had set aside the site, even though it contained enough land for 17,000 lots. It was a strange thing to do, but New York was socially ambitious, and, if London had Regent's Park and Paris had the Bois de Boulogne, then New York had to have a fine park, too. Other American cities quickly followed. By 1871, William Hammond Hall was laying out San Francisco's Golden Gate Park, replete with free-form dells.

National parks are part of this story, too. The first was Yellowstone, set aside from the public-domain lands in 1872. Yosemite and Grand Canyon followed, carved at no expense from the public lands. Many national parks, however, have involved land purchases. Grand Teton, for example, was a tiny thing when it was congressionally designated in 1929. John D. Rockefeller, Jr., tripled its size by buying land in Jackson Hole and giving it to the government. Rockefeller also gave $5 million as a matching grant to help Tennessee and North Carolina buy the land that's now the Great Smoky Mountains National Park. Authorized in 1926, land purchases went slowly, and the park was not dedicated until 1940. There's a similar story in the Everglades, which only became a national park in 1947, decades after preservationists first spoke up in defense of the river of grass.

In more recent cases, such as Redwood National Park, Congress has authorized the use of federal funds for land purchases. The price tag for a national park nowadays can be huge, and Republican administrations have thought it an unjustified use of taxpayer money. Still, even they haven't proposed abolishing the national parks. That's a remarkable fact, when you consider the implicit heresy of setting aside blocks of land to remain undeveloped, beyond the reach of the Baconian transformation of the world. National parks, after all, say without words that nature has a value apart from its development value. Such an idea cuts at the taproot of a technological society. It also creates huge management challenges for a park service that had to deal with 277 million visitors in 2002, up from 80 million in 1960.

(Lest you think there's something uniquely American about these parks, the Canadians created a Rocky Mountain Park—now Banff—in 1887. New Zealand in 1899 created the Tongariro National Park, near Lake Taupo on the North Island. In 1925 the Belgians created the Albert National Park as a wildlife sanctuary in the Belgian Congo, mostly for gorillas. A year later, South Africa created the Kruger National Park, adjoining Mozambique. The United States gets credit only for being the first.)

There's hobby gardening, too. Over the last decade, the Produce Green Foundation of Hong Kong has rented tiny garden plots to office workers who live 15 miles away in high-rises and want—maybe even need—to dig in the dirt and grow a few vegetables. A new idea? Far from it: A century ago, Germany began providing plots to apartment dwellers. They were called Schreber Gardens after their promoter, Daniel Schreber, a psychologist who thought that gardening was good for the mental health of factory workers. Today, there are a million such gardens in Germany—85,000 in Berlin alone. They're extremely popular in Poland, too, where on weekends people leave their apartments and go to a shack on a 1,000-square-foot plot loaded with fruit and vegetables.

Hard to believe you're in the middle of the world's most important city: Central Park, New York City.

Americans are too suburban for many such gardens, though Long Island's Eeco Farm rents 20-by-20-foot plots to New Yorkers. Gardening apparently does wonders for New York's prison inmates, however. The Greenhouse Project, run on Rikers Island by the Horticultural Society of New York, puts 125 men and women each year through a flower-growing program. Two-thirds of the prisoners at Rikers are reincarcerated after their release, but the recidivism rate for graduates of this gardening program is 5%. The comparison isn't quite fair, because the project accepts only nonviolent offenders, but whatever the statistical significance of the numbers, working with plants seems to help the program's graduates stay out of jail, after they're released.

Americans meanwhile spend an amazing $2 billion a year feeding birds. It's not so great for the birds. Attracted in droves to backyard feeders, millions crash into windows, are caught by waiting predators, or succumb to diseases that spread quickly through congested populations. (Highly contagious conjunctivitis, for example, kills millions of finches. Their eyes become glued shut, and the birds starve to death.) Why do Americans feed the birds? If you ask them, they'll probably say something about wanting to be kind, especially in winter. Just maybe, however, there's another reason, one that has less to do with helping birds than helping themselves.

CHAPTER 16

POLITICAL REACTIONS

In the previous chapter I considered the romantic rebellion against a technological society and some of the many ways in which that rebellion expresses itself in the cultural landscape. Now I want to take up the political reaction to progress. The distinction is that the romantic rebellion comes from people who enjoy the benefits of progress but see its limits; political problems arise when people want those benefits but can't get them.

WHEN IDEOLOGY FAILS

Many pages ago, in Chapter 6, I stressed ideology as the key to the emergence and maintenance of civilizations. Later, while discussing ours, I said that progress was our ideology. I suggested, in other words, that we accept the inequality in our society in exchange for prosperity, or at least the opportunity to seek to prosper.

For many, this ideology works. They see the payoff in every new house, car, and brimming shopping bag. For many more, however, and for a majority of people around the world, there is almost no hope of prosperity. What then? How can a civilization endure when its ideology is not credible?

FEAR

The leaders of civilizations have always known that ideology must be reinforced by force or at least the fear of force. Napoleon had his secret police, headed by Joseph Fouché, sometimes considered the father of the police state. So did the czars and their Soviet successors, whose secret police one day arrested Nikolai Vavilov, the plant explorer I mentioned in Chapter 4. Hitler had his Gestapo, the *Geheime Staatspolizei* or secret state police, which was run, without appeal, by Heinrich Himmler. Since those bleak days, secret-police organizations have supported many other rulers, some of whom saw their regimes collapse but others of whom still cow millions of people.

Fear stabilizes the democratic nations of the world, too, although it usually is the fear not of a knock on the door in the dead of night but of economic insecurity. About 40% of the Israeli labor force, for example, works at the country's minimum wage. That's about $800 a month, in a country where gasoline is $4 a gallon. So much for the myth of rich

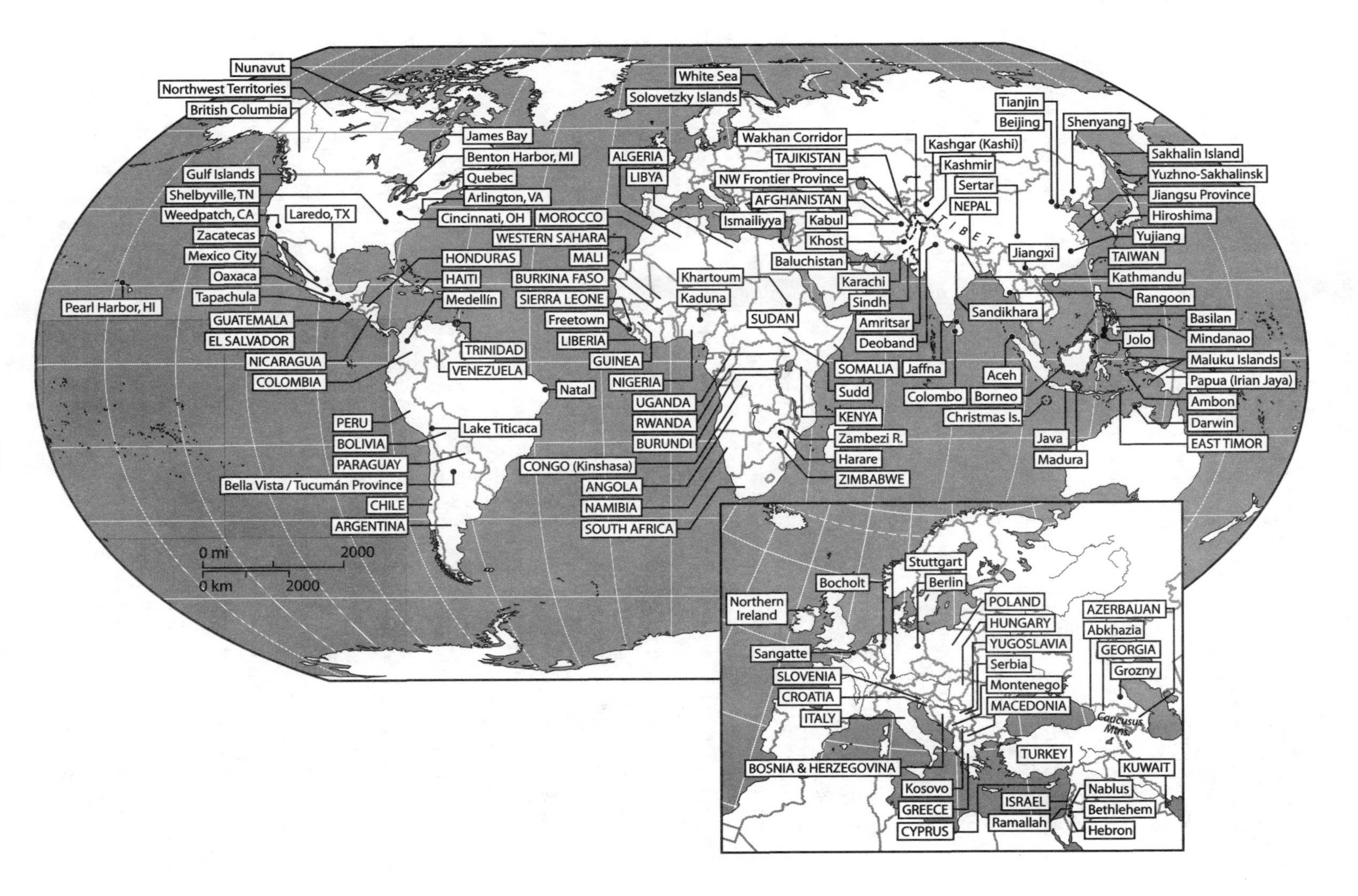

Nunavut
Northwest Territories
British Columbia
Gulf Islands
Shelbyville, TN
Weedpatch, CA
Laredo, TX
Zacatecas
Mexico City
Oaxaca
Tapachula
Pearl Harbor, HI
GUATEMALA
EL SALVADOR
NICARAGUA
COLOMBIA
James Bay
Benton Harbor, MI
Quebec
Arlington, VA
Cincinnati, OH
HONDURAS
HAITI
Medellín
TRINIDAD
VENEZUELA
Natal
PERU
Lake Titicaca
BOLIVIA
PARAGUAY
Bella Vista / Tucumán Province
CHILE
ARGENTINA
0 mi
2000
0 km
2000
White Sea
Solovetzky Islands
ALGERIA
LIBYA
MOROCCO
WESTERN SAHARA
MALI
BURKINA FASO
SIERRA LEONE
Freetown
LIBERIA
GUINEA
NIGERIA
UGANDA
RWANDA
BURUNDI
CONGO (Kinshasa)
ANGOLA
NAMIBIA
SOUTH AFRICA
Khartoum
Kaduna
SUDAN
Wakhan Corridor
TAJIKISTAN
NW Frontier Province
AFGHANISTAN
Ismailiyya
Kabul
Khost
Baluchistan
Karachi
Sindh
Amritsar
Deoband
SOMALIA
Sudd
KENYA
Zambezi R.
Harare
ZIMBABWE
Jaffna
Colombo
Kashgar (Kashi)
Kashmir
Sertar
NEPAL
TIBET
Tianjin
Beijing
Jiangxi
Sandikhara
Aceh
Borneo
Christmas Is.
Java
Madura
Shenyang
Sakhalin Island
Yuzhno-Sakhalinsk
Jiangsu Province
Hiroshima
Yujiang
TAIWAN
Kathmandu
Rangoon
Basilan
Jolo
Mindanao
Maluku Islands
Papua (Irian Jaya)
Ambon
Darwin
EAST TIMOR
Stuttgart
Bocholt
Berlin
Northern Ireland
POLAND
HUNGARY
YUGOSLAVIA
Serbia
Montenego
MACEDONIA
Sangatte
SLOVENIA
CROATIA
ITALY
AZERBAIJAN
Abkhazia
GEORGIA
Grozny
Caucusus Mtns.
TURKEY
KUWAIT
BOSNIA & HERZEGOVINA
Kosovo
GREECE
CYPRUS
ISRAEL
Ramallah
Nablus
Bethlehem
Hebron

Jews. Why do Israeli workers accept such conditions? Answer: "There's bad and there's worse." So it is in the United States. A worker on the General Motors assembly line in Dallas thinks his job is killing him but keeps at it. He explains: "When you're married with kids you have to make a living." Laid-off textile workers in the hard-hit towns of the lower Rio Grande Valley say, "I have to make enough to hold my house."

RELIGION

Fatalism can sustain civilizations, too. It's even a cliché, coming from an 1844 critique of Hegel in which Karl Marx wrote that religion is "the sigh of the oppressed creature, the heart of a heartless world, and the soul of soulless conditions. It is the opium of the people." Opium or not, many people do bear poverty more readily if they consider this life only a trial. Such a belief is commonplace among Christians, Muslims, Hindus, and Buddhists, all of whom have strongly antimaterialist traditions.

MIGRATION

For still other people, modern civilization's failure to deliver on its side of the bargain is the spur to move. I'm not thinking here of the approximately 10 million political refugees in the world today. (That number, by the way, excludes 4.5 million Palestinians, who, in a triumph of diplomatic fiddling, are technically not refugees, even though many live their whole lives in camps administered by a U.N. agency.) Nor am I thinking of another 5 or 6 million people who have been internally displaced in their own country. (Colombia ranks first in this category.) I'm excluding another 2.5 million returned refugees in need of resettlement help and another million seeking asylum. I'm overlooking still another million, who are stateless.

In short, I'm excluding the 20 million people of concern in 2003 to the U.N. High Commission for Refugees and am considering instead the tide of economic migrants who move—as did the ancestors of most Americans, Canadians, Australians, and New Zealanders—because of poverty at home. Approximately 175 million people fall in this vast category. Not all these people are desperately poor, nor are they all illegal. Many, however, are both, and it is against them that so many nations build futile walls. Why futile? Because permanent residents of the United States, for example, earn, on average, $20,000 a year more than they would make if they had stayed home. It's the same in Europe, which has more of these people than North America—56 compared to 41 million, by U.N. estimates.

This is a story of heartbreak. A half million illegal immigrants get to Europe each year. Among them are 30,000 Turks, mostly young men desperate to support their families. The men enter Bosnia, which allows Turkish entry without a visa. From there, they try to enter Croatia, Slovenia, and Italy. They are joined by Afghans, Kurds, and Pakistanis who have already got illegally to Turkey.

During the year 2000, the Turks caught almost 100,000 of these illegal foreigners, but others made it through and into countries beyond. Unable to legally enter Bosnia, they try getting into Greece and through the porous borders that lie beyond. Many succeed. In January 2001, however, the cargo ship *Pati*, full of illegal migrants, set out from a Turkish port and ran aground near a luxury hotel. Forty-seven people on board the vessel died, and the lucky but miserable survivors were returned to their countries of origin. About a year later, in March 2002, half a dozen Pakistanis trying to get from Bulgaria to Greece were spotted by Macedonian soldiers. It was night, and the men were resting along a deserted road. The soldiers shot them dead, and the Macedonian authorities proudly claimed that its military had captured and killed a group of terrorists. Two years later, after an investigation, a spokeswoman for the Interior Ministry said the episode "was staged, a monstrous killing of seven economic migrants" done "to demonstrate Macedonia's commitment to the war on terrorism." The Pakistani father of one of the victims says that he now tells people to "eat dirt but never send your children abroad."

The migrant tide has changed the ethnic composition of western Europe and the political climate of countries that, more than most, have been able until now to consider themselves nations in the sense of sharing a culture. Right-wing politicians have built careers deploring the cultural change that accompanies immigration. Millions of French voters, for example, have cast anti-immigrant ballots, especially along the Mediterranean coast and near the German border, the parts of France with the most foreigners. In Belgium, people in the smaller towns say that Brussels is no longer Belgian. They mean that it has been overwhelmed by migrants from other continents. They add that they aren't racists, but their hostility is directed toward newcomers with dark skin. White-skinned travelers in Asia and Africa today shouldn't be surprised by hearing well-off people they meet say that they don't even want to visit Europe or North America; there's too much hostility, too much suspicion.

Migrants have been heading east, too. Some land in Australia. If undocumented, they are promptly confined in immigrant detention centers. Three-year-olds have been put with their parents in punishment cells in these places, 8-year-olds have been handcuffed, and teenagers have sewn their lips together while on hunger strikes. The centers are run by security professionals hired by an Australian subsidiary of Florida-based Wackenhut Corrections Corporation.

In an attempt to stop the migrant flow, in 2001 the Australian government took the remarkable step of refusing permission to land to hundreds of refugees aboard *The Tampa*, a Norwegian container ship. This was so popular with Australian voters that it probably accounted for the reelection that year of the country's prime minister, John Howard. Some of Australia's centers have been closed, such as the one at remote Curtin, west of Darwin, but an even more remote one is being expanded on Christmas Island, a speck of Australia that lies off the south coast of Java. Denied visas, the detainees at Christmas Island took over the camp at the end of 2002, burned part of it, and forced the camp guards to retreat. The government stood firm, however, and Australia remains the only rich country that routinely jails asylum seekers who arrive without proper documentation. In its defense, Australia says that it faces greater assimilation problems than any other rich country. It may be true. A quarter of all Australians are foreign-born, while the comparable fraction for the United States and most of western Europe is a tenth. In different but related terms, foreign-born workers make up 16% of the American work force but 26% of the Australian. Still, the Australian defense is akin to the justification of cannibalism among shipwreck victims, which goes to show just how seriously nations can take the threat of unrestricted economic migration.

Millions of Chinese are meanwhile moving from China's interior to the coast. It's a classic rural-to-urban shift. The scale of this migration, however, is exceptional, with some 200 million migrants expected between 2000 and 2010. Beijing alone has 11 million people with Beijing *hukous*, or registrations. It also has 3 million *waidiren*, outside people who do the dirty jobs. That doesn't discourage new arrivals. Chinese migrants also try to go abroad, of course. Often they succeed, but trucks and shipping containers in England and the United States are also periodically found to contain suffocated corpses.

In Africa, children from Burkina Faso and Mali migrate to the Ivory Coast, where they hope to find work, particularly in cocoa plantations. Many become near slaves, but the flow continues because conditions back home are so bad.

And then there's the case of Latin America, where every nation except Brazil and Chile has more emigrants than immigrants. There has been a recent flood of Argentinians seeking residency in Spain, but the better publicized case is of migrants seeking to enter the United States. Nearly 1,000 enter the United States every day from Mexico alone. The number has risen slowly since 1993, which is surprising because during the same period there has been a doubling in the number of Border Patrol officers. "What are you doing?" shouts one officer to a Mexican swimming across the unfenced Rio Grande at Laredo. "Fishing," is the good-natured reply. Maybe he'll be luckier than the 400 people who died in 2003 while attempting to enter the United States illegally from Mexico.

There are so many Mexicans in the United States today—about 8 million—that a majority of the population of at least one Mexican state, Zacatecas, is said to live in the United States, not in Mexico. Mexicans are now living in Weedpatch, a town near Bakersfield once famous for housing Okies. For them, as it was 60 years ago for the Oklahomans, hope is a powerful lure. The old ads from the 1930s read as follows: "California. Cornucopia of the World. Room for Millions of Immigrants." You do not see such ads anymore, but the message is still drawing people like a magnet, and not just to the Southwest. Shelbyville, Tennessee, had fewer than 100 Spanish speakers in 1990. Then Tyson Foods arrived and attracted several thousand Mexicans, legal and not. Now the Shelbyville *Times-Gazette* every Friday carries a two-page supplement in Spanish. Conditions in Weedpatch,

Hidden by the reeds, a Mexican tries to wade into Texas. The railway bridge has sensors to catch anyone trying to stay dry.

Shelbyville, and a thousand other places fall very far short of what the newcomers hope for. They stay, in the words of one, because "what's worse than being exploited in Fresno is not being exploited in Fresno."

Back in Zacatecas, more than $1 million is received every day from relatives up north. It's a small part of the $14.5 billion annually sent home by Mexicans living in the United States. Some of those people have become wealthy, and a few are working to create job opportunities in Mexico. They fund projects to raise flowers in greenhouses, for example, or weave wool carpets. They hope that these projects will provide jobs so villagers can stay home, but there are many impediments. One is the chronically poor state of Mexico's rural roads.

Mexico is meanwhile trying to bar Central American migrants who enter through Guatemala. Rather than continue with the old practice of merely deporting illegal migrants to the Guatemalan side of the border, where they wait a bit then try again, Mexico has a *Plan Sur*. Migrants caught crossing from Guatemala are shipped all the way back to their home countries, most likely Honduras, El Salvador, or Nicaragua. In the year 2000, for example, Mexico under *Plan Sur* deported 168,000 illegals; in 2003, 155,000. Few if any came from the north, of course. That's why pedestrians on the Rio Grande bridge at Laredo are ignored so long as they're walking south.

How easily we ignore it all. Central Americans without money often try to ride the rails north. It's very dangerous—so dangerous that there's a philanthropic clinic at Tapachula, the southern railhead. The patients are all amputees who either fell off a train from exhaustion or were pushed off by local gangs. Fitted with artificial limbs, the patients are sent home to a family they were trying to help.

A mile from the French end of the railway tunnel under the English Channel,the Red Cross runs a refugee camp at Sangatte. Signs warn in four languages of mortal danger by electrocution or crushing, miles of razor wire surround the tunnel entrance, and 100 security guards are stationed there. In August 2001, however, hundreds of refugees overwhelmed the guards, and 44 managed to get 7 miles inside the tunnel before they were stopped. On Christmas Day that year, 500 asylum seekers forced their way into the tunnel, which was closed for 10 hours. People do sometimes get all the way through by clinging to the side of a train. Some have been apprehended in England. Others have died or been hurt in the attempt. In 2003 some 200 people were still trying to enter every day. That's when the government of France, under pressure from Britain, agreed to close the Red Cross camp. Where its 1,800 residents will go is another matter, but the Red Cross estimates that of the 50,000 who have used the camp an astonishing 42,000 finally did make it to the United Kingdom, mostly on ferries.

POLITICAL ACTION

Ideologically weak regimes can survive, in short, by keeping their people fearful, by encouraging religious fatalism, and by using emigration as a safety valve. Yet there is still another possibility, with governments failing to suppress their populations and with the people themselves forcing change on reluctant elites. In its soft version, this is social reform. The hard version is revolution.

There's no mystery here. Look at old photos of children working in mines and mills, then compare them to pictures of the children of the mine and mill owners. It's hard not to wonder at the injustice of a world where one child stands at a machine all day while another wears silk, visits France, and studies music. What kind of world is it where one child is born to poverty, another to wealth? Isn't it reasonable to assume that human beings can build a society where such inequalities do not exist? Think of William Blake's "The Chimney Sweeper" (1790–1792), with its bitter condemnation of a society that abuses its children both physically and socially. Think of Robert Burns and his poem (he called it a bagatelle) of 1795, "A Man's a Man for A' That." The last stanza is this: "For a' that an' a' that,/ It's coming yet for a' that,/ That Man to man, the warld o'ver,/ Shall brother be for a' that" (the *a'* stands for "all"). Beethoven took Schiller's similar words as the text for the finale of his ninth symphony. Think of Francisco Goya (1746–1828) and his "3 May 1808," with humanity helpless before faceless oppressors. Think of Dickens in the 1840s, with novels that make his England seem monstrously unjust.

REFORM

Without reformers, slavery wouldn't have been abolished in England in 1835—or in the United States in 1861. Without reformers, women and children would have continued working in English coal mines long after 1842, when their labor was banned by a parliament prodded by Anthony Ashley Cooper, the seventh Earl of Shaftesbury. A few years later, Shaftesbury helped enact a then-progressive 10-hour work day. Bismarck introduced old-age pensions to Germany in the 1880s, not because he was a humanitarian but because he wanted to weaken Germany's socialists. Lloyd George brought pensions to Britain in 1910, and Franklin Roosevelt brought them to the United States in the 1930s. By now, Americans are used to minimum-wage laws, a 40-hour workweek, paid vacations, pensions, and social security. They think they're well off, but Europeans have socialized medicine and maternity leaves that make the United States seem barbaric in its devotion to the mercies of the marketplace. Swedish parents, for example, can take a total of 480 days of parental leave for each child, and they'll draw 80% of their pay while they do it.

The average American takes 13 vacation days annually. The average Japanese or Canadian takes twice that many. The French and Italians take 37 and 42 days. Even the punctilious German official packs up at Friday noon—maybe a trifle earlier—for a long, energizing weekend. Over the course of a year, the German works 1,400 hours, but the American works 1,800. The Americans will also keep working longer and shake their heads in disbelief on learning that 60% of all Europeans aged 55–64 are retired.

Reform seems less successful if we consider the declining membership in American labor unions. It's down from about 35% of the American work force in the 1950s to about 13%. (The sign over the entrance to the now closed Fruit of the Loom plant in Harlingen, Texas, read "Wear the Union Label: Unemployed.") There's trouble on the political front, too. Democrats used to wear the label "Liberal" proudly, as an emblem of their belief in the power of government to cure social ills. Today, the label is an electoral death sentence, routinely prescribed by conservative political consultants. Britain's Tony Blair hugs the middle of the political spectrum and shuns the radical roots of his own Labour Party, with its historic ties to labor unions and support for state-owned corporations.

This near death of liberalism can be explained by the brutal fact that social-welfare legislation is a drag on economic growth. Germany, for example, is stagnating because it refuses to abandon hard-won social reforms. It pays high wages, for starters. That sounds good, but Bosch, which has made automobile starters for many years in Stuttgart, moved to Hungary because workers there make $4 an hour instead of $26. The same logic leads hotel operators in Berlin to send their sheets to Poland for laundering. But there's more at stake than wage levels. Volkswagen builds all its new Beetles and Jettas in Mexico because German labor law makes it almost impossible to fire a worker. And more still: Paying for the benefits mandated by reform legislation is so costly that many German manufacturers are going out of business. They prospered when their customers bought only from German suppliers. They're dying now, as foreign suppliers undercut them at every turn. That's why the workers at the Siemens plant in Bocholt voted in 2004 to go from a 35-hour workweek to a 40-hour one, at no pay increase. They knew that Siemens has found Chinese engineers capable of making German-quality products. They hear their own chief executive say, "We have to ask ourselves what we have to offer."

So far, the German electorate is hanging tough, even though the cost of Germany's social-welfare programs is bringing the economy to its knees. In 2004 the German chancellor, Gerhard Schröder, froze state pensions and began charging Germans about $50 a year for doctor visits. He explained that "our policies are necessary and correct," but that didn't stop his party's approval rating falling to 24%, its lowest ever. If the United Kingdom is any guide, change will be delayed until a moment of crisis. That's what happened when Margaret Thatcher arrived at Downing Street with her axe.

The United States is well ahead of continental Europe in undercutting its own relatively modest reform achievements. Franklin Roosevelt introduced a minimum wage, in his words, to "end starvation wages." (Today, such a phrase coming from a Democrat would be condemned by Republicans as an incitement to class warfare.) Dwight Eisenhower supported that minimum wage in the name, he said, of fairness. The minimum wage continued to rise to a peak during the Nixon administration, when it was worth $6.82 an hour, measured in 2001 dollars. The minimum wage came under attack during the presidency of Ronald Reagan, who said it was responsible for "more misery and unemployment than anything since the Great Depression." It's recovered from the low of his time, but while Democrats say it's too low, conservatives argue that a low (or no) minimum wage will increase the rate of economic growth. Oddly, this position isn't persuasive for minimum-wage workers. It falls almost as flat for the quarter of all working Americans who make less that the $8.70 an hour needed in the 1990s to keep a family of four above the official poverty line. The hard-nosed solution, of course, and a very common one, is to have a working spouse. It's a strategy that works wonders for toddlers packed off to daycare.

Why has Europe been so much more generous than the United States in funding social-welfare programs? One answer, of little consolation to American liberals, is that Europe's population has historically been ethnically homogeneous, so that the middle class could identify with the poor much more easily than in the minority-rich United States. In the words of one study, "racial animosity in the US makes redistribution to the poor, who are disproportionately black, unappealing to many voters." It's an argument that helps explain the weakening of the Democratic Party in the American South after Lyndon Johnson's Great Society programs.

REVOLUTION

And how has the hard form of political action fared? The answer is badly, at least if one considers the fate of the nations that set out to follow Karl Marx's advice. Marx himself was driven by outrage over the gross inequalities embedded in every civilization, but in practice his corrective dictatorship of the proletariat turned out to be dictatorship, period. It didn't take long. Stalin's purges in the 1930s put an end to the idealism of many of Communism's strongest supporters. For many, including some who kept the faith, it put an end to their lives. Even survivors who did well by Communism came to see its flaws. Werner Klemperer, for example, a professor who lived comfortably in East Germany, by 1958 wrote that Communism could "pull primitive peoples out of the primaeval mud, but it pushed civilized people back into it." Inconceivable as it would have been to people of his generation, the Soviet Union was abolished in 1991 by the stroke of a Russian pen. In China, the government insists that it remains staunchly Communist, but its "socialism with Chinese characteristics" turns out to be capitalism and, in the words of James Kynge, a "ferociously competitive brand of it." Soviet-style Communism carries on only in Cuba and North Korea, where everything is in decay except the totalitarian apparatus.

It does seem that a technological civilization is firmly entrenched, dogged in its defense of economic efficiency and resistant to palliatives both soft and hard. Still, social upheavals are a daily occurrence on this planet. They cause great suffering and lead at best to small gains and often not even that. They will continue to do so until the promise of universal prosperity becomes a reality or until the repressive hand becomes omnipotent. I want to sample them in enough detail to justify my use of the phrase "daily occurrence." Then I'll venture into a few speculative generalizations.

ASIA

A decade after the dissolution of the Soviet Union, things are getting better for the residents of Moscow. (Things are usually better in the national capital.) They're much better for the oligarchs who grabbed the Soviet Union's most valuable assets, but they're also better for the many shoppers lined up at the 68 cash registers ka-chinging at Moscow's Auchan hypermart. It's part of a French chain.

Money's coming to Yuzhno-Sakhalinsk, too, the largest city on Sakhalin, an island 4,000 miles east of Moscow—9 hours flying time on a good day. There's oil nearby, which is why the Megapolis shopping center now has an escalator. It's the first on an island that in czarist times was a place for political exiles and until 1990 was closed to foreigners. Russians with jobs in the Sakhalin oil business make over $500 a month, three times the Russian average.

To redress the balance, visit Koltsovo, once a collective farm about 125 miles southeast of Moscow. The land now belongs to UGRA, a private company that farms 70% of the land and has let the rest slide back to trees. There were 30 people here in 1991; now there are 17. "The collective was so rich," one woman says. Her husband adds that when the Soviet Union collapsed, "everything was stolen and disappeared." Someone even took the village pump, so people have to carry water in buckets from a nearby spring.

Farther down toward the nadir of the scale, there's Grozny, the capital of Chechnya. This is an area that caused the Russians trouble when they seized it in the mid-19th century. In Soviet times, however, Grozny became a city of 400,000. It was known for its oil refineries and chemical industries. Today, 200,000 remain. They struggle without public services. Some inhabit almost deserted apartment buildings, where plywood sheets separate living quarters from rooms whose walls and floors have been blown away by Russian artillery. Many Chechnyans depend on bread lines, and none dare go on the streets after dark, for fear of being shot by Russian troops fighting Chechnyan separatists. The Chechan hatred of Russians has been compared to the human hatred of rats, and the Russian troops reciprocate with almost incomprehensible atrocities. One Chechen mother is consumed by hatred. "The bombs destroyed my house," she says. "Soldiers killed many of my relatives, and this war has turned my children into cripples. . . . I hate this war, and I hate what the Russians are doing to us." Another says, "How can people live like this? How could we live like this? But we live."

Drift east to what is now Tajikistan. A woman who spent her working life in the irrigated cotton fields raised 10 children and received a Heroine Mother medal from the Soviet government. Today, she and her husband have a television, a refrigerator, and a stove, but the first two are broken, and there's no fuel for the stove, so they cook over an open fire. The fuel is stolen coal. The menu is potatoes.

Look farther east. The Chinese invaded Tibet in 1950 and forcibly incorporated it into China. Ethnic Chinese have settled in Tibet by the thousands, and thousands of Tibetans now live in exile, including about 10,000 in the United States. One lama, Khenpo Jigme Phuntsog, set up a monastery at Sertar, a very remote village in northwestern Sichuan. With 7,000 monks and nuns, it became the leading center of Tibetan Buddhism. In August 2001 the Chinese destroyed it, presumably because they saw it as a political threat. Meanwhile, at least 100 members of the Falun Gong spiritual movement have died in police custody since the Chinese authorities began treating it as a political threat.

It's not just minorities who feel the heavy hand of the Chinese government. In April 2001, 600 Chinese troops attacked the unarmed villagers of Yuntang, a village in Yujiang County, Jiangxi Province. The farmers had dared to refuse to pay what they called impossibly high taxes. Payment had been demanded even in 1998, when floods destroyed the farmers' crops. It's not the first incident of farmers protesting such taxes, but in this case Prime Minister Zhu Rongji promised relief. In 2004, Premier Wen Jiabao announced that the government would abolish its farm tax by 2007. This would leave farmers responsible for the local taxes that are intended to support schools, clinics, and other public services, but in theory at least it would reduce their tax burden by about a third.

For the moment, the Chinese government has permitted tens of thousands of villages to elect not only village heads, an empty position, but their Communist Party secretaries, a much more important one. Elections are also being held for urban neighborhood councils. Budgets are becoming more open, and elected officials are scrambling to help their constituents raise incomes, for example, through encouraging the production of profitable crops like peanuts and sesame and garlic, instead of the traditional rice and wheat. Elections may be held to elect officials at the higher-echelon township level. The Ministry of Agriculture has even been considering a plan that would replace township governments with elected boards. A contrary idea under consideration, however, is to replace them with branch offices of the county government.

Nobody knows how fast democracy can emerge in China without initiating a spiral of increasing instability, but everyone knows that the Chinese authorities don't like being hurried. One elected village head in Hejian township, about 100 miles south of Beijing, dared in 1999 to lead his villagers in a protest in front of township officials. He was jailed, charged with attacking state institutions, sentenced, then released though not pardoned. Qiu Guojun moved to Tianjin to pay off his legal debts. His teenage daughter quit high school because her family could no longer afford the fees.

Since China's economic reforms of 1997, 25 million workers have been permanently laid off from unprofitable state-owned companies, and in the particularly affected Northeast such layoffs have pushed unemployment rates to 40%. Some 20 million Chinese now receive a state living allowance—$36 a month in rustbelt Shenyang—not because of a Chinese reform movement but because the government doesn't want restive ex-workers threatening the flow of foreign investment, now running at an immense $50 billion annually. The risk of instability is so great that in 2003 the Chinese state council began to take apart the old system of *hukous*, those household registrations whose origins can be traced back to imperial times. Until now, getting a permit to live in the booming coastal cities has been almost impossible for peasants. Even so, the promise of a better life was so strong that people were willing to pay $10,000 for a forged Beijing *hukou*. Now, in an effort to provide a safety valve for desperately poor people, the authorities are prohibiting discrimination based on legal residency. Chinese workers who get a job, in other words, are now automatically entitled to a residency permit for the city where they work.

It doesn't help moderate social tensions that, 50 years after the Chinese Communists exterminated one generation of landowners, a new economic elite has arisen. By 2002 a mere 2.4 million Chinese held two-thirds of the country's liquid assets. Rich Chinese were driving S-class Mercedes to Gucci stores in Shanghai. The country's new premier and president had served in backwater provinces and knew that such a gulf was dangerous.

In the background hangs the status of Taiwan, to which the Nationalist government of Chiang Kai-shek retreated in 1949 and which the government of China views as a renegade province. It's been a huge issue for decades, although it may be slowly declining in importance as both sides see the benefits of the economic integration taking place before their eyes. Half a million Taiwanese now live in China, mostly in connection with Taiwanese factories there.

There's another Taiwan story, too. It goes back 400 years, to a time when there were no Chinese on Taiwan. Instead, there were indigenous tribes. Some have been destroyed, but about 400,000 members of these groups remain, mostly in the rugged eastern mountains. They constitute only 2% of the island's population, but they are struggling to regain the use of their languages and even their non-Chinese names, which were forcibly stripped from them in a rehearsal for what is happening today in Tibet. The largest tribe, the Ami, has begun catering to busloads of tourists, for whom they perform traditional dances. Such displays may be degrading, but at least the Ami are no longer hiding their identity.

Burma (or Myanmar, as the present military government calls it) is ruled by generals who refuse to accept the result of an election held in 1990. The winner of that election, Aung San Suu Kyi, has spent most of the time since then under house arrest in Rangoon (or Yangon, as the government calls it). The only enemy in sight is the country's 50 million people, but to control them the government maintains an armed force of 400,000. A million Burmese meanwhile live as refugees in Thailand, while the regime prospers with payoffs from Thai and, more recently, Chinese merchants. The Thais have had a longstanding interest in Burma's valuable teak forests, while the Chinese have invested particularly heavily in Mandalay, the city at the other end of the Burma Road from Kunming. The Burmese government is an international pariah, but frequent scoldings haven't softened its brutal treatment of both the majority Burmans and the many ethnic minorities in the country. A Burmese living in the United States will call home to speak to his parents but, certain that the line is tapped, will ask "how are you?" only from politeness and not expecting a serious answer.

Offshore, Muslim separatist movements have been fighting for decades on Mindanao, the southernmost of the large islands of the Philippines. The Moro Liberation Front made peace with the government in 1996, and the Moro Islamic Liberation Front agreed to a cease-fire in 2001. A splinter group known as Abu Sayyaf renounced that cease-fire, however, and has become internationally notorious by kidnapping and occasionally beheading foreign tourists. Abu Sayyaf has ties to Osama Bin Laden's al Qaeda, so the United States has given the Philippines $100 million to fight the group. In 2002 American military advisors began assisting the Filipino army on the southern islands of Jolo and Basilan, historically a pirate haven. In 2003 that involvement was stepped up with the deployment of 3,000 American troops sent to engage the few hundred Abu Sayyaf fighters on Jolo. In the central Philippines, the New People's Army operates with far less international notice, because its victims are Filipinos. And what do Filipinos hope for? The last elected president of the country, Joseph Estrada, was forced from office for corruption, but he remains popular with many poor voters. His successor is an economist who has dismissed the protestors outside the presidential palace as a "subculture of the urban underbelly."

Like most Muslim countries, Indonesia has been unable to find an alternative to authoritarian government. The regional economic crisis of 1998 forced the resignation of Suharto, Indonesia's military ruler for 32 years, but people were so fearful that year that Javanese mobs murdered people accused of sorcery. After September 11, 2001, outsiders began paying much closer attention to Indonesia, particularly the threat posed by the Jamaah Islamiah, or Islamic Group. It seeks to create a Southeast Asian caliphate—a unified religious and civil authority—that would rule Indonesia, Malaysia, Singapore, and the southern Philippines. To help reach that goal, the group tried shortly after September 11th to bomb several American targets in Singapore. Singaporean authorities prevented those attacks, but despite repeated warnings from foreign intelligence agencies, the Indonesian government took little action until the lethal bombing in October 2002, of a crowded nightclub in Bali. The Indonesian government, however, has yet to close the Educational Institution of Indonesia–Saudi Arabia. It's a Saudi-funded school advocating Wahabism, the puritanical Islam that supports and, in a mutually beneficial alliance, is supported by the Saudi government. Nor has the Indonesian government closed the Pesentren al Makmin, a Solo school that has graduated many terrorists. (One current student says, "It's not hatred, but it's what God says. The Koran says we don't want Jews and Christians to live with Muslims.") Perhaps the government should not close such schools—freedom of religion is an important element of democratic regimes—but Al Haramain, a Saudi charity funding such schools, has also pumped money into the Jamaah Islamiah, and that group was probably responsible for the bombing in 2003 of the Marriott Hotel in Jakarta. Both the Saudi and American governments now insist that Al Haramain be shut down.

Java, by far the most populous island of Indonesia, is encircled north and east by secessionist movements on Indonesia's outer islands. One of those movements recently succeeded in winning independence for East Timor, but only at the price of at least 60,000 lives. Other parts of the country have taken note. The elderly leader of the Free Aceh Movement (GAM is the Indonesian acronym) now lives in asylum in Sweden, but Hasan Tiro's followers continue to fight both the government and Chevron. It's no surprise: A century ago, the militant Acehnese of northern Sumatra were among the last holdouts against Dutch control over the archipelago.

On Kalimantan, which is Indonesia's part of Borneo, the indigenous Dyaks have been slaughtering migrants from Madura who arrived decades ago, after the Dutch colonial government tried reducing the population of Java and its near neighbors, including Madura, by shipping residents to outer islands. To the east, the population of the Maluku Islands is almost half Christian, and many deaths have been reported from clashes between them and the Muslim majority. On Ambon, one of those islands, a half-million people are said to be homeless.

In an effort to quell the push for autonomy in Irian Jaya, or western New Guinea, the

Indonesian government has changed the province's name to Papua and agreed that the province can retain 70% of the royalties from Papuan mines, including the American-owned Grasberg mine. (The pit is almost 3,000 feet deep, and copper production—metal, not ore—runs close to 2,000 tons daily.) Such concessions have not satisfied the leaders of the Free Papua Movement, and in 2002 unknown assailants shot to death two American teachers at a school serving mine employees.

Now Indonesia is finding that overseas buyers are reluctant to become dependent on Indonesia for liquified natural gas (LNG). Indonesia is the world's biggest producer of this product, but BP can't find buyers for LNG from its new Tangguh field in Papua. Customers instead are considering Qatar and Malaysia's Sarawak, which they hope will be more stable.

On Sri Lanka, the island on the western side of the Bay of Bengal, the secessionist Tamil Tigers have been fighting since 1983 for independence from the Sinhalese majority. Sixty-four thousand lives later, the Tigers control most of the far north from their center at Jaffna. Until very recently they used suicide bombers to devastating effect even in Colombo, the capital. The country's president lost her father and husband to the war, along with one of her own eyes. In 2001, the prime minister scolded his fellow citizens for having so few children that the army was running short of recruits. A year later, a cease-fire was followed by peace talks, and the Tigers agreed to autonomy rather than independence. Perhaps autonomy will come, but the Tigers are anything but democratic, and it's unclear how minorities like Muslims will fare under them. Meanwhile, the Sri Lankan government is itself so riven that the talks may fail. Blocks of downtown Colombo remain fenced off and under round-the-clock military guard; merchants within the cordon may as well be out of business.

India remains locked in its bitter and long-standing dispute with Pakistan over the status of Kashmir. Internal conflicts between castes and between Hindus and Muslims flare periodically, too. For most Muslims, caution is the order of the day, but ardent parents send their boys to a school called the Dar-ul-Uloom, or House of Knowledge. It's north of Delhi, in Deoband, and it's been in business since 1867. An official manual from 1911 calls it "a famous Mohammadan theological school," but times have changed. The school denies that it trains terrorists, and its head says that Muslims can live peacefully in India. But the government of India, apprehensive that the school will produce terrorists, no longer allows foreign students to enroll. No problem: The Dar-ul-Uloom has established branch schools in Pakistan. Some of their graduates in the 1990s moved west and joined the Taliban in Afghanistan.

And India faces other political challenges. In the 1980s the government of Indira Gandhi crushed a separatist movement in the Indian Punjab, where the Sikhs had sought an independent state of their own, which they had presumptively named Khalistan. Indira Gandhi, the prime minister, ordered an attack on the militants holed up in the Golden Temple in Amritsar—the holiest Sikh shrine. In turn, her Sikh bodyguards shot her to death in her Delhi garden. Now the government is fighting a group of secessionist organizations in the northeast, groups such as the United Liberation Front of Asom, the National Democratic Front of Bodoland, and the Kamtapur Liberation Organization. The names sound comical to outsiders, but the groups are no joke. A venerable group of Marxist guerrillas known as Naxalites also makes periodic trouble. Naxal camps raided in the southern state of Andhra Pradesh in 2003 contained precise instructions in Telugu, the local language, about how to make crude but lethal mortars of World War I vintage. The Naxalites periodically pick off an unlucky policeman or minister.

Neighboring Nepal since 1996 has faced a particularly virulent group of 5,000 Marxist rebels. Under the leadership of a former schoolteacher, the rebels have been responsible for

9,500 deaths. In November 2001, the group staged an attack that killed at least 75 officials, police, and soldiers, along with 200 of its own members. In the run-up to elections they had vowed to stop, 1,000 of the rebels the next year attacked a police post in Sindhuli, 100 miles west of Kathmandu, and killed 50 policemen. The next day they attacked Sandhikhara, 200 miles west of Kathmandu, and killed at least 65 soldiers and police. A cease-fire was declared in 2003, partly because the insurgents were no longer getting support from either India or China. Even so, the rebels control most of the countryside, and the government believes that a military victory is impossible. Landlords have been obliged to move to Kathmandu and, for the first time in their lives, work for a living. In an ironic twist, however, the rebels have organized themselves along caste lines, so that low-caste peasants who join the movement are stuck in menial positions.

Pakistan remains a deeply feudal country, particularly in the province of Sind, famous for its big rural landlords and cowed villagers. Half the country's arable land is owned by only 4% of the country's households. That helps explain why most boys attend school for only 2 years; girls average less than 1. You could hardly construct a better formula for keeping a people in servitude, yet the government of Pakistan is so riddled with corruption that no Pakistani government, elected or military, has seriously challenged the social order.

Free speech is meanwhile muffled because in Pakistan the wrong words can lead to charges of blasphemy, a capital offense. How far can such things go? In 2001 a professor in Islamabad was sentenced to hang for pointing out in class that the Prophet Muhammad's parents couldn't have been Muslims. Meanwhile, violence has plagued Karachi, Pakistan's biggest city. The trouble centers on the country's 30 million refugees from India. These are the *mohajirs* ("refugees" in Urdu), the descendants of the Indian Muslims who pushed the British to create Pakistan and then moved to the new country. Ironically, the children and grandchildren of the elite who created the country have become targets of the Muslims who already lived there. Ironically, too, a Muslim who wants to live in a South Asian country with a sustained history of free elections has to move to India.

Ironically, finally, Muhammad Ali Jinnah, the father of Pakistan, was himself thoroughly secular, as anglicized as Jawaharlal Nehru, his counterpart in independent India. Jinnah would be aghast to see Pakistan today and learn, for example, that in 2003 a group of Islamicist parties won control of the North West Frontier Province and called for Taliban-style *sharia,* Islamic law to be enforced by a new department of vice and virtue. Their next electoral goal is neighboring Baluchistan. The Baluchis may join up, but they have no interest in Pakistan. "Why must Punjab be my destiny?" one tribal leader asks. That's why there's a serious joke to the effect that Pakistan is a crowd, not a nation.

Adjacent Afghanistan in the 1990s became the source of more refugees than any other country on earth. (I'm following the United Nations here in declaring that the millions of Palestinians scattered in refugee camps in the West Bank, Gaza, Jordan, Syria, and Lebanon are in fact not refugees.) Two million Afghans fled to Iran, almost a million to Pakistan, and others north to Tajikistan. They did so partly to escape the Taliban government, which forced men to grow beards and women to hide behind cloth. They did so to escape a countryside blasted by drought and the legacies of war, including ruined cities and a countryside riddled with landmines. Twenty-five years ago, there were traffic jams in Kabul, but Kabul in the days of Taliban rule became a city of horse-drawn carts. An estimated 70% of the city's labor force was unemployed, and charities distributed bread or wheat daily to 400,000 people. The only visitors, it seemed, were journalists and terrorists in training. Kabul's changed a lot since then. Beer's back; so are photo and video shops. So is traffic, with car dealers offering vehicles fresh from Dubai. Behind the main streets, 80,000 homes are still rubble.

Many of the teachers of Afghanistan's terrorists were trained by the United States during the 1980s, when anyone fighting the Russians was a good friend of Washington. Though hardly known to the public before September 11, 2001, al Qaeda, "the base," took shape in that decade. Its name comes from the database that Osama bin Laden created to keep track of the thousands of freedom fighters, or *mujahideen*—both Afghan and foreign—who fought the Soviet army in Afghanistan. In response to the presence of American troops in Saudi Arabia during the Gulf War, al Qaeda became much more than a list of names: It organized the bombing of the American embassies in Kenya and Tanzania in 1998, the bombing of the *USS Cole* in 2000, and the destruction of New York's World Trade Center. During the late 1990s, the organization ran primitive camps near Khost, in southeastern Afghanistan. Some 70,000 people trained in those camps. They then settled in loose networks abroad, frequently with ties to organizations such as Egypt's Islamic Jihad, Palestine's Hamas, and Lebanon's Hezbollah. Meanwhile, and until late 2001, hundreds of Arabic speakers lived with their families in upscale neighborhoods of Afghan towns like Jalalabad. They had very little contact with the Afghans. Some of them have retreated now to Pakistan, where they have regrouped.

Iran is slowly emerging from the rule of Islamic *mullahs*. Since the Shah was deposed in 1979, literacy rates have risen and economic inequality has been reduced; that's the upside. The down is that per capita income has fallen from $600 to $150 a year. An inconclusive but devastating war with Iraq in the 1980s cost each side more than a half-million men. Even Iranian religious leaders such as the grandson of Ayatollah Khomeini have become opponents of the regime. He, in particular, says that if a government claims to represent God's will and yet cannot deliver prosperity, there must be something wrong with the religion. It's a good argument for keeping religion out of politics. What *is* surprising is that there's so little recognition across the Islamic world today of the unpopularity of the Iranian government. You'd think that the millions of people eager to live under an Islamic regime would wonder why so many Iranians want to get rid of the one they have.

Then there's Israel, whose Jewish pioneers fought for and got an independent state in 1948. In 1967 they extended their control over the West Bank of Jordan. It was a great military triumph—a Six Day War over almost before it began—but it was also a curse, leaving the Israelis controlling 2 million Arabs who, like any occupied people, vacillate between fatalistic acceptance and burning hatred. During the 1990s, when peace seemed possible, the Palestinians invested a lot of money in the West Bank, especially Nablus, Ramallah, Bethlehem, and Hebron. The Israelis, however, continued to invest (with loan guarantees from the U.S. government) in hundreds of settlements on the West Bank, some of which now have thousands of people—in a couple of cases, tens of thousands.

Since the outbreak of the second Palestinian uprising or *intifada* in late 2000, there has been almost no work on the West Bank or Gaza and almost no way to get to what work there is in Israel. Moshe Ya'alon, the Israeli Defense Force's chief of staff in 2003, said that he intended to make Palestinians realize "in the deepest recesses of their consciousness that they are a defeated people." Perhaps he can do it, but the Palestinians find strength in many places. Family networks are extremely important to them. For some, Islam itself is fortifying. For some, the words of the Syrian poet Nizar Qabbani help. In one of the many political poems that won him enemies in high places across the Arab world, Qabbani wrote: "Ah, this generation of betrayal,/ of surrogate, indecent men,/ this generation of leftovers,/ will be swept away—/never mind the slow pace of history—/by children bearing rocks." To Palestinian ears, those words are as encouraging as anything Churchill said to the British in 1940.

Israelis have their own burden. Immigration is down. The tourist business crashed from 283,000 arrivals in May 2000, to 112,000 a year later. Those numbers are terrible, but they pale alongside the silent fears that Israelis repress every time they put one of their children on a bus.

AFRICA

Could things be worse? The short answer is yes, a lot.

In a spate of postcolonial venom, the government of Robert Mugabe has been evicting the white farmers who are among Zimbabwe's few economic assets. The story can be traced back to the 1980s and 1990s, when the country gave land to over 70,000 families. Then it began to resettle some 300,000 families under a fast-track program. The new program was ruled illegal by the country's senior courts, but the government responded by packing the Supreme Court. A presidential order then confiscated the land of almost all the country's white farmers. The government ordered some 2,900 white farmers to stop cultivating by June 24, 2002, or go to jail. The farmers were then told to vacate their lands, without compensation, by August 8, 2002. Although their land is nominally destined for distribution among the country's many landless people, it is actually being turned over to government supporters with little knowledge of farming. The especially fine Iron Mask Estate—2,500 acres with a 29-room house and a swimming pool—was summarily assigned to Mugabe's wife, who wanted a nice place close to Harare.

Forty-two thousand square miles have been confiscated, and only 500 white farmers are still on their land, out of 5,000. The contribution of agriculture to Zimbabwe's gross domestic product has fallen sharply; cereal production, for example, fell by over 60% between 1999 and 2002. By January 2004 the United Nations estimated that 7 million of the country's 11.5 million people were "food insecure" and dependent on handouts from the World Food Program (WFP). The dietary staple in Zimbabwe is *sadza*, boiled ground corn mixed with vegetables and eaten twice a day. The WFP's ration in Zimbabwe is 10 kilograms per person per month, with added soy for those most in need of protein. The government continues to defend itself by saying that the land was stolen by the whites during the colonial period. Conveniently, it overlooks the fact that it has also confiscated the land of many black-owned farms. The leaders of Namibia and Mozambique continue to support the Zimbabwe government because they, too, find it convenient to blame their troubles on the colonial past. They all forget the moment in 1980, the year of Zimbabwe's independence, when the first postcolonial leaders of Tanzania and Mozambique told Mugabe: "You have the jewel of Africa in your hands. Now look after it."

There are many other heavy hands in Africa. The United States considers Egypt a friend, but during the 1990s President Hosni Mubarak ruthlessly crushed a movement aimed at creating an Islamic state in Egypt. Tens of thousands were jailed indefinitely before the Islamic Group in 1999 officially suspended its terrorist operations. Mosques came under government control—every preacher with a government license. Mubarak of course remembered the fate of his predecessor, Anwar Sadat, who was assassinated by Islamic Jihad for making peace with Israel. But Mubarak has crushed dissent as well as terror, and his government strikes not only at Muslim extremists but at secular opponents who dare to have liberal views.

To the west, the military government in Algeria called off in 1992 an election that would have brought to power the Islamic Salvation Front. Fearful of a democracy of the one man, one vote, one time variety, the military took over. (It wasn't an idle fear. Democracy has been compared to a streetcar: Ride it to your destination, then get off.) Since then, 100,000 Algerians have been killed, often savagely. Per capita annual income, at $1,600, is half what it was in 1986. In 2000, President Abdelaziz Bouteflika came to power through a rigged election, but he offered an amnesty to rebels who laid down their arms, and things have grown much more peaceful. In 2004, Bouteflika was reelected overwhelmingly, not so much because he was widely supported but because, in the words of one observer, "people are just worn out from so many years of violence, and they didn't want to risk a change."

Next door, Morocco is probably the most democratic country in the Middle East and North Africa, with the possible exception of Turkey. Yet it continues to control neighboring Western Sahara, which it invaded in 1975. The Polisario Front has been fighting for independence there and has been supported by Spain, which formerly held this area as the colony of Spanish Sahara. The Front hasn't been strong enough to push the Moroccans out, however, and Morocco is meanwhile encouraging Moroccan settlement. If a plebescite is ever held—one's been promised by the United Nations since 1991—the Saharawi may be outvoted by Moroccan newcomers.

Shall we make an equatorial traverse? During his 24 years as president of Kenya, Daniel arap Moi miraculously transformed himself from a country schoolteacher to one of his country's richest men. He stepped down in 2003, a risky step for any autocrat. Somalia, after all, disintegrated into warlordism after the ouster in 1991 of Siad Barre, a general who had seized power after the assassination of his predecessor 20 years earlier. Whether Moi manages to stay a free man, enjoying his wealth in retirement, remains to be seen; certainly there are Kenyans who think he should be arrested.

Sudan, Africa's biggest country, is barely cohering. It's torn between an Arab Muslim North and a non-Arab, non-Muslim, but oil-rich South. Since a fundamentalist military regime seized power from an elected one in the late 1980s, there has been such an exodus that 3 million Sudanese are now said to live in Cairo. The southerners still in Sudan are worst off, caught between the brutal Sudanese army and the equally brutal Sudan People's Liberation Army. As the southerners say, "When two elephants meet, it is the grass that suffers." Just as the north and south seemed about to make peace early in 2004, atrocities were reported in Darfur, along Sudan's border with Chad. This is the home of the non-Arab Fur, and they, like the southerners, have been marginalized by the government in Khartoum. The government response to their rebellion has been to allow Arab militias to burn defenseless villages, rape and murder their occupants, and drive 100,000 Fur to camps in Chad.

Back to the Equator. In 1994 the Hutus of Rwanda killed a half-million fellow Rwandans who happened to be Tutsis. A Canadian general had the misfortune to command a U.N. force that was ordered to sit on its hands through the butchery. He came home but could not forget what he had seen. In dreams, "I would literally be up to my waistline in bodies. I would have my hands spread out, to the sky, and my whole body would be red with blood." The Belgians in colonial times had consistently favored the Tutsis, and despite the losses they suffered in 1994, the Tutsis again regained power and sent a million Hutus fleeing to Zaire.

That country, now the Democratic Republic of the Congo, or Congo (Kinshasa), had been devastated during the long rule of Mobutu Sese Seko, who fled in 1997 and was replaced by Laurent Kabila, a rebel who was himself shortly assassinated by his own bodyguards. Before his death, Rwanda, Burundi, and Uganda were committed to his overthrow.

After it, their soldiers continued to stay in eastern Congo and stir tribal animosities. The Congolese government was supported by troops from Zimbabwe, Angola, and Namibia, all of which hoped not to bring peace to the country but to make money from the sale of Congo's gold and diamonds. The result has been more than 3 million dead Congolese since 1998. Mostly, they have been civilians who flee their homes, seek refuge in the forest, and die there of malaria, measles, and diarrhea. A Final Act, signed in 2003, was supposed to lead to a constitution and democratic elections, but the day after it was signed 966 villagers in the northeastern Ituri forest were slaughtered by unknown assailants. The International Rescue Committee points out that there's been very little press coverage of this awkwardly remote conflict, which now ranks as the world's deadliest since World War II. Ten thousand U.N. peacekeepers are in the country, but killings continue.

Back in Rwanda, 10,000 children have been born from the gang rapes a decade ago of a quarter-million Tutsi women. Some of those women now have AIDS as well, and they struggle to answer children who ask why they have no fathers or other relatives. All the mother's relatives were killed, but what can a mother say to a 10-year-old who asks: "Why don't we have any aunties? Why no uncles? No cousins? If a dad dies, does that mean everyone dies also?" The mother struggles to find a way to answer these questions before she herself dies for lack of the medical treatment she cannot afford.

A cluster of countries in West Africa have recently fought abominable civil wars, marked by machete amputations of the arms and legs of civilians of every age and both sexes. Rebel leaders have made fortunes by selling so-called conflict diamonds, now estimated to supply one-tenth of the world's $10 billion diamond market. International terrorist groups, ready to do anything to further their cause, are in the thick of the business. The British and French forcibly restored order to Sierra Leone and Ivory Coast—former colonies of theirs—but though Liberia was established by Americans the United States has kept its involvement there to a bare minimum and left peacekeeping as much as possible to African and U.N. forces.

What about Africa's most populous country? Since the sudden death of Sani Abacha, a barbaric and usurping general, Nigeria has returned to civilian rule. Many say that the elected government is incapable of addressing either the country's notorious corruption or the cleavage between the Islamic north and the Christian or animist south. Nigerians in the oil-rich southeast think of the wealth that could be theirs if the country broke up—wealth siphoned off to the European bank accounts of the country's rulers.

In the north, courts since 2000 have begun adopting *sharia*. It's been a very popular move, partly because everyone is aware of the country's endemic corruption. Yet there's nothing to prevent the judge in an Islamic court from being corrupt, which perhaps explains why only poor Nigerians have been tried and punished. They have lost their hands for stealing a goat, a cow, a bicycle. Women (Amina Lawal is the best-known) have been convicted of adultery, for which the punishment is death by stoning after burial to the neck. No man has been convicted, because *sharia* savants declare that DNA tests are prohibited by Islam. So far, the sentences against the women have not been carried out. Riots between Muslims and Christians, however, have made many Nigerian cities terrifying places for their residents. Kaduna, for example, is a city of about 2 million people. Its Muslims have sought safety by moving to one side of the Kaduna River, while many Christians have abandoned their homes and churches and sought safety on the other.

In this savage context, South Africa is all the more remarkable for its peaceful transition from white to black rule. Unemployment, however, has risen from 17% in 1995 to about 30%, and violent-crime rates have risen sharply. Some poor South Africans look enviously at Zimbabwe's land reform program and say that liberation without land is not liberation. They have on their side the cruel historic fact that the government of South Africa in 1913 passed a Land Act that reserved 92% of the country for whites. On the other hand, however, is the equally hard fact that smallholdings are a recipe for poverty. Some South African land reformers praise Zimbabwe's Mugabe for having had the courage to say "enough is enough, we are taking back our land," but the South African government understands that a radical program like Zimbabwe's would disrupt not only the agricultural sector but foreign investment in all sectors. Instead of foolishly copying Zimbabwe, the South African government set out in 1994 to move slowly on a program to transfer a third of South Africa's white-owned farmland to blacks. A decade after the end of apartheid, however, 87% of the country's agricultural land was still owned by whites. The Restitution of Land Rights Amendment Act of 2004 seeks to speed up the process by allowing the government to seize farmland when no willing sellers can be found; even so, equitable compensation must be paid, and the government has pledged to use the law only when white farmers are "unreasonably opposed" to selling. The government also knows that it must do more than transfer land to blacks: It must train the new owners in modern farm production methods, and it must make capital available to them. Huge as the issue is, it's overshadowed by even more dramatic problems. AIDS is projected to kill between 4 and 6 million South Africans by 2010.

THE WESTERN HEMISPHERE

Across Latin America voters are looking for leaders to lift them from poverty.

Begin at the bottom, with the poorest nation in the Western Hemisphere. This is Haiti, where life expectancy is 51 years, where most people earn less than $1 a day, but where high school costs $145 a year, plus $20 a month and books. One woman says, "I only care about whether we can eat. It doesn't matter who's in power." Often there's so little food that parents sell their children into virtual slavery in the hope that the buyers—families with a bit more money—will give the children something to eat in exchange for their household labor. There are 300,000 of these *restaveks* (from *rester avec*) as young as 4; they rarely go to school or have a bed to sleep in. Like so many Haitians, when they grow up they are likely to try to get out. Many flee to the United States, but others try the Bahamas—which responds, as America does, by sending them back home. Jean-Claude Duvalier, Haiti's president for life, fled the country in 1986 after stealing, with family and friends, some $500 million in the last decade of Duvalier rule. Now he's in Paris, broke, and heir to a regime that killed some 50,000 political opponents. His own assessment: "Perhaps I was too tolerant."

There's a weekly march in the Haitian capital by relatives of the victims of the post-Duvalier regime that ruled between 1991 and 1994. The march is modeled after one in Argentina, where the Mothers of the Plaza de Mayo seek punishment for the murderers of *their* children by elements of the Argentinian military regime of the 1980s. Most Argentines, however, don't have a lot of time for history: They're preoccupied by an economic collapse more severe than that of the 1930s. Per capita income fell from $9,000 in 1998 to $2,500 in 2002, by which time half the population was living on less than $3 a day. The job of bank teller became dangerous, while some 40,000 *cartoneros* pushed carts around the streets in a search for paper and cardboard to sell for recycling. Students in formerly middle-class schools grew school gardens and raised rabbits not for science projects but for lunch.

Says one man: "There are places that are much worse off. What's hard for us is that we've known something better. We've lived well." Out in the province of Tucumán, "the garden of the republic," doctors report seeing children with kwashiorkor, a protein-deficiency disease. "We never thought that this could happen in Argentina," they say.

Poor neighboring Paraguay has never been democratic. Its long-time dictator, Alfredo Stroessner, was kicked out in 1989, but his Colorado Party remains in power. The country is ranked the most corrupt in the Western Hemisphere, but most Paraguayans are too poor to care when it was revealed that their new president and his wife were driving stolen cars. The cars in time became an embarrassment and disappeared, but no legal action was taken because everyone understands that Paraguay's courts are hopelessly corrupt.

The embattled president of Venezuela, Hugo Chavez, has said that in the name of social justice he will not evict people who squat in empty office buildings. This has earned him many supporters—the *Chavistas*. They say that Chavez has given land to people who otherwise would have died of hunger, and they point out that tens of thousands of squatters in many illegal settlements now own the land under those houses. A *Chavista* says "this is the first time in my life that the government has done something for me." The business community, however, is outraged. Late in 2002 it launched a 2-month-long general strike that cost the economy $6 billion. Even generals and admirals spoke at protest meetings in Caracas' Plaza Altamira. They warned that Chavez was another Fidel Castro, as indeed he seems to be.

Or think of Colombia, with a record of violence unequaled in Latin America. Coca acreage has declined in Peru and Bolivia to about 100,000 and 20,000 acres, respectively, but meanwhile rose in Colombia to about 300,000 acres, concentrated along the eastern flanks of the Sierra Oriental. The country now produces about 500 tons of cocaine annually. That's 90% of the world supply. American military aid has included helicopters and herbicides that killed 180,000 acres of coca in 2001. For a time, it seemed that the growers were staying ahead of the sprayers, but the balance may be tipping. The crop in July 2003 was reported by the United Nations as 170,000 acres, down from 251,000 in December 2002. Other estimates don't show such a big drop, and it's possible that the drop, however large, is caused less by spraying than by declining consumption in the United States.

Colombian coca is grown in areas controlled by guerrillas, chiefly those of the Revolutionary Armed Forces of Colombia (FARC in the Spanish acronym). The FARC has grown wealthy by filling the vacuum created by the destruction of the Medellín drug cartel. It collects a $200 tax on every kilo of coca base grown in the areas it controls. This amounts to about $1 million a day, supplemented with ransoms collected from several thousand annual kidnappings—counting only those that are reported.

Colombians risk getting caught between the guerrillas, the army, criminal gangs, and the country's rapidly growing right-wing paramilitary forces. One of these paramilitary groups, the United Self-Defense Forces of Colombia (AUC in the Spanish initials), now has 14,000 combatants. In the first half of 2001 it was reported to have killed almost 1,300 people, including not only rebels but people thought to be helping rebels. The AUC is becoming especially powerful through collusion with the Colombian military and through seizure of FARC territory, where the AUC inherits the tax collection business and reportedly assesses a modest $50 per kilo. One demonstration of AUC power took place in November 2001, when 15 mayors negotiated a cease-fire with a secondary guerrilla group, the National Liberation Army. One mayor said, "Our people are being shot, our roads are being blocked, we are being kidnapped." The paramilitaries promptly kidnapped six of the mayors. All told, 58 Colombian mayors were assassinated over a 4-year period beginning in 1999, and many mayors dare not set foot in their own towns. The fortified state capital of Florencia, east of the Andes, is the refuge of a dozen mayors—remote-control mayors, they're called. You can hardly blame them.

In May 2002, the FARC threw a bomb into a church where hundreds of people had sought sanctuary. This happened in Bella Vista, west of Medellín. The FARC apologized—said it was a mistake in the heat of battle with the AUC—but 117 people died, including many women and children. The Colombian military was nowhere in sight.

With 5,000 hostages in rebel hands at the start of 2004, and with rebel forces estimated at 30,000, Colombians were speaking with their feet: more than a million have emigrated since 1995. Efforts to encourage coca producers to grow other crops have failed because no other crop is as profitable as coca. This fact leads American strategists back to aerial spraying. Some people say that the large-scale coca growers find new sites for their crops, while the 100,000 peasant cultivators, who collectively produce about 15% of the crop, see their food crops killed, along with their coca. Others say that the spraying program has weakened FARC so much that it has turned from battling the military to petty thievery and urban terrorism. What's to choose? The daughter of one elderly victim caught in crossfire between FARC and the army says, "I don't believe this war will ever end. Every day it causes the death of more innocent people." Her mother had moved to remote La Montanita 50 years earlier to find a peaceful place to live. No luck. "She fled the violence," a neighbor says, "but the violence found her." The journalist Alma Guillermoprieto has speculated that the only hope lies in establishing an effective judiciary, in offering an alternative to Colombia's thousands of teenage rebels and paramilitaries, in creating truth and reconciliation commissions, and—she herself wonders if this is quixotic—in the worldwide legalization of drugs.

All along the Andean spine from Venezuela through Bolivia, indigenous peoples are demanding a larger role in the governance of countries dominated for centuries by people of European origin. One Bolivian leader, Felipe Quispe, says that the indigenous peoples—70% of all Bolivians—no longer want "to be governed by whites who have robbed and stolen our natural resources." A miner is more eloquent: "Globalization is just another name for submission and domination. We've had to live with that here for 500 years, and now we want to be our own masters." A local journalist writes, "Indians no longer want to watch Bolivia from the mountains. They want to watch it from the balcony of the government palace."

It's easy to understand the anger: More than half the people in all the Andean countries, except Chile, live below the poverty line. At the other extreme, Bolivia's great mineral wealth, especially in tin, supported the Patiños, who mostly lived in Europe. But the populist programs of income equalization call for nationalized industries and high tariffs on imports. Such measures will discourage foreign investment and impede economic growth. Bolivians got rid of their last president, as well as his plans for a gas pipeline. (Along with his defense minister, Sanchez de Lozada now lives in the United States.) Rather than see the gas exported to Chile and points beyond, many Bolivians want the gas to be used in new factories built in Bolivia. It's not going to happen. As a diplomat asks, "who in their right mind is going to be willing to invest in a country that is so unstable and hostile to foreign capital?"

Meanwhile, the United States managed to cut the coca harvest in Bolivia, but its efforts to introduce alternative crops like pineapple, hearts of palm, black pepper, bananas, and passion fruit failed, as usual, because the profits weren't comparable to those from coca. The result has been the near collapse of the Bolivian economy. Coca acreage is rebounding. In a few towns, including Sorata, just east of Lake Titicaca, the police and army have been driven out and replaced by a Peasant Union Police.

Move north to Guatemala, a Central American country emerging from 35 years of civil war and 200,000 deaths. That's more than the deaths from the wars in El Salvador and Nicaragua, combined with the victims of military governments in Chile and Argentina. During the war, Guatemalan soldiers conducted hundreds of massacres, killing even kindergarteners who they considered sympathetic to leftist rebels. Bones are still being exhumed by organizations like the Forensic Anthropology Foundation of Guatemala. If possible, they're identified and given a simple burial. One woman in the village of Ziquin Sanah came to such a service in 2003. She was there for her 15-year-old cousin, whose bones were in an envelope when she arrived. They were removed and placed in a box for burial, but she intervened, stepping forward to cover them with a white handkerchief. She explained that she didn't want her cousin to be cold.

Think of Mexico, where the army and secret police in the 1970s disappeared hundreds of people suspected of involvement with leftist movements. It was an episode of Mexican history that began on the night of October 2, 1968. Hundreds of student protestors had gathered in Mexico City's Tlatelolco Plaza. They threatened to embarrass a country busily preparing for the Olympic Games. The army shot hundreds of them that night—and then suppressed all knowledge of the event.

More recently, Indians from Oaxaca have been fighting to secure rights to land. It's complicated, because land ownership across Mexico is extremely confused, with boundaries in most cases neither surveyed nor demarcated. President Vicente Fox took the unprecedented step of inviting the leaders of the Zapatista National Liberation Army to address the national congress. They did so, marching proudly to Mexico City. Many congressmen boycotted the event, and the Zapatistas soon broke off negotiations with the government. Fox's party, the National Action Party (PAN), had itself come to power when Mexicans turned against the long-reigning but thoroughly corrupt Institutional Revolutionary Party (or PRI). Now, it appears, the PAN is succumbing to corruption and violence. In 2002, María Tames, a city councilwoman in the Mexico City suburb of Atizapan, was gunned down and killed on her doorstep, apparently by officials who thought she was getting too close to exposing party corruption. Police corruption is another matter, particularly evident in Mexico's new growth industry, kidnapping. The families of victims do not seek police help. They know the police will be ineffective. Besides, they fear that the police are already involved in the worst way.

The United States remains crippled, or at least injured, by the legacy of slavery. Race riots erupt from time to time—famously in Watts in 1964, more recently in Cincinnati and Benton Harbor, Michigan. Benton Harbor in 1960 had been 75% white, but now it's 92% black. The median household income is $17,000, one-third of the $51,000 in nearby St. Joseph, which is 90% white. A Benton Harbor resident says, "jobs are low. Our kids have nothing to do. You know what's the highlight of our day? Standing on the four corners right here." How did she explain the riots that broke out in June 2003? It was simple: "We're sick of them killing us," she said. It's no surprise to Americans to hear that blacks have the highest rates of incarceration of any ethnic group in the country and that they are disproportionately subject to capital punishment. It was no accident 50 years ago that the nation's Pullman porters were black. It's no accident today that the supervisors at the world's biggest pork packer, Smithfield, are white. The men doing the gruesome work on its butchering floor in Tar Heel, North Carolina, are black. Fifty years have passed since *Brown v. Board of Education* held that separate but equal schools are unconstitutional, yet public schools in the United States are often as segregated today as they were then.

Not all aggrieved Americans are black. In the last days of the Clinton administration, the Assistant Interior Secretary for Indian Affairs said that it was time for contrition. On behalf of the Bureau of Indian Affairs, he apologized for policies that, he said, had amounted to ethnic cleansing. What should be done, besides apologizing? The Ute tribe is regaining some 84,000 acres taken from it in 1916, and the 1,400 Southern Utes have struck it rich with natural gas from their Red Willow Production Company. Most of the proceeds have been invested for the day when the gas stops, and tribal assets now exceed $1 billion, with annual payouts of over $50,000 to every elder. Other tribes have won rights to fish in closed waters. Most dramatically, Indian-owned casinos have become cash cows: California alone has 51. They are regulated by the state, but regulation is nominal because, in the words of one observer, "in Sacramento the tribes never lose. They never lose. They always get their way." Some of the casinos are in bucolic countryside, but protests against them are in vain: "When we try to organize against the casino, it is like facing a juggernaut," says a resident of tiny Capay; "what ultimately happens is that we are labeled as racists." The only salvation in sight is lifting the restriction that keeps Indian gaming confined to tribal lands; if Californians could hit a casino around the corner they wouldn't bother driving 2 hours to place their bets.

Casinos have generated so much money that Indians joke that when they're talking about the past and use the phrase "B.C." they mean "Before Casinos." Casino profits, however, go mostly to tribal bosses and their backers, who include not only established American casino owners but foreign ones. Malaysia's Genting Corporation, for example, bankrolled Connecticut's huge Foxwoods Casino. Meanwhile, only 15 of California's 107 tribes have casinos. About 80 students graduate each year from South Dakota's Pine Ridge High School—and that's from a class of 200. A happy side to the story is that some of this new income is being used to buy back ancestral land. That's what the Miwuk and several bands of Mission and Cahuila Indians are doing in California. Some of the land will be farmed, but some will be left as is. One member of the Morongo Band of Mission Indians explains: "This canyon and everything in it, including that rusting barbed-wire fence, tells our story."

Canada has gone further than the United States toward making reparations to its native peoples. The Nisqa'a Treaty, implemented in 2000, gives that tribe rights over some 22,000 square kilometers of British Columbia. In Quebec, the Cree have succeeded in forcing major changes in the provincial government's plans for hydropower dams on the rivers flowing into James Bay. Most dramatically, the new territory of Nunavut has been carved from the Northwest Territories and made into an ethnic homeland. Nunavut in this way has become a model for groups as far away as Argentina. Across the American tropics, similar ethnic homelands are being created. From now on, indigenous peoples will fight to be stakeholders in development projects. Canada meanwhile continues to struggle with a separatist movement in Quebec. It's quiescent at the moment—the Parti Québécois was soundly defeated in provincial elections in 2003. The demographics of continued immigration and a falling Québécois birth rate will tend to keep the party weak. Still, these things run in cycles, and it's a mistake to dismiss sovereignty as a dead issue.

EUROPE

Finally Europe. If one had asked a decade ago about social upheaval, the first response would have been the conflict between the Irish Republican Army, dedicated to a united Ireland, and the government of the United Kingdom, adamant in protecting Northern Ireland's Protestant majority. A second response might have been the long-running insurgency by

ETA, the Basque separatist organization in northwestern Spain. Other answers might have included the problems of racial minorities like North Africans in France, especially around Marseilles, or perhaps South Asians living in Great Britain. Someone might have mentioned Eastern Europe's despised Gypsies, or Roma.

There was also Cyprus, shared by Greeks and Turks since 1574, when the Ottomans took the island from Venice. In 1974 the Greeks tried to incorporate Cyprus into Greece. They failed, but the Turks in response calved off the northern third of the island, which then became the Turkish Republic of Northern Cyprus. This used to be the wealthiest part of the island, but it quickly became a pariah semistate, recognized only by Turkey. Per capita income by 2000 was about $4,000 a year, compared to $14,000 in the south. Those numbers explain why most Turkish Cypriots want unification, but they don't want Greek Cypriots moving back and reclaiming abandoned homes and properties—which of course is the Greek Cypriot price of unification. In 2004, the Republic of Cyprus was scheduled to join the European Union, and the fear among Turkish Cypriots of being left out—and the fear in Ankara that such a thing would delay Turkey's own entry—finally led to a vote on reunifying the island. Under the terms of the Annan Plan, Turkish troops would have remained for some time in the north, and the number of Greeks allowed to return home in the near future would have been limited. The Turks voted yes, but the Greeks voted no, derailing the process.

Grievous as they are, these troubles pale next to the breakup of Yugoslavia. There, several nationalities were dominated by the Serbs, who were supported by the Russians, fellow members of the Orthodox Church. With the collapse of the Soviet Union, only Serbia and tiny Montenegro wanted to stay in Yugoslavia. For Catholic Slovenia and Macedonia in 1991, the break was relatively easy, but Croatia and Bosnia had large Serbian populations and had to fight civil wars to gain their independence in 1995. Civil war came the next year to Kosovo, a part of Serbia inhabited chiefly by ethnic Albanians. An uneasy truce between the Kosovar Albanians and the very few remaining resident Serbs is now maintained by U.N. forces that can't leave without setting off another spiral of violence. A decade after the troubles began, none of these pieces of the former Yugoslavia are doing well. Slovenia is the best off, though it has trouble attracting foreign investment because it will not allow foreigners to own land. The others, including Serbia, are physically, economically, and socially traumatized. The name Yugoslavia was finally abandoned in 2003, but even as Serbia and Montenegro, the country is likely to split again, this time with Montenegro seeking independence.

FROM THE GREAT GAME TO THE GLOBAL WAR ON TERRORISM

So much for a survey of the social upheavals around us. If the nightly news ever attempted this kind of thing, millions of remote controls would hunt for a sitcom. Fortunately, we can retreat into comfortable generalizations. I'll offer three.

First, the world is awash in ethnic conflicts. One people, typically distinguished by language or religion but also by its control of economic and political institutions, abuses another past enduring. The result is a struggle for a homeland, whether in Chechnya, Tibet, Sri Lanka, Mindanao, Palestine, the Sudan, South Africa, Quebec, or Ireland. Although these are political movements, their ethnic or religious roots infuse them with a zealotry that statesmen a century ago were ready to dismiss as a thing of the past. For them, political conflicts seemed likely to arise in the future from a cold-bloodedly rational competition among nations. The governments of Europe would sit down, for example, and divide Africa as it suited them. If Germany wanted its Southwest Africa (Namibia) to have access to the Zambesi River, then the Caprivi Strip was attached to it like a tail. It's still there today, giving the Namibians purely theoretical access to a navigable waterway to the Indian Ocean. Britain and Russia meanwhile agreed on the importance of a buffer between their empires and so created the Wakhan Corridor in Afghanistan. It touches China and separated British India from the Russian Empire. Today, it pointlessly separates Pakistan from Tajikistan. In North America, Britain and the United States could not agree on whether the Canadian boundary should stick faithfully to the 49th parallel when it reached the Gulf Islands on the West Coast. They invited Kaiser Wilhelm to arbitrate. He did, with results that the United States respects today, even though the Kaiser's decision favored Canada. It was all very gentlemanly, this great game, as it was called in the case of the British–Russian rivalry.

The game is still being played. We're reminded of it every time an appeal is made to someone's national interest.

Now, however—generalization number two—the fans are rushing the field. That's what's happening with the butchery in the Caucasus, in Southeast Asia, in South Asia, in the Middle East, in Central Africa, in Southern Africa, in West Africa, in Colombia, in the Balkans, and in Ireland and Spain. The diplomats of a century ago would be appalled to see the governments of our time so impotent.

The nation-state is losing the capacity to shape its own future. Partly this is because those states are disintegrating, especially in Africa. Partly it is a matter of the emergence of supranational institutions, such as the European Union, which very, very gradually may become the political entity to which Europeans give their allegiance. Mostly, though, it's a matter of governments relinquishing part of their sovereignty to attract capital. Almost by definition, this capital is international and goes where returns are large and safe. Governments might like to shape unique policies, in other words, but they can't alienate investors without joining the wretched losers of the global economic race. Thomas Friedman discusses this dilemma at length in *The Lexus and the Olive Tree* (1999), but the newspapers supply plenty of other examples.

Half of Brazil's income, for example, goes to one-tenth of its people. President Lula da Silva pledged in his inaugural address at the start of 2003 that his top priority would be ending hunger, and late that year he told a crowd of landless workers that "I want to leave the presidency of the republic looking you in the eyes and saying I did what was possible." Yet Lula soon said that "creating jobs and distributing money to the poor is not easy." The people who voted for Lula's Worker's Party heard that the time was not right for radical social reform. Anything more aggressive than incremental change would jeopardize the country's standing with investors, they heard. The government managed to shelve plans to buy a dozen fighters and redirect that money to social programs. It announced plans to build 230,000 houses for low-income families and to grant land titles to squatters in Brazil's *favelas,* or slums. It announced its intention of ending debt slavery, which holds thousands of Brazilians prisoner on guarded ranches. It proposed to give tax breaks to private universities who admitted a certain number of poor—usually black or Indian—students. But when it dared to propose reforms to pensions and taxes, which would hurt many Brazilians, howls of protest were heard even from the party's supporters.

Or consider Vietnam, where foreign investment fell from $8.6 billion in 1996 to $2.3 billion in 2002. This dropoff was partly a consequence of regional and global events, but it was also the result of Vietnam deciding in 1996 to tell investors to build roads, schools, and hospitals. The investors packed their bags and opened factories in countries that didn't attach such foolish conditions. Now Vietnam is backpedaling, but it takes time to regain investor trust.

The third generalization is that new technologies are improving communications, heightening envy and resentment among the poor, and giving the poor and angry the power to launch deadly attacks on thousands of people. Assassinations and general strikes seem pallid compared to the weapons available today.

THE POLITICAL NECESSITY OF ECONOMIC DEVELOPMENT

What to do? The United States has known since World War II that it's not enough to crush an insurrection, declare victory, and walk away. Do that and you only guarantee that a new and perhaps more dedicated generation of opponents will emerge.

Instead, you must fix the problems that turn people into revolutionaries. A reporter visits Pakistan's garbage dumps and sees Afghan refugee children sorting trash. One of them says, "I want America to be finished." Another says Americans "are very cruel to us. They kill our people." The brightest future the boys can imagine is to join a *madrasa*, a religious school. The acting headmaster of one such school near Karachi teaches the *Qur'an* even though he neither speaks nor reads Arabic. Instead, he has committed to memory a text he cannot understand. The real headmaster has gone into hiding because the authorities are looking for him.

There's no end to such stories, which are disseminated now not only by the Western media but across the Arab world by Arab media, especially satellite television. The affable owner of a restaurant in Amman tells a reporter that "if an American soldier comes in to my place, I will poison his pizza. I will kill him." Ahmed Kamal Aboulmagd, an Egyptian intellectual, says that "to most people in this area the United States is the source of evil on planet earth." Over on the other side of Asia, Abu Sayyaf rebels kidnap two American missionaries, Gracia and Martin Burnham. The Burnhams are told that the rebel goal is to create an Islamic state like that of the Taliban. The rebels have a fallback plan, though. If they can't create an Islamic state they'd like to go to America and find work.

These are the conditions that lead a Saudi political scientist to say "the more poverty you have, the more fundamentalism you have." The same idea has been expressed more genteelly by James Wolfensohn, president of the World Bank: "unless we pursue equitable development the pursuit of peace is likely to be evasive. . . . An environment in which people don't have any hope or don't have any expectations are places in which terrorism can flourish."

Morocco, which prided itself on a moderate brand of Islam, now confronts the same truth. Fourteen suicide bombers killed themselves and 45 others in attacks in Casablanca in 2003. Police made many arrests, including two 13-year-old girls who admitted that they were about to attack the parliament and a shopping mall in Rabat. Like the other suspects, the girls came from Sidi Moumen, a huge Casablanca slum. "What kind of future awaits a person who lives in this filth," asks a Moroccan reporter, pointing to men foraging in piles of rotting garbage. "What do these wretches have left except for a belief in God?" Many suspected terrorists had spent time in the one clean, safe place they knew: the mosque. A mother of sons either dead or in hiding says, "All my troubles began on the day that Mohammed, my eldest, burned his jeans and put on an Afghani robe and leather sandals. I saw my son change before my eyes, without being able to do anything."

Back in the distant days when the United States was locked in a global battle with Communism, every administration knew that economic development was essential if desperately poor people were to be diverted from Communism. That's why the United States offered foreign aid through the State Department's Agency for International Development. That's why it supported the World Bank and regional banks in Asia, Africa, and Latin America. That's why it encouraged the growth of privately run nongovernmental organizations, such as CARE and Save the Children. It sought to give the poor a stake in the system.

Foreign aid has never been very effective, however. There are lots of reasons for this. For starters, Americans have been stingy, in recent years increasing their aid programs by 50% to a magnanimous 0.14% of gross domestic product. On a per capita basis many European countries make the Unired States look like a nation of tightwads. A nation that wears its Christianity on its sleeve turns out to be much less Christian in its willingness to help the poor than several European countries that many Americans would judge wickedly secular.

Foreign aid has failed also because aid money has often been directed to the support of inefficient parastatal or quasi-governmental corporations. Through the 1980s, aid agencies acted as though they thought these corporations could foster economic growth without the huge inequalities of capitalist economies. Like socialist enterprises everywhere, however, such state-run corporations were sheltered from competition and became grossly inefficient. Gurcharan Das's *India Unbound* (2000) is a rompingly enthusiastic critique of that era. Firms were wildly overstaffed, and low-paid workers would wink and say, "They pretend to pay us, and we pretend to work."

Another cause of aid's failure is that investments are crippled by unsound macroeconomic policies. Egypt, for example, has absorbed at least $50 billion in foreign aid over the last 25 years. Despite all the investment to rebuild its sewers and power stations and communications, the country is still desperately poor, largely because it continues to stifle private investment. Egypt grows some of the finest cotton in the world, but production has been cut in half since 1980 because the government insists on directing two-thirds of the crop to the domestic manufacture of rough cloth. Such incompetence leads to unemployment or petty jobs. They in turn reduce young men to desperation. It's no accident that the Gama'at Islamiyya, or Islamic Movement, was founded in Egypt in the 1970s.

Compare Sudan and Singapore. During the 1980s, Sudan was a major recipient of aid, which bred dependency, sapped initiative, and spawned corruption. Singapore, in contrast, governed by authoritarian but smart leaders, has pulled itself up in the space of a few decades so that its standard of living exceeds that of Britain, its former ruler.

Whatever good has been accomplished through foreign aid has meanwhile been negated by protectionism. Tariffs are one example. Global steelmakers complain about American barriers, but there are less publicized barriers facing the owners of Mexican sugar and Pakistani textile mills. American tariff walls reach down to Argentinian honey producers, excluded from the American market while the United States claims to support free trade.

I touched on farm subsidies earlier, when discussing American agriculture. They, too, are part of the reason why foreign aid has been so ineffective. The United States, Europe, and Japan together pay their farmers about $350 billion in annual subsidies. The Japanese are the most lavish: Their average full-time farmer collects $23,000 a year in subsidies, compared to $20,000 in the United States and $16,000 in Europe. Viewed on an acreage basis, the Japanese are more generous still. They pay their farmers about $4,000 an acre in subsidies. The European figure is about $300 an acre. In the United States it's a measly $50.

These subsidies drive down commodity prices, including the prices received by the very farmers targeted by aid programs. Americans ignore this contradiction, but Africans don't.

The world price for cotton, for example, sagged from $1 a pound in 1995 to 40 cents in 2002, but the United States continued to guarantee its cotton growers 70 cents, regardless of the market price. This kept them busy growing cotton. There are American cotton farmers whom the government pays $750,000 annually. Nobody subsidizes the producers in Mali, however. In 2004, the World Trade Organization ruled that the U.S. subsidies were illegal, but with appeals and compliance procedures it will take years for this decision to have much practical effect.

There's a similar story with sugar. The world price is probably 20% below what it would be without subsidies in America and Europe. Europe guarantees its sugarbeet producers three times the world price for sugar, so a farmer there with a 30-acre sugarbeet quota receives a subsidy of about $20,000. What to do with the sugar? Europe dumps 6 million tons into the same world market into which unsubsidized farmers in poor countries try to sell their crop.

The United Nations estimates that as a result of these subsidies the world's poor countries lose $50 billion annually in agricultural sales. By a bitter coincidence, that's the amount they receive in foreign aid. No wonder that Africans hear about a European Union booklet called "Fighting Poverty." It was distributed at the 2002 World Summit on Sustainable Development, and Africans in attendance said that a better title would have been "Creating Poverty." One African agricultural minister offered this advice: "If you want to do an agricultural experiment in Africa, experiment with taking away subsidies in the West."

Africa isn't alone. Mexico has 25 million people dependent on agriculture—almost a quarter of its labor force. NAFTA will soon remove the tariffs that protect them against American growers. Both American and Mexican agricultural exports have approximately doubled since 1994, when NAFTA began coming into effect, but Mexico ran a $2 billion farm-trade deficit with the United States in 2001, and that figure is unlikely to decline. Some Mexican producers are large and efficient enough to prosper under the new regime, but 70% of Mexico's farmers cultivate less than 12 acres each. Within this group, farm abandonments are likely to accelerate rapidly, especially when the tariff on corn is removed in 2008. The livelihoods of 3 million corn producers are at stake. Together, those farmers produce a crop that covers more than half of Mexico's cropland. Still, as one Mexican expert puts it, Mexico's farmers have long been the caboose on Mexico's economic train. Now the caboose is about to be uncoupled.

One more case: Vietnam has a booming export business in catfish. Shipments to the United States rose from half a million pounds in 1998 to 20 million a few years later. That's a fifth of the American market. Catfish are cheap to produce in Vietnam, because labor costs are low and the Mekong delta is a good place for fish pens. American producers have responded not by cutting their own costs but by persuading Congress to forbid the labeling of the Vietnamese fish as catfish, although that's what they are. Now the American producers are accusing the Vietnamese of dumping the fish, which could lead to exclusion from the American market. For the Vietnamese, it's a hardball lesson in free-market protectionism, American style. Among global hypocrisies, this one's up there with "socialism with Chinese characteristics."

POLITICAL OBSTACLES TO DEVELOPMENT

It's a mistake, however, to see aid as a panacea, because economic development depends not only on donors but on recipients. Many—perhaps most—governments think first, second, and third of their own survival. For them, the turmoil accompanying economic change can be a threat. Nothing but the fear of change explains why the Egyptian government has adopted policies that have depressed the number of books published in Egypt since 1960 by 90%. Nothing but fear explains why the title of Disney's "The Lion King" was changed when the film was dubbed for the Arabic market: a movie with the words "lion" or "king" in its title would have been banned as an insult to the self-proclaimed but none too secure lions and kings who rule most of the Arab world.

By this measure, the Chinese government looks good, because while it unhesitatingly crushes dissent, it also stimulates economic growth. Out in Xinjiang, for example, Muslims have been executed for supporting the Hizb-ut Tahrir, the Freedom Party, dedicated to the establishment of an Islamic state across Central Asia. At the same time, however, the Chinese have encouraged trade and travel, and Kashi (Kashgar) has become a bazaar for goods from as far away as Turkey and Japan. That trade is bringing a degree of prosperity to Kashi, and the Chinese encourage it as a damper on upheaval.

ROMANTICISM AGAIN

There's another cultural flaw in prescribing economic development as the way to world peace. Development, after all, means progress in the Baconian sense. And as we have seen over and over again, people have mixed feelings about that kind of progress. They want it, but they also want tradition and the meaning that it gives to their lives. Americans condemn materialism all the time, and their history, like Europe's, is full of groups opposed to the construction of railways and factories or, more recently, urban freeways and high-rises. Getting the right balance between progress and tradition, in other words, is a balancing act for everyone, but it's especially difficult in poor societies that want to grow quickly without cutting their traditional roots. Aid agencies have no better understanding than anyone else of how to achieve this balance. Typically, their professional staffs are evaluated by how much money they push, yet those same people eventually retire—these are real examples—to write children's books or open antique stores.

At this point social upheaval bumps into romanticism. Plenty of terrorists come from middle-class or even wealthy backgrounds. These aren't people who have nothing to lose, yet they're prepared to lose everything. To a degree, it's a case of Muslims helping fellow Muslims, most obviously the Palestinians. But there's more than altruism involved here, more than the quest for social justice. An Indonesian journalist says, "People like me are looking to religion to help us improve our lives." He's not looking for Islam to make him rich; he's looking for something beyond the material wealth promised by modernity. His hunger is the hunger that leads so many Americans to become born-again Christians, Sufis, Buddhists, artists, or radical environmentalists.

You can see this same romanticism shaping the formation decades ago of radical Islamic movements. One of the oldest is the Muslim Brotherhood, established in 1928. Its founder was Hassan al-Banna, an Egyptian schoolteacher who worked at Ismailiyya, the Suez Canal's headquarters town. "Weary of this life of humiliation and restriction," as he wrote, al-Banna began to preach the importance of virtuous communities, from which virtuous government would arise. This kind of reform from below has become known as *da'wa*, usually the call to Islam but by extension the call to social service as part of a Muslim's social obligations. Many of the organizations that Americans label terrorist, like Hamas and Hezbollah, are known on their own turf as generous providers of vital social services available nowhere else. Some of these organizations grow into political parties, and at least in the case of Turkey's Refah Party they win elections.

The Brotherhood gradually became impatient for change. Fearing a coup, in 1948 the government of Egypt disbanded it. A few months later the prime minister was assassinated. Al-Banna was soon shot dead in Cairo, presumably by a government agent.

Al-Banna's most important successor in the Brotherhood was another Egyptian schoolteacher, Sayyid Qutb. In a kind of study-abroad program, Qutb was sent to the United States in 1948. He hated it, as did plenty of Americans of the time who lampooned, along with H. L. Mencken, *Boobus americanus*. Qutb sent home a postcard on which he wrote that "if all the world were America, it would undoubtedly be the destruction of humanity." What did he object to? Qutb listed first the lack of "human sympathy and responsibility for relatives." Second came America's "materialistic attitude which deadens the spirit."

That's straight from an assigned essay on Romantic poetry, and it has many echoes. A Tunisian journalist explains that Islamism arises from "the emptiness people feel—they seek refuge in religion." And just as some Americans in the 1950s and 1960s became Beatniks, then Hippies, then revolutionaries, so Qutb turned to radical politics. He ended on Nasser's gallows in 1966 because he set out to overthrow not only the Egyptian government but every other government that claimed to be Muslim but which, in the Western tradition, tried to separate religion from political institutions. Insisting on that separation, in Qutb's view, was a return to the ignorance of the *jahiliyya*, the pre-Muslim era of human history.

Qutb's brother would become a professor in Saudi Arabia, where Osama bin Laden was among his students. Osama, in short, is part of a chain that goes back to men educated in Egypt during its most liberal period, when the educational curriculum was dominated by the study of European culture. Even today, you can find Palestinian agricultural-extension officers who can recite Wordsworth by the pound—can drive through the West Bank and recall from memory not only "Daffodils" but those more somber lines I mentioned in the last chapter—the ones where Wordsworth warns that we lay waste our powers by "getting and spending."

The deeply romantic root of this insistence on a spiritual value in life isn't the only thing that the Brotherhood's leaders found in Western thought. Both al-Banna and Qutb believed that true freedom could only be found in a society where actions are governed, dictated, by infallible authorities. Sound familiar? It should: It's the kernel of totalitarianism. Hitler said of his Hitler Youth that never in their lives would they be free, but Lenin was ahead of him. The Soviet Union's first concentration camp, established on the White Sea's Solovetzky Islands in 1923, had a sign that read: "With an Iron Fist, We Will Lead Humanity to Happiness." The Muslim Brotherhood sounded the same apocalyptic note in its founding manifesto: "God is our purpose, the Prophet our leader, the Koran our Constitution, Jihad our way, and dying for God's cause our supreme objective."

Putting Islamic extremists in this context provides an alternative to the apocalyptic view that Islam and the West are inevitably locked in a state of war that can only end with the destruction of one or both sides. It allows us to believe that Muslims are not chained to a literal interpretation of the *Qur'an*, any more than Christians and Jews are to the literal interpretation of the Bible. Instead of encouraging secular education, however, we offer a War on Terrorism and, as a side dish, more getting and spending. We're like wealthy parents who want to connect with an estranged child but don't know what to do except discipline the kid or offer money. It's ridiculous, but you can see how we come to make this mistake. We can barely balance progress and tradition for ourselves, and we don't have a clue how others should do it.

A SUPERPOWER'S DILEMMA

Faced with the challenge posed by al Qaeda, the United States took the comparatively easy steps of destroying the governments of Afghanistan and Iraq. Now it is trying to rebuild them as democracies. Is this a realistic goal?

It's discouraging to remember America's long and ultimately failed efforts to establish stable governments in the Philippines and, 50 years later, Iran. More currently, at the end of 2002 the United Nations had a force of over 6,000 civil servants and 30,000 troops in Kosovo. The U.N. had spent over $2 billion helping the 2 million Kosovars, but unemployment remained at about 60%. The Kosovars were divided between the majority Albanians, who wanted independence, and the minority Serbs, who wanted to preserve Kosovo's status as part of Yugoslavia. Unmik—the United Nations Mission in Kosovo—was increasingly seen as a colonial ruler with no way either to make things right or to get out without triggering a return to war. Many Albanians meanwhile cherished the illusion that the United States, having rescued them from the Serbs, would welcome them as the 51st state.

The United States is faring no better in Afghanistan. A flood of Afghan refugees moved back home after the fall of the Taliban, but a half million of them are now homeless in Kabul, desperately waiting for aid that hasn't come. Regional warlords are back in charge almost everywhere outside Kabul, and American influence in the countryside is minimal. New international relationships are evolving outside the control of either Kabul or Washington. Afghanistan's traditional exports, for example, were carpets and dried fruit; its imports were manufactures. Historically, these things moved through Karachi, but Pakistan is suspicious of the new Afghanistan; after all, elements of the Pakistan government created and sustained the Taliban. Iran, on the other hand, is keen on developing trade, not only with Afghanistan but through Afghanistan to Central Asia, so it's building a port at Chabahar, on its own Indian Ocean coast. India, no friend of Pakistan, is paving a road north from that port through Iran to Afghanistan. From there, other roads lead north to Uzbekistan. Think that the United States can control these developments? Not likely: It can't even prevent the resurgence of opium. Say what you will about the Taliban, in 2000 they eliminated the cultivation of opium poppies. Now Afghani growers are back in business. In 2003, Afghanistan produced 3,600 metric tons of opium, with a value of $2.3 billion, equal to Afghanistan's gross domestic product and accounting for three-fourths of the world's opium. Meanwhile, the United States is busy rebuilding the same Kabul-to-Kandahar highway that an earlier generation of Americans built in the 1960s. The job's a lot more dangerous now than it was then. Local Taliban commanders tell their men: "Go burn a school. We will give you money. Go rob a house. We will give you money." Highway engineers are fair game—and very hard to protect.

Then there's Iraq. The United States has moved quickly to repair and renovate the country's infrastructure and oil fields, but investors are needed to get the economy moving, and they won't come without trust that their investments will be safe. They need security, in short, and the United States hasn't been able to provide it. Fundamentally, security depends on trust, and America has been unable to win the trust of Iraqis. Trust is always hard to build when you have a gun and the other fellow doesn't. In Iraq's case, prisoner abuse scandals have made it doubly or triply difficult. Then, in the background, there is America's continuing and unwavering support for Israel. It's "the great poison in the region," in the words of a senior U.N. official. U.N. spokesmen dashed in to say that Lakhdar Brahimi was speaking only in his personal capacity, but he was saying no more than what every Arab believes.

And there's yet another reason for Iraqis to distrust America, because it calls for democracy but won't tolerate an Iraqi government hostile either to it or to Israel—even an Israel that is supporting the Kurds in what may become a bid for an independent Kurdistan, at the price of a dismembered Iraq.

One Saudi official says that the idea of building democracy in the Middle East is "the most preposterous, idealistic statement I have heard in a long time." Perhaps remembering Algeria, he insists that free elections in his own country would promptly lead to the election of a government as repressive as the Taliban. It's no different in Egypt, Jordan, Pakistan, or Syria. It's already happened in Kuwait, where parliamentary elections after the fall of Saddam Hussein went overwhelmingly in favor of Islamic rather than liberal parties. The only thing that kept Kuwait from adopting Islamic law was that real power in Kuwait rests with the emir, not the debating society called parliament.

Supporters of American policy dismiss such hesitations. Maybe America can't fix the world, they argue, but if it doesn't try there will be no end to terrorism. "What is almost impossible," they think, "turns out to be indispensable." They have no patience with anyone who thinks that the best course is to cultivate the liberal elements already within Muslim societies, elements that can help develop a respect for human rights, especially the rights of minorities. They're not interested in finding out why Muslim countries like Senegal and Mali are democratic and why the imam at Timbuktu's mosque says, "I am neutral and I will vote for no one." They lack the patience for the Chinese solution, which seems to be the very gradual introduction of democracy. "Been there, done that," is the dismissive reply. One American historian expresses this ferocity very well: "Those who find militant Islam terrifying have clearly never seen a militant democracy."

In all this, the United States is almost bowling alone. It has broken treaties governing nuclear arms, refused to sign international agreements on the environment, and boycotted new international organizations like the International Court of Justice. America's allies conclude that the United States is destroying the very stability it claims to seek. They see the United States embarking on neocolonialism or perhaps democratic imperialism. They compare it to Germany in 1914, when Victor Klemperer wrote that "we, we Germans, are a truly chosen people." Jason Epstein writes that Americans seek "to Americanize the world, as previous empires had once hoped with no less zeal to Romanize, Christianize, Islamicize, Anglicize, Napoleonize, Germanize, and communize it." The European reaction is obvious. In the words of the newly elected Spanish prime minister, "We must not allow one country to decide the future of our planet."

LANDSCAPES OF VIOLENCE

What is there to see of all the upheavals we've looked at in this chapter? Are they visible in the cultural landscape?

The old British governor's palace in Khartoum used to be protected by a decorative chain, draped from post to post around a large rectangular lawn. British ministries in Rangoon used to have a buffer of grass separating them from the street. Now, there's a high wall around the lawn in Khartoum and there's chain-link and barbed-wire fencing in Rangoon. The governments of both nations are proud to have thrown off the colonial yoke, but they now need more protection from their own people than the British ever did.

As late as the 1970s, a visitor to Washington could stroll without anyone's permission into almost any federal building. There were no guards at the doors, no metal detectors, no required visitor badges. You could drop by an office, introduce yourself, and ask whatever question you liked. That's gone now, and not just in Washington: Even city halls and state office buildings are often rimmed with Jersey barriers, those low concrete walls that were designed as highway medians but which have found a new use. Stopping a car in central London is just about impossible now: There must be hundreds of signs warning motorists off. Such is the price paid by a global civilization whose ideology teases millions of people.

Meanwhile there's the void along the line of the now-destroyed Berlin Wall. There's a world of difference between North and South Korea, between Haiti and the Dominican Republic, and between the cities on the American and Mexican sides of the Rio Grande. Those differences all reflect the differing abilities of governments to serve their people and win their allegiance.

There are famous military cemeteries, like those at Arlington and Normandy. There are military cemeteries in unexpected places, too, like the British ones in Jerusalem and Khartoum.

And then there are the unintended memorials. Twenty years ago there were plans for the 200-mile Jonglei Canal, which was going to bypass the Sudd, Sudan's great papyrus swamp on the White Nile. A huge German-built bucket-wheel excavator was brought in from Pakistan, where it had dug canals in the Punjab. It went to work in 1978. Burning 10,000 gallons of fuel daily, it excavated 3,000 cubic meters hourly and had completed about half the canal before civil war forced the French crews to abandon the machine in 1984. Vandalized, it hasn't moved since.

Or try Waller Field, an abandoned airfield in Trinidad. It's a relic of World War II, when an unending stream of planes flew here from the United States. They landed briefly

Independence brought brutal suppression to the Burmese, and the government found it advisable to ring its ministries with barbed wire.

Waller Field, east of Arima, Trinidad. It's been quiet for 60 years, but a concrete runway doesn't go away overnight.

before continuing to Natal in Brazil and then making the 1,800-mile jump across the Atlantic to Freetown, in Sierra Leone. They crossed Africa. Along the way, they stopped at airports that, like the one at Waller Field, were designed by Pan American. Who had more experience in setting up airports in remote locations? Eventually the planes flew to Egypt and the campaign in North Africa and Italy. Back at Waller Field today, you can have the runway all to yourself—you and the grass growing in the expansion joints.

CHAPTER 17

CONSERVATION, NATURAL RESOURCES, AND POPULATION

So far, I've considered two problems of our technological civilization. They've both been social, one chiefly affecting people who have been the beneficiaries of progress, the other chiefly affecting those who have not. Both problems have shaped the cultural landscape and are visible in it. Both can be inferred from it. Now, however, I want to consider two other problems that are more physical than social. One is the relatively new concern with environmental quality; I'll consider it in the next chapter. The other, which I want to discuss now, is the long-standing fear that our society is such a voracious consumer that it is bound ultimately to collapse for lack of the resources on which it depends.

Water is an obvious example. The Rio Grande is a roiling river in northern New Mexico but only a sandy streambed at its mouth. The same impoverishment is visible on the Colorado and farther afield on China's Yellow River, the Ganges, and the Nile. During the winter of 2002 Lake Kinneret, or the Sea of Galilee, was 20 feet below its normal level because Israeli withdrawals were greatly exceeding the lake's inflow. (Twenty feet may not sound like much, but it's equivalent to a 2-year supply of water for Israel.) The threat was so acute that in 2001 the Israelis cut their farmers' irrigation supplies in half; they even seriously considered turning off the drip-irrigation systems in their prized city parks. Farther east, the marshes that once teemed with wildlife at the mouth of the Tigris and Euphrates are parched salt flats. Saddam Hussein drained them to deny rebels a refuge, but even if he hadn't, the marshes would have died as diversions are made upstream for irrigation developments in Turkey, Syria, and Iran. A similar but less well-known story can be told about the great Indus, so reduced by diversions for irrigation that the last 80 miles of the stream is now brine from the Arabian Sea.

No need to single out water: most American oil wells are strippers, producing no more than 10 or 15 barrels a day. Often they produce only one barrel, which is mixed with 100 barrels of water. Wells like these are run by independent producers; the big companies can't be bothered with such marginal properties. It's a discouraging sight: old separating tanks and pumping jacks standing idle most of the day. Does it portend a day without gasoline, a night without lights?

Or fly out of Seattle on a clear day and look down at the patchwork of clear-cuts draped across the Cascades. Will people a century from now look back at us as wantons, butchering the planet for cheap lumber and paper? If that sounds improbable, go down to

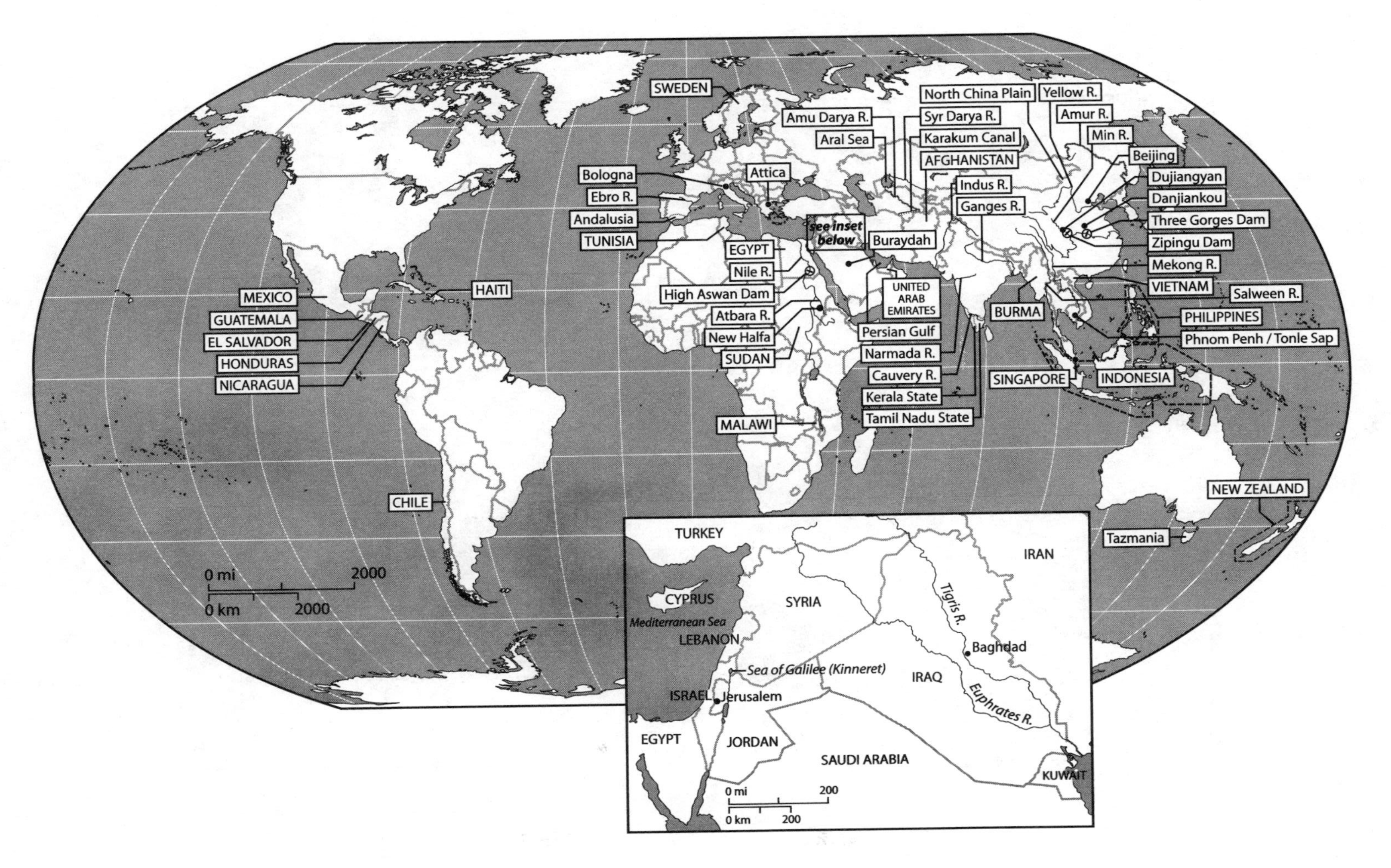

SWEDEN
North China Plain
Yellow R.
Amu Darya R.
Syr Darya R.
Amur R.
Aral Sea
Karakum Canal
Min R.
AFGHANISTAN
Beijing
Bologna
Attica
Indus R.
Dujiangyan
Ebro R.
Ganges R.
Danjiankou
Andalusia
see inset below
Three Gorges Dam
TUNISIA
Buraydah
Zipingu Dam
EGYPT
Mekong R.
Nile R.
VIETNAM
Salween R.
MEXICO
HAITI
High Aswan Dam
UNITED ARAB EMIRATES
BURMA
PHILIPPINES
GUATEMALA
Atbara R.
Persian Gulf
Phnom Penh / Tonle Sap
EL SALVADOR
New Halfa
Narmada R.
HONDURAS
SUDAN
Cauvery R.
SINGAPORE
INDONESIA
NICARAGUA
Kerala State
MALAWI
Tamil Nadu State
CHILE
NEW ZEALAND
Tazmania
0 mi 2000
0 km 2000
TURKEY
IRAN
CYPRUS
SYRIA
Tigris R.
Mediterranean Sea
LEBANON
Baghdad
Sea of Galilee (Kinneret)
IRAQ
ISRAEL
Jerusalem
Euphrates R.
EGYPT
JORDAN
SAUDI ARABIA
KUWAIT
0 mi 200
0 km 200

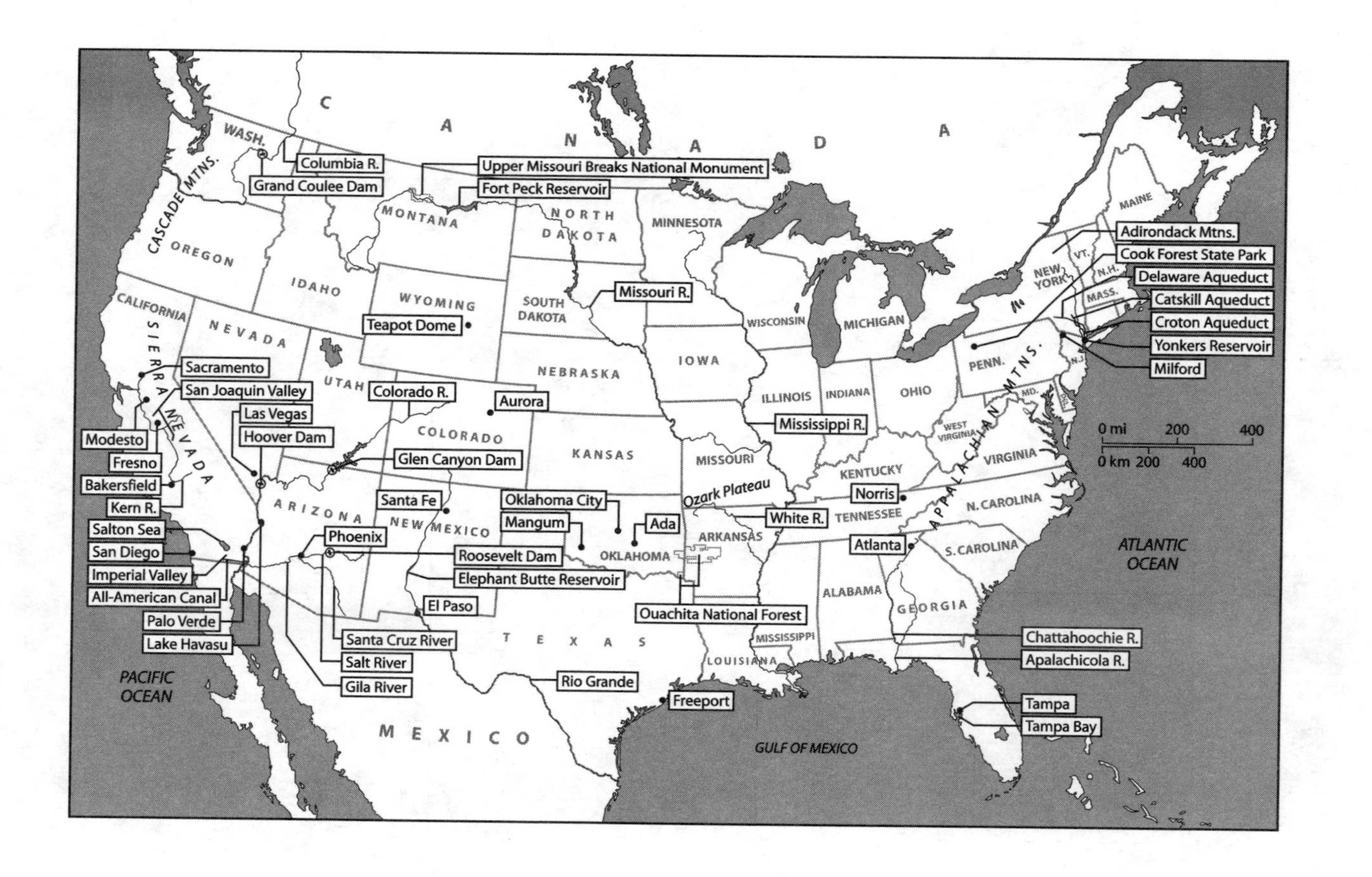

C A N A D A
M E X I C O
PACIFIC OCEAN
ATLANTIC OCEAN
GULF OF MEXICO
0 mi 200 400
0 km 200 400
CASCADE MTNS.
SIERRA NEVADA
APPALACHIAN MTNS.
WASH.
OREGON
CALIFORNIA
IDAHO
NEVADA
MONTANA
WYOMING
UTAH
ARIZONA
COLORADO
NEW MEXICO
NORTH DAKOTA
SOUTH DAKOTA
NEBRASKA
KANSAS
OKLAHOMA
TEXAS
MINNESOTA
IOWA
MISSOURI
ARKANSAS
LOUISIANA
WISCONSIN
ILLINOIS
MICHIGAN
INDIANA
OHIO
KENTUCKY
TENNESSEE
MISSISSIPPI
ALABAMA
GEORGIA
S. CAROLINA
N. CAROLINA
VIRGINIA
WEST VIRGINIA
PENN.
MD.
N.J.
NEW YORK
VT.
N.H.
MASS.
MAINE
Columbia R.
Grand Coulee Dam
Upper Missouri Breaks National Monument
Fort Peck Reservoir
Teapot Dome
Colorado R.
Glen Canyon Dam
Aurora
Missouri R.
Sacramento
San Joaquin Valley
Las Vegas
Hoover Dam
Modesto
Fresno
Bakersfield
Kern R.
Salton Sea
San Diego
Imperial Valley
All-American Canal
Palo Verde
Lake Havasu
Phoenix
Santa Fe
Santa Cruz River
Salt River
Gila River
Roosevelt Dam
Elephant Butte Reservoir
El Paso
Oklahoma City
Mangum
Ada
Rio Grande
Freeport
Ozark Plateau
Ouachita National Forest
Mississippi R.
White R.
Norris
Atlanta
Chattahoochie R.
Apalachicola R.
Tampa
Tampa Bay
Adirondack Mtns.
Cook Forest State Park
Delaware Aqueduct
Catskill Aqueduct
Croton Aqueduct
Yonkers Reservoir
Milford

When the right hand is oblivious of the left: a Forest Service sign posted in a fresh clearcut praises beauty. Sadly, this is the south side of the Olympic National Forest, Washington.

Haiti and watch coffin makers at work sawing brittle planks of avocado. The mahogany they used until a few years ago is gone, along with the island's forests.

THE CONSERVATION MOVEMENT

Fears like these are deep and old: They go back more than a century to George Perkins Marsh, who was Abraham Lincoln's ambassador to Italy. Marsh was a classicist. He looked at the Mediterranean Basin and remembered the *Critias*, in which Plato writes that Attica once was forested, with plenty of water, but now is "a skeleton of a body wasted by disease." Marsh thought of his own New England and of Wisconsin, where forests were being cut and burned at a terrific rate. In a sober and heavily footnoted book called *Man and Nature* (1864), Marsh warned that the United States was on the verge of repeating Greece's sorry environmental history.

Marsh's warning was ignored for many years, and today there are only a few scattered remnants of the eastern forest as it was before European settlement. There is the Five Ponds Wilderness in New York's Adirondacks, for example, and there is Cook Forest State Park, on Pennsylvania's Allegheny Plateau. Congress was sublimely oblivious, but cartoonists began drawing Uncle Sam bald, his head dotted with stumps. Finally, in 1891, a clever participant in a legislative markup session tacked a rider to an appropriations bill. It gave the president the power to set aside forest reserves on the public lands. Presidents Cleveland, Harrison, and Roosevelt all exercised this new authority, and the dusty, leather-bound volumes of *Statutes at Large* from those years are full of proclamations listing the townships being set aside, page after page. Roosevelt exercised his authority with special gusto, and in 1907 Congress angrily woke up and stripped him of it. Too late. By then, nearly all the national forests that exist in the American West today had been set aside.

What was the government to do with them? For a decade the answer was nothing. They were to be a reserve, period. In Germany, however, foresters were looking at trees as a crop. This thinking was now imported to the United States, and in 1905 the forest reserves were given the name of national forests. In keeping with the new philosophy, administration was shifted from the Interior Department to the Department of Agriculture. Sustained-yield logging was now the goal: a flow of forest products in perpetuity.

The key figure here was Gifford Pinchot, the first head of the Forest Service. Pinchot (his finely understated title was Forester) endowed that agency with a charisma that lasted long after Pinchot's ideal ranger—smart, dedicated, tough—had nearly sunk beneath bureaucratic mire. For the decade in which he held office, Pinchot was the undisputed leader of what became internationally known in those years as the Conservation Movement. Its avowed purpose was the wise use of natural resources, with wise use defined as the greatest good for the greatest number. You can quickly become tangled up in arguments over the definition of the greatest good, but Pinchot wasn't a philosopher; he just wanted to be sure that the United States had plenty of forests forever. To ensure that it would, he recommended government ownership, as in the national forests. Contrary to what you might expect, however, Pinchot was a silk-stocking aristocrat. Grey Towers, his fine home in Milford, Pennsylvania, is now a museum.

Pinchot's power base collapsed during the Taft administration, but the conservation idea endured and was especially important to Franklin Roosevelt, whose presidency Pinchot lived to see. It's still around, too, though "sustainable development" is probably now a more stylish formulation of the same idea.

Pinchot left an account of his Washington years. It's biased, but *Breaking New Ground* (1947), written many years later, shows what fun Pinchot and Roosevelt had. They, and the men around them, wanted to develop national forests in the East, for example, particularly in the Appalachians and Ozarks, but there was little public-domain land in these states. Still, there are national forests in those mountains today—and even on coastal lowlands, for example, in the Florida Panhandle. That's because the conservationists were able to secure passage in 1910 of a law giving the federal government the authority to buy private land for inclusion in national forests. There was no direct constitutional authority for such action, but the conservationists argued that forest management was the only way to prevent downstream floods. Floods affected navigation, of course, and the constitution specifically mentioned navigation as a federal responsibility. Voilà! The claim that forests actually reduce floods more than do scrub or grass proved embarrassingly difficult for the forest service to demonstrate, but with the sponsorship of Rep. John Weeks (R-MA), the legislation became law. The Geological Survey made formal findings that the proposed forests would reduce flooding. Still, it wasn't until the 1930s, when another Roosevelt was in the White House and when land was available at a dollar or two an acre, that the program went into high gear. Cheap as the land was, the forest service haggled for the best deal, even in negotiations with owners who were selling small, failed farms. The government was determined to stretch its dollars.

Although Pinchot was a forester first and last, he and his associates were interested in much more than forest conservation. They weren't able to deal with floods directly, because they were the jealously guarded domain of the Army Corps of Engineers. Nor did they get involved in urban water supply, usually a local matter. They did, however, look at the arid

West, where from the utilitarian perspective water was wasted when it flowed into the sea. This was the impetus for the creation in 1902 of the Reclamation Service, later renamed the Bureau of Reclamation. Among its earliest projects was Roosevelt Dam, on the Salt River east of Phoenix. Later, the bureau built several world-class dams, including Hoover (Boulder), Glen Canyon, and Grand Coulee. From most of the bureau's dams, canals brought irrigation water to desert lands that were soon green with alfalfa and other crops. It was an image that would inspire engineers from Spain to Israel and from South Africa to Australia.

The Corps of Engineers, embarrassed by huge floods on the Mississippi around 1930, decided that its unswerving faith in dikes should be supplemented, after all, with a little devotion to dams. Fifty years after *that*, the Corps decided that maybe dams weren't such a great way to prevent floods: Maybe people should stop building in floodplains. By then, however, the Corps had built a series of immense earth-fill dams on the Missouri River. Fort Peck was probably the best known, but it was upstaged by the contemporary and internationally renowned Tennessee Valley Authority. Created in 1933, the TVA was geographically restricted but thematically audacious. It was conceived as a master conservation program, incorporating not only dams and hydroelectricity but improved agriculture that used fertilizer produced by TVA electricity. The TVA was also interested in—big jump here—model towns. It built a few; the best known is Norris, Tennessee, named for Sen. George Norris (R-NB) and the senator who led the push to create the TVA. Norris was eager to bring the TVA concept to other American river basins, such as the Missouri, but the utility industry easily convinced Congress that the TVA was an exercise in socialism that should be curtailed, not expanded. The agency was reduced to being little more than a federal utility, generating power at coal- and nuclear-fired power stations, as well as at its dams. As such it survives, still taking pride in a fame that it no longer deserves.

Oil caught the conservationist eye. Shortly after 1900 the Navy had converted its fleet from coal- to oil-fired boilers. Meanwhile, Americans were beginning to drive automobiles and pave streets. What would happen when the nation's finite oil reserves were depleted? Nobody at the time knew what lay under the ground in East Texas, or off the Gulf Coast, or up on the North Slope of Alaska—or, for that matter, in Venezuela and the Middle East. (In the 1920s, one American oilman asked T. E. Lawrence, "Lawrence of Arabia," if it made sense to spend money looking for oil in Arabia. "No," was Lawrence's one-word reply. Perhaps Lawrence was actually expressing his opinion, but it's hard for me not to believe that he deliberately misled the man to prevent Americans doing what, a generation later, they would in fact do. In any case, the oilman said thanks and went home, I assume to Lawrence's relief.)

The Navy, with conservationist support, had meanwhile urged the creation of naval petroleum reserves. The most important ones, measured by proven reserves, were west of Bakersfield, in California's San Joaquin Valley. A small one, however, was in Wyoming, and ironically it became the best known. It was located at a place called Teapot Dome, and an oilman in the 1920s bribed the Secretary of the Interior so his company could drill it. The secretary, Albert B. Fall, went to jail.

Fall was corrupt, but that didn't make the reserves a good idea. They weren't, any more than forest reserves were a sound way to prevent timber famine. Keeping the oil reserves locked up until the nation ran out of oil might help the Navy, but when that day came the country would have much bigger problems than battleship fuel. Mostly from congressional inertia, however, the reserves sat more or less intact until the 1980s. By then, it was clear that the way to avoid shortages was to have an active program of exploration, to develop substitute fuels, and to use fuels more efficiently. With one exception, the oil in the naval reserves was then auctioned off to the highest bidder. The exception is the 23-million-acre Naval Petroleum Reserve No. 4, which lies on the Alaska North Slope. Oil companies covet it. Paradoxically, the United States in the 1980s began building a strategic petroleum reserve. By 2002 that reserve held almost 600 million barrels of oil—about a 50-day supply. Unlike the early reserves, which were intended to guarantee the Navy an oil supply when other resources were exhausted, the new reserves are intended to buffer the country against short-term supply disruptions.

The federal government took two other steps to conserve the nation's oil. One was to abolish the application of the mining law of 1872 to oil on the public lands. Under that law, conceived primarily with gold in mind, the only way to secure rights to oil on the public domain was to file a claim, drill, and find oil. It was tremendously wasteful, because claims were limited to 160 acres, which meant that if a prospector found oil, he was suddenly surrounded by a scrum of eager neighbors. Oil, as I mentioned in Chapter 11, is no respecter of property lines and merrily flows to the nearest well, so the application of the General Mining Law created a free-for-all. The waste, was huge, especially when the lack of transportation facilities meant that the oil wasn't going anywhere. There are places in the hills west of Bakersfield, California, where you can still find traces of old dams hurriedly built to store the oil.

The Mineral Leasing Act of 1920 substituted for this chaos the leasing of blocks covering thousands of acres that could be developed less wastefully. Under its terms, a portion of the value of the oil produced from the lease was paid to the government as a royalty. It was the first time that the U.S. government took a direct share of the mineral wealth produced from the public lands. In this, it was no different from the practice on private lands, where owners were given a share of the proceeds. (The classic royalty was one-eighth, though

A hurriedly constructed dam, built a century ago to store oil, near Taft, west of Bakersfield, California.

petroleum landmen liked to joke about the grizzled farmer who demanded a tenth, and not a penny less.)

The other conservation measure was federal leasing of offshore lands. For a time, the states claimed jurisdiction of these lands, but in the 1950s the Supreme Court settled federal title to everything beyond a coastal fringe. Since then, the continental shelves have become a major source of oil and federal revenue. They have also pitted the federal government against states like California and Florida, which from time to time have feared the environmental consequences of offshore development.

Two other resources attracted conservationist attention. One was soil, which during the Dust Bowl of the 1930s famously blew from the Great Plains all the way east to dramatically darken the sky outside congressional hearing rooms. In response, Congress in 1933 created the Soil Conservation Service (SCS, originally the Soil Erosion Service) under the leadership of Hugh Hammond Bennett. A worshipful national press called him Big Hugh, and the SCS became world famous. Today it exists as the Natural Resources Conservation Service—a case of a bureaucracy figuring that a bigger name must mean a better agency—but it continues to work as it always has with farmers, on a cooperative rather than a compulsory basis. Rather than force farmers to adopt conservation measures, in other words, the agency subsidizes the development of farm ponds, the construction of terraces, and the establishment of shelterbelts. (The first of those, by the way, is off State Highway 34, north of Mangum, Oklahoma. It's a linear forest.) Later droughts, such as one in the early 1950s and one on the Great Plains today, have been as climatically severe as the drought of the early 1930s, but their impact on soil has been much less severe. The SCS should get much of the credit.

Finally, there was range land. By 1930, the public domain suitable for farming had gone into private ownership. Forest lands were either in private ownership or reserved as national forests. A huge amount of land, however, remained vacant, an open range stretching from the Rocky Mountains on the east to the Cascades and Sierra Nevada on the west. It was good for grazing, but it was being severely overgrazed, so in 1934 the federal government closed it and initiated a permit system. Local ranchers were now allowed to use specified lands within specified limits—so many cows and calves allowed on a given pasture for a certain number of months each year. The grazing districts are administered today by the Bureau of Land Management (BLM), the successor to the venerable General Land Office. The government went further and bought several million acres of private but degraded farmland, mostly on the Great Plains. This land was put into national grasslands, similarly subject to controlled grazing by local ranchers but administered by the Forest Service.

CONSERVATION'S SIGNIFICANCE

I go into this conservation history for three reasons. First, you cannot travel very far in the western United States without encountering national forests, BLM lands, and the dams built by the Bureau of Reclamation and its sister agencies. In other words, the conservation movement was a potent force shaping the cultural landscape of the United States today. Many people in rural communities of the western states wish that it wasn't so: They see the federal land management agencies as their sworn enemies. For many other Americans, however, the lands administered by the forest service and BLM are national treasures, supplementing the much smaller systems of national, state, and local parks. Without the surviving public lands, in truth, Americans would be very nearly trapped between fences. They'd be confined to

The United States's first shelterbelt; there's another field on the far side of the line.

whatever they owned, whatever they were invited to use, to public roads, and to the few places—mall, school, workplace, parks—to which they had regulated access. Ironically, this romantic legacy of the conservation movement is very far from the utilitarian goals of the conservationists themselves. Still, they might not have objected to the way things have turned out. Pinchot once flabbergasted a hardnosed congressman who wanted to harvest some fallen redwoods in California's national parks. Pinchot testified that he wouldn't touch even one of them. They were treasures, he said.

The conservation movement also left a much more questionable legacy, however, in helping to embed the utilitarian attitude toward nature. Despite Pinchot's defense of old redwoods, his Forest Service generally prided itself on treating trees like cabbages. The U.S. Forest Service at least officially no longer thinks in those terms, but other forestry agencies around the world do. Tasmania, for example, has large areas of a giant eucalyptus, *E. regnans*. About 16,000 acres of these trees—almost as tall as coast redwoods—have been set aside in reserves, and the government wants to log the rest. The trees it wants to cut aren't much good for anything except pulp, because the heartwood of the giant trees is largely rotten. The plan, however, is to clear the land and replant it in fast-growing species, cabbages that will be harvested just as they come into maturity. Has enough old growth been set aside? Forestry Tasmania's general manager of operations thinks so. The state, he says, has set aside "more than required by international standards for preservation of the gene pool."

An even more controversial expression of conservationist utilitarianism is in the construction worldwide of giant dams.

The United States, ironically, has been out of the business now for a generation. The Bureau of Reclamation came under attack in the 1920s for projects that subsidized uneconomic agriculture, but the mystique of big dams survived until the 1960s, when the Sierra Club argued very effectively that dams did more harm than good. Since then, the bureau has built nothing very large. It continues to subsidize farmers in other ways. There's talk, for example, of a $200 million pipeline to bring water from the White River to rice farmers in southeastern Arkansas. The farmers rely on groundwater from failing aquifers, which the pipeline would replace at a cost of $300,000 per farmer.

More typical of the bureau's new role—its retreat from the reclamation mystique—was its decision in 2002 to pay $100 million to farmers in the Westlands Water District of California's San Joaquin Valley. Their land had been damaged by poor drainage, a bureau responsibility. Sued by the farmers, the bureau decided to settle by buying out the farmer's right to irrigate some 34,000 acres. There's talk that the program may expand to 200,000 acres.

Meanwhile, many Americans have come to regret the conversion of the Missouri River into a string of ponds. The government in 2001 made partial amends by setting aside the last remaining 150 miles of free-flowing water as the Upper Missouri Breaks National Monument.

Overseas, it's a different story. In the 1960s, Egypt's High Aswan Dam converted the Nile Valley from seasonal to perennial irrigation at the price of waterlogged soils, increased fertilizer requirements, and damaged fisheries. Abdel Gamal Nasser was the pharaonic president of Egypt at the time, however, and in the best pharaonic style he commanded work to begin, despite warnings about these environmental hazards. There were social consequences, too. As Lake Nasser filled, it flooded the world of the Nubians. Some lived in Egypt and were moved to slums at Aswan. Others lived near Wadi Halfa in the Sudan and were shipped south to a grim place mockingly called New Halfa. Most of them eventually abandoned it and drifted elsewhere. The Arabic novelist Idris Ali gave voice to their loss in *Dongola* (1998). He writes: "Nubia, my homeland/ Land of my fathers, my palm trees/ Secret of my sorrows! My land, farewell."

It's a story that's been repeated many times. The Soviets diverted the waters of the Amu and Syr Darya into thousands of miles of canals built to irrigate cotton. One consequence was the drying up of the Aral Sea, which put 40,000 fishermen out of work. Another, following the breakup of the Soviet Union, has been the near collapse of the irrigation system: Without maintenance, more than half the water in the 800-mile Karakum Canal is now lost to seepage. Head-end farmers—those near the start of the canal—use water lavishly to grow 160,000 acres of rice in the desert. Downstream, tail-enders grow dust.

Yet China, India, and Turkey have all insisted in recent years on moving forward with huge dams. China is investing $25 billion in its Three Gorges, or Sanxia, Dam on the Yangtze. In 2003 the gates closed and water began rising behind the dam, 600 feet high. Everything about the project is huge, including the set of locks to permit seagoing container ships to pass through the reservoir on their way to and from Chongqing. The 370-mile-long reservoir floods the Qutang, Wu, and Xiling gorges and will displace more than 1 million people by 2008. Late in 2003, 720,000 had already moved. The displaced families were offered compensation for their homes and the opportunity to buy new ones, but many said the compensation was inadequate to buy the new ones.

The Chinese government tells the people that they should "forsake the small home. Support the big home." One man, looking at the reservoir, says, "When you build the biggest dam on Earth, people know you have a great nation." It sounds very Chinese, but the Three Gorges dam site was selected in the 1940s by Bureau of Reclamation engineers assisting the Nationalist government. One new departure, and one that would have surprised them, is that the Chinese have budgeted $120 million for archaeological rescue work. Many relics will be lost, but some historically important features will be saved. They include White

Crane Ridge, with thousands of carvings on a limestone outcrop in the middle of the river, and the Shibao Block Pagoda, on the side of a gorge cliff. Some 37 historic buildings will also be moved from Dachang, a town rich in historical buildings, but more will be lost along with the townsite.

Indian engineers continue to study the feasibility of a canal system linking India's rivers from the Ganges in the north to the Cauvery in the south. It's anybody's guess what will come from the Task Force on Interlinking of Rivers, but in the meantime work continues on the huge Sardar Sarovar dam on the Narmada River. Here, too, as on the Yangtze, the World Bank has withdrawn financial support because of concern about the ecological and social impacts of the project. It's becoming a familiar story. European contractors in 2001 withdrew from Turkey's Ilisu dam on the Tigris, because they feared lawsuits brought by displaced villagers and environmentalists. They knew there was a chance, too, of hostilities breaking out between Turkey and the downstream riparians, Syria and Iraq.

Still, dams keep coming. In 2001, the Chinese began building the Zipingpu Dam on the Min River of Sichuan. The dam and allied works threaten one of China's most famous irrigation works, the Dujiangyan Diversion, probably the only irrigation work in the world that's on UNESCO's list of world-heritage sites. Since the third century B.C., the diversion has sent about half the Min's flow through a cut made in the river's mountain wall. On the other side, the water irrigates a block of land that's now about 40 by 50 miles square on the Chengdu Plain.

China is now also planning dams on the Jinsha, or upper Yangtze. It's already begun building dams on the nearby Nu, China's part of the upper Salween of eastern Burma. A farmer says that "if the government wants to go ahead with dams, there's nothing peasants can do about it." He's right. The Nu, along with the Jinsha and Lancang, or upper Mekong, are a UNESCO world heritage site called Three Parallel Rivers, but an official of the Yunnan Development Planning Commission says that if it weren't for the Chinese government "these people would still be living a primitive way of life, like monkeys or ape-men." It came as a shock when, in 2004, Prime Minister Wen Jiabao halted the dams, at least temporarily. A professor at Yunnan University said, "I don't think I've ever heard of anything like this ever happening before." The Nu might continue to flow, unconfined, through the Grand Canyon of the Orient.

Cambodian farmers along the Se San are suffering from fluctuating water levels caused by the upstream Yali Falls dam, in Vietnam. Fishermen who depend on the Tonle Sap, a Cambodian backwater of the Mekong, are worried by talk of dams that would stabilize the Mekong's flow but put an end to the annual rise of the Tonle Sap. The great lake now expands annually from 1,000 to almost 5,000 square miles and supports over a million people dependent on its fishery and shoreline agriculture.

So much for the juggernaut of dam construction and the unfortunate parts of the conservation movement's legacy. The third reason for my dwelling on the movement is to suggest that fears about resource depletion have been around a long time and, without exception, have always been exaggerated. Theodore Roosevelt, for example, called together a conference of state governors in 1908. Appearing before them and speaking about the need for the conservation of natural resources, he said that it was "ominously evident that these resources are in the course of rapid exhaustion." The country, he continued, was "on the verge of a timber famine." Experts, he said, were convinced that "the end of both iron and coal is in sight."

Roosevelt was wrong on every count, and all the actions that the conservationists took in response to fears like his have had only minimal impact on American patterns of resource consumption. Americans get the overwhelming preponderance of their timber from private lands, not the national forests. They get only the tiniest sliver of their food and fiber from federal reclamation projects. They get next to none of their oil from the old naval reserves and only a very modest amount of their electricity from federally owned dams. How has the sky not fallen? Why have we not run out? The answer brings us back to the discovery of more resources, the substitution of new resources for old ones, and the more efficient use of resources. Remember this when you hear stories of impending doom. Fears of resource shortages have circulated for longer than anyone now alive can remember, and they have always proven alarmist.

So are we running out of resources? Are fears today better grounded than they were a century ago? I think the answer to both questions is no, but this is such an unconventional—some would say irresponsible—position that I'll take some time to support it. I'll look at water in detail and then touch briefly on energy and forest resources. Meantime, ask yourself why we're so inclined to doomsday scenarios. I'll circle back to that question before too long.

WATER RESOURCES

About 2% of the water on our planet is fresh. Two-thirds of that freshwater is permanently frozen, mostly in Antarctica. Almost all the rest is in groundwater. Odd as it may seem, we rely primarily on the still smaller quantity of freshwater that is in rivers and lakes and which, unlike groundwater, can be rapidly replenished.

Even though we rely on a very small percentage of the world's water, even that sliver provides a theoretically abundant source of water for almost all the Western Hemisphere, except chiefly Mexico, the Caribbean islands, and Peru. (I'm using a World Bank definition of abundant as more than 5,000 cubic meters of water per person annually.) The Eastern Hemisphere's supplies are abundant in Northern Europe, Russia, tropical Africa, and Southeast Asia; supplies are short in a broad band stretching from Morocco through Central Asia to northern China and Japan. They are short as well in Great Britain, Poland, South Africa and New Zealand. Nobody thinks of Great Britain as arid, but these are per capita calculations, and Britain is crowded.

Using these gross figures, most countries have a water surplus. Indonesia, for example, uses only 1% of its renewable water. Russia uses 2%; China, 16; India, 18; and the United States, 19. A few countries do use much higher percentages: They include Belgium, which uses 74% of its water; Israel, 84; Iran, 85; Egypt, 95; and New Zealand, a surprising 100%. In the Middle East, many countries import or desalinize water and in this way consume more than their natural freshwater supplies: The figures for Libya and Saudi Arabia, for example, are 700% of natural supply; for the United Arab Emirates, 1,400% ; and for Kuwait, 2,700%.

Such generally reassuring numbers can be very misleading because they mask huge variations within countries. Canada has a vast surplus of freshwater, yet water is often in acutely short supply in its southern Great Plains. Brazil has rivers carrying immense quantities of water to the sea, but its northeast is notoriously prone to ruinous droughts. The figures are misleading in another way, too, because they skirt the issue of water quality, of freshwater that's not.

As misleading as these numbers are, they suggest that water shortages are an economic problem, not a physical one. This is no consolation to villagers too poor to pay for an adequate water supply, but it means that a solution is possible. We're not, as the stock phrase has it, running out of water. The challenge is to become wealthy enough to develop available water.

Hard to believe? A billion people lack even primitive water-supply systems, and this number is projected to rise to 3 and perhaps even 5 billion by 2025. Africa is worst off. A third of its population, or 300 million people, rely directly on ponds and streams, rather than a piped supply or wells. Many of these sources are erratic or provide poor quality or polluted water. It's a huge problem, but the United Nations calculates that everyone on earth could have clean water for an annual investment of $10 billion.

Give an engineer enough money, after all, and he'll work wonders. Engineers in 1842, for example, finished the Croton Aqueduct, which brings water 30 miles south to New York City. It was a godsend, but by 1892 it was supplemented by the much bigger Catskill Aqueduct, which brings water to a Yonkers reservoir via a tunnel that passes under the Hudson River. In the 1930s, the city added the Delaware Aqueduct, bringing water from that river to the Yonkers reservoir. Even today engineers are tunneling under Manhattan as they build a third tunnel to supplement the two that now bring water from the Yonkers reservoir into the city.

Engineers helped Denver reach west of the Continental Divide to bring Colorado River water eastward through a tunnel under the Rockies. They helped California's major cities reach to the Sierra Nevada and to the Colorado River. An Oklahoma company, PESA, is working to tap the Arbuckle-Simpson aquifer, near Ada, to feed a pipeline supplying water to towns on the west side of Oklahoma City. Oklahomans think it's a big deal—especially the people in the source area, who fear major changes in the local hydrology—but it's a pipsqueak alongside a scheme proposed in the 1960s by the Ralph M. Parsons company of Los Angeles. That pipedream proposed bringing Alaska water through Canada to the American Southwest. Acronymized to NAWAPA, for North American Water and Power Alliance, the plan never got off the ground. It generated a tidal wave of environmentalist opposition, especially in Canada.

Similar schemes were proposed at about the same time to bring Siberian water to the deserts of Central Asia. Nothing came of them, either, but Libya built an immense project to bring underground water to the settled and coastal part of the country from the Nubian Sandstone aquifer in the country's empty southeastern desert. Until it was recently cancelled, Spain's national hydrological plan aimed to divert water from the Ebro River south to the dry but agriculturally rich region of Andalusia. The rice growers at the Ebro delta weren't happy, but the greenhouse owners down south were pushing hard. Now, they're likely to get locally desalinized water.

Entrepreneurs want to help. Cadiz, Inc., wants to mine water from its California desert aquifer. Cleverly, it also wants to use the aquifer as an underground storage basin to hold Colorado River water in wet years for resale in normal or dry ones. Boone Pickens is trying something similar in Texas, where he's set up Mesa Water to mine and export groundwater from the northern Panhandle. Pickens plans to build a $1.2 billion pipeline to Dallas. It will be 8 or 9 feet in diameter and over 300 miles long. He wants to send 200,000 acre-feet through the pipe each year until the aquifer is half depleted. He's got a permit to do it, too, provided he leaves the other half in place. The water will come partly from some 27,000 acres he owns but also from land owned by ranchers who will be paid $350 an acre for their water—at least twice the value of the land itself. They're happy to sell. Still, there are objections from critics who don't like the idea of exporting groundwater. Pickens's tart reply is that he's doing nothing fundamentally different, in principle, from farmers who pump water onto crops later sold to distant consumers. He's got a point. Pickens figures that by the time he's removed his half of the groundwater, the project will have paid for itself and the purchasers will have found other supplies. All he needs to get started is a couple of big cities to sign a contract.

So much for the engineering approach. With the decline of the reclamation mystique in the United States, however, we've come to realize that there are usually much better solutions to water shortages than massive construction schemes. The better way is to couple those schemes, in reduced form, with the more efficient use of water—a phrase that is very close to meaning: make users pay more for water.

California, for example, is required by interstate compact to reduce its withdrawals from the Colorado River. To everyone's surprise, the federal axe fell on January 1, 2003, when three of the eight giant pumps at Lake Havasu on the Colorado were shut down by federal order. Southern California has temporary alternatives to this reduction in the water it gets through the 242-mile Colorado Aqueduct, but it needs a long-term agreement. The problem is that farmers in the Imperial Valley irrigate a half million acres with Colorado River water to which they have senior water rights. The farmers say that Los Angeles must reduce its consumption before they are forced to reduce theirs. Such is the hoary rule of appropriation, where first in time is first in right.

Closer to the Colorado, there are about 80 farmers in the Palo Verde Irrigation District. They agreed in 2004 to fallow between 7 and 29% of their land annually. The water saved would go to Southern California's cities, which every year would pay the farmers $600 for every fallowed acre. From the farmers' perspective, this was a good deal: They rarely earned even $400 an acre from land in cultivation. Unlike their Palo Verde cousins, however, the Imperial Valley farmers aren't at all thrilled by the prospect of selling water to city slickers. One Imperial Irrigation District director said in 2002: "This is a hill we're prepared to die on." The board president explained that "without water, the Imperial Valley is nothing." When the Bureau of Reclamation's regional director ruled in 2003 that Imperial Valley farmers used water wastefully and should have their allotment cut by 9%, Imperial farmers denounced the ruling. "Wow, how generous," said one. Another muttered about a "kangaroo court." A third said, "the deck is stacked against us. . . . But we're going to see this thing to the end. We have to."

Lawyers will do well on this battlefield, but in the long run some Imperial Valley land will be taken out of production so the limited supplies of the Colorado River can be directed to higher-value uses. There won't be any resulting shortage of food and fiber in the United States; there will only be a group of people whose livelihoods have changed. That's a political and social problem that will rouse passions and sympathies, but society accepts such changes everyday. Just talk to the Americans who used to make Life Savers or Etch A Sketches.

(As the regional director's ruling suggests, the Imperial Valley could irrigate more land with less water if it irrigated more efficiently, for example by making more use of parsimonious drip-irrigation systems. That's trickier than it sounds, though. The Imperial Valley depends on water transported from the Colorado River by the All-American Canal. A long stretch of the canal is unlined and loses a lot of water to seepage. The canal managers want to line the unlined section. Sounds good, but hundreds of wells have been drilled in Mexico to capture the seepage water, and farmers there will be severely hurt if the canal is lined. The water that's presently "wasted," in short, isn't wasted at all; it's transferred to users on the other side of the border. There's another way, too, in which the waste of the water in the Imperial Valley isn't so bad. Used irrigation water from the Valley flows to the Salton Sea, a lake created when the Southern Pacific railway in 1905 accidentally and temporarily diverted the Colorado River onto what had been dry land. During the 20th century, the Salton Sea has become a very important habitat for migratory birds whose habitats elsewhere in California have been lost to agricultural and urban development. If the sea evaporates in the name of efficient irrigation, those birds will die. Actually, they will die before the Salton Sea disappears, because the birds depend on the sea's abundant fish—mostly tilapia and croakers. Those fish will die if the Salton Sea become much more salty. It's already 25% saltier than the ocean.)

Drought, of course, intensifies the sense of crisis and makes it harder to think clearly about how to manage water resources intelligently. Across the Rocky Mountain West from Canada to Mexico, the land in 2002 was as dry as anything seen in the 1930s or 1950s. Many of Colorado's 16,000 cattle ranchers were forced out of business as 2 million of the state's 3 million cattle were sold at distress prices that year. The same drought extended north into Canada, where eastern farmers shipped free hay to the prairie in a gesture of support. The next year, the drought was so bad in northeastern Kansas that even drought-resistant sorghum failed, along with tenderer crops of corn and soybeans. Farmers were reduced to relying solely on the winter wheat they hoped to plant. Europe, too, was hit by an exceptionally hot summer that year. Wheat production in France fell 25%. Production in the Ukraine fell 75%.

The Western drought affected cities, too. El Paso, for example, depends heavily on New Mexico's Elephant Butte Reservoir. The reservoir in 2003 was so low that the Bureau of Reclamation brought in dredgers to dig an 18-mile ditch in the reservoir bottom so water in the 85%-empty reservoir could reach the dam. Farther downstream, Mexican and American farmers were getting desperate for lack of irrigation water. Supplies were short partly because of the drought upstream and partly because of rapid economic growth on both sides of the Rio Grande. Mexico and the United States agreed in a 1944 treaty to take equal shares of the Rio Grande, but Mexico has run up a debt of about 500 billion gallons. There are plans to modernize the Mexican water system so less water is lost to seepage, but that solution will take time to implement and will not solve the problem if drought and growth continue. Conditions may get so bad that American farmers go out of business, their lives turned inside out. Shall we view this as a water crisis or as an economy that needs to be reorganized? The press tends strongly to the former interpretation, which is dramatic and offers the opportunity to use arresting phrases like "water wars," but the sensible answer is the latter.

Eventually, Americans will stop using water in the arid West as though it's the humid East. Residents of Modesto, Fresno, and Sacramento, for example, don't have water meters. Instead, they pay a flat monthly fee for water. That will stop. Home builders who want a permit for a new house in Santa Fe are required to find an existing building with low-efficiency toilets. They have to replace those toilets—8 to 12 per permit—with high-efficiency models that use water sparingly. Then they can build something new. Meanwhile, the suburban yard is under attack. In 1870 the landscape architect Frank Scott wrote in *The Art of Beautifying Suburban Home Grounds of Small Extent* that "a smooth, closely shaven surface of green is by far the most essential element of beauty" in a yard. No longer: Southwestern cities now commonly discourage lawns. The Denver suburb of Aurora in 2003 prohibited the planting of trees and shrubs, not to mention the filling of swimming pools. In Nevada, the city of Las Vegas pays homeowners $1 for every square foot of turf they remove: The city estimated that it would spend $25 million this way in 2003. Xeroscaping has become commonplace, too, though it's sometimes carried to extremes with gardens, in the words of one Nevada official, composed of "two tons of gravel and a cow skull." Another trend, popular in Southern California, is restoration horticulture, which uses native plants to simulate presettlement conditions.

Even people in the humid East will learn to use water more sparingly. Atlanta's major source of water, for example, is the Chattahoochee River, which is likely to be fully appropriated within the next 30 years. People downstream along the Apalachicola River in Florida are already worried about the impact of low flows on wildlife, including the oyster harvest in Apalachicola Bay.

What about overseas? We might look to Singapore as a model. Determined to reduce its dependence on water imported from Malaysia, Singapore plans to supply consumers with recycled sewage, thoughtfully renamed NEWater. The immediate task is to persuade Singaporeans that the water is perfectly potable and actually better tasting than the usual, chlorinated supply. As is often the case, however, Singapore is so far ahead of its neighbors that for the moment it has little to teach them.

The big overseas water users are the big irrigators, starting with China, India, and Pakistan. (Globally, the United States comes third, after India—and that's because it, too, is a big irrigator.) None of these three countries uses irrigation water efficiently. They know it. China, for example, is extracting unsustainable quantities of groundwater from the aquifers under the North China Plain. It plans to spend hundreds of billions of dollars on a North–South Transfer Project, three canals that will bring 150 million tons (45 billion cubic meters) of water every year 800 miles north from the Yangtze. Like the Three Gorges Dam, this transfer project will displace many people. Some 330,000 will have to move so the Danjiangkou Dam, which is on the border between Henan and Hubei and is part of the project's central route, can be raised 45 meters. Critics say that there won't be enough water in the canals to alleviate the shortages faced by North China's farmers: project water will only replace groundwater, which is sinking to levels that can't be economically tapped. In order to irrigate more land, the Chinese will have to apply water more efficiently, so that an irrigated acre requires less water than it does now. One way to do this is to develop drought-resistant plants, and the Chinese are doing that. In 1983 they also began making farmers pay for irrigation water. Since 1995 they've organized about 1,500 water users' associations to cut waste and costs. India has done the same thing, though big landowners there tend to control the associations and retain control over a disproportionate share of association water.

Even with better allocation and management of water, there will still be droughts with devastating impacts. By 2001, for example, a drought that began in 1999 had driven almost a million Afghans off their land. A U.N. undersecretary for humanitarian affairs visited a camp and reported that he "saw a sea of people living in unbelievable misery."

Such crises come even to countries that usually have plenty of water. Nicaragua and Honduras had a corn-killing drought in 2001. City residents weren't much affected, and the nation's forests and rivers weren't visibly altered, but 700,000 families from Nicaragua to Guatemala lost at least half their crops. Families subsisted on mangoes gathered from the forest. "Everything is lost. There is nothing left to hope for," said one Salvadoran. In Malawi, conditions in 2002 were perhaps even worse, because drought came to a population already devastated by AIDS.

What can be done in such cases, besides provide relief? The straight answer is that famine is the sister of subsistence; putting an end to one means putting an end to the other. That's a painful prescription for anyone who believes that the West has no business leading anyone into the joys of consumerism, but it's the blunt truth, and few subsistence farmers are likely to disagree. It also explains why droughts are most ruinous in countries like Afghanistan, where the economic and political systems are themselves in ruin.

Even in the Middle East, where water is often described as the likeliest cause of military conflict in the future, shortages are the result of a political inability to reallocate water away from low-value uses. Israel sells water to farmers so cheaply that the water in an orange is said to cost the country more than the price of the orange in the market. Rather than wage the political battle to restructure its agriculture to semidesert conditions, Israel is preparing instead to import Turkish water in refitted oil tankers.

It's not the most absurd case. Saudi Arabia is presently mining groundwater at the rate of about 6 trillion gallons annually. It uses that water to produce irrigated wheat at highly subsidized prices. It produces so much wheat that some is exported and some discarded. Around the old oasis of Buraydah, a landscape of center-pivot sprinklers now ranks high on the list of global water follies.

Technological optimists believe that desalination may yet allow us to continue using water lavishly. Saudi Arabia already operates the world's largest desalination plant and gets about 70% of its water this way; it plans to spend $50 billion on new desalination plants over the next 20 years. Extravagant? Maybe not. The cost of desalinized water in Israel today is $2 per cubic meter, down from $6 a decade ago, and a plant now under construction in Israel should produce water for about 50 cents a cubic meter. That's about twice the price Israel charges for water. It's also half the price that Israel is paying for that Turkish water coming by tanker. Thames Water, which is German owned, proposes to desalinize water for London. The United States already has a major desalinizing plant at Tampa Bay. It consumes about 50 million gallons of seawater daily, forces it through membranes, extracts 25 million gallons of freshwater, and returns the rest to the bay as very saline brine. Several towns in eastern Massachusetts, including Brockton and Braintree, are weighing the merits of building desalination plants in local tidal rivers. Other plants are under way at San Diego and Freeport, Texas. Like the one at Tampa Bay, they're being built by Poseidon Resources, which proposes to build in southern California plants twice the size of the Tampa one. Desalination has been held out as a technological savior for such a long time that many people dismiss its ever becoming a major source of fresh water. The Tampa plant, they'll point out, has had problems with clogged filters. Like much conservationist thinking in the past, however, they're probably wrong.

Even with cheap desalination, water still has to be distributed, which gets expensive. Shall it be subsidized? That's easy to say, but hard for many poor governments. Shall users then pay? Suppose they refuse. Shutting off their water creates an instant crisis. Rather than do that, South Africa has tried prepaid water meters, like prepaid phone cards. It has also tried lifeline service, where people who can't afford even minimal supplies can take water from taps set to drip, not pour. Either way, users are unhappy. The head of the Orange Farm Crisis Water Committee tells people to "destroy the meters and enjoy the water." Once again, as in the countryside, we're led back to the importance of economic development, not the absolute insufficiency of water.

ENERGY RESOURCES

At current rates of production, the world has a 250-year supply of coal. If only the best grades—anthracite and bituminous—are counted, the supply will last for more than a century. The figure for petroleum, excluding unconventional sources like tar sands, is more than 50 years. For natural gas, it's about 70 years. For uranium, 100 years. If you're still a dedicated believer in dams, you'll be pleased to learn that only 18% of the world's waterpower resources have been developed, and the only areas where a majority of those resources have been developed are North America, Europe, and Japan.

All these numbers will drop precipitously if technology doesn't change and if everyone on earth uses energy as Americans do. The average American—we've all heard the numbers—uses the equivalent of 8 tons of oil annually, while the average Chinese uses less than 1 ton and the average Indian only half a ton. But it is also true that energy efficiency is almost certain to rise and that exploration will continue to increase known reserves. Better production methods will increase recoverable reserves, too. There are fields in the San Joaquin Valley—places like the Kern River and Midway–Sunset fields—that were moribund in 1960 but which have been rejuvenated by advanced recovery methods, chiefly the injection of steam that decreases the viscosity of heavy oil. Daniel Yergin estimates that such methods, if put in place globally, would raise the world's oil reserves by the equivalent of Iraq's proven reserves, about 125 billion barrels.

Use this new and newly available energy to extract hydrogen from reformed natural gas, or to extract hydrogen from water by electrolysis, and you have a clean replacement for gasoline-fueled engines. It won't be easy. You have to build a hydrogen-distribution system or install reforming machines at service stations that are already on or near a natural gas pipeline. Also, it takes energy to produce hydrogen, which means that on the way to making clean fuel you're producing carbon dioxide unless—and this is a very big if—you get that energy from solar or wind power. Still, the federal government announced plans in 2002 to sponsor a research program (Freedom CAR) to develop a hydrogen vehicle. General Motors

Steam generators and pipes for the steam that, injected into the ground, have given a second life to Midway-Sunset, a giant field near Taft, California. In 1960 the field was senescent.

promises to sell hydrogen-powered cars by 2010. California and Illinois are planning networks of hydrogen filling stations. BP says bravely that its initials stand for Beyond Petroleum. Shell says that half of the world's energy in 2050 will probably come from renewable sources. These are brave words and probably unrealistic. For decades to come, it will probably make more environmental sense to use gas to replace coal, not oil. Eventually, however, we'll make the transition away from gasoline.

When we talk of energy crises, in short, we usually are talking about disruptions in our energy delivery system. Two such disruptions occurred in the 1970s. The first was the Arab oil embargo of 1973, and the second was the Iranian Revolution of 1979, which brought a tripling of oil prices. Other domestic spikes have been caused by the temporary shutdown of even a single refinery. But that's not the same thing as running out of energy. The biggest energy problem the United States faces is probably the power of Saudi Arabia to keep world oil prices just below the point where Americans would move to alternative fuels. Americans shouldn't blame the Saudis for pursuing Saudi interests. The United States could cure itself of this addiction by taxing conventional fuels, but congressmen who like their jobs won't do that, so the transition to alternative fuels will be much slower than it could be. Along the way, watch for energy producers to warn loud and often about impending shortages. That's the best way for them to gain access to lands from which they're excluded for environmental reasons.

FOREST RESOURCES

One more case, one that echoes the timber-famine fears of George Perkins Marsh and the early conservationists. About 3 billion tons of wood are cut annually. Do shortages lie ahead?

Wood cut in the temperate latitudes, particularly in North America, Europe, Japan, Chile, and New Zealand, goes to saw, pulp, or paper mills. Wood cut in the tropics, however, is mostly burned as fuel or is converted to charcoal. The quantities cut from these two regions are roughly the same, but the land-use consequences are very different. The United States, for example, has a negative deforestation figure of –0.3% annually, partly because cutover land is replanted and partly because farmland is reverting to forest. Canada, though a major producer of forest products, also has a negative figure—and for the same reasons. Sweden, Europe's chief forest-products producer, shows no net gain or loss of forest lands.

Annual deforestation rates, however, exceed 2% annually down the spine of Central America, in West Africa, and in Southeast Asia from Burma through to Indonesia and the Philippines. Most of the cutting is the result of timber theft or of villagers seeking fuelwood; neither group replants. In the tropics, cutover forest land often becomes rough pasture stocked with the notorious *Imperata* grass, whose roots are so tough that farmers without machinery find it impossible to cultivate the land.

There's no foreseeable shortage of timber or pulpwood except for particular species like redwood and teak and pernambuco wood, which is used in the making of bows for violins and other stringed instruments. Fuelwood's a different matter, because villagers are already having to go on longer and longer collecting trips. In many cities in the semiarid tropics, including the Sahel and India, the fuelwood trade reaches many miles from major cities. In southern India it's no strange sight to see a bundle of branches strapped to the outside of a third-class railway carriage. Firewood for sale is increasingly expensive, too. An armful in Sri Lanka will sell for $1, a lot of money in a country where unskilled workers make only $3 a day. Gas is available, and wealthier households use it. So would the poor, if they could afford it. The cure for the fuelwood shortage, then, is the same as the cure for the droughts affecting peasant farmers: economic development. Recognize it as such, and you face a challenge instead of inevitable defeat.

One country that doesn't fit this tidy division into temperate/managed and tropical/unmanaged is Russia. Its most productive forest region is Karelia, bordering Finland, but afforestation is uncommon; loggers just keep moving to virgin areas. There aren't many left, and some of the country's most pristine forests are ironically those so close to the Finnish border that they're in a no-man's land.

POPULATION

In all these cases I've been suggesting that we are unlikely to run out of resources. Asserting something is not the same as proving it, of course, and many experts would be incensed by what they would consider my near-criminal optimism. Just because people were wrong about shortages a century ago, they will say, doesn't mean we won't have shortages in the future.

On top of this, they will say that I'm overlooking the catastrophic consequences of population growth. Fair enough: I can even pitch in and help them make their argument. Take Pakistan, with 30 million people at its creation in 1947. Projections show it in 2025 with an almost inconceivable 250 million. The projections for Burkina Faso, Mali, Niger, Somalia, and Yemen show those countries quadrupling their populations between 2000 and 2050. Drive through the streets of cities in these countries, and it's easy to think that we're already in dire trouble from overpopulation.

Our mistake—if that's what it was—lay with the colonial regimes that imposed peace in places accustomed to near-perpetual conflict. It lay with colonial transportation systems that made it possible to ship food into stricken areas. It lay with the belief of colonial powers that they were morally obligated to undertake famine-relief measures, such as organizing road-building crews in times of famine, so farmers would have some money to buy food. It lay finally with colonial policies that stimulated agricultural production. A classic case was the *kultur* system introduced by the Dutch to Java in the 1830s. Out went a crushing 40% tax on the value of all crops. In came a requirement that farmers give to the government only the produce from a fifth of their land, which had to be devoted to export crops. The system worked well, maybe too well. Coffee and sugar production doubled, which was what the Dutch wanted, but subsistence crops were no longer taxed and so they grew quickly too. So did the island's population. Between 1815 and 1850, it more than doubled.

It's time to remember Thomas Malthus, a professor of political economy at the East India Company's Haileybury College. In a famous essay of 1798, Malthus wrote that "the great question is now at issue, whether we shall henceforth start forward with accelerated velocity toward illimited, and hitherto unconceived improvement; or be condemned to a perpetual oscillation between happiness and misery." Malthus predicted the gloomy alternative and justified his choice by saying "that the power of population is indefinitely greater than the power of the earth to produce subsistence for man." Population tended, in other words, to increase geometrically, while food supplies increased arithmetically. Could we have a society "all the members of which, should live in ease, happiness, and comparative leisure?" Malthus thought not. "The argument," he wrote, "is conclusive against the perfectability of the mass of mankind." Population growth would continue until checked by famine and disease.

In a later version of the essay, Malthus in 1826 softened his position by admitting that parents might limit the size of their families. If they stopped at two children—technically 2.1—population would gradually stabilize.

This, of course, is exactly what has happened in many countries. Typically, it's because children in a modern economy are an economic burden, not an asset. There are other reasons, though. For some parents, personal freedom is more important than family. "Kids aren't at the top of my list." For others, peer pressure and government inducements are a powerful nudge toward fewer children.

This fundamental change—the demographic transition, to give it its proper name—is far advanced among the populations of the world's wealthiest countries. In Canada, the average woman will have 1.52 children during her lifetime. In Japan, the figure is 1.33. In Italy, it's all the way down to 1.2, and in the city of Bologna, it's 0.8. Walk that city's streets on a summer evening and you'll see lots of adults—young and old—but very, very few kids. It's eerie, a bit like a science fiction film where nobody notices that something's dreadfully wrong.

A country with a birth rate of 1.5 and without immigration will see its population cut in half over the space of a century. Toyota and Nissan are building new plants in the United States but none back home, where population is expected to peak at 126 million in 2006 and be cut in half by 2100. The Japanese even have a word for a society tending toward fewer children—*shoshika*—and it's used by politicians who want, probably without much luck, to reverse the decline. If Italy keeps its present birthrate, and if it excludes immigrants, its population will shrink by 75% over the next century. Should it plan for a smaller future? Its schools have to downsize or consolidate. Retirement programs have to be modified as the pool of workers grows smaller and smaller, while the pool of retired workers grows larger and larger. On current projections, Italy in 2030 will have two pensioners for every worker; it already has more people over 60 than under 20. The United States, where women are having 2.08 children, is not much better off. In 1950, it had 16 workers for each pensioner. Now it has three for every two, and it's headed for two to one in 2030. You can raise the retirement age, of course, and it's very likely that governments will do so, but that won't bring the kids back.

In the rich countries, population is growing so slowly today that at current rates it will take 800 years for those countries to double their present population. The poor countries, on the other hand, will double in 42 years at their current rate of growth. Another way of saying this is that the population in the poor countries is growing from about 4.9 billion today, out of a world total of 6 billion, to a projected 7.8 billion in 2050, out of a world total of 9 billion. An obvious solution to the problems of a declining population in the rich countries is to welcome people from the poor ones, although this is a political third rail. Another solution is to follow Norway, grant mothers a whole year of maternity leave at 80% of their salary, grant fathers 4 weeks' leave, give parents a child allowance in excess of $100 per month per child, and add in 10 days for each parent to stay home each year if they have a sick child. With policies like that, Norway has resisted the European tide and kept its fertility rate at 2.0. Still, it's hard to imagine many other rich nations being willing to pay the taxes necessary for such programs, or choosing to have more children even with such incentives.

Will the world's poor countries pass through the demographic transition? It used to be assumed that this would not happen until they became at least moderately wealthy, but women in the South Indian states of Kerala and Tamil Nadu are now having on average only two children. These are relatively literate states, but it appears that birthrates are falling below two per family in Andhra Pradesh, too, which has a more typical population. India no longer forcibly vasectomizes men, as it notoriously did in the 1970s. Instead, the government actively rewards sterilization, both male and female. It also puts more subtle pressure on would-be parents. The state of Haryana, for example, says that adults with more than two children cannot be elected to village councils. The supreme court of India in 2003 upheld the law with the startling explanation that India's population crisis was so severe that the country should not put "undue stress on fundamental rights and individual liberty."

Predominantly Islamic countries are usually opposed to contraception, which is why they typically have very high birthrates. Yet Tunisia, under the long-term leadership of Habib Bourguiba, proved that birth control programs could succeed in Muslim countries. In the 1950s, Tunisia outlawed polygamy, despite the *Qur'an* allowing Muslim men four wives. In the 1960s Tunisia legalized abortion, and it's still the only Muslim country to have done so. Meanwhile, reproductive health clinics were set up. Many are mobile and bring three-person teams of a nurse, social worker, and midwife to villages. The teams distribute free condoms. Sex education is offered not only to women but to men and college students. Sometimes it's the topic of Friday sermons at the country's mosques. Bourguiba stepped down in 1987, but his successor has maintained the program, which has brought the nation's birthrate down from 7.2 in the 1960s to less than 2.1 today. With 10 million instead of 15 million people, per capita income in Tunisia has risen over the last decade by 50%.

Even China is coming around to seeing that persuasion works better than draconian methods. In 1980, when it passed the billion-people mark, China implemented a soon-notorious one-child policy. Officially, the policy merely encouraged parents to stop at one child, but encouragement could take the form of posting on public bulletin boards the menstrual schedules of village women, of forcing women to have abortions, and of tearing down the houses and confiscating the land of parents who were labeled extra-birth guerrillas.

Within a few years, however, most rural counties allowed parents to have a second child if their first was a daughter, and today about 80% of the children born since 1980 are from families with two or more children. Will truly voluntary programs be more successful? In 1995 a pilot program was tried. Instead of coercion, officials provided information about birth control, then let couples make their own decisions. The results were so encouraging that the program was officially expanded to 32 counties, without any rise in the number of births. In 2002, the old policy was replaced by a law requiring couples to pay a social compensation fee if they have more than one child. From now on, there will be a cost associated with large families, but no stigma. That's not so different from the reality facing American or European families.

IDEOLOGY ONCE AGAIN

Will we squeak through? Will a combination of increased supplies, more efficient use of resources, and a stabilized population be enough to avert the kind of resource collapse envisioned by the early conservationists? My presentation suggests a positive outcome. Once again, however, "suggests" isn't the same as "proves." The truth is that nobody knows. What we do know, for sure, is that intense, localized, and periodic shortages of every resource will be interpreted as signs of a wider and catastrophic failure. Why, then, are we so inclined to fear the worst? Why, for that matter, were the early conservationists so fearful? This is the question I asked earlier and said I'd come back to. I think the explanation lies in our romantic doubts about progress. We don't know how to stop this train, and our minds aren't quite made up about the wisdom of stopping it, so we're idly amused to think of things that will stop it for us. Meanwhile, let me mention Israel's Lake Kinneret again. At the beginning of this chapter I said that its level at the start of the winter of 2002 was 20 feet below normal. I didn't mention that heavy rains in the following few months brought the lake up over 10 feet. The next year the lake overflowed. Feel a twinge of disappointment? Prefer a real cataclysm? That's the romantic spirit. You can't shed yourself of it, but you can see it for what it is.

CHAPTER 18

POLLUTION, BIODIVERSITY, AND CLIMATE CHANGE

The conservationists of Teddy Roosevelt's time, who feared that the United States would run out of resources, would have a hard time believing that a century later the nation's biggest resource producers habitually portray themselves as enthusiastic conservationists. The early conservationists had been seen—had seen themselves—as radicals. No longer. Every resource-production company claims to be conservation-minded, and it's more or less true. If it wants to stay in business—at least its historic business—it must have trees to cut or coal to dig or oil and gas to extract.

A century on, environmental reformers have shifted ground. There is still a concern with resource depletion, though it tends to focus on how indigenous peoples and peasants can develop in environmentally sustainable ways. But there are also three relatively new concerns. One is the grim fear that we may be contaminating the earth with poisons so toxic that the only creatures likely to keep us company in the future will be roaches, crows, rats, and squirrels. Another is the sense that we are on a biological demolition derby, killing off species at a rate at least 100 times faster than the natural rate of extinction and along the way jeopardizing the stability of the biosphere. The third is that we are changing the earth's climate in ways we do not understand and in ways that many people do not want to hear about, for fear that they might be forced to change the way they live.

What is the label that embraces these concerns? Environmental quality does the job, but a shorter label is ecology. The word was coined in 1866 by Ernst Haeckel, a German biologist, from the Greek *oikos*, for home or habitation, and etymologically it denotes the study of the home—in this case humanity's home, planet Earth. A hard-edged scientific term until about 30 years ago, ecology became a household word when the Apollo 8 astronauts in 1968 sent home pictures of a gibbous earth, emerging in white and blue from behind a barren moon. The photograph for a time was a popular poster, the sort of thing that went up on thousands of classroom and dormitory walls. It's been largely forgotten now, perhaps because in a media-saturated culture even the most powerful images have a short shelf life. The ecology movement, too, has lost some of the intensity it had in the early 1970s, when Earth Day was a major event on many campuses. The complexity of the management issues it raised was off putting, and ecology raised unsettling questions about the morality of a consumer society. Still, pollution, threats to biological diversity, and climate change haven't gone away.

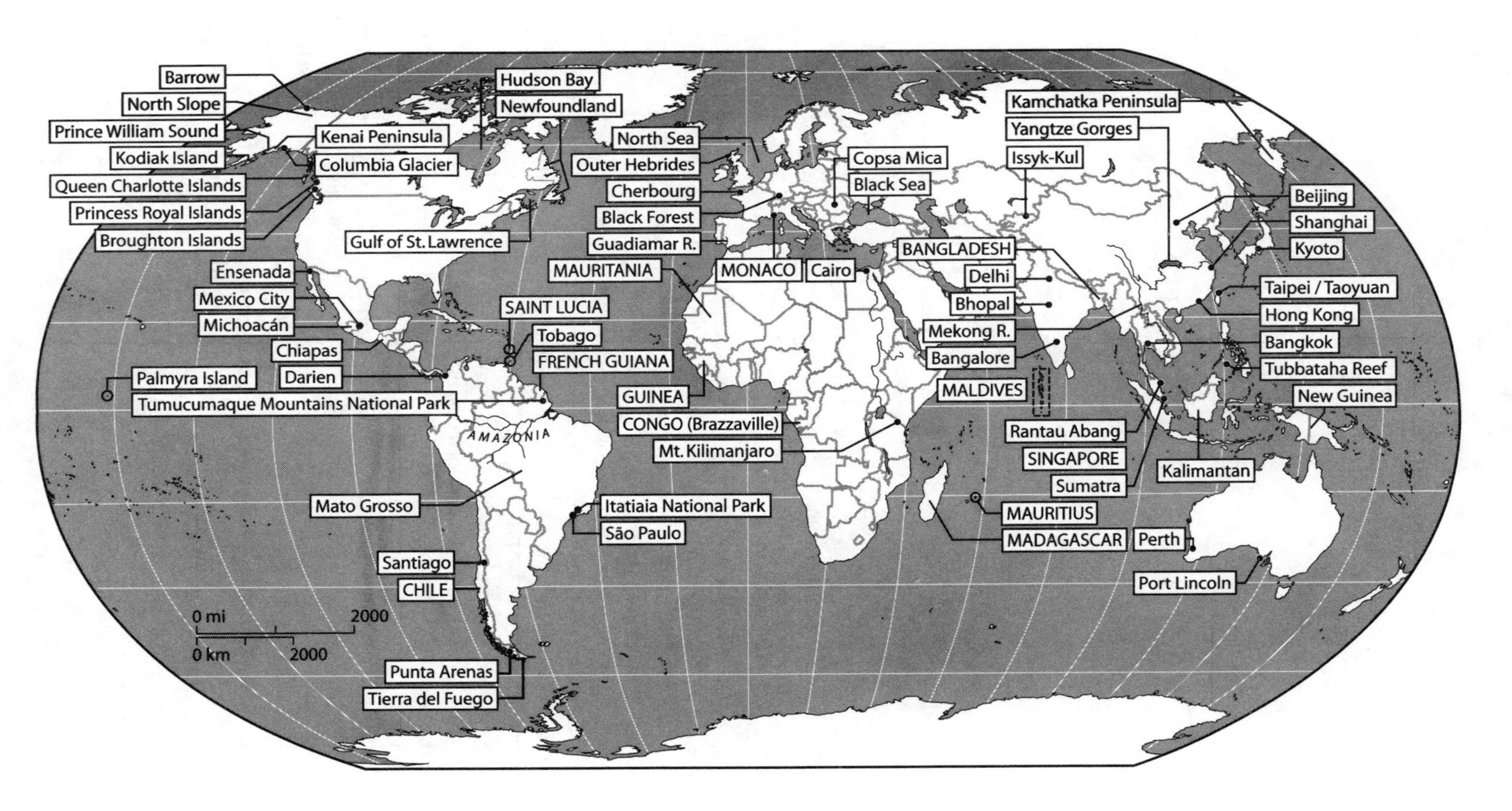
Barrow
North Slope
Prince William Sound
Kodiak Island
Queen Charlotte Islands
Princess Royal Islands
Broughton Islands
Hudson Bay
Newfoundland
Kenai Peninsula
Columbia Glacier
Gulf of St. Lawrence
North Sea
Outer Hebrides
Cherbourg
Black Forest
Guadiamar R.
MAURITANIA
MONACO
Cairo
Copsa Mica
Black Sea
Kamchatka Peninsula
Yangtze Gorges
Issyk-Kul
BANGLADESH
Delhi
Bhopal
Mekong R.
Bangalore
MALDIVES
Beijing
Shanghai
Kyoto
Taipei / Taoyuan
Hong Kong
Bangkok
Tubbataha Reef
New Guinea
Ensenada
Mexico City
Michoacán
Chiapas
Darien
Palmyra Island
Tumucumaque Mountains National Park
SAINT LUCIA
Tobago
FRENCH GUIANA
AMAZONIA
GUINEA
CONGO (Brazzaville)
Mt. Kilimanjaro
Rantau Abang
SINGAPORE
Sumatra
Kalimantan
Mato Grosso
Itatiaia National Park
São Paulo
MAURITIUS
MADAGASCAR
Perth
Port Lincoln
Santiago
CHILE
0 mi
2000
0 km
2000
Punta Arenas
Tierra del Fuego

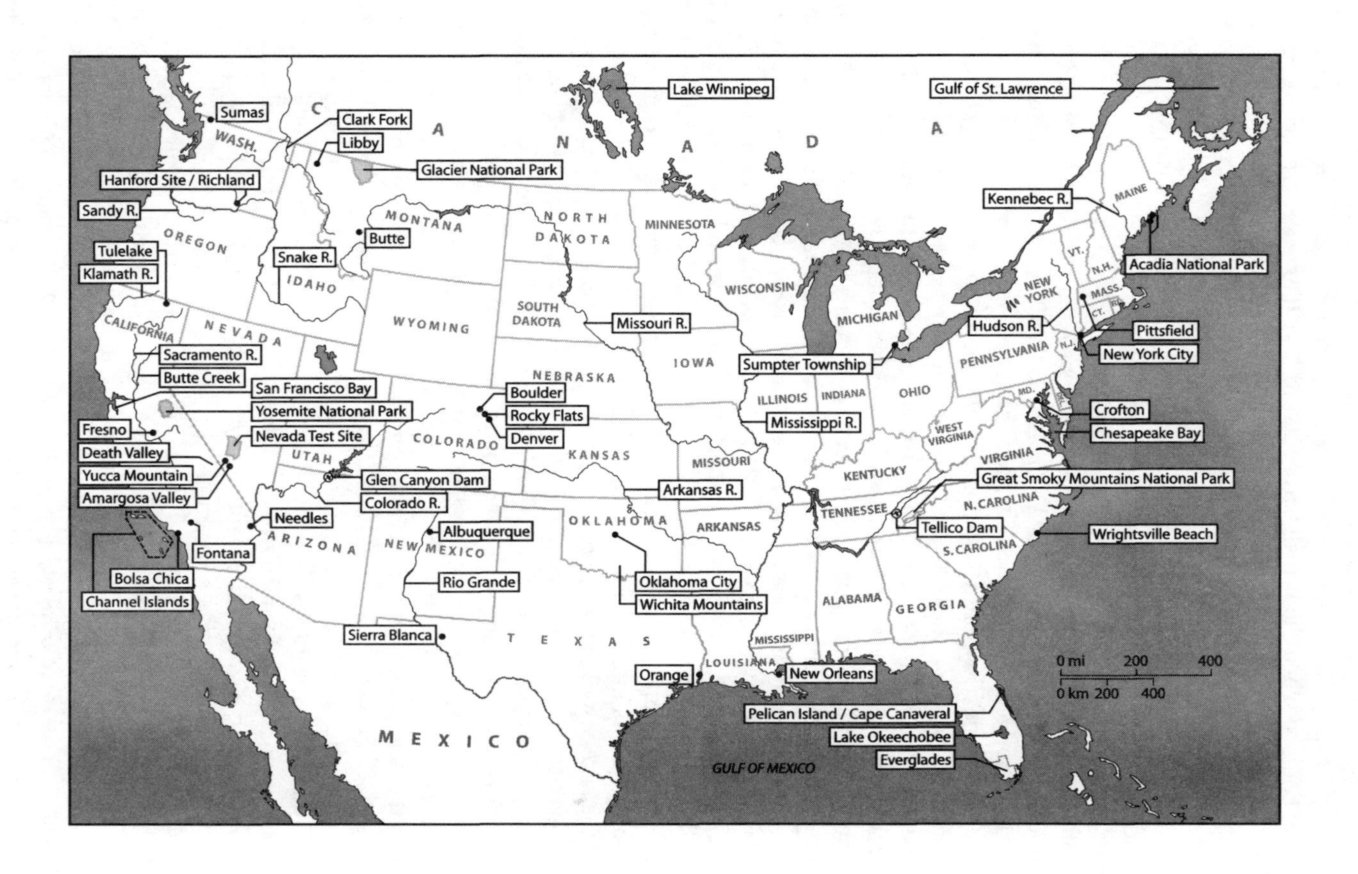
Lake Winnipeg
Gulf of St. Lawrence
C A N A D A
Sumas
WASH.
Clark Fork
Libby
Glacier National Park
Hanford Site / Richland
Sandy R.
OREGON
MONTANA
Butte
NORTH DAKOTA
MINNESOTA
Kennebec R.
MAINE
Acadia National Park
VT.
N.H.
MASS.
CT.
RI.
NEW YORK
Hudson R.
Pittsfield
New York City
Tulelake
Klamath R.
Snake R.
IDAHO
WYOMING
SOUTH DAKOTA
Missouri R.
WISCONSIN
MICHIGAN
CALIFORNIA
NEVADA
Sacramento R.
Butte Creek
San Francisco Bay
Yosemite National Park
Fresno
Nevada Test Site
NEBRASKA
IOWA
Sumpter Township
PENNSYLVANIA
N.J.
Boulder
Rocky Flats
Denver
ILLINOIS
INDIANA
OHIO
MD.
DEL.
Crofton
Chesapeake Bay
Mississippi R.
WEST VIRGINIA
VIRGINIA
Death Valley
Yucca Mountain
Amargosa Valley
UTAH
COLORADO
KANSAS
MISSOURI
KENTUCKY
Glen Canyon Dam
Colorado R.
Arkansas R.
Great Smoky Mountains National Park
N. CAROLINA
TENNESSEE
Needles
ARIZONA
NEW MEXICO
Albuquerque
OKLAHOMA
ARKANSAS
Tellico Dam
Wrightsville Beach
S. CAROLINA
Fontana
Bolsa Chica
Channel Islands
Rio Grande
Oklahoma City
Wichita Mountains
ALABAMA
GEORGIA
Sierra Blanca
TEXAS
MISSISSIPPI
LOUISIANA
Orange
New Orleans
0 mi 200 400
0 km 200 400
Pelican Island / Cape Canaveral
Lake Okeechobee
Everglades
MEXICO
GULF OF MEXICO

POLLUTION

Of the three, pollution seized the public imagination first, before biodiversity and global warming were even in the dictionary. Smog, a portmanteau joining of smoke and fog, appeared in England when the British before World War II experienced sickening, coal-laden fogs. After about 1950, the term jumped the Atlantic. Los Angeles was becoming a freeway city, and cars were loading its air with nitrogen oxides. Sunlight acted on the gas, and the resulting photochemical soup was trapped by thermal inversions. Millions of people began seeing some kind of bitter joke in America's most glamorous city having a glorious mountain backdrop that on many days was invisible.

Since then, intricate regulatory programs have been devised to deal with the many sources of air pollution. In the United States, these programs are administered chiefly by the Environmental Protection Agency (EPA), created in 1969. (In retrospect, the Nixon administration did a lot for the environment. Maybe its environmental activism was a sop or diversion, but maybe it was also a response to the public mood. In any case, later Republican administrations look spineless and cavalier in comparison.) The Clean Air Act was passed the next year. It charged the EPA with setting national ambient air-quality standards for several key pollutants. There are now six of these criteria pollutants, as they're called: carbon monoxide, nitrogen oxides, ozone, lead, particulates, and sulfur dioxide. Originally, Congress assumed that the nation's air would be in compliance with these standards by 1975, but today, almost 30 years later, noncompliance is still commonplace. Cars emit 70–90% less pollution today than they did in 1970, partly because of reformulated gasoline and partly because of cleaner-burning engines, but there are more vehicles now and 150% more vehicle-miles. That's why New York, Los Angeles, Houston, Chicago, and Dallas continue to violate emission standards for ozone, which along with particulates has proven to be one of the two pollutants most difficult to control. Denver is the only major American city that meets all the criteria-pollutant standards. It's done it by mandating special gasolines and by shifting power-station fuel from coal to natural gas. Such programs probably require a community more concerned than most with environmental quality. Colorado, it seems, has one.

The EPA is still struggling with power plant emissions, which were recognized a generation ago as the cause of the dying forests of New England. Trees were being blanketed by acid deposition put into the atmosphere by coal-burning Midwestern power stations. A few days later, the acid fell downwind as acidic rain. (A similar problem became notorious in Germany's Black Forest, where it was spoken of as *Waldsterben*, or forest death.) The 1990 amendments to the Clean Air Act capped sulfur-dioxide emissions at an annual 8.95 million-tons, and during the 1990s there was a dramatic decline in the production of the gas in southwest Indiana and western Pennsylvania, the sources of most of the acid falling on New England. Across the eastern states, sulfur-dioxide emissions declined 25% between 1982 and 2001. Nationally, emissions were halved.

In 2003 the EPA warned that coal-fired power plants were also responsible for about a third of the mercury emissions in the United States. The emissions were linked to health problems, especially in children, but a proposal to cut the emissions in half by 2010 was dismissed as unrealistic by the Edison Electric Institute, speaking on behalf of the utility industry. One solution was to import electricity. That's why in 2002 some 16 plants aimed chiefly at the American power market were under construction in Mexico, where emission standards are less strict than in the United States. Call them energy maquilas. The biggest was being built by InterGen, a joint venture of Shell and Bechtel. Some of the pollution will drift or blow back across the border, compounding the already severe air pollution in California's Imperial Valley, but the plants satisfy Mexican law, and the EPA has no control over them. In Washington State, meanwhile, a power plant is planned at Sumas, near the Canadian border; its emissions will drift toward Vancouver, instead of the Seattle residents who will use the power. The Canadians may protest, but they'll have a hard time getting rid of the offending plant.

Programs to control power plant emissions have become so complex that, no matter how important they are, the public ignores them. Consider the law that requires the upgrading of pollution-control devices in older power plants whenever significant investment is made in them. It's called new source review, and the Clean Air Act was amended in 1977 to mandate it. The law was ignored for over 20 years, however, in what one head of the EPA's enforcement division has called "the most significant noncompliance pattern EPA has ever found." The Clinton administration took aim and filed suit in 1999 against seven major utilities violating the law. The second Bush administration, however, set out to scrap new source review and replace it with pollution trading, in which a cap is set and polluters who clean up their plants more than the law requires are allowed to sell the surplus savings to companies that cannot or choose not to clean up. The approach is a clever one, but its effectiveness depends on where the cap is set. The Bush administration intended to set caps so high that air would be dirtier under its proposed law, dubbed "Clear Skies," than under the Clean Air Act. The measure failed to become law, but the administration decided to achieve the same goals using EPA's rule-making authority. New rules were announced that combined pollution trading with a toothless new source review. The Bush EPA proposed, in other words, to raise the threshold so high that a smart plant operator would never have to upgrade pollution-control facilities. Officials in the Clinton EPA had proposed a threshold of 0.75%, so that an investment of $7.5 million in a billion-dollar plant would trigger upgrade requirements. The Bush administration raised the threshold to 20%, or $200 million. In the words of one watchdog, this was "such a huge loophole that only a moron would trip over it and become subject to N.S.R. requirements." The American Lung Association signed a statement declaring the rules "the most harmful and illegal air-pollution initiative ever undertaken by the federal government." They are now being tested in the courts, where several states have brought an action arguing that such major changes require congressional approval.

Hydroxyl radicals are meanwhile declining. They're short-lived but important as natural destroyers of pollutants such as methane and sulfur dioxide. The reason for their decline is unclear but may be because there's so much haze over American cities. Haze blocks the ultraviolet light whose action produces the radicals.

Another air pollutant whose danger has only recently been discovered is soot from power plants and diesel engines. The particles are so small that they pass through the lungs into the blood stream. They are calculated to account for 50,000 premature deaths annually in the United States, but no control measures are yet required.

Fractured as American policy may be, and poor as air quality may be in the United States, air quality overseas is often much worse. Places like Mexico City and Bangkok are notorious for their dangerously foul air. It's not just an urban phenomenon, either. India is now creating an annual mountain of haze over the Indian Ocean. The Asian Brown Cloud is as big as the United States. It's created by many sources, ranging from power plants to millions of sooty dung fires. The dry winter monsoon pushes the cloud south over the Indian Ocean. Come the wet summer monsoon, the cloud is pushed back over India, where some of it falls as acid rain.

China, like India, relies heavily on coal for electricity. One consequence, epidemiologists say, is that China now has 178,000 premature deaths annually from air pollution. The Chinese are trying to import cleaner fuels from Russia and Australia and elsewhere. They're trying to relocate their generating stations to be near coal mines, not cities. They're insisting that cars be equipped with catalytic converters. Still, the Chinese standards for vehicle emissions are much lower than American ones, and the number of cars on Chinese roads is rising extremely rapidly.

Brazil, to take a third case, is one of the top 10 air polluters in the world, but it's not because of motor vehicles: It's because of fires in Amazonia. This is as much a problem of biodiversity as of pollution, and measures are belatedly being taken to control it. In the soybean-growing state of Mato Grosso, big landowners cannot clear more than 20% of their land without risking fines and imprisonment.

The United States isn't the only country concerned about air quality, however, and it is by no means the one most aggressively combating it. Motor vehicle licenses in Singapore grow more, not less, expensive year by year, because the government is determined to discourage the use of older, more polluting cars. That's partly why Singapore's air is much cleaner than Hong Kong's. Private investors in Singapore have meanwhile set up car clubs. Instead of owning a car, members pay an annual fee and take any car they like from any of the company's many lots. The cars are all clean-burning hybrid Hondas, and hourly fees are set to favor short rentals. (A similar undertaking, Zipcar, has sprouted in Boston and operates in several U.S. cities.)

The Thai government banned leaded gasoline in 1996, and new Japanese cars sold there are set to European emission standards. In addition, Thai motorcycle dealers sold their last two-stroke bikes in 2002. The shift was justified, because the old engines emitted about half of all the particulates in Bangkok's air.

In 1999, India's supreme court ordered Delhi's 10,000 buses to switch to natural gas. In the best Indian fashion, the bus owners and the city authorities ignored the order. Two years later, push came to shove and the court threatened to shut down the city's bus fleet. The city caved, and its buses, and even the thousands of motorized rickshas in Delhi, now run on natural gas. The city's air is still appallingly dirty, but the improvement for people stuck in traffic—and there are millions of them daily—has been very noticeable. Delhi also now has the beginnings of a subway system that is supposed to replace 3,500 buses. A parallel story has occurred in Cairo, Shanghai, and Beijing, where buses run on natural gas and subways have been built.

Despite these impressive accomplishments, air pollution remains an immense global problem. In 1981, the ozone hole over Antarctica measured less than 1 million square miles. Now it's over 17 million, despite the 1987 Montreal Protocol that phased out the production of ozone-destroying chlorofluorocarbons (CFCs). If the hole gets big enough, inhabited areas could be showered by high levels of ultraviolet light. It may already be happening, because skin cancer rates in Santiago doubled in the 1990s. Farther south, in Punta Arenas, a "solar stoplight" is published each day, like ozone alerts in the United States.

A similar story can be told for water pollution. The Clean Water Act of 1977 charges the EPA with developing and administering a program of national standards and effluent levels. Federal grants help cities and towns finance sewage systems, and some 200,000 industrial polluters whose waste enters municipal sewer systems must secure permits issued as part of the national pollutant-discharge-elimination system. About half the nation's water pollution, however, comes from nonpoint sources such as parking lots, feedlots, and croplands. These are regulated by 1987 legislation, but a third of the nation's streams and lakes remain unfit for swimming.

Overseas, there are plenty of dirty rivers. The reservoir forming in the Yangtze Gorges, for example, will be heavily contaminated by waste left on the reservoir floor. Measures are being taken by many countries to deal with similar problems, but whether they'll be successful is an open question. The Asian Development Bank helped Thailand build the $750 million Samut Prakarn wastewater-treatment plant south of Bangkok, at a place called Klong Daan. Fishermen protested that the plant would damage the shrimp and mussels on which they depend. The bank then commissioned an external review, whose conclusions it rejected. Perhaps the fishermen's fears and those of the consultants are groundless, but environmentalists say that the bank—like its big sister, the World Bank—is incorrigibly arrogant.

Water pollutants tend to be handled on a pollutant-by-pollutant basis. After many years of study, for example, the standard for arsenic in American domestic water supplies is to be cut from 50 to 10 parts per billion. Things are much worse in Bangladesh, where arsenic from over a million tube wells is now responsible for the biggest mass poisoning in history, with perhaps 20 million victims. Bangladeshis were persuaded to stop using polluted surface water and to start using new, shallow wells. Nobody at the time realized that many of those wells tapped groundwater that was highly charged with naturally occurring arsenic. The aquifer contamination is spotty, with dangerous wells close to healthy ones. The bad ones are to be closed, but for people who have drunk the water from bad ones for many years, the damage is done. On their behalf, a London law firm sued the British Geological Survey in 2001 for negligence in not testing for arsenic before it recommended drilling the wells. In 2004, however, London's High Court threw the case out.

Then there are oil spills. The early hopes of the second Bush administration to expand drilling on Alaska's North Slope, as well as off the coasts of Florida and California, were battered by legislative and judicial opposition—odd in a way, as there have been no major domestic spills from offshore wells since one near Santa Barbara in 1969. It's a remarkable statistic, because 3,700 rigs operate without major incident off the Gulf Coast alone. The federal Minerals Management Service reports that there were three significant oil spills in the Gulf during 2002, but the largest was less than 1,000 barrels.

Perhaps it's guilt by association, because the public is acutely aware of the damage caused by tanker spills. The notoriety of these accidents began with the *Amoco Cadiz*, which broke up off the coast of Brittany in 1978. The next year there was an even larger, but much less reported, spill from the *Atlantic Empress* near Tobago. Americans, of course, are most familiar with the *Exxon Valdez*, which in 1989 ran aground in Prince William Sound, at the southern end of the Alaska Pipeline. This media-saturated accident involved a 984-foot ship spilling 257,000 barrels (11 million gallons) of oil. As spills go, the *Valdez* ranks 20th, well below the *Atlantic Empress* and *Amoco Cadiz*, both of which lost six or seven times as much oil. Still, the *Exxon Valdez* killed a quarter million birds, along with 2,800 otters, 300 seals, and 22 killer whales. Exxon spent $2 billion on the cleanup, and it says that the area has recovered. Residents disagree. A permit in Valdez for a salmon seiner was worth $160,000 before the spill. Now it's worth $50,000. The *Valdez* is barred from Prince William Sound, even though it's been cryptically renamed the *SeaRiver Mediterranean*.

There's solid waste, too. Until the 1930s, American cities piled and burned their trash. Then, starting at Fresno in 1937, trash began to be placed in trenches and buried. This helped control rats, but it didn't prevent toxic seepage. (Ironically, the pioneering Fresno dump is now a toxic-waste site precisely because it has polluted the local groundwater.) The EPA has published elaborate standards for sanitary landfills, which now include liners intended to keep drainage from entering groundwater. Sanitary or not, such dumps are very unpopular with local residents, who, when a new dump is announced in their neighborhood, are likely to parade outside government offices and carry signs with acronyms like LULU (locally unwanted land use) and NIMBY (not in my back yard). Many cities consequently ship their solid waste long distances. Toronto exports 130 truckloads of garbage every day to Sumpter Township, just south of Detroit. A million tons of the city's waste are buried there annually at a landfill with the wonderful name of Carleton Farms. (The landfill, by the way, is owned by Republic Services, whose chief stockholder is Bill Gates.) You can export sewage, too. New York City ships 400 tons of sludge every day to Sierra Blanca, 90 miles southeast of El Paso.

Recycling could reduce the pile of solid waste that Americans produce, but it's a hard sell. Imagine what would happen to the American politician who tried to copy Bangalore, India, where stores are required to charge for all their plastic bags and where they may no longer stock bags made of film so thin (less than 20 microns) that it's good for only a single use. For lack of such policies, the United States has become the Saudi Arabia of scrap. Sound implausible? Perhaps, but in 2002 the United States exported $1.2 billion worth of scrap and waste—mostly metal, plastic, and paper—to China. That was enough to make scrap and waste America's third most valuable export to China, right after aircraft and semiconductors.

What about toxic waste? The Resource Conservation and Recovery and the Toxic Substances Control acts of 1976 were supposed to monitor and control hazardous wastes in the United States. A few years later, the Superfund was created to treat the nation's most dangerous toxic-waste sites by raising $1.6 billion (later raised to $8.5 billion) through taxes on the nation's refineries and chemical plants. Many of the sites remain untreated, however, because the cost of cleaning them exceeds the funds available. Superfund tax collection stopped in 1995, and the fund is now almost exhausted. The second Bush administration proposes to continue work on a reduced level, with funding from general revenue.

The biggest Superfund site is at Butte, Montana, where Anaconda mined copper for generations. By 1955 the ore was so lean that the underground mine was converted to an open pit, the Berkeley Pit. In 1978 the property was sold to ARCO, the old Atlantic-Richfield oil company, which closed the mine in 1982. The pit then flooded with 30 billion gallons of water so acidic (pH 2.5) that migratory birds have landed and died on the spot from throat burns. The polluted water from the pit flows downstream into the Clark Fork, where bones from animals that drink regularly from the river are tinged blue–green. Birds are now kept away from the pit by a sound system that imitates predators, but cleanup of the river remains a costly challenge. ARCO, now part of BP, says it made a mistake when it bought Anaconda.

Big Sky Montana doesn't like to advertise such catastrophes. Butte isn't its only one. There's the vermiculite, for example, mined at Libby from 1940 to 1990. The material contains a form of asbestos that lodges in the lungs. Scar tissue builds up, slowly stiffening the lungs until breathing is difficult and finally impossible. Some 200 Libby residents have died of asbestosis and 20 to 30% of the town's 2,700 people show lung abnormalities. The mine has belonged to W.R. Grace Chemical since 1963, and the company has paid $20 million in medical claims. Now, however, the company is in bankruptcy, and the EPA may declare Libby a Superfund site. It's part of a much broader asbestos problem in which thousands of lawsuits have pushed many companies into bankruptcy. A trust fund on the order of $100 billion is likely to be created by the federal government to replace this legal mountain, but it won't satisfy the 100,000 Americans projected to die of asbestos poisoning during the next decade.

Toxic wastes aren't a problem only at remote locations. Between 1946 and 1977 General Electric released into the Hudson River, with government permission, a million pounds of polychlorinated biphenol (PCB), an oily liquid used in electrical transformers. Some of these PCBs are still on land, but most are at the bottom of the Hudson, where they work their way into fish that become unfit for consumption. The EPA has proposed a dredging plan that would cost at least $500 million, but GE, which would bear most of the cost, strenuously argues that the dredged material would become a hazard itself and should be left on the river bottom, where fresh sediment will gradually bury it. Supporters of the dredging plan say that GE is afraid that the plan, if implemented, will become a precedent for action at other company PCB sites, such as one at Pittsfield, Massachusetts.

Again, the crises seem to rain upon us. The Clinton administration bought the land of a proposed gold mine at the edge of Yellowstone. It also proposed a mining ban on 400,000 acres of national-forest land in the northern Rockies, just south of Canada. Mercury levels continue to rise, however, despite a sharp cutback in the use of mercury in mining. The cause is trace quantities in coal burned in power plants.

The best-known pollution disaster involving toxic materials overseas occurred at Bhopal, India, in 1984. A Union Carbide plant there released a cloud of cyanide gas that killed at least 5,000 people. Five years later, Union Carbide paid India $470 million. The families of people who died in the accident have received about $1,300 each, while almost a half million people injured in the accident have received payments of about $500 each. Union Carbide has since merged with Dow Chemical, but survivors continue to fight for more compensation. They say that the payments they received were absurdly low because the government of India wanted to show foreign investors that they could build a factory in India and not fear huge legal judgments.

More recently, the waste pond of the Los Frailes zinc mine in Spain breached in 1998, releasing a toxic slurry into the Guadiamar River. It flowed downstream, burying farmlands. The Canadian mine owner, Boliden, spent $30 million on a cleanup, but the arsenic residue has left some 10,000 acres of cropland permanently toxic. Plans call for the government to buy the land and make a greenbelt of it. As so often happens in such cases, the local people have supported reopening the mine, because 500 of them work there.

Many other pollution disasters have received little attention. Consider Copsa Mica, a Transylvanian town between Bucharest and Cluj. During the Communist era, Copsa Mica had a lead smelter and a carbon-black plant. The town and rural neighborhood turned black with soot. When the Ceausescu regime fell in 1989, the carbon-black plant was shut down and the smelter was privatized. Things are green once again, but the soil is so deeply contaminated with lead and other heavy metals that it's dangerous to drink the local water, eat vegetables grown in the local soil, drink milk from cows eating the local grass, or eat cheese made from that milk. The 5,000 townspeople do all those things because local jobs are scarce. The health hazards won't decline for decades.

In Italy, organized crime is deeply involved in the illegal disposal of toxic wastes at some 700 sites. Italian doctors say that these sites are responsible for rising rates of cancers and tumors. Ironically, the landowners are sometimes themselves implicated. Salesmen with a company called Ecoverde, for example, have offered a free initial shipment of fertilizer. It's too good to be true, and the farmers who have said yes have watched in horror as their land died from unknown poisons that they themselves had spread. Ecoverde has also been charged with dumping waste in Italy's national parks.

High-tech industries, famous for their clean rooms, can also be serious polluters. Endicott, New York, used to be the site of an IBM plant that used toxic solvents in making circuit boards. A lot leaked, and today, Endicott—alias "Emptycott"—has 337 homes with plastic tubing, paid for by IBM, to vent the toxic chemicals that rise from the groundwater into the basements of homes that nobody will buy. There's another story at Taoyuan, some 10 miles west of Taipei. In the 1960s, the Radio Corporation of America (RCA) built a television manufacturing plant there. It was one of the earliest instances of American manufacturers going abroad, and the Taiwanese were proud of the facility. The plant relied heavily on organic solvents, however. They were discarded casually, polluting the local groundwater. RCA was sold to General Electric in 1986, and GE sold its consumer electronics division to France's Thomson a year later. Thomson now supplies bottled water to Taoyuan residents, who believe that their cancer rates are above normal because of the plant, which closed in 1992.

Finally there's nuclear waste. It's currently stored at 150 sites around the United States. At about 100 of them, the material will be dangerous for centuries. Some of these sites are close to major cities. Rocky Flats between Denver and Boulder is a good example. Others, such as the Hanford Works in central Washington, are already leaking waste into aquifers from which the waste cannot be retrieved. Leakage is occurring at the Nevada Test Site, too. It will percolate at an unknown rate under the town of Beatty, then under alfalfa and pistachio farms in the Amargosa Valley. Poetically, it will come to rest under Death Valley. A permanent storage site for high-level waste has yet to open, but $4.5 billion has been spent developing one at Yucca Mountain, near the Nevada Test Site. The state of Nevada is objecting to the project on the ground that leakage to the perimeter of the site may occur in as little as 50 years, far less than the 10,000 during which the site was supposed to provide safe containment. In response, the federal government is considering loading Yucca Mountain with waste stored in special containers of titanium and steel. The outcome of the battle between the state and federal governments is very much in doubt. Both sides say they are confident of victory.

France already has a concrete vault for such materials; it's near Cherbourg and is designed to permit monitoring for 300 years. Germany has a site, too, but the Germans have decided that the whole problem of nuclear waste storage is so immense that over the next 25 years they will phase out all nuclear power plants. As for Yucca Mountain, the plan is to keep it open for 300 years, then seal it forever. A small but nontrivial problem is how to post signs on the site. What language do you use when you're warning people who won't be born for 10,000 years? Pictographs seem to be the way to go, primitive as we may think they are.

BIODIVERSITY

So much for polluting our environment. We're also exterminating the creatures we share it with.

We've been doing this for a long time, and we've been slow to change. Think of the Dutch, indifferent to the death in 1681 of the last *walgvogel,* or ghastly bird. That was how the Dutch saw the dodo, a defenseless and flightless pigeon unique to Mauritius. It was inedible, so killing it was wanton. The Dutch killed it anyway, with help from the rats that accompanied them from Europe and which hungrily ate the dodo's undefended eggs.

When did we begin to change our ways? The Bible is always a good source. Deuteronomy 22:6-7 permits taking eggs and young but not parents. That warning to protect animals of breeding age underlies America's early game laws, with their closed seasons and restrictions on size and number. Iowa in 1878 limited hunters to 25 prairie chickens a day. Oregon in 1901 limited them to 100 ducks a day. Enforcement of such laws was lax, which is why the last passenger pigeon died in 1914. By then, at least some Americans were ashamed at having almost exterminated the American bison for the sake of robes and tongues. Teddy Roosevelt set up and stocked the first bison reserve, down in the Wichita Mountains north of Lawton, Oklahoma.

Since then, we've learned to talk about ecosystems. It's a term that goes back to about 1935, when a sober British ecologist named Arthur Tansley wanted a less emotive term than "biological community," a phrase introduced earlier by Frederic Clements, an American. Ecosystems are used popularly now in discussing the web of life, or the interaction of species.

Aware that we don't understand this web, we're making some effort to put the brakes on species destruction. The United States—not one of the world's richer biological domains—has listed some 500 animal and 700 plant species as endangered. About a third of all its plant species are threatened by loss of habitat that could lead to their endangerment. Hawaii is a particularly vivid case. It was only settled 1,800 years ago, but it has already lost 80% of its bird species. Part of the damage was done by rats escaping from European ships; part by mongooses released in 1883 to kill the rats. A third of Hawaii's plants are gone, too, a transformation unnoticed by tourists enjoying what they think is a pristine tropical paradise. The Center for Biological Diversity calculates that 114 species have become extinct in the United States since passage of the Endangered Species Act; of that total, more than 50 were on Hawaii.

What are the costs of such extinctions? There are two kinds of answers to this question. The utilitarian one says that the world is full of species whose value we do not yet recognize. Tamoxifin, for example, is a tumor-inhibiting drug that was developed from a chemical discovered in yew bark. AZT, a drug used in treating AIDS patients, is based on chemicals from a Caribbean reef sponge. The hoodia plant, a succulent used by Bushmen to stave off hunger, is being developed by Pfizer into an appetite suppressant that might be valuable in treating some of the world's 150 million obese people. (The active chemical has been patented by the government of South Africa, which has agreed to pay 6% of its royalties to the Bushmen.) The Chinese have discovered that sweet wormwood (*qinghao*) yields a chemical called artemisinin, which is very effective against malaria; health agencies around the world have ordered millions of doses. Even leeches, used by European doctors long ago but then discarded as barbaric, are on the verge of being farmed to produce leech saliva, which is a more powerful blood thinner than anything currently available to physicians. A broader utilitarian argument is that extinctions may tear the fabric of nature in ways we do not understand, with consequences we cannot anticipate.

There's a romantic answer, too. The 19th century produced not only nature-loving poets but the Society for Prevention of Cruelty to Animals (SPCA), founded on the anti-Cartesian belief that animals are capable of suffering. Come down to the 1920s and meet Bambi, the fawn created not by Walt Disney but by Felix Salten, whose *Bambi: Lebensgeschichte aus dem Walde* appeared in 1923. The English translation, with a less clomping title, appeared in 1928 as *Bambi: A Life in the Woods*. Walt Disney's movie, released in 1942 with a still simpler title, brought the story to millions. It played an important role in making every American child believe that killing things is bad. Shopping malls meanwhile post signs next to their Christmas trees. Shoppers at Los Angeles' The Grove read in December 2002 that the mall's tree was cut before it succumbed to disease—"an ignominious end for a noble tree." The sign didn't say that in January the tree would be ground to mulch.

The Monterey Bay Aquarium in California publishes a seafood guide advising visitors to avoid eating lobster, shark, prawns, swordfish, and other species in decline. Furs are now almost taboo, with the paradoxical result that many Inuit have been forced to abandon a traditional occupation and take up driving trucks for the oil industry. Or, consider the annual Newfoundland seal hunt, so effectively denounced as an exercise in bloodlust that since 1972 the United States has barred imports of Canadian seal products. So does Western Europe. The annual take fell from 250,000 to 15,000 in 1985. No longer were the helpless seals, too young to swim to safety, clubbed on ice floes and, if not quite dead, skinned alive. Activists were pleased, but many Inuit who had made a living selling pelts went on welfare and were bitter about Americans who, as one hunter said, "treat their pets better than human beings." Those hunters should be happier now. Eastern European consumers don't share the sympathies of Americans and Western Europeans; they're eager to buy sealskin. The population of seals has meanwhile grown dramatically in the last few decades. That's why Canada in 2004 authorized the killing of 350,000 seals by the tried-and-true methods, although it's now illegal to kill any seal less than 3 weeks old.

The two arguments—utilitarian and romantic—often fuse. Mountain lion populations, for example, are growing fast these days, and there have been about a dozen fatal maulings in the last decade. Yet suburbanites in Boulder who see mountain lions in their backyards often do not call animal control. On the one hand, they don't want to disturb the balance of nature. On the other, they don't want to see an animal hurt.

These arguments have been powerful enough to elevate the protection of world wildlife to center stage in discussions of environmental quality.

Fisheries are one important aspect of the debate. Since the 1950s, the annual global catch has risen from about 19 million to 80 million tons. A lot of the credit (or blame) goes to the Japanese, who after World War II pioneered new fishing techniques. There are other causes, however. An obvious one is electronic gear for finding fish. Less obvious are the better weather forecasting and communication systems that permit fishermen to operate in situations that once would have been foolhardy. Also, better transportation has opened new markets for the catch.

Production worldwide is now on a plateau, but though level it's not stable. That's because the fish caught today are increasingly juvenile, which implies declining catches in the future. Cod, for example, naturally live 40 years and grow to 6 feet, but fishermen don't catch cod like that any more: 90% of the catch by British fishermen is of fish less than 2 years old, too young to have bred. It's a problem reaching up the food chain all the way to great white sharks, whose numbers in the Atlantic have been halved since 1986.

For the number of fish caught, there are also too many fishermen, perhaps by a third or a half. There are so many that the industry suffers huge economic losses, about $50 billion a year worldwide. Many European fishermen survive only because their government subsidizes them. The subsidies are unlikely to go away, especially since the European Union decided in 2003 to reduce the take of cod by 45% and to permit fishermen to fish for cod only 9 days a month. Severe as it sounds, this decision had been watered down from the recommendation of the International Council for the Exploration of the Sea, which called for a total ban on fishing for cod in the North Sea, Irish Sea, and the waters west of Scotland.

Things are no better in North America. Since 1995, the federal government has paid fishermen $200 million to stop fishing and give up their boats. The hope has been that the remaining fishermen would catch enough fish to make a living. On the East Coast, however, the take of cod has plummeted from over 200,000 tons annually in the 1980s to about 18,000 tons. The Canadians experimented with a moratorium from 1992 to 1998, but their cod stocks didn't regenerate. Now it appears that the fishery will be shut down indefinitely, at least in the Gulf of St. Lawrence. Once again, there will be subsidies, as there were during the moratorium, when Canada paid 26,000 unemployed fishermen $1 billion.

Cod stocks are down 90% in North America since 1980, and they're down 75% in Europe. If you want cod now, you head to the Barents Sea, north of Norway. That's where half of all the cod caught now come from, but you'd better hurry: the Barents Sea is being overfished.

The cod crash isn't unique. Most major fishing nations have not ratified the 1995 U.N. Agreement on Straddling Fish Stocks, governing fish that live within no single jurisdiction. Internationally prohibited drift nets, 20 miles long and hanging like curtains from floats, still trap anything that comes their way, including sea turtles, dolphins, and sharks. Another hazard for those animals is long-line fishing, with thousands of hooks on a line running for miles, like oceanic barbed wire. Some 100,000 miles are laid daily by fleets based in China, Taiwan, Korea, Japan, and Spain. One fisheries expert says, "everything is going in the Pacific." Says another: "Fisherman used to go out and catch these phenomenally big fish. But they cannot find them anymore. They're not there. We ate them."

Even waters under national sovereignty can't be monitored perfectly, although fishermen are occasionally prosecuted vigorously. This was the case in 2001 with the *South Tomi*, a Spanish trawler loaded with $800,000 worth of Patagonian toothfish, a species marketed in the United States as Chilean sea bass. The boat was intercepted off Western Australia by an Australian inspection vessel. The Spanish vessel sailed away, but the Australians followed it all the way to South Africa, where with South African Navy help they boarded the vessel and forced it to Perth for prosecution.

Freshwater fisheries are in trouble, too. Giant catfish grow in the Mekong to 600 pounds, but they're almost extinct now, a victim not only of fishermen but of dams that interfere with migration. Or think of the sturgeon, source of caviar. After the collapse of the Soviet Union, limits on the sturgeon harvest disappeared. Production spiked, then crashed. An annual catch of over 20,000 tons in the 1970s became a catch of barely 1,000. As a result, the Convention on International Trade in Endangered Species (CITES) came close to prohibiting caviar from entering the United States, Europe, and Japan. Facing the unhappy prospect of tons of unmarketable caviar locked away in warehouses, the Russians agreed to a 10% quota cut, and CITES approved the export of 142 tons of caviar for 2002. Now, however, the U.S. Fish and Wildlife Service may declare sturgeon an endangered species, which will make the importation of caviar illegal. In search of substitute products, poachers are descending on Kamchatka, where they gut salmon for the roe. (Don't be too quick to condemn the poachers. Although some are wealthy, many are villagers who lost their jobs with the collapse of the Soviet Union. One says, "if you get enough caviar, you sell it. We have to survive, and that's the only way.")

Is fish farming a good alternative? Pioneered in Scandinavia in the 1960s but now a major industry also in British Columbia and Chile, fish farms raise 50,000 or more fish in a wire cage 100 feet on a side. The fish are doctored with antibiotics to resist the diseases that can decimate animals in close confinement. Waste settles to the sea floor, and there it produces bacteria that use oxygen needed by nearby shellfish. Fish sometimes escape from the cages, too. In the case of British Columbia, this means that Atlantic salmon—the fast-growing farmed species—now swim in most Pacific waters. In addition, farmed fish are fed on a diet of fish meal, and it takes at least 2 pounds of meal, which is made from sardines, anchovies, and herring, to make 1 pound of salmon. To make the flesh pink, by the way, the feed is tinctured with canthaxanthin, synthesized by Hoffman-La Roche. More worrisome to consumers, most of the feed comes from fish caught in the polluted North Sea, so PCB and dioxin levels are much higher in farmed than wild salmon. The levels are so high that some scientists recommend eating farmed salmon no more than once a month. (If you want a real horror story, try the levels of PCBs and mercury in the breast milk of Arctic natives. Their diet in heavy in animal fat, which concentrates these substances to levels that make mother's milk toxic.)

Critics compare salmon farming to feedlots and floating pig farms. Alaska in 1990 banned fish farms as a threat to the existing fishery. British Columbia is catching up. In 2003 it ordered the closure of 11 farms near the Broughton Islands so pink salmon would have a safe route to the sea. Not content, local Indians have sued to close two large operators, Heritage Salmon and Stold Sea Farm. A chief explains: "Clams, prawns, crabs, salmon—we had it all at our fingertips. . . . It's all being depleted because of all the algae, the sea lice and the contamination the fish farms are putting in our territory." Despite these problems, commercial fisheries are in such poor shape that a start's been made to farm cod and other white fish, such as halibut. Australian fishermen near Port Lincoln have become very wealthy by corraling wild tuna and fattening them for sale to the Japanese. The fish double their weight over 3–6 months in captivity.

Land species aren't doing much better. Ape populations in the Congo were halved between 1983 and 2000 because there's a widespread market in Africa for bushmeat, including not only apes but monkeys and antelope. Gorillas fetch about $60 in Cameroon markets, which is why the next decade may see their virtual extinction in West Africa. There's a similar trade in Indonesia, where forest animals like macaques are captured and sold as food. Populations are declining so sharply that, for lack of prey, tigers on Sumatra are turning to humans.

Plants are under attack, too. The world's most biologically diverse forests are the lowland forests of Indonesia. During the Suharto regime, there was central control over Sumatra and Kalimantan. Not now. The virgin forests of Sumatra will probably be gone by 2005 and those of Kalimantan by 2010. Over in the Brazilian Amazon, 2,500 sawmills are at work in the largest of the world's tropical rainforests. It is being destroyed by cutting but also by fires that start in cutover areas and spread.

The Amazon forest absorbs huge amounts of carbon dioxide and in this way counteracts our production of it through the burning of fossil fuels. That's why preservation of the Amazon is often described as vital to the control of global warming. In 2002, however, some 10,000 square miles of Amazon forest were cleared. So much forest was burned that the Amazon that year may have added as much carbon dioxide as it sequestered. Burning in the Amazon that year certainly produced more than four times as much carbon dioxide as Brazil produced by burning fossil fuels.

Brazil's Itatiaia National Park, atop the Serra da Mantiqueira west of Rio de Janeiro, is meanwhile the prime habitat of the *jucara* palm, which over a century grows into a 65-foot tree. The *jucara* is also the source of the tastiest heart of palm, whose removal kills the tree. One tree yields enough of this prized delicacy to fill two 14-ounce cans that sell in the United States for $3.99 each.

Farther north, the monarch butterfly migrates annually to forests in Michoacán, Mexico, where it overwinters. Those forests are being decimated, though seedlings are being distributed to villagers in the hope that the forest can be stabilized before the monarch's habitat is lost.

The introduction of exotic or weed species to both marine and terrestrial environments has meanwhile been as damaging as the direct destruction of native species. Some of these introductions are very well-known, like Europeans bringing horses to the New World. Others aren't. Glaciers removed earthworms from everywhere north of a line running from the Missouri River across to Cape Cod. The worms are returning now, but we're hastening the wave and adding species previously unknown in North America. That's why if you buy a fishing license in Minnesota you're told not to discard worms you don't use. Worms are great composters, it's true, and we usually think of them as beneficial, but they feast on forest duff and eliminate habitat for native animals, like salamanders.

All told, about 7,000 species have been introduced to the contiguous United States since 1500. This is a small number compared to the 150,000 native species, but 700 of the aliens have become pests. Among them are kudzu and leafy spurge, yellow star thistle, Norway maple, melaleuca, Brazilian pepper, and purple loosestrife. People who deal with these plants groan at the difficulty of getting rid of them.

Hawaii is a dramatic example. Until the 1990s, it had no snakes or terrestrial amphibians. Now it does, courtesy of the nursery trade, which apparently is responsible for the inadvertent introduction to Hawaii of the *coqui* frog, native to Puerto Rico. In Hawaii, the frog lacks predators and is expanding rapidly as it competes for food with local birds. Hawaii now has ants, too, although historically it had no social insects. The islands have plant invaders as well. One is Miconia, a tree whose canopy shades everything beneath it to death.

There are invasive marine plants. One is Caulerpa, an inedible algae that escaped, ironically, from the Oceanographic Museum of Monaco. It has spread across the Mediterranean, displaced native vegetation, and caused local fish to starve. Somehow, Caulerpa has now made its way to California.

Weed species are a problem with aquatic animals, too. Highly destructive jellyfish have appeared in the Black Sea. The United States has zebra mussels, introduced by Russian freighters in 1986. The mussels cause billions of dollars of damage annually to boats and other man-made structures in the Great Lakes. Another threat to the Great Lakes is the Asian carp, now working its way upstream from Arkansas. The fish was introduced in the 1970s by fish farmers eager to control algae. The carp are now within 20 miles of Chicago, whose mayor says, "If we don't stop them now, they will destroy the Great Lakes." Asian eels are a newer problem, apparently destined to invade the Everglades and as unstoppable as fire ants, yet another invader. In 2002, the northern snakehead was found in a pond near Crofton, Maryland. Apparently some had been released by a cook who decided not to kill them, but the sympathy may have been misplaced. One owner of a marina says that he kept some snakeheads in a tank. No longer: "They scare me," he said. "They just kill and eat and eat and kill." The authorities poisoned the pond in hopes of stopping the invasion. No luck: The snakehead can crawl on land between bodies of water, and in 2004 snakeheads began appearing in the Potomac. Poisoning the Potomac was out of the question, and posters were put up asking fishermen to kill snakeheads whenever they caught one.

Land animals, too, can become pests. The most famous case is probably that of rabbits in Australia, but there are plenty of others. American gray squirrels have pushed Great Britain's native red squirrels to northern refuges. In Tierra del Fuego, beavers were introduced in 1946 as a fur-growing venture. When the business failed, the animals were released. Fifty years later, 120,000 beavers are destroying the island's forests, and there are fears that some will cross to the mainland. In New Guinea, crab-eating macaques, perhaps escaped pets, are now eating so much of the food used by the island's native species that those native species are likely to disappear unless the macaques are removed.

The impact of invading animals has been especially severe on bird populations. Guam's birds were driven to extinction by the introduction of brown tree snakes after World War II. The seabird population of British Columbia's Queen Charlotte Islands has been reduced by introduced raccoons. A similar story has taken place in the Outer Hebrides of Scotland, where pet hedgehogs escaped in 1974. The hedgehog population expanded to 5,000 animals, which cut the bird population in half. Plans to cull the animals were stalled by the British Hedgehog Preservation Society but got under way in 2003 with battalions of sniffer dogs trained to find hedgehogs, which were then killed by lethal injection.

Perhaps whitetail deer should be seen as another weed—not exotic but out of control for lack of predators. The population in the United States has risen from a low of 500,000 in 1900 to perhaps 20 million today, about as many as were present when the first gun was carried ashore. The prospects for limiting the population aren't good, even though the deer cause about a million vehicle accidents annually in the United States. Another weed animal may be the many gaggles of resident Canadian geese, which do not migrate. They are descendants of farmed geese or of geese trapped as living decoys, and their population now exceeds the number of migratory geese using the Atlantic Flyway.

Efforts to exterminate the invaders are undertaken from time to time, and not just with hedgehogs in the Hebrides. A recent case is the proposed destruction of 60,000 ruddy ducks, native to America but introduced to Britain in the 1940s. They are now interbreeding so aggressively that they threaten to eliminate Spain's rare white-headed duck. Another approach is to bar the introduction of weed species. New Zealand has taken especially aggressive measures. Each year, it intercepts almost 5,000 organisms in transit. Such programs overlook plant domesticates, however. Considered ecosystemically, they—especially the grains—are weeds that have gone a long way toward eradicating native plants, especially grasses. Those lovely amber waves can be seen as biological freaks.

Just as we've acted to reduce pollution, so too we've acted to slow down this biological killing spree. There have been many success stories. Tiger populations are growing, for example, even though 20 years ago biologists had almost given up on the tiger's survival. In the United States, the Endangered Species Act of 1972 has helped the bald eagle, peregrine falcon, and California condor. Floridians may be less thrilled by the dramatic resurgence since the 1960s of the then-endangered alligator.

A century ago, 20 million bushels of oysters were taken annually from Chesapeake Bay. By 1995, the number had crashed to less than 100,000, but now the oysters are reviving with artificial reefs built by volunteer labor. To save salmon, the U.S. government in 1999 breached McPherrin and other dams on Butte Creek, a tributary of the Sacramento River. The spring salmon run promptly jumped from 1,000 fish to 6,000. Edwards Dam on Maine's Kennebec River has been deliberately breached. There have been discussions about doing the same thing to much larger dams on Idaho's Snake River. Portland General Electric apparently will take down the fair-sized Marmot Dam, built in 1902 on Oregon's Sandy River. Removing it won't be easy, because breaching a big dam can release a huge load of impounded sediments that wash downstream and destroy habitat.

Saving land species can also be tricky. The leatherback turtle, which can weigh over 1,000 pounds, lays its eggs on beaches. Until recently, the second most important site, after one in French Guiana, was the 12-mile-long beach at Rantau Abang, on the northeast coast of Malaysia. When hunting threatened to eliminate turtles there, the Malaysian government set aside protected nesting grounds. They seemed to work well, but they were all in the sun.

Nobody thought to ask why the turtles themselves always laid some eggs in the shade. It turns out that all the eggs hatched on sunny beaches yield females. Now turtles have almost vanished from Rantau Abang. Two came ashore in 2002, but they laid no eggs.

Protection programs can also be extremely controversial. American tourists in Africa are virtuously shocked to hear villagers speak of lions and elephants as pests. The tourists, of course, don't depend on the crops and livestock that these animals destroy, but when the tables are turned Americans are just like Africans. A few years after Congress in 1972 passed the Endangered Species Act, for example, it revised that law so that the Tennessee Valley Authority could go ahead and build Tellico Dam, which the courts had blocked because it threatened the endangered snail darter.

Since then, there have many cases in the United States where the protection of endangered species has collided with people. There are angry North Carolinians who can't go to the beach at Wrightsville because it's fenced off as nesting habitat for the endangered piping plover. There are angry Texans on the High Plains whose water supply may be jeopardized by the need to protect the Arkansas River shiner. Albuquerque is depleting the groundwater it relies on and has built a reservoir behind Heron Dam. Now the city's being told that it must release water from that reservoir into the Rio Grande, because the endangered silvery minnow needs it. Out on the West Coast, there are angry Californians whose vineyards are threatened by the presence of the tiger salamander. Farther south, the proposed Empire Center at Fontana is blocked because it is to be built on part of the 1,200 acres of sand dunes that are the only breeding ground of the endangered Delhi sand fly. The shopping center's promoters think the species should be allowed to go extinct, but opponents say that the fly is part of the dune ecosystem, which may depend on the fly in ways we don't understand. Down in Ensenada, the biggest fish-processing plant in Latin America was put out of business in 1990 by an American embargo on Mexican tuna, which were being caught by nets that also trapped dolphins. Forty big tuna boats are idle, along with hundreds of unemployed and unhappy fishermen.

Typically, these situations pit one set of our principles against another. We want to help Mexican fishermen, but of course we want to save dolphins. Out by Needles, California, there's only one surviving ranch in the Mojave National Preserve. It's the Blair 7IL Ranch, and it operates in such an arid environment that it spreads its 400 cows and their calves over more than 200,000 acres of public land, to which the Blair family has a legal permit. That land is also used, however, by the endangered desert tortoise. Environmental groups have sued to ban grazing in the reserve, and Howard Blair knows what's coming: "We just don't know when. Whether it is 10 years or next year, we don't know." Shall we support the tortoises or the Blairs?

A particularly bitter case arose on Oregon's Klamath River. To save salmon and sucker fish, the Bureau of Reclamation in 2001 reduced by 90% the water it released to 1,400 farmers around Tuleluke, on the Oregon–California border. As a result, 220,000 acres turned to dust. Ironically, this is the same land that had been reclaimed after 1907 by farmers the bureau had welcomed as settlers. In 2002, the Secretary of the Interior came to Oregon in person to open the gates and release water to the fields. The biologists of the Fish and Wildlife Service had been given a few days to attest that this would not hurt the aquatic resources of the Klamath River. Downstream, there were plenty of unhappy fishermen, who had watched the salmon yield of the Klamath fall from 4 million pounds in 1980 to 58,000 pounds 20 years later. When a group of scientists argued in 2004 that a group of Pacific salmon species should not be removed from the endangered species list, the federal Fish and Wildlife Service disagreed. It explained that large numbers of these salmon were successfully raised in hatcheries. This position overlooks the fact that when hatchery fish are released into the wild their behavior is different—often fatally different—from that of wild fish. One of the rebuffed scientists called the government's explanation ridiculous: "If you have animals in the zoo," he said, "you don't call the natural populations restored."

A similar contest may emerge on the Missouri River. Its dams are managed to minimize seasonal variations in flow. This is an unnatural hydrological regime and damaging to wildlife, but simulating the natural regime by increased spring releases would at least occasionally flood downstream farms. Local farmers point out that adaptive management has already been tried on the Colorado, whose regime was profoundly disrupted by Glen Canyon Dam and the introduction of rainbow trout. In 1996, releases from Glen Canyon Dam deliberately simulated the Colorado in flood, and there have been follow-up experiments. So far, these experiments have failed to move sediment and build beaches as the river used to do. Tamarisk trees are invading the remaining beaches, because they're no longer flooded out. River managers now hope to release water to kill the eggs of the rainbow trout, which so thrives here that native species are being eliminated. Guides and other people dependent on the introduced trout are strongly opposed. All in all, says a retired National Park Service employee who ran the Grand Canyon science center, "the Grand Canyon river corridor is getting nuked."

Back on land, the annual cut on the national forests has been declining for years as environmentalists fought to protect endangered species. The most famous case was that of the spotted owl in the Northwest. That controversy was settled by the Northwest Forest Plan in 1994, but logging authorized under the plan has now been blocked by a court order designed to protect the region's still declining salmon and trout fisheries.

A completely different approach to species preservation involves real or threatened boycotts of retail stores. Home Depot and Lowe's, under pressure from environmental groups, now say that they will not buy wood from old-growth forests. Home Depot has agreed to sell no wood products without certification by the Forest Stewardship Council, a small organization based in Mexico and dedicated to saving forests, wildlife, and indigenous peoples. Under pressure from Kinko's and L.L. Bean, Boise (formerly Boise Cascade) announced in 2002 that it would no longer cut old-growth trees growing on patches exceeding 5,000 acres. It was an easy concession for the company to make, because less than 1% of its wood comes from such sites. Office Depot announced in 2004 that it would join with conservation organizations to form a Forest and Biodiversity Conservation Alliance whose purpose, according to the company's chief executive, was "to incorporate conservation science into our procurement decisions."

Parallel steps are being taken in Europe. Environmental groups in Paris and London have argued that tropical woods proposed for bridges across the Seine and Thames should come from sustainably managed forests. Ikea, the Swedish home-furnishings company, now buys wood only if it's grown in plantations, not natural forests. Doubtful that it can really control the sources of teak, Ikea is experimenting in Laos with furniture made from young eucalyptus. It's tricky, because young eucalyptus logs have to be dried and cut with special care or they almost explode under a saw. On the other hand, it appears that furniture can be made from eucalyptus logs only 7 years old. It's a development with the potential to take a lot of pressure off Southeast Asia's forests. Similarly, the Swiss supermarket chain Migros announced in 2002 that it would no longer buy palm oil (it buys 3,000 tons of it every year) unless the oil is certified as coming from plantations that have not been recently established on previously forested land.

One problem with these success stories is that they target single species or only a few species. There are too many species at risk, however, for this kind of approach to succeed, especially in the tropics, where many species are not yet even known. That's why a different approach is needed, one that emphasizes habitat, not species.

How does one protect habitat? The obvious way is to set aside large blocks of land as sanctuaries. This is an idea whose roots go back to medieval hunting preserves, but its modern history begins at 5-acre Pelican Island, near Cape Canaveral, Florida. There, the federal government created the first wildlife preserve in the United States. Near its centenary, Pelican Island today can be seen only by boat; landings are prohibited.

A century after the creation of Pelican Island, the American refuge system covers almost 100 million acres, an area the size of California. Every now and then, the system gets bigger. The grandest example was the addition in 1980 of about 50 million acres of Alaska. The creation of a new large refuge, however, is a rare event. In the year 2000, the Nature Conservancy, which owns about 2 million acres, bought the whole of Palmyra Island, 15,000 acres of reefs and islets 1,000 miles south of Hawaii. Two years later, Brazil made the news when, in conjunction with the World Summit on Sustainable Development, it announced the creation of the Tumucumaque Mountains National Park in the state of Amapá, north of the Amazon's mouth. At almost 16,000 square miles, this is now the world's largest tropical park. At the same meeting, Gabon announced the creation of 13 new national parks covering a total of 10,000 square miles.

Several other big land reserves are planned or dreamed of. Perhaps the biggest is in Canada, where the Boreal Forest Conservation Network, led by the World Wildlife Fund, hopes to put over 1 billion acres—more than a third of Canada—into a mix of ecological reserves and areas to be developed under strict environmental controls. About a third of all the land birds in North America, including huge numbers of ducks, warblers, and sparrows, use this forest at some time in the year, and so its preservation has impacts reaching well beyond Canada. The Network has the support of many resource-based companies, as well as environmental groups, and the prospects are good that it will actually be created.

Conservation International meanwhile wants to spend $6 billion protecting 400 million acres of tropical forests. The sites it's chosen are some 25 hotspots. This is the term introduced in 1988 by Norman Meyers to call attention to the 1 or 2% of the earth's surface that provides habitat for 60% of all terrestrial species. The hottest of these hotspots are Southeast Asia, including Indonesia and the Philippines, Madagascar, the Caribbean, and coastal Brazil. To protect them and other hotspots, Conservation International received $250 million from one of Intel's founders. It hopes to leverage that money into a lot more.

In the search for win–win solutions, compromises are sometimes worked out between private landowners and government agencies. A good example involves some 400,000 Weyerhaeuser acres in Oregon that are now subject to logging controls in exchange for a federal guarantee not to bring lawsuits during the life of the management agreement.

California is trying another approach and is acquiring hundreds of corridors between reserves. The hope is that mountain lions and many other animals will use them to range over larger distances than the state can provide in massive blocks. A grander corridor project is the Mesoamerican Biological Corridor, mile-wide strips unifying reserved blocks in a belt running from Mexico's Chiapas to Costa Rica's Darien. Progress has been slow, impeded by economic development projects. What comes first? The Montes Azules Biosphere Reserve is the surviving fragment of a Chiapas rain forest decimated by commercial logging and peasant farming. You might think that locals would want to save it, but the Zapatistas see the reserve as "a war of extermination against our indigenous communities." They go further and condemn ecotourism as a threat to their existence as "indigenous peasants with our own ideas and culture." Their culture has certainly been transformed by the 40,000 tourists who come each year to nearby Frontera Corozal. Hear an echo of desert tortoises and Mojave ranchers?

Protecting habitat without disrupting economic activity is a problem in the United States, too, but it at least has the money to experiment with heavy-duty engineering solutions. A fine example is taking place in the Florida Everglades, where in 2000 the Army Corps of Engineers announced an $8 billion recovery plan. The swamps are shrinking as their water supply, from Lake Okeechobee, is diverted by irrigation and drainage works. Ninety percent of the swamp's wading birds are already gone, but the plan seeks to restore the habitat and revive the populations of at least 68 endangered species by drilling 333 deep wells through which freshwater will be injected into saline aquifers to create huge bubbles of freshwater, 2,000 feet in diameter. The water will be pumped to the surface and put into Lake Okeechobee when the lake is low. A seminatural system can in this way be created without threatening either urban growth in Florida or the continued operation of farms growing federally subsidized sugarcane. As long as those uses continue, however, the Everglades will remain dependent on elaborate and wholly artificial environmental plumbing. Other complex wetland restoration projects have been proposed in California. One is at Bolsa Chica, down south; another is on San Francisco Bay, where some 26 square miles of evaporation ponds used for commercial salt production are to be purchased for $135 million and gradually restored to wetlands.

There are marine reserves, too, in countries as diverse as Canada and tiny Saint Lucia. Fish stocks recover in these reserves, with the result that fishermen now fish intensively just outside the boundaries. An early example was the 120-square-mile Tubbataha Reef reserve, created by the Philippine government in 1988. California has now proposed no-take reserves on the waters it controls around the Channel Islands. If the federal government extends the reserve to include waters under federal control, the project would be the second largest in the United States. The biggest is the giant Hawaiian Islands National Wildlife Refuge, which extends some 800 miles northwest from the main islands.

Australia in 2003 announced a no-take policy on about 150,000 square miles, or about a third of the area in its marine park covering the Great Barrier Reef. Protected areas are also being set up off the coast of West Africa by a group of nations stretching from Mauritania to Guinea. Promising as they are, such protected areas cover less than 1% of the oceans. Perhaps that explains why the director of the U.N. Environment Program has written that teachers in years to come "will have the tough job of explaining what a fish is." Acknowledging that "the oceans are in trouble," the chairman of the U.S. Commission on Ocean Policy released his commission's report in 2004. It recommended, among other things, that transponders be attached to commercial fishing boats so satellites could monitor them and authorities act if the boats entered marine reserves.

CLIMATE CHANGE

The ecological stakes have risen still further in the last few years with the threat of global warming. How real is it? Nine of the warmest years of the past 140 have occurred since 1990. True, the change hasn't been huge: 2001 was a trifle less than 1 degree Fahrenheit warmer than the 57-degree average of the last 40 years. Nor do the expected changes sound very great: perhaps an additional rise of 3 or 4 degrees in this century. Still, this change is faster than any other temperature trend in the last hundred centuries, and there's only a 9-degree spread from temperatures today and those of 20,000 years ago, at the peak of the last Ice Age.

There's no end of dramatic anecdotes about the impacts of recent climate changes. Siberian hunters along the Arctic coast report thunder and lighting, which they've never seen before. They also report open water in winter. Farther south, the level of Kyrgyzstan's Lake Issyk-Kul has risen about 10 inches since 1998—this, after decades of decline. Increased precipitation is the apparent cause. The lake's core temperature, meanwhile, is rising.

In North America, Glacier National Park is unlikely to have any glaciers by 2030. They've already melted drastically since the park was created, and many tourist attractions of the 1920s no longer exist. Alaskan cruise ships can no longer nose up to the tip of the Columbia Glacier to watch bergs break off, because the glacier's snout has retreated 8 miles in 16 years. The glaciers that are still accessible by sea, like the Hubbard, are so popular that cruise ships line up to get a look, with disappointed tourists allowed only a glimpse before their ship has to back off to make way for the next one. Polar bears on Hudson Bay are meanwhile losing weight and having fewer pups, because with less ice the bears are trapped on land longer than usual and are unable to get to the seals that are their main food supply.

Robins have been seen in Canada's Northwest Territories, turtles have been seen on Kodiak Island, and the migratory guillemot now winters in the Beaufort Sea, 500 miles north of the wintering grounds the birds used in the 19th century. Mosquitoes now live at Barrow, Alaska. On the Kenai Peninsula of southern Alaska, warmer temperatures have led to an explosion in the population of spruce bark beetles, which in turn have killed 38 million trees in a 4-million-acre spruce forest.

There are dramatic changes in the tropics, too. The Andean glaciers are shrinking fast. The icecap atop Mount Kilimanjaro, Africa's highest mountain, has lost 85% of its volume in the last century and is apparently going to vanish within the next 15 years. In the South Pacific, there have been repeated instances of coral reefs losing their color, a bleaching effect linked to rising sea temperature. If the bleaching is prolonged, the coral dies.

The consequences may be dramatic for Americans, too. In the future, it has been said, summers in New York City will be like Atlanta today, and summers in Atlanta will be like summers in Houston today.

Agriculture at high latitudes may benefit from a longer season but suffer from a drier one. At lower latitudes, crops may suffer from excessively high temperatures, impeding plant growth. Moister midlatitude air will support increased populations of insects such as the malaria-bearing *Anopheles* mosquito. It will also produce more flooding, which will lead to more frequent outbreaks of cholera, carried in sewage-contaminated floodwaters. Counterintuitively, there may be more frequent droughts in these latitudes. Higher temperatures should permit more evaporation and therefore more precipitation, but global warming seems to be linked to El Niños, the periodic upwelling of cold water in the tropical Pacific. El Niño, in turn, is linked by the mechanisms of global atmospheric circulation to widespread drought.

A change of several degrees is also likely to melt enough polar ice that the oceans will rise between 1 and 2 feet. That doesn't sound like much, but it will increase coastal erosion and have very serious effects on the world's mangroves and corals. It will also affect the Dutch, two-thirds of whom live below sea level. They are already planning on a 2-foot rise of sea level and are planning some $5 billion worth of coastal protection measures. Raise sea level 3 feet and there are major impacts in the densely populated Nile and Ganges deltas, as well as in several low-lying island nations, such as the Maldives. In the longer term, there are more ominous sea-level-rise scenarios, such as the melting of the West Antarctic Ice

Sheet. Over the space of several centuries, that sheet could melt, as it has done previously. This would raise sea level by 10 to 20 feet, enough to flood many coastal areas. Paradoxically, the heart of Antarctica is actually growing colder. It's a continent, after all, with variations to match.

Climate change may not be gradual and steady. On the contrary, it may occur suddenly, as global circulation systems are pushed beyond a critical tipping point. One such mechanism might be the disruption of the Gulf Stream, which, if it happened, would drastically cool Europe. The outflow of cold, deep water from the Arctic Ocean east of Greenland is, for unknown reasons, apparently already declining. Another tipping point might be reached with the release of methane trapped in the world's slowly melting permafrost. The released gas might be enough to accelerate warming dramatically. The melting of permafrost would also have severe and direct consequences in Siberia, where thousands of miles of pipelines are now supported on frozen ground.

The key question is no longer whether temperatures are rising but what is responsible for the rise. Natural fluctuations have occurred throughout the earth's history, and some of them have been large enough not only to produce ice ages but to permit crocodiles to live in polar areas. We, on the other hand, may also contribute to global change through the production of greenhouse gases such as carbon dioxide, which trap solar heat in the planet's gaseous envelope. Opinion is divided, but the Intergovernmental Panel on Climate Change, which before 1995 took no position on this question, now states that there is "new and stronger evidence that most of the warming observed over the last 50 years is attributable to human activities."

If we're partly responsible for climate change, then in theory at least we should be able to manage it, or part of it. In the long run, we will do so by moving away from carbon-rich fuels like wood, coal, and oil. In their place, we'll use alternative energy sources, ultimately hydrogen. That's already a long-term trend. In 1860, when people relied on wood for most of their energy, they produced three times more carbon dioxide than we do today for the same amount of energy.

Want something sooner? That was the intent of the Kyoto Protocol of 1998, where delegates from most of the world's countries agreed that by 2012 their countries would, on average, cut greenhouse-gas emissions to 95% of their 1990 levels. Early in 2001, however, the second Bush administration announced that the United States would not be bound by the protocol. The cost of emissions reduction was too high, it said, and the need for reductions was unproven. Glancing at the EPA's list of criteria pollutants, it then was pleased to note that carbon dioxide, though a greenhouse gas, was not a pollutant. By early 2004, the administrator of the EPA had resigned in private frustration. So had her head of regulatory enforcement. In his letter of resignation, he wrote that the White House "seems determined to weaken the rules we are trying to enforce."

The administration was particularly unhappy that the United States was being asked to reduce its emissions by 7% below 1990 levels, while no sacrifice was asked of India or China. This view was severely criticized at the time as yet another demonstration of American selfishness, but this criticism is much more muted now, because the Delhi Ministerial Declaration on Climate Change and Sustainable Development in 2002 revealed that economic growth was everybody's top priority. Rather than curb production of greenhouse gases, the declaration emphasized the need to adapt to climate change. In a masterful demonstration of how to make a problem go away, India pressured the U.N. Environmental Program to stop funding research into the Asian Brown Cloud and then persuaded it to call the cloud the Atmospheric Brown Cloud. Voilá! The next year, Russia said that it would not sign the treaty; only two European signatories—Sweden and the United Kingdom—were on schedule to implement it. Since then, Russia has changed its mind.

When curbs do come, as they likely will, the United States will probably want to engage in international emissions trading so American companies can buy gas reductions from foreign companies paid not to emit whatever quantity the American company buys. Though this strategy smacks of smoke and mirrors, Dupont has already shown how it can work. The company voluntarily changed its operations at Orange, Texas, in a way that reduced its production of nitrous oxide. Rather than claim that it was reducing its production of greenhouse gases, for $600,000 it instead sold a certificate to Entergy, a power generator in New Orleans. The certificate allowed Entergy to claim that it had cut global carbon dioxide production by 125,000 tons. Dupont in turn used the money to pay for the gas reduction, which it could achieve more cheaply than Entergy.

Kyoto won't be binding until it is ratified by 55 countries, including countries responsible for 55% of the emissions from industrialized countries in 1990. Even then, it's likely to bog down. That's why some jurisdictions are moving ahead on their own. In August 2001, for example, the New England states signed a compact with several Canadian provinces to reduce their greenhouse-gas emissions to 1990 levels by 2010. Many major companies have already elected to do so, because they see this as an inevitability. Among them are Dupont and BP.

Where do things stand? It's easy to exaggerate here. Don't blame one year's drought or another year's rainy winter on global change. Instead, think of naturally occurring variations superimposed on a climate that's changing both on its own and with our help. It's prudent also to remember that climatologists a few decades back thought we were heading toward another ice age. In that sense, a little warming might be a good thing. But, again, things aren't so simple. More precipitation associated with global warming might mean more snow cover. More snow cover would mean more sunshine reflected back to space. In this way, global warming might paradoxically push us to another ice age.

IDEOLOGY AGAIN

Such cautious conclusions are unsatisfying. Once again, it's almost as if we're primed for catastrophe and in a perverse way disappointed when we don't get it. You can see our ambivalence in Rachel Carson's *Silent Spring* (1962). Carson was arguing that pesticides were killing birds, and she substantiated her argument very carefully. Yet her title has nothing to do with science. It strikes the romantic note and talks not of agricultural shortages but of the tragedy of a world without birdsong. No wonder that the chemical companies fought back with everything they had. They had an enemy whose secret weapon was ancient human sentiment.

PART V

READING LANDSCAPES

CHAPTER 19

AMERICAN CITIES

We've come at last to pure geography, to the character of places rather than the processes shaping them. In other words, we're looking now at the visible consequences of all the processes we've traced.

Where do we start? How do we proceed? I'm going to operate in the simplest way I know, starting with the familiar and ending with the remote. That means I'll begin with American cities and end in faraway countrysides. Just saying this makes me chafe, but I'll restrain myself, stay put, start downtown, and work to the suburbs. Meanwhile, keep an eye out for that ambivalence we have encountered so many times. It's visible in the American urban landscape. I'd go further and say that you can't understand American cities without understanding the contradictory values shaping them. On one side are the high-rise and the automobile, both emblems of progress in the Baconian sense of controlling or dominating the environment. On the other are the suburbs, a genuine but failed attempt to escape progress and make a home in nature.

THE CENTRAL CITY

The London and Paris skylines are generally low, and until recently smart builders haven't even bothered trying to disrupt them with high-rises. Gustave Eiffel in the 1880s wasn't so smart, but he was lucky. Parisians bitterly resented the intrusion of his tower, but over the course of time it became an icon. A century later, tall structures are still forced to the periphery of Paris, as with the complex known as Le Défense, on the west side of the city. So, too, in London, with Canary Wharf on the Thames a few miles below the city. Only now are the barriers coming down, at least in London.

Americans do a bit of this themselves. Until about 1990, builders in Philadelphia respected as an unofficial height limit the statue of William Penn atop the city hall. That limit has been breached now, but Washington, D.C., still reflects Thomas Jefferson's hope that the capital would follow in the steps of Paris and keep its buildings "low & convenient, and the streets light and airy."

Tall buildings weren't really an issue in Washington until 1894, when the 14-story Cairo apartment building went up near the intersection of 16th and Q, some blocks north of the White House. It's not as appalling now as it was then; if you like exotic design, you'll even admire the façade's Moorish ornament. In 1899, however, the building provoked an

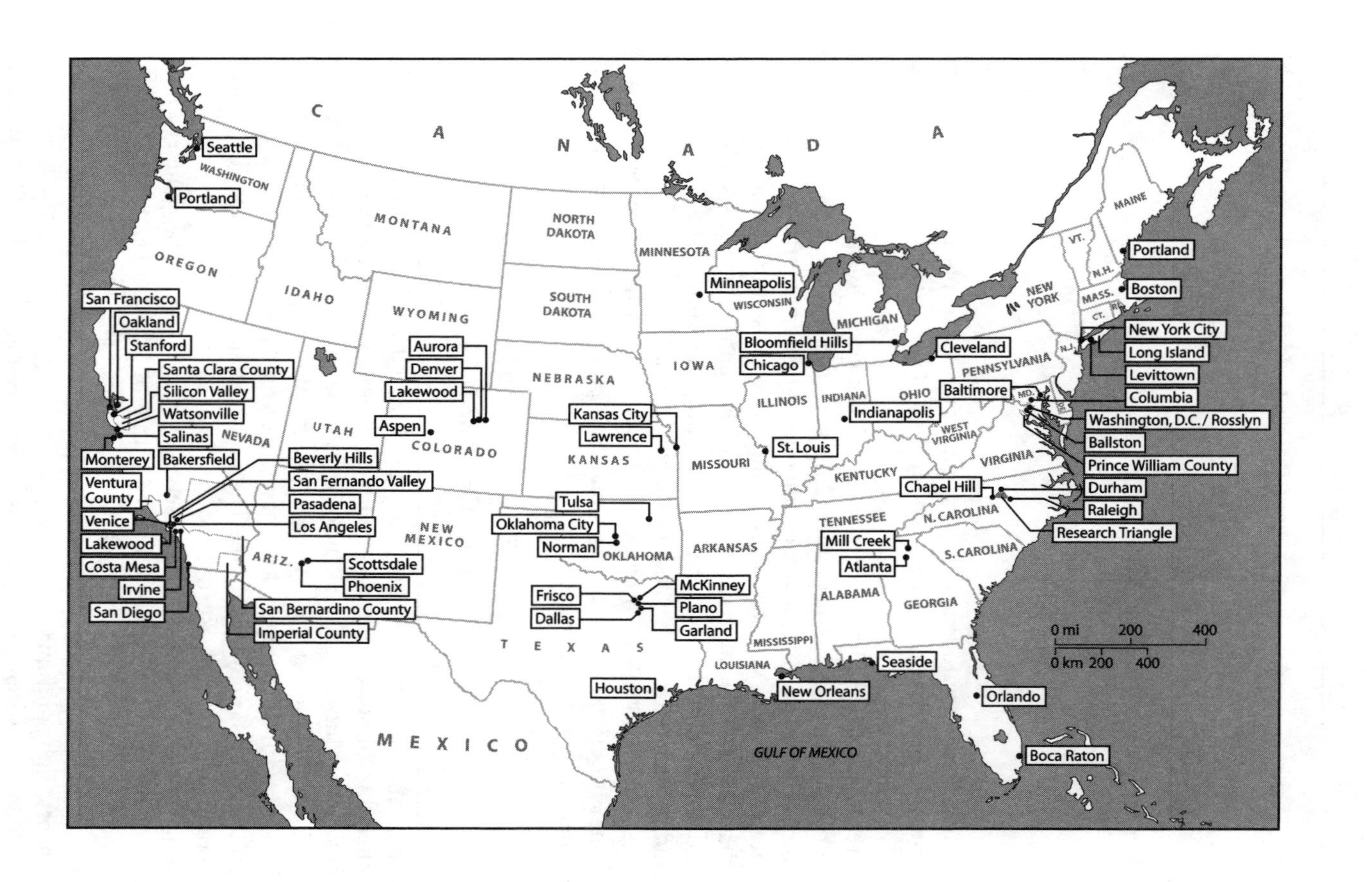

C A N A D A
M E X I C O
GULF OF MEXICO
WASHINGTON
OREGON
IDAHO
MONTANA
WYOMING
NEVADA
UTAH
COLORADO
ARIZ.
NEW MEXICO
NORTH DAKOTA
SOUTH DAKOTA
NEBRASKA
KANSAS
OKLAHOMA
TEXAS
MINNESOTA
IOWA
MISSOURI
ARKANSAS
LOUISIANA
WISCONSIN
ILLINOIS
MICHIGAN
INDIANA
OHIO
KENTUCKY
TENNESSEE
MISSISSIPPI
ALABAMA
GEORGIA
S. CAROLINA
N. CAROLINA
VIRGINIA
WEST VIRGINIA
PENNSYLVANIA
NEW YORK
MD.
N.J.
CT.
MASS.
VT.
N.H.
MAINE
Seattle
Portland
San Francisco
Oakland
Stanford
Santa Clara County
Silicon Valley
Watsonville
Salinas
Bakersfield
Monterey
Ventura County
Venice
Lakewood
Costa Mesa
Irvine
San Diego
Beverly Hills
San Fernando Valley
Pasadena
Los Angeles
Scottsdale
Phoenix
San Bernardino County
Imperial County
Aspen
Aurora
Denver
Lakewood
Kansas City
Lawrence
Tulsa
Oklahoma City
Norman
Frisco
Dallas
McKinney
Plano
Garland
Houston
Minneapolis
Bloomfield Hills
Chicago
St. Louis
Indianapolis
Mill Creek
Atlanta
New Orleans
Seaside
Orlando
Boca Raton
Cleveland
Baltimore
Chapel Hill
Portland
Boston
New York City
Long Island
Levittown
Columbia
Washington, D.C. / Rosslyn
Ballston
Prince William County
Durham
Raleigh
Research Triangle
0 mi 200 400
0 km 200 400

Shades of Granada: the entrance to Washington's tallest residential building, now a condo.

act of Congress "to regulate the height of buildings in the District of Columbia." You see the consequences of that act and its amendments when you look down central Washington's streets and see ranks of high-rent but tedious, apparently truncated office buildings.

Elsewhere, Americans have been building skyscrapers for a century. More often than not, the buildings are referred to now as high-rises, perhaps because they've become so clustered that a passerby has a hard time seeing their tops at all, let alone seeing them scraping the sky. Even if the metaphor is in retreat, there are staunch environmentalists, ardent in the defense of redwoods, who live in the Berkeley hills and are happy to show guests the panoramic view of San Francisco, lit at night like a fabulous jewel. Forget the icons of Manhattan or San Francisco. Every American city has its tallest building, and a visitor can no more ignore it than ignore a man 7 feet tall.

Up close and especially at night and weekends, it's a different story. When the streets are deserted and it's just you and the buildings, it's easy to fantasize that the buildings despise pestering bipeds like yourself. Even if you're in a car driving by, the buildings can easily look like giant tombstones on a dead planet. Pity the trees struggling to grow through heavy iron grates in wind-swept plazas.

Consider New York's now destroyed World Trade Center. Its architect, Minoru Yamasaki, was kept on a short leash by the New York Port Authority, his employer. It wanted the world's tallest building and—oh, by the way—10 million square feet of rentable space. Those were Yamasaki's marching orders, the so-called Program. Like a pedestrian lost at the base of an unfamiliar skyscraper, Yamasaki went where he was told. David Childs, one of the architects most involved with the reconstruction of the site, sums up the old Center as "anti-urban—cold, harsh, hard to get to."

Aftermath of the Cairo controversy: height-limited buildings line Washington's down-town streets, in this case 12th Street at Pennsylvania.

Every high-rise—even the few that draw the attention of architectural critics—is somebody's property. In the midst of all the design chatter, it's good to remember that Chicago's Sears Tower, the tallest office building in the United States since September 12, 2001, has been the object of complex real estate deals. Skidmore-Owings-Merrill designed the building for Sears, Roebuck, which saw the tower completed in 1973 but lost it in 1990. Metlife took over. Seven years later, Metlife signed a deal that would have gradually transferred ownership to Trizec Properties. After September 11, 2001, the value of the building declined from $911 million to $826 million. Lead tenants including Goldman Sachs, which rented over 200,000 of the 3.5 million square feet in the building, announced they were leaving, and rents fell 25%. Trizec no longer had the stomach to ante up $766 million as, according to the 1997 deal, it was supposed to do to complete its purchase. Metlife, however, was lucky enough to find investors with steadier nerves, and early in 2004 it announced that it would sell the building. It did not release the purchase price or the names of the owners, but reportedly they included some of the same private investors who had held the lease on the World Trade Center.

Such heartburn is nothing new in the world of commercial real estate. When the Empire State Building was completed in 1932, its occupancy rate was so low that the building became known as the Empty State Building. When Rockefeller Center opened the same year, tenants were bribed to come so it wouldn't flop. Of course there have been good times, too: when the General Motors building, at the southeast corner of New York's Central Park, was sold in 2003 to Macklowe Properties, it fetched $1.4 billion, a record $800 a square foot.

Even as Americans build skyscrapers, they try to escape them. Probably the best known critique is Tom Wolfe's *From Bauhaus to Our House* (1981). It's a typically speedy Wolfeian polemic against the modernist architects who surrounded Americans with high-rise office buildings and condominiums that few Americans actually liked but which they meekly tolerated. In the long run, however, clients buy what they like, and the lordly architects have capitulated, abandoned the egalitarian austerities of the Bauhaus, and given clients what they wanted. The first sign was Philip Johnson's AT&T (now Sony) Building. A student of Mies topping a building with trim? Was he serious? Was it ironic humor? No matter. Skylines today are full of these postmodern structures. And it's not just high-rises. Stores, motels, churches, houses—all are in costumes hinting at distant times and places.

There are other ways to tame the high-rise. In 1916, upset by the 42-story Equitable Building, New York passed its first zoning law: Buildings were limited so sunshine would strike the street at least a few hours a day. The consequences of that law are visible in the stepped-back high-rises, like ziggurats or wedding cakes, that were built in the 1920s and 1930s. A 1961 revision allowed extraordinary height if a portion of the lot was left open. That's why Manhattan in the last four decades has been dotted with a very watered-down version of the "towers in a park" proposed 70 years ago by Le Corbusier. Recent proposals would have prevented the construction of buildings higher than existing nearby ones, but developers successfully fought the change, and it seems now that New York City will live a while longer with the 1961 law.

A more radical alternative is to abandon the high-rise altogether. When Sprint built a new headquarters, it chose to build a campus at Overland Park, in suburban Kansas City. There it houses 14,500 employees on a 240-acre site with 18 office buildings, none over five stories. The company's explanation: "skyscrapers chill communication." It's not unique. Microsoft's headquarters, after all, are in Redmond, in suburban Seattle.

Never more emotive than after its destruction.

More often than not, high-rise office buildings cluster in a plausibly named financial district; after all, the owners and lead tenants are typically banks. Nearby, there's a downtown shopping district, often dead or dying. A few blocks away, there is also a warehouse and manufacturing district. Almost certainly it's partly derelict, although some sections may have been converted to other uses, both residential and commercial. North and west of this core, upwind of the old factories, there will be residential neighborhoods in a patchwork of slums and expensively rehabilitated homes.

Slums have been popularly recognized as an integral part of American cities since at least 1890. That's when Jacob Riis published *How the Other Half Lives*, a muckraking exposé of tenements in New York City. Until 1880, the city's tenements had been notoriously dark, ultimately because most of Manhattan is covered by long and narrow lots, 25 feet wide and 100 deep. Thousands of apartment buildings stood on these lots: Consisting of a hall and a simple line of rooms, they were called railroad tenements because the rooms were lined up like cars in a train. The only rooms with windows were the front and back ones until, in 1879, the city required airshafts.

By 1900, a decade after Riis's book, the city had 60,000 improved tenements known as dumbbells, because of the profile of two tenements sharing a common airshaft. Conditions were still terrible, and the city's building laws were revised again in 1901. True courtyards were required, as well as running water in each apartment. This spelled the end of the New York tenement, because you couldn't fit a courtyard on such a narrow lot. Instead, big apartment buildings appeared, straddling several lots. Many of them became slums, too. Others became the standard housing unit for the central city's middle class. A good example is the work of the Lefrak Organization, which over the course of the 20th century built 200,000 houses and apartments in greater New York City. Their projects include Lefrak City, population 15,000. It's in Queens and consists of 5,000 apartments in 20 18-story buildings clustered on a site that was once a Waldorf estate.

A later generation of reformers nudged the federal government into paying for public housing projects. The overtones of the word "projects" hints at the fact that these places in the 1970s became notorious—poorly maintained and riddled by gangs and drugs. The poster child was St. Louis's Pruitt-Igoe. In 1955 it was a model public housing project. Seventeen years later, on July 15, 1972, the government blew it up. (Ironically, Pruitt-Igoe had been designed by Minoru Yamasaki, designer of the World Trade Center.) The city with the greatest number of people in projects, however, was Chicago. Between 1955 and 1968 it built 20,000 high-rise public housing apartments. One of them, the notorious Cabrini Green, was demolished in 1995. By 2003, half of all of Chicago's projects had been demolished. The surviving ones have turned out to be reasonably popular with tenants. The problem seems to have been not the buildings themselves but city governments overwhelmed with delayed maintenance and unable to help the people living in these buildings find jobs.

In the 1950s, the federal government fought slums with urban renewal. Between 1958 and 1960 Boston alone evicted 7,000 poor people from houses that were knocked down and replaced by residential high-rises for middle- and upper-income people. The poor had to go somewhere, and so the city's decayed neighborhoods moved but did not disappear.

Typically the poor in any case stayed close to downtown but were deliberately separated from it by new freeways. Chicago and Atlanta are good examples. In *Here's the Deal: The Buying and Selling of a Great American City* (1996), Ross Miller writes that Chicago's Mayor Daley "secured the perimeter of the downtown with multilane highways as forbidding as moats." On one side was the mayor's new civic center. On the other were public housing projects.

The most famous critique of such midcentury American city planning must be *The Death and Life of Great American Cities* (1972). In it, Jane Jacobs attacks the "marvels of dullness" created by planners. At the same time, she celebrates the messy but vital neighborhoods—places like Greenwich Village and Boston's North End—that evolved with minimal government planning. Her eloquence is persuasive for many readers, although in a review Lewis Mumford pointed out that what worked at a neighborhood level could not work for a metropolis as a whole. Jacobs seems to have been right at least to the extent that people do like complex, pedestrian-friendly neighborhoods. Chicago, for example, malled State Street in the 1970s, only to find businesses leaving in droves; in 1996, when the same street was demalled, it rapidly returned to commercial success.

Developers in many cities have meanwhile created successful downtown commercial centers in neighborhoods of decayed warehouses and factories. In the trade, they're known beguilingly as festive retail projects. San Francisco's Ghirardelli Square came first, in 1964, but it's been followed by Boston's Faneuil Hall, Dallas's West End, and Cleveland's Warehouse District. Portland, Oregon, has its Pearl District; Portland, Maine, its Arts District.

Restoration goes beyond old warehouses. New York City's Grand Central Terminal was rejuvenated at a cost of $200 million. Radio City Music Hall is being restored so it will look the way it did when it opened in 1932. There's a plan to restore the nearby Pennsylvania Station, which is partly housed in a post office building converted to a rail station when the real Penn Station was demolished in the 1950s.

The same implicit condemnation of 1950s planning echoes in the controversy over what to build on the site of the World Trade Center. Hallowed ground that site has become, but it's also prime real estate, still under lease to developers. One of those developers is the Australian company Westfield Holdings, which owns the rights to the Center's underground shopping center. It wants to maximize the site's commercial potential by building a mall. Arrayed against it are civic organizations arguing that the World Trade Center was always an eyesore. They want to restore the streets that existed on the site before the World Trade Center was built. Doing so, Westfield argues, will damage the site's commercial value, because pedestrians don't want to shop next to delivery trucks. A compromise is likely, with streets emerging once again on the site but with some of them closed to traffic and perhaps even enclosed.

Despite the conviction of many suburban Americans that their central cities are falling apart, developers and business owners think otherwise. Ralph's has opened a supermarket near the Staples Center in downtown Los Angeles, where there's been no supermarket for many years. Home Depot, Kmart, and Giant Foods have opened stores in long-depressed northeastern Washington, D.C. The site of Chicago's demolished Cabrini Green has a new supermarket, a Blockbuster, and a Starbucks. Logan Square, in Chicago's Hispanic West Side, has a Target that's among the busiest in the country. Pathmark has opened a market near New York's Kennedy airport, an area that hasn't seen a supermarket in 30 years. It's also got one on 125th Street in Harlem, close by a new center called Harlem, U.S.A., which also has a nine-screen theater and an Old Navy. Come to think of it, Manhattan has never had big supermarkets, but some are arriving now: Whole Foods is putting a big one, for example, in the Time Warner Center on Columbus Circle.

There's new investment in central-city residential development, too. Mostly these developments are small and built on land that falls between established neighborhoods. It's a kind of infill, in other words, but in the aggregate it's substantial. People think of Brooklyn as densely settled, for example, but in 2003 the borough issued 6,000 building permits, mostly for duplexes and apartments in spots that earlier developers had skipped. They're part of the exceptionally successful New York City Housing Partnership, which in 1982 scrapped the idea of public housing and instead seized 10,000 properties for taxes. It then transferred these properties for a nominal charge to small developers who built or restored 200,000 units of affordable housing for which the city paid about one-fifth of each unit's price. Now the mayor has announced a $3 billion New Housing Marketplace that seeks to add 65,000 more units over the next 5 years, probably packed more densely than in the past but often located on waterfronts no longer used by industry.

Developers have built 2,200 apartments in Washington's East End, around MCI Center. They're hoping that the prime location will attract residents, though their fingers are crossed because a lot depends on whether shops and restaurants come. Houston, too, whose residents are used to grinding commutes, is getting new downtown apartment buildings. In an effort to make them not only functional but attractive, the developers are consciously emulating Manhattan. They're putting up buildings with brick exteriors, baroque trim, 19-foot ceilings, and building names like Manhattan, Gotham, and Metropolis. (They've forgotten that the name Gotham started out in the English language as a derogatory allusion to a town famous for fools.) A resident of one of these buildings, the preciously named Renoir, says "here it just seems natural to want something that feels a little old." That sentiment explains why old office buildings and hotels are being recycled in Dallas, Phoenix, and San Diego. In some cases, these developments will once again displace the poor, but in others the goal is a mix of conventional and subsidized housing.

There are megaprojects in the works, too. This is nothing new. The Rockefeller Center

Near Washington's MCI Center, the façade of the Atlantic Building is all that remains, as a new and enlarged Atlantic Building is built around it (2004). Gifford Pinchot maintained the central office of the Forest Service in the original building.

in Manhattan began when the Rockefeller family cleared three blocks between Fifth and Sixth avenues. A cluster of buildings went up, including the Radio City Music Hall (1932) and the adjacent, 70-story RCA Building (1933). The center later expanded into adjoining blocks. Seventy years later, Mayor Bloomberg announced a $10 billion plan for Lower Manhattan, which he would reshape into a cluster of urban villages and parks—Fulton Market Square, Greenwich Square—linked by direct subway connections to J.F.K. and Newark Liberty airports. In the separate Hudson Yards Master Plan, which includes 28 million square feet of office space and 12,000 apartments, the mayor hopes over the next 30 years to renew the lower West Side between 27th and 43rd streets.

Two Dallas developers, Hillwood and Southwest Sports, are planning Victory, a 72-acre project next to the American Airlines Center. It's on land that's been used for freight yards, silos, a powerplant and a dump, but the developers see a future of retailing, a hotel, a theater, office space, and some 500 residences. Out on the West Coast, the president of the Los Angeles Central City Association says that her city needs a strong downtown. You can't have a great city, she says, "defined by two theme parks, a beach, and a sign." A first step is the Sports and Entertainment District, which is being built around the Staples Center and is planned to have a 1,200-room hotel, a 7,000-seat theater, a quarter-million-square-foot convention center, and 800 apartments. It's part of the even bigger City Center Project, which calls for a municipal investment of $2.4 billion over 30 years. Staples Center is also a reminder of the increasingly popular entitlement deals coming to America's cities. Dallas has its American Airlines Center, while the New England Patriots play in Gillette Stadium. More of these deals are probably on the way, and not just for stadiums. Museums, concert halls, parks—donors rarely miss an opportunity to brag.

The picture, in short, is a complex mosaic. Investors bid $3.2 billion for a 99-year lease to the World Trade Center. The deal was done late in July 2001, less than 2 months before the destruction of the towers. That's a powerful vote of confidence. In *America's New Downtowns* (2003), Larry Ford goes beyond developments in a single city and describes what he calls a dumbbell model, having nothing to do with dumbbell tenements but roughly applicable to many downtowns. There's the old central business district, perhaps fitted out now with new museums and renovated theaters. There's also a high-rise-dotted spine stretching toward a new central business district, with more office space, shopping, and residences. Along with the traditional downtown functions, these new downtowns typically include convention centers and sports arenas. The Dallas Cowboys, for example, are planning a $1–$1.5 billion complex around a new stadium that they hope to build in the next few years; the present one opened in Irving in 1971 and is obsolete. The city that gets the new stadium—it will probably be Dallas itself—will be expected to contribute to the cost of the stadium, which is likely to run about $650 million. That's not too far from the $400 million being spent on NFL stadiums for the Seattle Seahawks and the Houston Texans.

Like highways, these huge buildings can also insulate a downtown from the nearby poor. The city's ability to help the poor, meanwhile, has declined drastically, because cities almost everywhere have outgrown their political boundaries. The mayor of a central city is no longer elected by the majority of the metropolitan area's wealthy or middle-income people. Safely suburban, they have no wish to merge with the central city they've abandoned. This political dilemma goes a long way toward explaining why New York City's leaders shudder to contemplate the pending $900 billion to upgrade their subways, water system, schools, and bridges.

THE SUBURBS

We'll return to the city center, but let's look now at the periphery, spreading out for miles in every direction.

We know the story here: Everyone wanted a country house, but only the rich could afford them. That's why cities were compact, with most people walking to work. Frank Sprague, who about 1900 more or less invented the electric streetcar, made it possible for average people to undertake longer commutes. By World War I, most cities were spreading radially along streetcar and interurban rail lines. Los Angeles had Pacific Electric, with an 1,100-mile network. The one in Oklahoma City provided such good service that surrounding towns like Norman, 20 miles to the south, were commercially atrophied, with a permanently undersized downtown.

Then came Henry Ford and a car in every household. The spokes between the radial lines of streetcar tracks infilled, and the city continued to spread. It's easy to distinguish the automobile suburbs from the streetcar ones: Houses in the automobile suburbs are detached from one another, and population density in the newer neighborhoods is roughly 5,000 people per square mile, instead of 10,000. Too abstract? Then think driveways. You can't get simpler than that. The diagnostic test is the presence of curb cuts for residential driveways.

What's not so easy to grasp is that the car is more than a vehicle that just happens to be everywhere in American cities. It's an intrinsic part of the city, like buildings and people. Don't be fooled by the fact that the vehicles could all be driven off a cliff. We celebrate those rare days when a heavy snowfall leaves the streets entirely to people, but the permanent disappearance of cars would cripple American cities as surely as the disappearance of water mains and electrical systems. Not that it's likely to happen: As much as Americans hate traffic, they love their cars more than they love their churches, schools, parks—often even their own homes and a good number of relatives. After all, the car gives them the power to escape.

At first, state and local governments took the lead in helping them get out. The first urban freeway was probably Manhattan's Henry Hudson Parkway, begun in 1934 under the imperious Robert Moses. Los Angeles wasn't far behind. In 1938 it began the Arroyo Seco Parkway, which provided a quick link to Pasadena. Federal funding of highways came in the 1950s, however, and has paid by far the largest share of the cost of America's urban freeways. Only now is that story coming to an end.

San Francisco in 1959 fought to stop the Embarcadero freeway from connecting the Bay and Golden Gate bridges; eventually, the elevated structure was torn down. Work continued elsewhere. In 1993, for example, California completed the 17-mile Century Freeway from Norwalk to El Segundo: Building it had taken 30 years, cost $2.2 billion, and required the demolition of 8,000 homes. The last urban freeway likely to be completed in California is the 28-mile, $1.1 billion Foothill Freeway between La Verne and San Bernardino; other freeways in the state will probably never be completed. Since the late 1960s, for example, a 6-mile link on Interstate 710 has stopped a couple of miles short of South Pasadena. Residents don't want it going any farther, and they will likely have their way, because the cost of the missing link has risen to over $300 million a mile. Texas is one of the few places in the country still building urban freeways in earnest. Dallas, for example, has its new President George Bush Turnpike, a toll road under construction on the booming northern fringe of the city.

As freeways allowed Americans to escape the central city, builders got busy. One of the first was David Bohannan of California, who adopted assembly-line methods to residential construction. The most famous of these developers, however, was William Levitt. Back from World War II, he began building 800-square-foot houses on Long Island, at a place then called Island Trees. These were four-room houses (kitchen, living room, two bedrooms, bath) built by assembly-line methods on a 25- by 30-foot concrete slab. At first Levitt rented the houses for $60 a month, but he soon started selling them for about $8,000. By 1951, he had built 17,000 in the place now called, at his insistence, Levittown. He subsequently built two other Levittowns near Philadelphia.

Ticky-tack tract housing became a cliché of postwar America, but its economics were unbeatable. Upgraded, tract homes still define the American suburb. Just north of the San Fernando Valley, for example, the Newhall Ranch is being subdivided into 21,000 housing units. Still farther north and about halfway to Bakersfield, the quarter-million-acre Tejon Ranch is ready for development. Its owners intend to build a city with 23,000 units. Feeling fine, they plan to call the place Centennial.

There's opposition, of course. Next door to the Tejon ranch there's a 95,000-acre property that was once a cattle ranch run by the Kern County Land Company. KCL sold it in 1967, and the property became a gleam in the eyes of developers. Then, in 1996, the Wildlands Conservancy bought the land for $140 an acre. The Conservancy wants a free, open-access wilderness only 50 miles from Los Angeles; the place will be called Wind Wolves, something to send shivers. Now the owners of the Tejon Ranch are talking about selling 270,000 mountaintop acres to the Trust for Public Land. Along with Wind Wolves, this would create a biological corridor linking the Sierra Nevada to the Coast Ranges, and it might win Tejon Ranch the public support it needs to build Centennial.

Used to winning only temporary or partial reprieves, environmentalists are doing surprisingly well stopping developers around Los Angeles. For a decade starting in 1992, for example, Washington Mutual of Seattle struggled to get permission to chop up Ventura County's 2,800-acre Ahmanson ranch into 3,000 homes and a golf course. Finally, in 2003, the state of California agreed to buy the property for $150 million and keep it as open space.

Opposition in the middle of the country is less ardent. That's partly why the built-up area of metropolitan Denver expanded by a phenomenal 66% between 1990 and 1996. The biggest suburban development there is Highlands Ranch. It's a project of Shea Homes, which plans on housing 90,000 people on a site a dozen miles south of the city. Dallas has some big deals, too. One piece on the block is the 2,400-acre Clements Ranch, which will provide about 7,000 units in Crandall, 20 miles east of downtown. Dallas' north side is meanwhile so built up that developers are reaching north along U.S. 75 to land that is closer to Oklahoma than downtown Dallas.

We're in the Boomburbs now, where life is interesting for mapmakers. Every 2 years, a Dallas company called Mapsco publishes a new road atlas for Collin and Grayson counties, out on Dallas's north fringe. The 2002 edition included 1,307 new streets. The one for 2004 added another 2,000. Collin County has meanwhile seen its cotton acreage plummet since 1972 from 55,000 acres to 1,100.

With Plano full and Frisco getting there, real estate scouts in Dallas are moving to the next town up the line, the aptly named Prosper. Land there runs $20,000 to $40,000 an acre, half the price in Frisco. Perhaps that sounds high, but it's nothing compared to southern Nevada. Las Vegas is building homes for 65,000 new residents every year. There's lots of land around the city, but it's mostly federally owned and off-limits to builders. The result is bidding wars that recently raised the price for one 1,000-acre tract 8 miles from the Strip to $160 million. That's $160,000 an acre for the tract to be developed as Mountain's Edge.

Not surprisingly, new homes in Las Vegas are being crammed onto land at the rate of 13 or more per acre.

Who's doing the building? The home-building industry in the United States has historically been extremely fragmented. About 1 million single-family homes are built each year in the country, in other words, but the hundred largest builders in the United States together build only a third of them, for an average of about 3,000 houses per builder. Nationwide, the average builder puts up only 20. It's atavistic, compared to other sectors of the economy. In Metro Dallas, to focus a bit, about 40,000 new homes are built annually, but the biggest builder in the area, Centex, builds only about 3,000 of them. Meanwhile Dallas has 950 builders each putting up one to four homes annually.

The kind of consolidation that has happened in so many other sectors of the economy, however, is finally coming to home building—and it's coming fast. In 1993, the five biggest builders in the United States were (from the top down) Centex of Dallas, Pulte of Bloomfield Hills (Michigan), Ryland of Columbia Park (Maryland), Kaufman and Broad of Los Angeles, and U.S. Home of Houston. Together they built 34,000 homes, or about 3% of the nation's total. Eight years later, Pulte was still in second place, but Centex had slid from first to fourth, Ryland from third to seventh, KB from fourth to fifth, and U.S. Home was gone. This was more than just the usual shuffling of winners and losers. The top builders in 2002 (again from the top) were D.R. Horton, Pulte, Lennar, Centex, and KB. Lennar had swallowed U.S. Home and bounced up from eighth place, and Dallas-based D.R. Horton had soared to first place from 24th place in 1993 by the simple strategy of swallowing a dozen companies. The five together had built 118,000 homes, a tenth of the national total. And Horton was on a tear: Its sales rose from 19,000 in 1999 to 39,000 in 2003. The company predicted 50,000 sales in 2004.

The driving force for this consolidation is access to land. "The opportunity to grow," one Horton executive explains, "is directly related to the amount of land you can control." The big builders have a huge advantage here, because it takes a lot of time nowadays to buy land and get the necessary permits to build on it. "The difficulty in buying and entitling is one of the key drivers right now," says another executive. He means that even medium-sized builders can't get financing for land that may not be built out for years. The big builders, on the other hand, are often corporations. "The public builders," a developer explains, "have a variety of sources, and the private guys are stuck with the banks." Small builders can perhaps find niches for themselves, but medium-sized ones can't. That's why some analysts predict that by 2011 the 20 biggest builders in the country will build three-quarters of America's new houses. Maybe things won't move that quickly, but however it plays out we're a long way from the Little House on the Prairie.

Built by big or small companies, today's houses are much bigger and more light-filled than those of Levittown. Insulation is better. There's also much more reliance on wood-particle members, because wood has become so expensive. Cement has become expensive, too, so in some parts of the country builders are turning away from slab foundations and returning to pier-and-beam construction.

Despite technical innovations, romanticism pervades suburbia. It was there from the beginning, especially in the once popular bungalow. The name derives from Bengal, where buildings are low, with deep wrap-around porches for shade and protection from rain. The bungalow was especially popular around World War I in California, and it still suggests gracious living in a time before television, radio, and a welter of electronic wonders.

The bungalow came under attack after 1945, however. The garage was competing for porch space. Even more important, air conditioning was eliminating porches, period. The typical suburban house adopted the semimodern look of ranch-style homes. Another victim of air conditioning was the trees that had lined city streets and towered over residential yards. Their cooling shade was now obsolete, and cities and homeowners thought of trees as nuisances, shedding leaves and branches, breaking sidewalks and foundations, and penetrating pipes. During the 1990s alone, about one-fifth of the nation's urban trees were removed, frequently to be replaced by shrubs like crape myrtle. Famous for trees, Washington, D.C., has removed half its trees since 1980.

In the simplicity of its shape, the ranch-style hinted at Le Corbusier's machine for living. But it was not to last. The customer wanted to pretend that his house was a castle, or at least a chateau. That's why new houses, though full of electronics, are trimmed with fake Elizabethan cross-timbering or Romanesque turrets. In a triumph of eclecticism, one Scottsdale builder calls his tall narrow windows Mission and Tuscan. It's our old quest for life before clocks, plus our desire for status. What better way to show your standing than to drape your house with the ornamentation of dead kings? Why else roof the house with a huge, unusable attic, its many gables suggesting a mansion?

Can't afford it? Try a 20-foot-high foyer: they're *de rigueur* for even middling new homes—and you don't even have to pronounce it the French way. If you think a two-story foyer wastes space, or if you just can't see climbing a ladder to dust a chandelier, then pave your driveway with concrete impressioned to look like slate. If even that's too expensive, you'll have to settle for a subdivision with a fancy name. How about Ashton Grove, with its overtones of Jane Austen? It sounds very nice, though given the choice I'd pick Sandringham.

Stores followed housetops. At first, they did so in commercial strips, independent businesses lining arterial streets for 10 miles or more. Deplored as eyesores, people still use them

American Italianate, with a foyer to die for. Beverly Hills? Palm Beach? Not a chance: Try Laredo, Texas. In the driveway, for proof, there's a Chevy Suburban, the Texas station wagon.

because they're convenient. With the passage of decades and the construction of new highways, however, many have become run-down, with cheap motels and hubcap shops.

By the 1920s, developers were experimenting with shopping centers, characterized by single ownership, architectural unity, and on-site parking. These centers are usually divided by size into three classes. The smallest is the neighborhood center. It serves at least 2,500 people, occupies 3 to 10 acres, and has a supermarket and half a dozen shops. The community center comes next, serving at least 40,000 people, occupying 10–30 acres, and having a wider range of stores but no full-line department store. Then there's the regional center. It serves at least 150,000 people, exceeds 30 acres, and has at least one department store. There's a subcategory here, the superregional mall, with three or more department stores.

Another way of measuring these centers is by gross leasable area, with the neighborhood center typically having 50,000 square feet, the community center having 150,000, and the regional center having 400,000. An alternative definition of a superregional center is that, besides having at least three department stores, it has at least 750,000 square feet of gross leasable area. These numbers are important because shopping centers generally have four to five parking spaces for every 1,000 square feet of gross leasable area. According to this parking index, a center with 750,000 square feet should have about 3,500 parking spaces. (The place to go for more is the Urban Land Institute's *Shopping Center Development Handbook*, 3rd edition, 1999.)

A final entrant into this taxonomy is the power center. Like the line of shops along arterials, the power center clings to a road, but in this case the road is an interstate freeway and the shops are big-box merchants. This is the empire of the national accounts: Home Depot, Barnes & Noble, and Bed Bath & Beyond. It's the land of megaplex theaters, which now account for about half of the 40,000 movie screens in the United States. (Remember the old multiplex theaters, with only half a dozen screens? More than 1,000 closed in 1999.) Power centers tend toward spartan simplicity, but planning agencies sometimes impose amenity requirements on them. Down in the Dallas suburb of Flower Mound, just northeast of Dallas/Fort Worth, Home Depot wanted to build a store in the Lakeside Business District. Permission was granted, but the company was told to forget orange. The sign had to be bronze, the walls had to be limestone, and the building had to have windows. The goods had to be inside, not piled outdoors, and there had to be trees in the parking lot. Home Depot said yes.

The archaeology of the shopping center reaches back to Baltimore, where in 1907 Edward Bouton created Roland Park, a subdivision that included a shopping district complete with a unified architectural scheme and parking for carriages. Bouton hired Frederick Law Olmsted, Jr., to design part of Roland Park, and such high standards inspired J.C. Nichols of Kansas City, who went on in 1922 to develop that city's Country Club Plaza. It spreads over an existing street network but to this day boasts the Spanish motif that Nichols borrowed from Seville, right down to a copy of that city's icon, the Giralda—a minaret turned cathedral spire.

Dallas's Highland Park came next. The residential portion was designed in 1907 by Wilbur David Cook, who had designed Beverly Hills, but the shopping center came later, in 1931. Like Country Club Plaza, Highland Park Shopping Village adopted a Spanish style. Unlike it, Highland Park provided off-street parking for stores that faced parking lots, not city streets. It was innovative enough that in 2000 Highland Park Shopping Village became a National Historic Landmark.

The demand for centers was stifled by the Depression and World War II, but then the lid came off. Los Angeles's Broadway–Crenshaw Center opened on 13 acres in 1947. Three years later, Northgate opened in Seattle and boasted a mall. The term comes from London's Pall Mall, a promenade near Buckingham Palace; under Northgate's mall, however, there was a tunnel through which freight deliveries could be made. Some other centers, like San Francisco's Stonestown, adopted this idea, but freight tunnels were expensive, and developers soon dropped them.

In 1952, the City of Lakewood's Lakewood Mall opened south of Los Angeles. It had no tunnel, but it covered 154 acres and had parking for 12,000 cars. Two years later, the world's biggest center was Northland, outside Detroit. Shoppers still had to walk outside to get from store to store, but that ended with Southdale, a center that opened in 1956 outside Minneapolis and was the nation's first fully enclosed, climate-controlled center.

Traditionally, the mall has been anchored by department stores that are expected to bring in the traffic, while intervening inline stores pay for the mall through high rents. The department stores are given long leases—typically a minimum of 20 years—while small tenants get leases no longer than 10, with rents usually set as a percentage of gross sales, with a guaranteed minimum. To verify sales, the mall owner has full access to the tenant's books. Generally, rents for the small tenants are set so high that about one-tenth of the stores fail annually. That's the way the mall owners want it: They want to keep shoppers trekking between anchors, and to do that they need a changing smorgasbord of retail diversions. That was the theory of the mall, at least, and it worked until people grew tired of department stores.

Many of America's 1,800 regional malls are owned by a handful of families. With 172, Simon Properties of Indianapolis comes first. In 1996 it absorbed the properties of "the king of malls," the pioneer regional-center developer Edward DeBartolo. More recently it's tried to take over the 20 upscale malls owned by the Taubmans of Bloomfield Hills (Michigan). The closest rival of the Simons, however, is General Growth Properties, which owns 160 regional malls and is controlled by the Bucksbaum family of Chicago. One tiny example of the difference between the companies: Simon rents corridor space to vendors who sell from carts. Taubman doesn't. It misses out on rent for the cart space but says that its lessees pay premium rents in exchange for the tonier atmosphere.

Simon also controls the biggest of them all, the Mall of America, which opened in 1992 in suburban Minneapolis. This giant, with 2.5 million square feet of leasable space, has 12,500 parking spaces, 520 stores, an occupancy rate of 99%, and 43 million visitors annually. Forty percent of them have traveled more than 150 miles, but the mall attracts locals, too, especially in winter. It includes an amusement park, an aquarium, and concerts. "What else are you going to do?" locals ask. "This is Minnesota." Appropriately, the mall was developed by a Canadian firm that had already built the even bigger West Edmonton Mall. That behemoth has 800 stores, a dolphin lagoon, the world's biggest indoor wave pool, indoor bungee jumping, and a full-scale replica of Columbus' *Santa Maria*, in case you're looking to get married someplace unusual. It also has 20,000 parking spaces. Disney World, just so you don't get carried away by such numbers, has 46,000.

Competition between malls is intense. A textbook example comes from North Dallas, which has long been served by Northpark and the Galleria, both developed by Texans. In 2000, General Growth Properties opened the 1.3 million-square-foot Stonebriar Centre in Frisco; it's anchored by a Nordstrom. Five miles to the south, in Plano, Taubman spent $300 million developing The Shops at Willow Bend, which opened the same year. Taubman followed its usual high-end strategy, and the initial anchors included Lord & Taylor and Neiman Marcus.

Within 7 miles of Stonebriar there are 770,000 people with a median family income of

$70,000. Some of these people, however, overlap with the 444,000 people within the 7-mile circle for Willow Bend. Malls usually draw most of their shoppers from within an 8-mile circle, so the immediate question was whether both Stonebriar and Willow Bend could survive. They weren't far from Prestonwood Center, either. That mall in the 1970s had had both Neiman Marcus and Lord & Taylor, but the opening of the Galleria had slowly killed it.

Stonebriar and Willow Bend would compete not only with each other and with Northpark and the Galleria but with an open-air shopping center at Preston Road and Park Boulevard. Many of its customers liked being able to shop without having to enter a mall, and in an effort to draw such tough customers, Willow Bend was designed without an atrium. Instead, it resembles a hotel lobby furnished with arts-and-crafts-style furniture. Taubman chose shops to attract visitors and was in no hurry to lease space. Even the center's name, which makes no mention of mall, suggests an effort to replace a tired image with the more attractive one of a traditional town center.

The result? The Galleria, which had been built and opened by the Hines family in 1982, was sold at the end of 2002 to Europe's UBS Realty for about $300 million. It will be managed by General Growth, which is spending $25 million remodeling it. Northpark meanwhile announced a $200 million upgrade and expansion. Willow Bend, however, struggles with a 72% occupancy rate. In the first 3 months of 2003, it lost 14 shops. "This is not going to be a Rodeo Drive environment," one observer explained, referring to the very upmarket neighborhood in Beverly Hills. Another analyst explains that shoppers want entertainment. Chastened, Willow Bend whispered that it would rent space to doctors and dentists.

With a carnival atmosphere, Stonebriar on the other hand is a huge success, not only for its developers but for Frisco, whose population has skyrocketed over the last decade from 6,000 to 60,000. Back in 1996, the city worked hard to get Stonebriar—competed for it against Plano and wound up offering a set of incentives worth six or seven times the city's entire annual budget. The payoff came with spectacular growth, not just for the mall but for peripheral stores. Collectively, these merchants constitute the biggest concentration of retailing anywhere in Texas.

The former Frisco mayor has credited the city manager for getting the mall. "If George was nervous," he has said, "you couldn't tell it. This was his deal; I knew it was in good hands. He kept the council informed, but mainly what we tried to do was stay out of his way." George Purefoy says, "It was our moment to win or lose the thing. We put our best deal on the table. We knew that was a deal we couldn't afford to lose." The vice president at General Growth says, "To be honest with you, it was very close. There were a lot of reasons for us to build that mall in Plano. . . . There was some indecision on our part. But in the end, George was very persuasive."

Meanwhile, the huge building at the old Prestonwood center sat vacant. For a while the owners, a subsidiary of Goldman Sachs, thought about making the building high-tech office space, but the telecom crash put an end to that. Finally, in 2004, Archon Group decided to knock it down and build an open-air shopping center. The demographics were apparently irresistible: within 3 miles there was a resident population of 130,000 people and a whopping additional 185,000 commuters. Who would rent space? Early rumors included a Wal-Mart Supercenter.

Like Willow Bend, hundreds of malls across the country are struggling in an overbuilt environment. That's why only two new regional malls opened in the entire country in 2003. Six were scheduled to open in 2004 and five in 2005. Of those 13, one was in the Bronx. One was in Des Moines. The rest were either in California, with three, or the South, with eight.

New malls, along with many more that are extensively renovated, often strive to be lifestyle centers. (That name comes from the improbable but very real International Council of Shopping Centers.) Seattle's University Village is an example. So is Costa Mesa's The Camp, which specializes in recreational sporting goods and even displays surfboards on water. The old Sherman Oaks Galleria, also in Southern California, was renovated as a lifestyle center. When it reopened in 2001, the old mall was gone and 1.25 million square feet of rentable space had been reduced to 700,000 square feet laid out like the downtown of a small city, with squares, restaurants, and performing spaces. What had been a Robinson–Mays department store was now Warner Brothers office space.

The Congress for the New Urbanism estimates that over 100 regional malls could be similarly redeveloped as town centers, with housing as well as shops and public spaces. Housing? You bet. The Villa Italia mall in Lakewood, on Denver's west side, was the biggest mall west of Chicago when it opened in the 1960s. It did poorly, however, largely because it wasn't on the freeway network, and it finally closed in 2002. Now it's been torn down and in its place there's a mixed-use development with 1,300 residential units—a mix of rental apartments, condominiums, townhouses, and lofts. Welcome to Belmar. (The name comes from the residential estate here in premall days.) Belmar is supposed to give Lakewood something it never had before: a downtown, with a plaza, a village green, and shops including a 16-screen multiplex.

Simon is embarking on a hybrid between the lifestyle center and the power center. Firewheel Center in Garland, on the west side of Dallas, will have trees and awnings but no roof, and you'll be able to drive right up to the stores, although there will also be garages for peak shopping times. Simon says it hopes to create a retail village with a nostalgic atmosphere. The implication of spectacle isn't overdrawn: a partner in California's upscale South Coast Plaza says that his mall is now a shopping resort, and pedestrians strolling around nearby Fashion Island tell reporters, "This is the only place we like to walk around."

In malls, too, architects have veered away from modernism. In the 1950s, strip malls were designed to look like the boxes they were. Now, in a kind of fakery that would have astounded and infuriated the pioneers of modernist architecture, commercial facades are changed every few years. It's advertising, a way of grabbing attention. For a while, Sonic drive-ins were dressed in red plastic triangles that evoked the family farm. It was homey, but now it's gone, replaced by a retro 1950s look. A third costume is meanwhile making its debut. Chili's has similarly experimented with a building that looks like a huge red pepper. It proved too expensive and never made it past the prototype at Aurora, Colorado, one of the chain's 800-plus locations. Such commercial designs often look crude up close, but they are meant to be seen by traffic moving quickly, when drivers don't notice how the detailing has been simplified, like Disney cartoon characters with four fingers instead of five.

Like advertisers, commercial developers sometimes know us better than we know ourselves. The Mall of Georgia is in Mill Creek, 35 miles northeast of Atlanta. It's a Simon three-level mall with 1.7 million square feet of leasable space. It cost $325 million. The mall also has five courtyards themed to suggest Georgia's geographic regions. They surround a new outdoor village, with four blocks around an imaginary Georgia town square with shops and dining. Like the Shops at Willow Bend, this is a mall for people tired of malls. Within the first 2 months of mall operation in 1999, 2.5 million people came to check it out.

In this shopper's paradise, it's easy to forget that most American manufacturing and much of its office space has also moved out to the suburbs. This is a story with deep roots, going back to England's Trafford Park, a 1,200-acre industrial estate built in 1800 along the Manchester Ship Canal. The 54-mile-long canal had been dug so oceangoing ships could sail to Manchester. Among the 20th-century occupants of this pioneering industrial park were Westinghouse Electric and Ford Motor, which opened its first foreign plant here. Procter & Gamble came; so did Kellogg. Eventually, heavy industry left the area, but since the 1980s Trafford Park has been redeveloped with 1,000 companies and almost 30,000 jobs. Another pioneering industrial park in England was Slough Estates, originally covering 600 acres just west of London. The estate was created in 1920 on the site of a World War I military depot, but the company that took over the depot now houses businesses that occupy over 7 million square feet.

Planned industrial districts came to the United States with two Chicago developments, the Clearing Industrial District of 1900 and the Original East District of 1902, which adjoined the Union Stockyards. The earliest occupant of the 530-acre Clearing district was the Corn Products Refining Company, which settled here in 1907 to make Mazola, Argo corn starch, and Karo syrup. By 1936 there were 111 factories in the District, and in 1940 the district was expanded to include an additional 1,300 acres divided in 40-acre superblocks to minimize traffic delays. The Original East District meanwhile provided rail sidings to each property, along with centrally generated electricity. Its Pershing Road extension of 1916 offered underground tunnels so railway traffic could be completely separated from road traffic. Factories in the extension also faced McKinley Park, a green space running the length of the development.

Railways played a major role in the development of industrial districts in the United States. A good example is the Los Angeles Central Manufacturing District, which the Santa Fe created on 280 acres of what became the industrial city of Vernon. The District was later expanded to 3,600 acres and for a time was thick with slaughterhouses. Between the wars, many other manufacturing companies located here, including U.S. Steel, Bethlehem Steel, Alcoa, Owens glass, American Can, and Studebaker. Most of these companies are gone now, but like Manchester's Trafford Park, the Los Angeles Central Manufacturing District has been renovated, in its case for use by companies working with fabric, film, and electronics.

As the American economy shifted away from heavy industry, new kinds of industrial parks appeared. Often, they look like college campuses. This is no accident. Perhaps the key technological innovation here was the development in 1912 of the vacuum tube by Stanford University's Lee de Forest. Forty years later a Stanford provost, Frederick Terman, pushed hard to build partnerships between the electronics industry and the university. The university had land—8,000 acres of it—and on part of it Terman set up an industrial park for electronics firms. The first occupant was Varian Associates, which arrived in 1953. A decade later, Stanford Industrial Park had 42 companies. In 1974 it became the Stanford Research Park, and today it has 150 companies on 700 acres. Stanford set the standard for university research parks around the world. North Carolina's Research Triangle Park was established very shortly after Stanford's, but it began to grow quickly only in 1965, with the arrival of IBM, which is still the largest employer of the park's 100 companies. The American Association of University Research Parks has 230 members, and there are many parks overseas. An early one was the Cambridge Science Park, created in 1970 and now housing 66 companies.

The campus-like design spread to private office parks, too. A good example is Las Colinas, a 12,000-acre development between Dallas and Dallas/Fort Worth. A hundred thousand people work at Los Colinas for companies like Abbot Laboratories, Microsoft, Citigroup, Nokia, and Verizon. There's a total of 20 million square feet of office space here, including room for ExxonMobil's world headquarters. Las Colinas is exceptionally large, but there are many other business parks, often on prime real estate at the intersection of interstate freeways. There, they are part of what Joel Garreau wrote about in *Edge City: Life on the New Frontier* (1991). They become part of the self-contained quasi-cities that have arisen on the periphery of metropolitan areas so large that the commute downtown is a nightmare. (For more on industrial parks, see the Urban Land Institute's *Business and Industrial Park Development Handbook*, 1988.)

THE NEW URBANISM

The suburbs were conceived as a combination of the best of two worlds, but they failed to deliver. That's a very sweeping judgment, but consider Irvine, California. It has a Buddhist temple, a Korean church, Jewish and Islamic schools, and a Greek Orthodox church. Maybe it's proof of religiosity, but maybe it's the result of a nodal position in the Southern California freeway network, combined with people desperate for social contact. After school, the kids at Colorado's Columbine High School went to the Southwest Plaza, with a video arcade. That was it. They may have enjoyed high standards of housing, but that's not enough. And it's not just the young who are in trouble. The suburbs deny access to anyone without a car. There's no public space, no street life, no accessible public activities—often enough, not even a sidewalk.

Not that having a car solves your problems. Average salaries in Research Triangle Park are twice the North Carolina average, but there's no residential space, and there are few buses. The result is an onerous commute, never anticipated by the developers of what was supposed to be a model development. Similar stories can be told about Columbia, Maryland, a planned community whose planners forgot that people age. Eventually, they don't want or can't handle a single-family home in a low-density, single-use neighborhood.

Such criticisms of suburbia underlie the so-called New Urbanism. It's really not so new. Ebenezer Howard's *Garden Cities of To-morrow* (1902) led to several experiments in urban design, famously including Letchworth and Welwyn Garden City in England and Radburn in New Jersey. Howard is also the intellectual father of planned communities like Columbia. Still, the New Urbanism is usually traced only as far back as 1980 and a Florida developer named Robert Davis. His family owned 80 acres of sandy beachfront west of Panama City, in the Florida Panhandle. Working with architects Andres Duany and Elizabeth Plater-Zyberk, Davis platted Seaside, a tiny subdivision—it has only 280 homes—that 20 years later stands as the original urban village.

Seaside is small enough that everyone can walk to the town center in 5 minutes. (That's a radius of about 2,000 feet). There's a mix of housing types. There are shops and offices. Traffic is deliberately slowed. Buildings are close to the street. The houses are of wood, with large enclosed porches, and they are wrapped (almost hidden these days) by native trees that have not yet reached mature heights. (Seaside was the set for *The Truman Show*, but the film's designers surrounded Truman's house with lawn, which is prohibited in ecoconscious Seaside.)

Seaside is now a tourist destination in its own right. Next door, St. Joe, a one-time Florida timber company and paper manufacturer, is building a 500-acre development called WaterColor. The aim is to preserve and extend the distinctive character of Seaside. Houses in WaterColor, according to promotional literature, have a "Cracker" look, with corrugated-metal roofs and rough-sawn planking. Some 15,000 bales of pine needles are sprinkled on the site each year to create a suitably rustic atmosphere.

Proponents of the New Urbanism praise these places as humane alternatives to mega-

The best community that money can buy: Seaside, Florida.

lopolis. They say they're especially good for children. But there's a snag. Only one-tenth of Seaside's homeowners live in Seaside year-round, and fully half the homes are on short-term rentals to vacationers. This goes a long way toward undermining the sense of community that is the Holy Grail of the New Urbanism. But jobs are scarce in Seaside, and its homes are too expensive for people on local salaries. A pleasant 2,000-square-foot house, for example, runs close to $900,000. Even a plot in the Seaside cemetery sells for $15,000, while the modest bungalows in WaterColor start at $750,000 and climb to $3 million. These are Cracker-style homes that no Cracker could afford.

Still, around the United States now there are some 400 Seaside imitators. They're in every region except the Great Plains and northern New England. Even Disney has tried its hand with Celebration, a modestly named town near Orlando that's planned for 20,000 people. Five thousand are there already.

Some urban villages have been built in cities. Washington Harbour, for example, is in Georgetown, along the Potomac. It's a mixed-use complex developed by Shorenstein Company, a San Francisco-based developer. (The British "u" in Harbour is complemented by the company's description of the central building as graced by "grande" columns. The extra letters exude leisured spaciousness, as though there's no need to rush anything, including spelling.) Boca Raton's Mizner Park was built in 1989 on the site of a failed mall, and it provides a combination of housing, shops, and offices.

Out in Southern California, Paseo Colorado stands on the site of the old Plaza Pasadena, an enclosed downtown mall built in 1980. The mall was demolished in 2000. In its place, three blocks of double-decked shops have been developed by Trizec. The New York firm sold the air rights over the stores to Post Properties of Atlanta, which built some 400 apartments on top of the stores. Rents are high, although not exorbitant by Los Angeles standards: apartments sized at 500–2,000 square feet rent for $1,500–$4,000 a month.

A much bigger development is going in between the Los Angeles International Airport and neighboring Venice. Playa Vista promises to house somewhere between 15,000 and 30,000 people, depending on how much wetland is preserved on the 1,100-acre site. Still another development, Long Beach City Place, is replacing the Long Beach Plaza Mall. Streets are being restored on the 12-acre site, and 258 apartments are being included. In Seattle, Paul Allen, one of the founders of Microsoft, plans to convert 50 industrial acres near South Lake Union to a mix of offices, shops, and apartments. Surface parking will disappear, replaced by a 1,000-car underground garage. McKinney, Texas, has plans for an expanded urban village that will house 25,000 people living and working in an integrated community. In Dallas itself, there's the West Village Complex on McKinney Avenue.

Even the suburbs are seeing an integration of land uses, so that residents no longer have to live in their cars. It's not exactly the New Urbanism, because nobody's talking about walking everywhere. Still, it's a step in that direction. The so-called Inland Empire of San Bernardino and Riverside counties, for example, has long been the home of long-distance commuters, but it gained almost 350,000 jobs between 1990 and 2002, while Los Angeles lost over 100,000. The people out there are finding jobs closer to home.

Another variation is emerging around the light-rail systems recently built in Los Angeles and San Diego. Los Angeles allows builders to increase residential densities by 35% in projects within 1,500 feet of a station. Parking requirements are reduced, too. House prices aren't any cheaper because special construction methods are needed to stand up to vibration and insulate against noise. Backyards shrink. But residents of these transit villages can take the train to work, while the village itself provides basic shopping requirements, so residents don't have to drive everyday. For many people, it's an attractive proposition. For Southern California as a whole, more such places are probably inevitable as commuting distances to conventional suburbs become unbearable.

ASSESSING THE AMERICAN CITY

A good case can be made that the exclusivity accompanying high prices is just fine with the residents of these new developments. Like most Americans, they think that downtown is dangerous and want no part of it or its people. When Footlocker and Walgreens close at night in Oakland, California, the manager doesn't just lock doors: He brings down metal gates. The city manager doesn't approve. He says: "It sends a message." He's happier with the local Gap store, which installed polycarbonate glass windows able to resist baseball bats. Oakland planners have lots of other ideas. Instead of razor wire around parking lots, they recommend cyclone fencing with a mesh too small to climb. Residents have a different idea. "Don't just make it *look* safer," one says, "*make* it safer." Easier said than done. Making it safer depends on finding jobs for all the people kept out by the gates and razor wire.

We're back now to the inner city and the intractable issue of racial segregation. Whites constitute only 44% of the population of America's 100 biggest cities, and they're a minority in 48 of them. In 1960, the Long Island Levittown had 82,000 people, and not one was black. Levitt explained that the country could build houses or try to solve its racial problems, but it couldn't do both. When plans to doubledeck the Long Beach freeway were unveiled in 2003, it turned out that about 800 residential buildings were under the enlarged road's footprint. Of the 10,800 people in the project area, fully 10,070 were minorities—mostly Latino or black. More surprisingly, three-quarters of the students at Hollywood

High qualify for free lunches. That's because kids from the wealthy parts of Hollywood—the hills, basically—are in private schools. Maybe Hollywood High is supposed to be weird, but 84% of the children in plain-vanilla Dallas's public elementary schools qualify for free lunches. Talk to a real estate developer in Oklahoma City, and you'll quickly see how bad things are. He'll say that he's on the lookout for property anywhere in the metro except inside the Oklahoma City school district. He can't sell houses inside the district, he'll say, because potential buyers believe that they'll be unable to resell the property.

There's no ocean to protect Americans from each other, so they buy plenty of guns and retreat to gated communities. These places are part of a broad trend that probably began with zoning laws. Zoning became popular about 1920 as a way of separating both classes and races. Public transportation systems were then laid out to avoid wealthy neighborhoods. Ever try taking the D.C. subway to Georgetown? It's no accident that it doesn't go there.

Gated communities themselves began with retirement communities in Arizona. They then spread widely, especially in California and Florida but also to Houston, Chicago, and New York. "Forting up" is the phrase Edward Blakely and Mary Gail Snyder use in *Fortress America* (1999). Styles are evolving to match. Windows are shrinking, especially in top-end houses with lots of art to hang. Turrets are back, with their impression of fortress-like security. Houses are increasingly hidden behind high walls—commonplace in Manila, where the poor are a perennial threat to the rich, but new in America.

Public housing and urban renewal didn't eliminate urban poverty. Some cities have tried rent controls. They're popular in New York City, where they are a relic of World War II. They have the unintended consequence, however, of forever delaying maintenance by owners who have no incentive to invest in their properties.

Cities have also tried inclusionary zoning, where developers are required or encouraged to include a number of housing units that are priced below market rates. Cambridge, Massachusetts, has such a program. Encouragement comes in the form of a density bonus, where the developer is allowed to build additional market-rate units in exchange for providing affordable ones.

There are bizarre variants. Housing is so expensive in California that school districts and state colleges provide subsidized housing to teachers and faculty. In Aspen, Colorado, $1 million buys a fixer-upper, and the official poverty line for a family of four is about $150,000 a year. Most of the people who can afford to buy in the city do so with cash but then live there only seasonally. To avoid becoming a millionaire ghost town, Aspen now buys and resells houses to families with incomes under $118,000. Developers are forced to set aside a majority of their units for this affordable housing program. When participants leave town, they can't sell their property on the open market; instead they are limited to making a small profit, while the unit goes back into the affordable-housing pool.

Americans in recent years have often been inclined to say that the country can do no better than rely on the marketplace. Go out to California and you'll see the results. Watsonville is now 75% Hispanic. Cheap motel rooms house a dozen farm laborers. An organizer for Teamsters Local 890, which represents lettuce workers, says that if you look around the east side of Salinas, "I'm willing to bet you money you can't find a garage with a car in it." She means that it's a bedroom. Is this a spur to government action? Not a chance: In 1979 the federal government spent $70 million on low-interest loans for farmworker housing; today it's spending $28 million. About 1 million of those workers are probably illegal, but another million aren't.

When the U.S. Army abandoned Fort Ord in 1994, it walked away from 45 square miles of prime California property near Monterey. The base included thousands of houses that are still empty. There's a huge demand for them, because real estate in this part of California is sky-high. At a public meeting, the local congressman talks about off-base house prices and says, "There isn't a single person in this room who could afford to buy the house he's living in, including me." But there are heavy pressures to knock the military housing down and replace it with expensive homes. A journalist puts it succinctly: "The newly arrived wealthy assume that living on the coast is their privilege. The laborers believe that being pushed elsewhere is their lot."

Or consider San Francisco's famous Chinatown. It's a major tourist attraction, but look above the street level with its tourist-oriented restaurants and shops. You'll see the city's most abominable housing—three generations sharing one 10-by-10 room, 60 people sharing one vermin-infested kitchen and bath. It's the same story in New York City's Chinatown, where population densities—190 per acre—are more than twice those of Manhattan as a whole. Out on Long Island, 100,000 Salvadorans pay top-dollar rents for miserable accommodation. Many are in the United States illegally, and they dare not complain.

What follows? If you're patient enough to wait for decades, you'll see that many of these minorities do well for themselves, just as so many earlier arrivals did. A third of the Latinos who came to the United States in the 1980s, for example, were living in poverty in 1990, but by 2000 the figure had declined to 23%. For Latino immigrants who came in the 1970s, the poverty rate in 2000 was 17%.

If you want to live in this life, not the next, and if you don't want poor neighbors, you'll likely join the tide of Americans moving away from the central city and hoping for safety on the fringe. Irvine, perhaps the country's biggest master-planned development, markets itself as the American Singapore and uses computers to control access to office-building restrooms. You can't be too safe. North Dallas grows fat with suburbs stretching out beyond Plano, while south of the Trinity River whites constitute no more than 12% of the population. Down there, you won't find a single mall or megaplex or university; the only growing commercial center in the southern part of the city is at the junction of highways 67 and 1382 in Cedar Hill. What do you think you'll find there? One corner has The Market at Cedar Hill, with Kroger's, Blockbuster, and Subway. Across the street is the Plaza at Cedar Hill, with Old Navy and Barnes & Noble. A third corner has Cedar Hill Crossing, with Home Depot, Staples, Starbucks, and Kohl's—pioneer of the freestanding suburban department store. The last corner has the new Cedar Hill Village, which has a Wal-Mart and will soon have a rare freestanding J.C. Penney. It's also the site of Cedar Hill's proposed new city hall.

Cities can be less generic than this. That's why affluent Americans by the thousand stream to Europe each summer. They marvel at the street life of even a second-string French city like Lille, where block after downtown block is lined with thriving small businesses. Many of the streets have been closed to traffic, but they're crowded with people who aren't driven away even by winter's chill or slippery cobblestones. The place is crowded after dark, although the shops close at 7 and only the many restaurants stay open. Gap and McDonald's invade these neighborhoods but do not destroy them.

A first-rate European city meanwhile just about makes Americans cry. Consider Venice. It has the extraordinary advantage of being built in water, and Americans wouldn't stand for having to cart their groceries home over rough pavements and bridges with stairs. They definitely wouldn't like living in the grim apartment buildings that most Venetians call home. But Americans like to think that their cities are pretty good places, and Venice takes a baseball bat to that pride. It forces Americans to acknowledge that motor vehicles have devastated the places most Americans call home. What price efficiency? Venice whispers that Americans pay much too high a price. Some will disagree. Many like their commute, as long as the traffic moves smoothly. Still, Americans go to Europe. There's something in those walkable cities that draws them. To a large degree it's the fact that these cities are shaped to serve people, not cars.

Instead of building pedestrian-friendly cities, Americans battle developers. Disney's plan for a national historic park in Prince William County, 35 miles from D.C., was scotched in the early 1990s by a campaign led by the Piedmont Environmental Council. Down in New Orleans, community organizers are suing to keep Wal-Mart from building on the 64-acre site of the former St. Thomas public housing project. Its 1,500 apartments were built about 1940 and knocked down in 2000. The protesters want them replaced with upgraded low-income housing. The mayor wants Wal-Mart, along with 180 public housing apartments, 270 affordable apartments, and 800 market-rate ones. Residents figure they've already lost: One says the displaced residents "have been treated like a crop on the plantation." It's not so different from the experience in Norman, Oklahoma, when Wal-Mart announced its plan to build a supercenter along Interstate 35. The public was overwhelmingly opposed to the deal, but nearby car dealers wanted the traffic. The average car dealership in the United States has profits exceeding $600,000 a year. Guess who won.

Compare that with Canada, where people are willing to limit urban growth in ways that Americans can scarcely comprehend. Vancouver has an Urban Design Panel that suggests design revisions. Going to the Development Permit Board without the panel's blessing is a recipe for permit denial. Why do Vancouverites accept such controls? The city's director of planning explains that development is a privilege, not a right. Good luck to the American city planner who tries that line. (For more on Vancouver, see John Punter's *The Vancouver Achievement*, 2003.)

In the United States, the best results come when a strong planning department finds common ground with the business community. Through such an alliance, Lawrence, Kansas, has preserved a thriving downtown. Or compare Rosslyn and Ballston, two towns on the Washington, D.C., subway system. Ballston insisted on underground garages, then encouraged dense development on the surface. Now there's life in the previously decaying town. The National Science Foundation is here; so is E-Trade; so is the Nature Conservancy. There's a Hilton Hotel. There are apartment buildings and lots of restaurants. There's a shopping center called (inevitably) Ballston Common. There are people around at night.

Then there's Rosslyn, the subway's first stop in Virginia. The office buildings empty out and leave behind a concrete desert. Because it's so close to Washington, Rosslyn has several big hotels, but there's nowhere for the people staying in them to walk at night. The developers may have made as much money here as at Ballston, but for the people who work in Rosslyn there's no hope of avoiding a commute. For the people in the hotels, the choice after dark is curling up with a book, watching TV, or climbing on the Georgetown shuttle.

CHAPTER 20

RURAL AMERICA

Rural America was once a place where men farmed, ranched, or worked in forests or mines. Those resources are still being produced—you have to hunt for the ones, like sheep, that aren't—but the jobs are evaporating, and rural communities have to find another mainstay. Many, especially on the Great Plains, won't. Others, particularly those near coasts or mountains, will not only find them but will prosper more than ever, with economies based not on resource production but on the recreational or esthetic value of nature and, to some extent, of America's cultural landscapes. You could say that the old Baconian drive to control the environment has paid off and that people can now enjoy the world instead of struggling to survive in it, but many of the people in rural America today are newcomers, city people looking for what they could not find even in the suburbs.

CHANGING POPULATION PATTERNS

Academics are fond of denying the distinction between urban and rural America. Fly at night from Boston to Washington, they'll say, and there's an unending quilt of lights.

They're right, at least about the lights. We've popularly recognized this emerging supercity since 1961. That's when Jean Gottman published *Megalopolis*, a book about America's northeastern conurbation. "Conurbation" had been coined by Patrick Geddes, a British biologist, sociologist, and town planner. In *Cities in Evolution* (1915), Geddes had described the emergence of supercities formed when existing ones grow so large that they fuse with one another. Geddes was thinking of half a dozen places in the United Kingdom, starting with Greater London, but he foresaw conurbations taking shape in the United States, too. He writes of "practically one vast city-line along the Atlantic Coast for five hundred miles." You can't get a clearer anticipation of megalopolis than that.

Today, the northeastern megalopolis has company. Southern California is the obvious contender, but you could argue for the Interstate 85 corridor between Atlanta and Richmond. A half million skilled manufacturing jobs have been created here since 1990, and there's hardly any open countryside along the route. Greenville, Charlotte, and Durham have almost fused.

Still, anyone who looks out airplane windows knows that a lot of the United States is mighty empty. Fully half of all Americans live in counties within 50 miles of the sea, if you count the Great Lakes as sea. The other half are spread out thin or thinner. East of the hun-

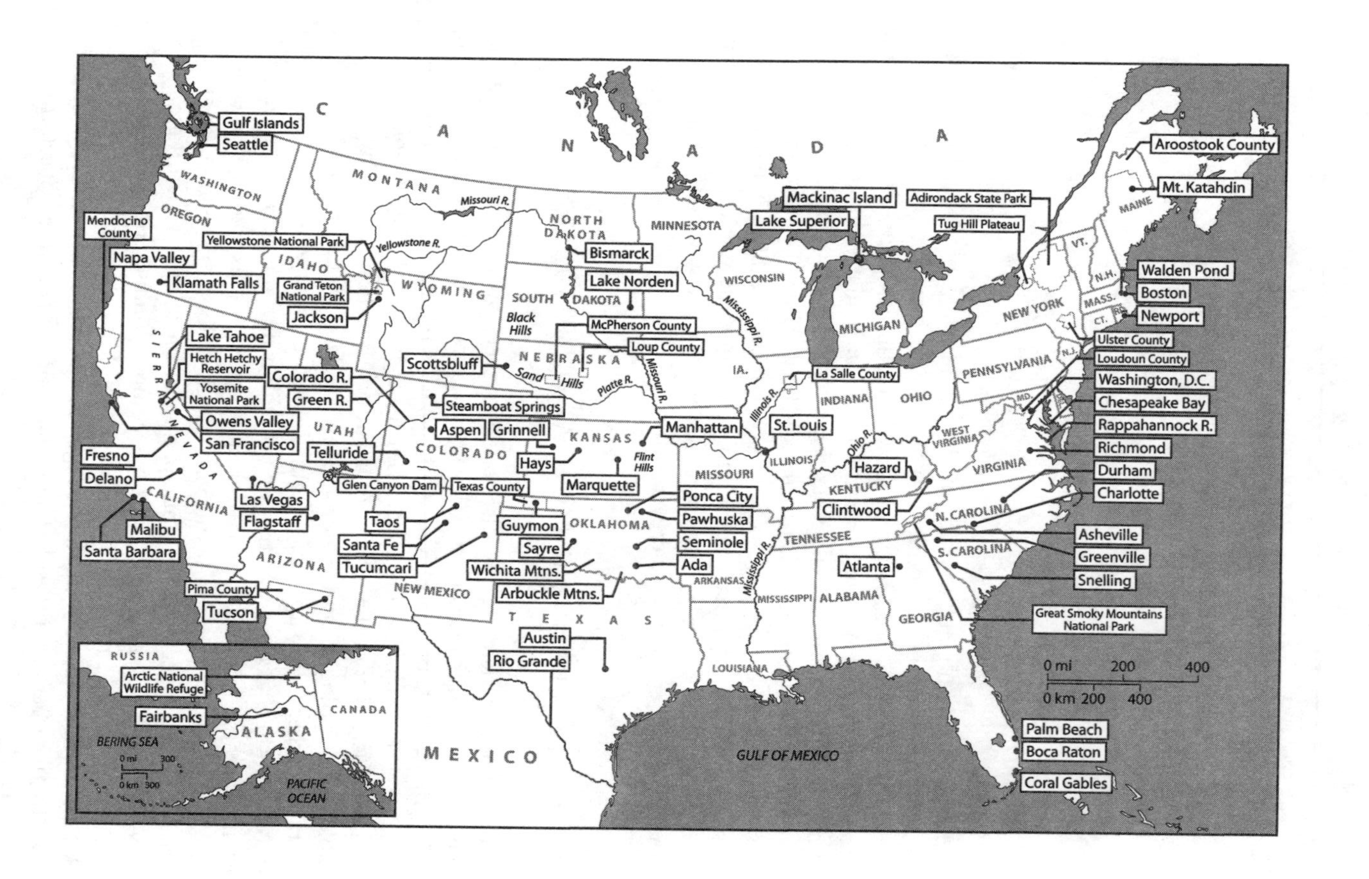

CANADA
Gulf Islands
Seattle
WASHINGTON
OREGON
Mendocino County
Napa Valley
Klamath Falls
Yellowstone National Park
IDAHO
Grand Teton National Park
Jackson
MONTANA
Missouri R.
Yellowstone R.
WYOMING
NORTH DAKOTA
Bismarck
Lake Norden
SOUTH DAKOTA
Black Hills
McPherson County
Loup County
NEBRASKA
Sand Hills
Platte R.
Missouri R.
MINNESOTA
WISCONSIN
Mississippi R.
Mackinac Island
Lake Superior
Adirondack State Park
Tug Hill Plateau
Aroostook County
Mt. Katahdin
MAINE
VT.
N.H.
MASS.
CT.
R.I.
NEW YORK
Walden Pond
Boston
Newport
MICHIGAN
IA.
SIERRA NEVADA
Lake Tahoe
Hetch Hetchy Reservoir
Yosemite National Park
Owens Valley
San Francisco
Colorado R.
Green R.
UTAH
Scottsbluff
Steamboat Springs
Aspen
Grinnell
COLORADO
KANSAS
Hays
Flint Hills
Marquette
Manhattan
Illinois R.
La Salle County
INDIANA
OHIO
PENNSYLVANIA
N.J.
MD.
Ulster County
Loudoun County
Washington, D.C.
Chesapeake Bay
Rappahannock R.
Richmond
Durham
Charlotte
WEST VIRGINIA
VIRGINIA
St. Louis
ILLINOIS
Ohio R.
Hazard
KENTUCKY
Clintwood
Fresno
Delano
CALIFORNIA
Malibu
Santa Barbara
Telluride
Glen Canyon Dam
Texas County
Las Vegas
Flagstaff
Taos
Santa Fe
Tucumcari
ARIZONA
NEW MEXICO
Pima County
Tucson
Guymon
Sayre
Wichita Mtns.
Arbuckle Mtns.
OKLAHOMA
MISSOURI
Ponca City
Pawhuska
Seminole
Ada
ARKANSAS
Mississippi R.
TENNESSEE
N. CAROLINA
S. CAROLINA
Asheville
Greenville
Snelling
Atlanta
MISSISSIPPI
ALABAMA
GEORGIA
Great Smoky Mountains National Park
TEXAS
Austin
Rio Grande
LOUISIANA
MEXICO
GULF OF MEXICO
0 mi 200 400
0 km 200 400
Palm Beach
Boca Raton
Coral Gables
RUSSIA
Arctic National Wildlife Refuge
Fairbanks
ALASKA
CANADA
BERING SEA
0 mi 300
0 km 300
PACIFIC OCEAN

dredth meridian, which corresponds to the eastern edge of the Texas Panhandle, there's usually enough rain to support agriculture without irrigation. (The 98th meridian is a more accurate boundary, but it's rhetorically awkward.) That's why maps showing population distribution in 1900 show a fine spray of population spreading eastward from that meridian and covering everything except northern Maine, the Adirondacks, the Great Smoky Mountains, and the Everglades. They still do. West of the 100th meridian, however, and except for the crowded West Coast and a few big cities like Denver and Salt Lake City, there are only threads of population along rivers like the upper Missouri, Yellowstone, Platte, Rio Grande, Green, and Snake. The rest is empty and in some places getting emptier.

Head west from Oklahoma City on old U.S. 66. With your tires thumping over the expansion joints in the old concrete, you'll pass abandoned fields and pastures, broken every half hour or so by a long-suffering town. Block after block of Tucumcari, New Mexico, is pockmarked with empty lots and old foundations.

The pain is greatest in the great flat middle of the country. Geologically this is the core of the continent, the ancient heartland battered by the immense collisions that created the Appalachians on the east and the Rockies on the west. South of Lake Superior, the ancient rock is buried under sedimentary layers that yield, among other things, the coal of Illinois and the oil and gas of Texas. On top of that layered rock is a veneer of soil that once sustained a grassy hide. A century ago, that soil was the apple of an agrarian nation's eye.

Those days are gone. St. Louis had 846,000 people in 1940. Sixty years later it had 348,000. In 2001, its population was officially estimated at 339,000. It sounds like one of those "last person turn off the lights" jokes, and farther west things just get worse.

During the 1990s, 676 of the nation's 3,141 counties lost population. Many of them were west and north of St. Louis. With declining population comes poverty, which is why 9 of America's 10 poorest counties are either in Nebraska or South Dakota. We can visit one: just go west on Interstate 70 from St. Louis to Grinnell, Kansas, then turn right on U.S. 83 and head up to North Platte, Nebraska. Don't stop: turn north on lonely State Highway 97 and go 20 miles to Tryon, the McPherson County seat.

We're in the Sand Hills now, great ranching country, with waving grass on old, stabilized dunes and plenty of underground water. The 2000 census, however, reported 533 people spread over McPherson County's 859 square miles. That's sparser than Alaska. The median household income was $25,000, compared to about $40,000 for the nation as a whole.

In 1910—I almost want to say "once upon a time"—McPherson County had 2,470 people. They were confident that their county would grow in the century ahead. They never imagined that their numbers would be cut by three-quarters. Anyone foolish enough to have predicted such a thing would have been considered downright inhospitable.

You can't blame the people of McPherson Country. Until the 1960s, most rural Americans saw their future in resource production. It wasn't a bad guess; it was a good one for a long time. It's just not right anymore, not when 1.7% of Westerners make their living that way.

COLD REALITY

We can get awfully sentimental about pioneer life—shades of our ambivalence toward progress—but except at speechifying time Congress has never been fooled. Throughout the 19th century it saw the public lands as investment capital. That's why it gave so much away, espe-

End of the road for a Great Plains town? There's lots more like Ringling, the boyhood home of Chuck Norris.

cially west of the 100th meridian. Homesteaders could file on a quarter section out here, but they could not survive on it unless they had irrigation. Usually they didn't. Ranchers took advantage of the situation and put together small kingdoms by bribing friends and relations—even strangers—to file homestead claims along the length of local watercourses, then deed them to the rancher. Whatever lay between the lines of claims was for all practical purposes the rancher's. The claims were fraudulent, because nobody had any intention of living on them, but this was a mere technicality. The most interesting account of such public-domain frauds is Stephen A. Douglas Puter's *Looters of the Public Domain* (1908). Puter, the "King of the Oregon Land Ring," was in the thick of organizing similarly fraudulent claims to public timberlands. He got caught—perhaps paid one too many schoolteachers to sign an affidavit. The frontispiece of his confessional book shows him at his desk in a prison cell.

Congress gave tens of millions of acres of Washington and Montana as a construction subsidy to the Northern Pacific. The railroad company sold the timberland in its grant to Frederick Weyerhaeuser, whose company owns it to this day. The same thing happened along the lines of the Union Pacific and Santa Fe, though there the buyers were mostly farmers and ranchers. There hasn't been a railroad land grant since Ulysses S. Grant was in the White House, but the grants are still a fact of life in the West. That's because in an effort to reduce the chance that the railroads would jack up the price of the land, the grants consisted only of the odd-numbered sections within a band straddling the track. Sometimes the band was 40 miles wide; in parts of the Northern Pacific grant, it was 80. The even-numbered sections in and west of the Rockies have mostly remained in federal ownership, so the railroad lands survive today as a checkerboard of private holdings complicating the management plans of the Bureau of Land Management and Forest Service. Sometimes the railroads were unable to sell their land, especially in the deserts. That's why the Southern Pacific throughout the 20th century was California's biggest private landowner.

The public lands are no longer given away, but the old view of rural America as a place to grow, cut, or dig things remains powerful. It's not always a matter of corporate barons making fortunes. Wander around in antiquarian bookshops and you occasionally come upon a pair of heavy volumes called *The Apples of New York*. There were published by the state as part of a series that also included volumes on the cherries, grapes, plums, pears, and peaches of New York. It was an agrarian world. Today that world is almost as remote as a Buddhist text written in Pali on palm leaves, but the books—heavy, bound in green buckram, full of handsome plates—fetch several hundred dollars each.

What to do instead of growing, cutting, or digging? Slaughterhouses are one possibility. They used to exist in every big city, but they're gone now, thanks to better transportation. As a result, most American children no longer even recognize that distinctively cloying, sweet, disgusting smell. The packers have moved to the Great Plains to be closer to the feedlots drawn there by the dry climate. It's not just cattle. Texas County, in the Oklahoma Panhandle, had 40,000 hogs in 1990. A decade later it had over 2 million, butchered at a processing plant in Guymon. Seaboard Industries invested $300 million in this operation, which can cut up 1,000 animals an hour. It has competition: Tyson Foods produces a quarter of the nation's beef and chicken, and now it's producing close to 20% of the pork, much of it from a new operation at Seminole, Oklahoma. As a result, trailer parks are growing fast in towns whose population has been static for decades. One side effect: Spanish is now a plus for anyone looking for a job as a salesclerk in these places.

Outside the towns, the population trend is still downhill. Congress continues to support agricultural prices, but this does nothing to stabilize the farm population. Go check La Salle County, on the Illinois River west of Chicago. Some 4,200 farmers there collect subsidies. According to data compiled by the Environmental Working Group, payments to county landowners totalled $136 million over the 1996–2000 period. That's an average of $7,000 annually, but nobody's average. The top 444 farmers in the county collected 51% of all the subsidies, for an average annual payment of $30,000. The bottom 80% collected 29% of the total payout, for an average of $2,000 annually. That's not enough to keep anybody anywhere.

It's a national folly. In 2001 the federal government paid farmers more than $49 billion. (That includes subsidies for dairy, sugar, and peanut producers, as well as payments made to the growers of wheat, corn, cotton, rice, and soybeans.) That's more money than the federal government pumped into elementary and secondary education, and it accounts for half of all the farm income in the country. As La Salle County hints, however, more than half the payments went to the top 10% of the nation's 880,000 participating farms. These 88,000 received payments averaging $40,000 a year.

Such a program will never keep most of America's farmers in business. That's why the Freedom to Farm Act of 1996 was referred to sarcastically as the Freedom to Fail Act. Over the objections of the White House and the Department of Agriculture—and to angry foreign protests—Congress in 2002 roughly doubled the amount of money going into the farm program. Sen. Pat Roberts (R-Kan.) said during early negotiations over the Farm Security and Rural Investment Act of 2002 that big farmers saw the pending legislation as their private ATM. Others observed that the bill would depress market prices by stimulating production. Millions of unsubsidized peasant farmers around the world would be severely hurt. The only defense anyone could offer, besides saying that Congress was properly putting the concerns of American farmers first, was that Europe and Japan were even crazier. Europe paid its farmers $93 billion in 2001. That's why food prices there, as in Japan, are three times what Americans pay.

No matter: the farm security law passed. Two-thirds of the money will go to one-tenth of the nation's farmers. There won't be any food shortage in the United States. There will just be fewer people growing it and collecting larger subsidies. Out on the Great Plains, visitors will continue to see towns with a surplus of old people and some incredibly cheap real estate. McPherson County has 200 households. Forty-five have somebody over 65. Twenty-nine have someone under 18.

Shall the United States try harder to stabilize or increase its rural population? That's what Abraham Lincoln did in 1862 with the Homestead Act, and Marquette, Kansas, is doing it today. It's on a smaller scale now: Instead of 160 acres, newcomers get about half an acre if they build a house on it. In a year the town—population, 600—has given away 20 lots, which means, it hopes, 20 more children in its school.

Marquette's just about in the center of Kansas and about a dozen miles west of Interstate 135. How are the newcomers to make a living? Rural advocacy organizations, like Nebraska's tenacious Center for Rural Affairs, suggest that small businessmen in rural America should be subsidized like farmers. It's hard to argue against that proposal if you accept the wisdom of subsidizing agriculture, and some legislators from the northern plains go further and propose that student loans should be forgiven for college graduates moving to the region. They also suggest that tax credits should be given to home buyers. Logical or not, these proposals are exceedingly unlikely to become law: Their constituency is just too small. Besides, in the optimistic 1960s Congress created an Appalachian Regional Commission to alleviate the notorious poverty in the backcountry of Kentucky and Tennessee. Yet if you define poor counties as those where the percentage of the population living below the poverty line is more than 25% above the national average, then there are more poor counties in Appalachia today than there were 30 years ago. Even with the best of intentions, in other words, the federal government may not know how to improve rural incomes.

There aren't many silver linings. The United States doesn't generate much power from the wind—only a fifth as much as Europe—but it could do much more. We're likely to see more, but the job prospects for wind-turbine maintenance workers are about like those for lighthouse keepers.

The United States could encourage immigrants and ask them to take up intensive farming on the plains. It's not impossible. Davisco Foods, for example, is building a mozzarella plant in Lake Norden, South Dakota. It's a big operation—65,000 cows to help keep Americans in pizza—and the state is trying to entice 150 dairymen from Western Europe. They attend meetings. They ask about the weather and the isolation. Some come.

South Dakota has also worked hard to attract biotechnology firms and has succeeded in attracting one, Hematech. Iowa and Missouri have also tried to play this game, but all these states are vanishingly far behind California, which has 450 publicly traded biotech companies. Hematech employs a few dozen people working on animal cloning and antibodies to cure diseases like anthrax, but there are dozens of biotech companies around San Francisco that each employ over 1,000. There's no reason why South Dakota has to have as many companies as California, of course, but the idea that one company will attract others is wishful thinking. The Howard Hughes Medical Institute is building the Janelia Farm Research Campus about 35 miles southwest of Washington, D.C., and though it will have about 300 scientists and engineers, experts doubt that it will have a snowball effect.

Until recently, poor states hoped that call centers might be their savior. Many welcomed Sykes Enterprises, for example, a company that handles 600,000 calls and messages daily for clients including SBC, Delta, and Microsoft. Sykes expanded quickly from 3 centers in 1994 to 40 by 2003, and it opened them in places like Hays and Manhattan, Kansas; Hazard, Kentucky; Eveleth, Minnesota; Scottsbluff, Nebraska; Minot and Bismarck, North

Dakota; Ponca City and Ada, Oklahoma; and Klamath Falls, Oregon. People in those towns were so hopeful that their depressed economies would begin to grow that they ignored the fact that Sykes was also opening centers in the Philippines and Costa Rica. In 2002, Sykes opened a center in Bangalore. The company's annual report to stockholders for that year speaks of a restructuring plan "to reduce costs and improve operating margins through a reduction of excess seat capacity in the United States and Europe." The centers in Kentucky, Minnesota, Nebraska, and North Dakota had all closed by early 2004.

Small consolation: These places weren't alone in their pain. In 2001 Travelocity opened a call center in tiny Clintwood, Virginia, close to the Kentucky line. Corporate executives said: "We plan to stay and be a part of the community." Less than 3 years later the center closed. The local Chamber of Commerce director says: "I was watching a TV program showing how customer service people in India were being trained to speak with a Southern accent. I didn't realize I was seeing our annihilation." Some of the 250 laid-off workers say they'll reluctantly go back to the coal mining they dread. Not all can. One observer says that, without fundamental change, "the working class will simply be ruined. They'll flip burgers, go on welfare, or sell drugs."

Private prisons are a possibility, of course, and one that the same Chamber of Commerce official calls "not quite as bad as being a nuclear waste dump site." Some 245 private prisons, each a "dumpsite for human misery," opened in the 1990s. They were spread over 212 counties whose population had risen only 1% annually in the 1980s. During the 1990s, the population in those same counties jumped 12%. The Corrections Corporation of America runs one of these prisons at Sayre, on Interstate 40 in western Oklahoma, where it houses 1,400 prisoners from jailed-up Wisconsin. In 1998, when the prison opened, unemployment in the county was cut from 3.2 to 1.7%.

Call it gray gold, but prisons don't always do the trick. Delano, in California's San Joaquin Valley, got a prison whose staff preferred to commute from Bakersfield. It was a second blow, because the San Joaquin Valley recently became the first part of the country to ask the federal government to designate it an area of extreme noncompliance with air-pollution standards. That designation will allow Valley residents more time to fix their problems, which is to say that it will allow developers and farmers to carry on, business as usual. The valley already has worse air than Los Angeles, not only measured by ozone levels but by particulates, mostly stirred up by agricultural operations. It's very abstract unless you were at the 2002 graduation ceremony at Malloch Elementary School in Fresno. Thirty of the 59 graduating sixth graders raised their hands when asked if they used asthma inhalers.

If all else fails, a community can volunteer to become a waste depository. "Whatever you do, don't call it a dump," says an executive of Chem-Nuclear, which runs the biggest nuclear-waste dump in the Eastern United States. It's in Snelling, South Carolina, a town that depends on the dump for 40% of its budget. The mayor says, "We're happy to have this stuff. It doesn't scare us one bit." A worker adds that "around here, this is probably the best job there is." An environmental lawyer counters that "your average county garbage dump has higher standards than this place," but the mayor isn't interested: "I've been drinking the water all my life and I'm still here," he says.

THE NEW WEST

What shall Delano and McPherson County do? You could ask the same question about most of Maine and West Virginia. One answer is to pack it in: Let the huge tractors, giant tree harvesters, and monster draglines produce the resources needed by an urban nation.

Quit struggling. Frank and Deborah Popper, who have made a cottage industry of promoting what they call the Buffalo Commons, know how such talk is received. Gentle souls that they are, on the Great Plains they speak with police protection.

There's another possibility, though. It's compatible with the Buffalo Commons but emphasizes the economic benefits of preserving and restoring natural ecosystems. It rests on the desire of city dwellers to live outside even the penumbra of cities, or to least to spend a lot of time out there. You could say that it's a determination not only to live well beyond that fringe but to live well, period.

This is a quintessentially romantic business, with roots all the way down to James Fenimore Cooper and Henry David Thoreau. Their admiration for Indians—red or white—helped Americans see Daniel Boone, Davy Crockett, and Josiah Gregg as heroes. Gregg is the least known of the three, but his *Commerce of the Prairies* (1844) describes the exhilaration of the Santa Fe trail, across whose 800-mile length Gregg made four round trips in the 1830s. They made him useless, he said, for city life. "I have striven in vain to reconcile myself to the even tenor of civilized life in the United States," Gregg wrote. He explained that "the wild, unsettled and independent life of the Prairie trader makes perfect freedom from nearly every kind of social dependence an absolute necessity." He wasn't speaking just for himself. "I have hardly known a man," he wrote, "who has ever become familiar with the kind of life which I have led for so many years, that has not relinquished it with regret." Gregg himself looked forward to "the plains again, to spread my bed with the mustang and the buffalo, under the broad canopy of heaven." By 1900, that kind of freedom was gone in the East, except perhaps for northern Maine and the higher elevations of the Appalachians. The West retained more of it; that's why John Muir quit Wisconsin for the Sierra Nevada.

There's another historical strand, too. It goes back to the town and country homes of the European aristocracy. The one was for society; the other was a link to the land that was, in preindustrial times, the source of aristocratic wealth. "Let Curzon hold what Curzon held" was the motto of one such family, and Kedleston, the family's immense home near Coventry, is still in the family, albeit run largely as a museum. Wealthy Americans were comparative parvenus, but that didn't keep them from copying the European nobility. George Vanderbilt ordered the construction of palatial Biltmore, near Asheville, North Carolina. Today, like Kedleston, it's a going business and advertises itself online with the phrase: "Visit Biltmore Estate and Escape from Everyday Life." That's the romantic impulse. Another Vanderbilt preferred the Atlantic Coast and ordered up The Breakers, a monster cottage at Newport, Rhode Island. Later on, the Rockefellers would build a magnificent retreat near the Grand Tetons. These were places where Creation could be enjoyed in splendid comfort.

Merely well-off Americans had to settle for vacation hotels. There was the Greenbrier in Virginia, which goes back to the 18th century but which was hugely enlarged after the Chesapeake and Ohio Railroad bought it in 1910. There was the Grand Hotel on Mackinac Island, near the strait between Superior and Huron. It opened in 1887 and still has its very grand, 600-foot-long front porch. These were hotels solely or primarily for the summer, but wealthy Americans sought out winter retreats, too, and in 1926 alone three luxury hotels opened in Florida: they were the Boca Raton Resort, the Coral Gables Biltmore, and the Palm Beach Breakers, rebuilt after a fire. A string of national-park hotels opened, too, including Old Faithful Inn at Yellowstone and the Ahwanee at Yosemite. For those unable to afford them, there was always camping. As a vacation activity it came later than country retreats and resort hotels, because it depended on the existence of both vacations for working people and affordable cars. With these, states began developing state parks with campgrounds.

Against this background, Congress in 1964 passed the Wilderness Act, creating reserves where human beings would be allowed only as visitors. Such a law would never pass today: it was a remarkable achievement in a remarkable decade. Forty years later, however, the appeal of wilderness is still strong. A good recent example occurred during the senate debate in 2003 over the fate of the Arctic National Wildlife Refuge. There was a Republican administration keenly devoted to the oil industry, there was a Republican majority in the senate, and there was a hopping-mad senator from Alaska who, in the worst way, wanted to have his way. As chair of the appropriations committee, Senator Stevens said on the senate floor that "people who vote against this today are voting against me. I will not forget it." Yet the bill died, 52–48, and Senator Boxer of California said, "There's something more powerful out there than any senator, even than any president, and that's God's gift to us." Thoreau and Muir had become mainstream.

Back in 1964, something else had happened, something that eventually would bring tens of thousands of affluent Americans out to live surrounded by nature. It was Sea Ranch, a recreational subdivision in Mendocino County, on California's north coast. The shore was left in common ownership, and Sea Ranch's wood-toned houses were so secluded behind lines of cypresses on marine terraces that motorists along the coast highway hardly saw them.

At the time, Sea Ranch seemed like just another exhibit in the gallery of Countercultural California, but it proved to be the start of something big. The diffusion wave was picked up sympathetically in Santa Fe, a community that has gone to extraordinary lengths to create a unique identity. This is a story that goes back to Isaac Rapp, an early 20th-century architect who copied and adapted for commercial use the Catholic church of the Acoma Pueblo. He would be appalled if he came back and saw how Santa Fe's building codes have produced gas stations and fast-food outlets with fake-adobe facades and make-believe *vigas*, those protruding, theoretically roof-supporting beams.

Santa Fe style replaced a territorial style imported from the East Coast. Clay veneer, in other words, replaced wood. That's why, even though Santa Fe style was conceived more for people than nature, the people who loved it instinctively shared the ecological awareness of Sea Ranch. They, too, wanted their community to look as though it had grown as organically as a seashell.

Forty years on, the first Sea Ranch houses seem monastic. They measure 1,000 square feet, hardly big enough for a garage in newer parts of the New West. It's not that the Sea Ranch idea has been lost. There's no dogma more secure in America today than the idea that nature should be preserved or ecological balance maintained. In the 40 years since 1964, however, America has become a much wealthier country than the one Lyndon Johnson knew. It's no longer the case, as a *New Yorker* cartoon once put it, that a cowboy hands his partner a loaf of French bread and warns that there aren't any decent baguettes west of the Pecos. People just want it all, environmental intimacy along with a 6,000-foot hideaway.

That's why they squeeze into an old town like Aspen, which instantly sprawls. That's why they come to an old town like Telluride and preserve it by building a new town hidden from view. That's why they start from scratch at a place like Lake Las Vegas.

Lake Las Vegas is worth a look. It's a 2,000-acre master-planned community in what a few years ago was utter desert 17 miles east of the Las Vegas airport. Its hard to get top dollar for cactus, so the Transcontinental Corporation of Santa Barbara built a lake, with help from the deep-pocketed Bass brothers of Fort Worth. The lake covers only half a square

Upscale Santa Fe, far from Wal-Mart but close to Canyon Road. The dry Acequia Madre is at the lower left.

mile, but that's enough for the Ritz-Carlton, which now spans it in an imitation of the Ponte Vecchio. If you're bored with Florence, try the nearby Moroccan-style Hyatt. Intrawest of Vancouver has meanwhile added a 50-acre shopping and residential center called MonteLago. Planned to its teeth, it's supposed to look like a fishing village that's grown, without planning of course, into a Tuscan town replete with clock tower and casino. Adjoining MonteLago are eight gated communities with romantic names like Monaco, Biarritz, and Siena. There are three golf courses whose water, like the water for the lake, comes from the Bureau of Reclamation's Lake Mead. That reservoir was named for Elwood Mead, the commissioner of the bureau in the 1920s, and if he saw Lake Las Vegas his jaw would drop. Whatever the failings of the Old West, at least it wasn't in drag.

The New West comes at many costs. One is ecological. Forty-acre ranchettes surround Santa Fe. The people buying them claim to be environmentally aware, but they don't like wildfires. By putting them out, they encourage the invasion of woody species in place of grass. Magpies are meanwhile attracted to the new sources of trash, and they kill native songbirds.

There's an esthetic cost, too. Half of Lake Tahoe's shoreline is now privately owned. Hard to believe, but a half-acre lot with 100 feet of waterfront is worth $3 million. There are plenty of people who can afford it, which is why there are almost 1,000 waterfront homes surrounding the lake. With land values so high, however, buyers build monster homes that leave the lakeshore looking like the high-rent urban neighborhood the buyers came to escape. That's why the Tahoe Regional Planning Agency finally imposed what amounts to a moratorium. Assuming that the new rule survives legal challenge, new lakefront homes are going to be less visually intrusive.

Then there are employment costs. The people moving to the New West are appalled at every tree that's cut, every hill that's dug, every range that's grazed. They're paying good money for nature, and they've got the money and lawyers to make sure nature is what they get. So loggers become nurse's aides, oilfield roughnecks go into computers, and cowboys become trail guides. Foreigners have preempted many of the seasonal jobs, however. Some 8,000 college students from Australia and New Zealand, for example, get visas each winter to work at Western ski resorts.

Even those Old Westerners who make the transition find it increasingly hard to live in the New West. They aren't likely to bid on the Teton Valley Ranch, which consists of 151 acres near Jackson, Wyoming, and which Sotheby's put on the market at $50 million. Developers near Steamboat Springs, meanwhile, are willing to buy ranches at prices that can't be remotely justified in terms of their agricultural potential. A ranch that yields $29,000 annually will have a market value of $6.5 million. If you're the owner and you want to move to Hawaii, well and good; if you're not, property taxes are going to kill you.

Then there are cultural costs. For the people of the Old West, the word "decadent" meant something bad. Not so in the New West. At Aspen, restaurants offer roast duckling with vanilla-scented potatoes and elderberry compote. Go over to the Ajax Tavern and try the oven-roasted porcini polenta with herbed ricotta. Too heavy? Try Rustique for sweetbreads and chanterelle mushrooms. If Aspen's too big, try Telluride. To avoid swamping the old mining town, its developers built an entirely new center called Mountain Village, and they put it on the other side of a mountain crossed by a free gondola. But watch out: At the top of the mountain there's a restaurant called Allred's. Diners enjoy seared *foie gras* with orange-honey sauce. Those who don't relish fat goose liver order lobster salad with grapefruit and macadamia nuts.

It's hard to imagine John Wayne eating here. It would be a Nicholson scene, as Wayne slowly goes ballistic explaining that first you cook the meat on one side and then you turn it over and cook it on the other. "Got it, sister?"

Easy as it is to mock the New West, the Old West had plenty of costs. Besides, the New Westerners are caught in a cultural trap. To really escape the city, they have to escape the grip of clock time. They have to escape the world of numbers that identify them and their property. They have to undo what Bacon and Descartes did. And they have to do it in a way that's fully reversible, lest they find themselves in serious trouble. It's not so easy.

It's like the dilemma facing rich Middle Easterners. They come from a society that has created some of the world's most beautiful buildings. Caught between the traditional and modern worlds, however, the descendants of consummate craftsmen hang velvet paintings on the walls of equally garish villas. In the most spectacular mosques they will hang fluorescently lit clocks next to the elegant *mihrab*, or prayer niche. Both Middle Easterner and American have to figure things out for themselves, and it takes time. Meanwhile, mosques fill with clocks and the New West spreads like a megalopolitan ski resort from Mammoth to Angel Fire, with race tracks in between.

EASTERN CATCH-UP

I keep calling it the New West, but it's a national phenomenon. New York State in 1892 added a forever-wild provision to its constitution, guaranteeing that the huge Adirondack State Park would never be developed. In the 1920s Benton Mackaye extended the recreation idea by initiating the Appalachian Trail, which runs from Maine's Mt. Katahdin down to the Great Smokies. Today there's a Hudson Valley renaissance, with heirloom potatoes and microproduction of fine (sorry, artisan) cheese. Way up in Maine's Aroostook County—so far north it's not even part of New England—people no longer see the woods just as a place for timber, black flies, and fishing expeditions. Instead, they see it also as a place where carefully groomed trails attract out-of-state cross-country skiers, especially now that southern New England has less snow than it used to. The transition comes none too soon, because the Maine potato industry is in near-terminal decline.

The one part of the country to miss the bonanza is the part that needs it most. It's going to be a while, maybe a long while, before an equivalent of Lake Las Vegas comes to western Kansas: Americans just refuse to believe that plains and muddy rivers can be beautiful. The entry points for development, if they exist, are likely to be the places where mountains rise from the plains. The Black Hills of South Dakota, the Sand Hills of Nebraska, the Flint Hills of Kansas, and the Wichita Mountains of Oklahoma won't attract the Santa Fe crowd, but they could attract people from Sioux City, Omaha, Kansas City, and Tulsa. If these places can be linked to Native Americans, they might draw Europeans, especially the Germans who still read Karl May.

CONFLICTING VALUES

Just as Los Angeles took water from the Owens Valley, so San Francisco wanted to dam the Tuolumne River in the Hetch Hetchy Valley. It was one canyon to the north of Yosemite Valley and it, too, was in Yosemite National Park. That's what made it so attractive, because once federal permission was granted the city would not have to pay for reservoir land or water rights. In signing the Raker Act, which gave San Francisco the permission it sought, Woodrow Wilson overlooked this economic rationale and based his decision on the conflict between preservation and progress. Preservation lost, and John Muir probably didn't help, because instead of looking to the money he castigated the city's supporters as barbarians. (His metaphor, comparing a dam at Hetch Hetchy with making a water tank out of a cathedral, was rejuvenated and put to more effective use by the Sierra Club in the 1960s, when it fought dams on the Colorado.) Wilson signed the bill, and O'Shaughnessey Dam still plugs the valley. Conservative politicians sometimes amuse themselves by suggesting that it be taken down. There's hardly anything they enjoy more than watching San Francisco's exquisitely liberal legislators spring to the defense of a dam.

Congressional hearings have also been held on breaching Glen Canyon Dam, on the Colorado River above the Grand Canyon. The residents of Page, who love this part of the world, also depend on the dam for their jobs, and every proposal to breach it leads to their cries carrying all the way to Washington.

Sometimes the contest becomes physical. Earthfirst!, famous for monkey wrenching or sabotaging development equipment, was once also known for spiking trees, so that a chainsaw hitting a buried spike would fly apart, killing or maiming the saw operator. (Spiked trees were supposed to be marked in warning.) Now it's more inclined to organize tree-sits, some of which go on for years. There has been a lot of cutting of ranchers' fences, too, and there have been cases where cattle were killed with automatic weapons.

The ranching debate isn't always so rough. Oklahoma's town of Pawhuska today welcomes visitors to the Tallgrass Prairie, whose creation a few years ago was bitterly resisted. The community gains a lot from tourists, however, perhaps more than from livestock production. On a grander scale, the Poppers' Buffalo Commons is a catchily named proposal to convert submarginal Great Plains cropland to open range stocked with bison. The regional economy stands to earn as much or more from ecotourism as it does from ranching.

The famous, or infamous, Hetch Hetchy Valley, a part of Yosemite National Park that was dammed so San Francisco could have cheap water.

On the one hand, there are roads and subdivisions to be blazed. It's big business. The demand for rural living has left Austin, Texas, struggling to control sprawl. In Arizona's Pima County, 40% of all houses are in wildcat or unregulated subdivisions, without paved roads or water mains. The appeal is understandable, because lots out here cost $15,000 instead of $200,000. Tucson's periphery, however, has become a rural slum whose residents suffer from respiratory infections caused by all the dust raised by traffic on the unpaved roads.

On the other hand, hundreds of land trusts are buying development rights to land. These trusts are increasingly important at a time when it's tough getting governments to buy land. In Connecticut, there's a land trust with an adopt-a-view program that clears trees from fields no longer farmed. (It's a regional problem: Forests in Abraham Lincoln's time covered a third of Massachusetts; now they cover two-thirds.) Or consider the apple orchards of Ulster County, 75 miles north of New York City. There's no money in apples, not with fresh apples from Chile and New Zealand and imports of very cheap concentrated juice from China. They've pushed prices down to $9 a bushel, $2 below the cost of production. What to do? Ulster County farmers can now sell their development rights to the Scenic Hudson Land Trust and use the money to stay in a money-losing business.

Upstate, some 45,000 acres of the Tug Hill Plateau on the west side of the Adirondacks were sold in 2002 by a logging subsidary of John Hancock Life Insurance. The purchasers included GMO Renewable Resources, a logging company that agreed not to remove more timber than grows on the land. The other purchaser was the Nature Conservancy, which announced its intention of giving a conservation easement on its land to New York State. The Tug Hill Commission would meanwhile attempt to maintain a forest economy while building a recreation one.

The conflict between preservation and development is most intense where there are the most people. During the 1990s, Loudoun County, west of Washington, D.C., grew from 86,000 to 170,000 people. It was one of the nation's 10 fastest-growing counties. Would the whole county be built up? Referring to the county between Loudoun and Washington, bumper stickers read "Don't Fairfax Loudoun." Developers were denounced as landscape rapers. They in turn damned the supporters of rural zoning as snob zoners and ecofascists.

The zoning law was changed to require a minimum lot size of 20 or 50 acres in the western two-thirds of the county; as a result, some 80,000 houses that could have been built under the old zoning were now prohibited. Building permits crashed from 104 in 2000 to 8 in 2002, but slow-growth advocates worried that irrevocable permission had already been given for another 35,000 houses, about a 7-year supply and about half the county's existing stock. Even that backlog didn't discourage would-be developers from filing by early 2003 150 lawsuits against the county's new restrictions. Would-be buyers found that the limited supply of new houses was forcing prices up. One of them sued on the grounds that the slow-growth policy was forcing him to make a dangerously long commute.

The *Washington Post* in 2003 tried to get some perspective on the issue and surveyed 14 counties between Chesapeake Bay and the Rappahannock River. It found that more than half the land in these counties was zoned to require at least 3 acres for every house. Was this land protection or was it, as the newspaper put it, "Gucci sprawl"? The authors did not say, although the obvious answer is both. A year later the newspaper reported the census finding that Loudoun County had grown faster in the previous 3 years than any other county in the nation. The reporters covering the story summarized the county's effort to control development as "ambitious, controversial, and failed." The development community had now succeeded in electing enough sympathetic members to the board of supervisors that the board had begun repealing the restrictions.

The same kind of battle is fought on the West Coast. There are 3,000 commercial vineyards now in the United States, for example; they're in just about every state. The industry's epicenter, however, is California, where the Napa Valley has awakened from a century-long lassitude to become Winery Central. Landowners want to convert forested hills into vineyards. There's money to be made. There's also strong opposition from preservationists who want to preserve the hillside trees. The irony is that the would-be vintners think of themselves as environmentalists. That's part of the reason they come to the Napa Valley.

A few miles to the west, the oceanfront itself has become a legal battlefield. In 1974, California voters passed the California Coastal Act, which declared that "developers shall not interfere with the public right of access to the sea." Sounds good, but all those proudly liberal property owners at Malibu have walled off several miles of beach in front of their cheek-to-jowl estates. Voila! *De facto* private beach, from which the unwashed public is excluded.

Up the coast a bit, just beyond Santa Barbara, the National Park Service undertook a feasibility study for a proposed Gaviota National Seashore. The park would have protected the area around Point Conception, which is the largest block of undeveloped land along the Southern California coast. Environmentalists were for it—there's hardly any such coast left. (This bit survives only because since World War II most of it has been part of Vandenberg Air Force Base.) There was powerful opposition, however. The owners at Hollister Ranch, which has been chopped into 100-acre ranchettes, monopolize 8 miles of coast. Hearing of the study, they hired lawyers to kill it. The lawyers failed, but bigger landowners in the neighborhood had even more to lose from a park that preempted their development plans. The American Land Rights Association describes the congresswomen who initiated the study as a "left-wing extreme enviro member." That's just one click shy of ecofascist, and perhaps it did the trick. The National Park Service in 2003 had good news for the landowners. Gaviota, it said, was "not a feasible addition to the National Park System. There are few willing sellers of land and land acquisition and operations costs are beyond what the NPS can undertake."

In 2003, an adventuresome group of middle-age Californians tried walking the length of the coast. The Coastal Act had decreed that "a hiking, bicycle and equestrian trails system shall be established along or near the coast," but the hikers found that only half the shoreline was actually accessible. Among the inaccessible areas, ironically, was the 7-mile strip owned jointly by the owners of Sea Ranch.

Compare that situation—or the situation in the San Juan Islands of Washington State—with beach access on British Columbia's Gulf Islands. There's only Kaiser Wilhelm's line in the water separating the two island groups, but thanks to public-access laws it's much easier to beach comb in Canada than the United States. Choose for yourself which is better, but remember that you're looking at two different legal regimes that rest on two different sets of values. The landscape is as good as a book to get you thinking.

Drive on back roads from New Mexico to Oklahoma. The roadside in the first is often unfenced. In Oklahoma, on the other hand, the right of way is lined with endless miles of three-strand barbed wire. Land ownership is part of the story. In New Mexico, most of the land is in public ownership, administered by the National Park Service, Forest Service, or Bureau of Land Management. In Oklahoma, it's private. But underneath that distinction is the question of how we should treat land. Barbed wire is a wonderfully rational response to managing livestock. Get some, along with a stack of T-bar, and you can just about bury your cowboys. Yet "don't fence me in" is as American as Daniel Boone.

So you drive from one of these states to the other and think about this contrast. If you want to get down to the Oklahoma rivers that flow this way—the lazy Canadian is a good choice—you'll fulminate at the idiocy that walls Americans off from their own country. If you think muddy water's disgusting, you may find nothing wrong with landowners who chop creation into hash. Either way, you'll be doing real geography. About time.

CHAPTER 21

CITIES ABROAD

I want to look abroad now, again first at cities, then the countryside. Here, too, the distinction is fuzzy but real, and it helps structure the discussion around the worldwide migration that is taking place as people move from the land to cities or towns.

Consider, first, the phenomenal growth in the population of the cities of the developing world. In 1925 the world's 10 biggest cities, in descending order, were London, New York, Berlin, Chicago, Paris, Moscow, Shanghai, Osaka, Leningrad, and Buenos Aires. Of these, only the first three had more than 3 million people. Today, the 10 biggest cities all have more than 10 million people, and the rankings have been radically revised. London, which topped the list in 1925, is no longer on it. Neither are Berlin, Chicago, Paris, Moscow, or Leningrad. The biggest city now is Tokyo, followed by Mexico City, Mumbai (Bombay), São Paulo, New York, Lagos, Los Angeles, Kolkata (Calcutta), Shanghai, and Buenos Aires. Look forward to 2015. Mumbai will be first, followed by Tokyo, Lagos, Dhaka, São Paulo, Karachi, Mexico City, Delhi, New York, and Jakarta. Only two cities, New York and Shanghai, will have remained in the top 10 throughout the period.

LIVING IN SQUALOR

The outstanding pattern in all this jostling is the shift away from cities inhabited chiefly by people of European descent. In 1925, 8 of the 10 biggest cities fell in that category, but by 2015 only two will. That's not because life is so good in these new megacities; it's because life in the countryside is worse. We'll look at that countryside in the next chapter, but rural conditions must be terrible there if people prefer the megacities. Even people with apartments, running water, and air-conditioning must breathe chronically unhealthy air when they step outside, and they find themselves in a grim world of highrises and traffic. The far more numerous poor can't shut the window. Their house can be a section of concrete pipe, not yet laid underground. It can be a shed of discarded cardboard. It can be a mat on the sidewalk. The latrine can be a curb, and the water supply can be a grossly polluted pond or stream.

Makoko is a Lagos district with roughly half a million people living without sewers, garbage collection, or electricity. Jobs are scarce; schools and medical clinics, a luxury. The Nigeria factor—corruption—is so debilitating that the population of Lagos as a whole can only be estimated. Estimates are hazarded from 7–20 million.

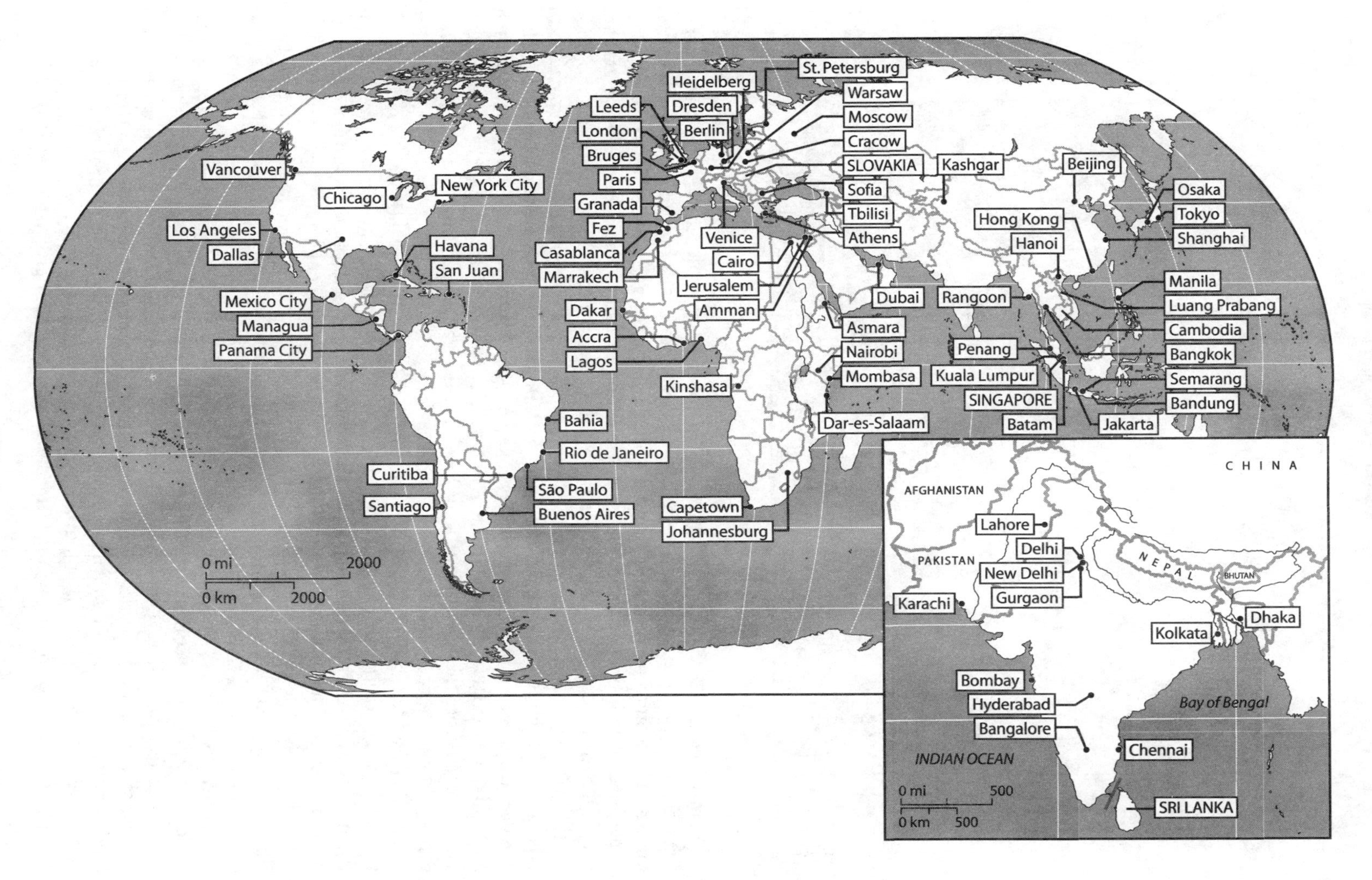

St. Petersburg
Heidelberg
Leeds
Dresden
Warsaw
London
Berlin
Moscow
Cracow
Bruges
Vancouver
SLOVAKIA
Kashgar
Beijing
Paris
New York City
Chicago
Sofia
Osaka
Granada
Tbilisi
Tokyo
Hong Kong
Los Angeles
Fez
Athens
Venice
Shanghai
Hanoi
Havana
Casablanca
Dallas
Cairo
San Juan
Marrakech
Manila
Jerusalem
Dubai
Rangoon
Mexico City
Luang Prabang
Dakar
Amman
Asmara
Managua
Cambodia
Accra
Penang
Panama City
Nairobi
Bangkok
Lagos
Mombasa
Kuala Lumpur
Semarang
Kinshasa
SINGAPORE
Bandung
Bahia
Dar-es-Salaam
Batam
Jakarta
Rio de Janeiro
Curitiba
São Paulo
Santiago
Buenos Aires
Capetown
Johannesburg
0 mi 2000
0 km 2000
C H I N A
AFGHANISTAN
Lahore
PAKISTAN
Delhi
New Delhi
N E P A L
BHUTAN
Gurgaon
Karachi
Dhaka
Kolkata
Bombay
Hyderabad
Bay of Bengal
Bangalore
Chennai
INDIAN OCEAN
0 mi 500
0 km 500
SRI LANKA

Since the 1940s, Cairo's garbage has been collected by children driving donkey carts. They clip-clop home to a shack in the middle of the dumps at Moqqatam, on the east side of the city. The women of the family then take over, sorting the garbage and separating plastic, paper, glass, cloth, metal, bones for glue—anything of value. These *zabaleen*, or garbage people, have attracted a lot of attention from aid agencies, but a tour of their neighborhood is still grim, with families living next to pens in which swine forage in a deep muck of organic waste. Pigs in Islamic Egypt? The *zabaleen* are Orthodox Coptic Christians. They antedate the Muslims in Egypt—and they'll let you know it. Even their name speaks of their autochthonous origin. It comes from *Aegyptios*, which is what the Greeks called Egypt 1,000 years before the founding of Islam. Another derivative from that root, by the way, is gypsy, although those people more likely come from India.

Embarrassed, Cairo officials contracted in 2002 with two Spanish companies to pick up 8,000 tons of daily waste with a mechanized system with dumpsters and a recycling center in the desert south of the pyramids. The materials in the garbage would become company property. What about the 30,000 *zabaleen?* Some of them hoped they would find a niche in the system, perhaps taking garbage to the dumpsters or working in the new recycling center. Others thought they were being pushed off the bottom rung of the social ladder.

Down in Nairobi, 700,000 desperately poor people are crowded into Kibera, East Africa's largest slum. People pay a third of their household income for rent on shacks that measure 8 feet on a side. There are no latrines, and householders use plastic bags, which they tie shut, then throw aside. "Stop flying toilets" has became the motto of a campaign to fund latrines, but progress will be slow, because these slums are very profitable for their nonresident landlords, half of whom are Kenyan politicians or civil servants.

I'll stop picking on Africa. A daytime landing in Mumbai gives you a view over extensive slums with thousands of one- and two-story, tin-roofed shacks packed so tightly that from the air you can see no streets, no yards, no parks, no trees—nothing except roofs. Nearby apartment blocks look positively elegant in contrast. If the people in these slums get sick, they can't afford a doctor. No problem: Indigenous practitioners are as common as real doctors in India. Prescribing medicine is no problem, either: pharmaceuticals are very loosely controlled in India.

That's true in Thailand, too, where you can walk into a pharmacy and get all kinds of things restricted in the United States. The clerk is likely to volunteer some little blue quadrilaterals, "if you have an older friend, sir." Isn't that why men visit Thailand? Village girls in Thailand are promised factory jobs by crooked recruiters who place them in brothels like those of Patpong, Bangkok's sex-trade district. There, a customer can take an escalator up to a hallway with a one-way plate-glass window. On the other side sit 20 women, each wearing a number pinned to her dress. They watch television until they see a customer's face behind one of the several circles in the window that permit two-way vision. A smile of sorts crosses their faces, only to drop as the face disappears and the stranger moves on.

Is Latin America better? Rio has 680 *favelas*, or slums, and 2 million of the city's 9 million people live in them. They can be big; the biggest is Rocinha, population 150,000. A woman resident of another, Rio das Pedras, says: "There's no better place to live. This is paradise." It's a strange thing to say, but Rio das Pedras has vigilantes who kill criminals

Boys and men soap up and rinse off by a street hydrant in downtown Kolkata.

and keep the slum safe. She continues: "We can put up with anything—rats, floods, trash—as long as we're spared drugs."

In neighboring São Paulo, there are an estimated 5 million people in illegal housing. Some of them are scattered in about 150 unauthorized subdivisions in the northern Sierra da Cantareira. Others are crammed into *corticos*, illegal and extremely overcrowded boarding houses. Some live in *favelas*. Still others have joined the *Movimento dos Sem Teto*, the Movement of the Roofless, whose 5,000 homeless families now occupy abandoned mansions, offices, even theaters and a hospital.

Up in Nicaragua, an earthquake in 1972 wrecked buildings that, for lack of money, have still not been repaired but whose shells are inhabited by squatters. It's not so unusual. The victims of an earthquake in Armenia in 1978 are still living in the wreckage.

True, you can find poverty in the rich countries of the world. Make a list of the 20 poorest wards in England, and you'll find that 14 of them are in London, where one out of every six people is on the British equivalent of welfare. These people mostly live in the eastern half of the city, away from tourists. If seen, they'd be an embarrassment, of course, but no more so than the poor of Paris or Berlin or Rome. Poor as they are, however, Europe's poor generally have clothes on their backs, clean water, and access to schools and medical help. The material conditions of their poverty are not to be equated with those of the developing world's destitute.

THE INEVITABILITY OF URBAN GROWTH

Despite these conditions and despite occasionally brutal efforts to exclude immigrants, cities around the world continue to swell. In the 1980s the Khmer Rouge drove millions of Cambodians into the countryside, but Cambodia's urban population today is growing at almost 5% annually, while its rural population is growing at less than 2%. Squatters flood Jakarta, and a government official explains that "these people are a threat to our city and our future. If we don't move them, then how will our investment climate improve?" It's a good point, but the resident whose house has been knocked down says, "We are human beings. Why are they treating us like animals?" No matter: Despite periodic demolitions, the urban population of Indonesia is growing at about 4% annually, while the country's rural population is declining. This is the rule, not the exception. Egypt has struggled to divert Cairenes into new towns in the deserts around Cairo, but Cairo continues to grow. Soweto (an acronym for Southwestern Township) continues to grow, too, but what started out as a satellite city has now been swallowed up in the continuing growth of metropolitan Johannesburg.

REFUGES AND POVERTY AMELIORATION

People with money can find luxurious niches in these cities. In Mexico City, the rich can retreat to the western suburb of Santa Fe, built on a former garbage dump. There's luxury housing here by the Centex Corporation of Dallas, and there's shopping at the Centro Comercial Santa Fe, a major Latin American mall. A Canadian company, Reichman International, has meanwhile built a $750 million office building in Mexico City. It's the 55-story Torre Mayor, and its occupants will scarcely have to deal with La Capital, the dysfunctional central city. Rents at Torre Mayor are even charged in dollars so there's no problem with peso fluctuation.

Agencies public and private, international and local, are trying to help the far more numerous urban poor. The United Nations has specialized agencies dealing with health, education, and jobs. The World Bank pumps money into the task. So do regional institutions like the Asian Development Bank. So do bilateral or country-to-country programs such as those generously funded by Japan and the Scandinavian countries. Private foundations are involved, along with hundreds of nongovernmental organizations. Wander around poor neighborhoods in the new megacities and you're likely to find signs crediting a sewer here or a clinic there to aid from Germany, Sweden, Spain, Japan, or the United States. Donors do expect recognition.

National governments sometimes undertake urban improvement projects on their own. The countries that have done the most are naturally those with the most money. In the 1950s, for example, Hong Kong's public housing consisted of tall apartment buildings without elevators and with shared lavatories on each floor. They were considered a big improvement on the shacks they replaced, but over the next 20 years 2,000 of Hong Kong's older apartment blocks are to be demolished and replaced.

This isn't an entirely happy story, because renovation brings higher rents. Many elderly poor who have lived in these buildings for decades fear that they will be forced out. The government recommends their moving to older and cheaper apartments on the city's periphery, but one man tried it, only to move out. "I was alone there," he said. "How would anyone know if I died?" In the year 2000, about 30 displaced people at Kennedy Town on Hong Kong Island took a drastic step and, in the Hong Kong idiom, became street sleepers.

São Paulo has undertaken its own program of residential highrises. They're called Singapores, but unlike the buildings in Singapore on which they're modeled, the Singapores of São Paulo are under the control of gangs and drug dealers. Conditions are so bad that some residents have moved back to the *favelas*.

Urban poverty is almost always too big a problem for governments to solve, either with their own resources or with foreign aid. That leaves the private sector.

Despite India's socialist tradition, Indian newspapers carry ads for new, privately built apartments. The developers have mastered the patter of promotion, too. Konark's Oasis, in Delhi's northern suburb of Bhiwadi, takes out ads saying in English, "Stop dreaming for Heavenly Homes! Let us do it for you!" Parsvnath Prestige, in the booming eastern suburb of Noida, advertises "enviable homes at heartwarming prices." Eldeco's Utopia, which is next door, says: "So far you've just dreamt of it. Now it begins to take shape. Presenting Utopia. Life that's picture-perfect."

On a sunny Sunday afternoon, the subdivision offices are almost like American suburbia in the 1950s. You come down a new freeway running through green fields—these places, too, have a wicked commute. Then there's a walled enclosure plastered with posters promising that the subdivision on its way will have gardens, a cafe, barbeque pits, a swimming pool, of course, and rock-climbing walls. There's nothing on the ground yet except some rebar, a busy sales office, and a model apartment with a steady stream of families. They've all arrived in their own cars, and wide-eyed wives slide kitchen drawers back and forth on roller bearings.

The apartments are often about the size of the original Levittown houses: 1,000 square feet. In Delhi, though, they're stacked up, sometimes a few stories and sometimes 10. Building materials are different, too. India doesn't build American-style balloon frames, with wall plates assembled from two-by-fours. It builds walls of solid brick, supported by a concrete frame. Perhaps that's why the apartments, even when floored in laminated wood, feel so solid: There's no flex in the structure.

Prices are low by American standards. At the upper end of the range you can buy a 2,000-square-foot apartment for about $100,000. Indian builders, however, demand full payment up front. You can take a couple of years to buy your apartment, but you'll finish paying before it's ready for occupancy. A penalty clause encourages the company to complete the units on time, and this seems to do the trick, because many units are sold years before there's anything on the ground.

Do such places help India's poor? Indirectly they do indeed, because the middle-class Indians buying them will vacate premises quickly taken over by someone with less money. And that person will be vacating a still cheaper place. It's like musical chairs, but housing standards rise across the board.

China, too, has turned to the private sector. It has an urban population of about 350 million people, squeezed on average into 80 square feet. That's why the government has privatized its urban real estate market and stopped providing free housing to workers in state enterprises. Some 60% of the people in China's big coastal cities now own their homes, typically in high-rise apartment buildings. There isn't much choice, because the value of land in China's major cities in 2003 averaged 1,166 yuan per square meter. That's about $400,000 an acre, and the figure was higher in Beijing and Shanghai.

The transition hasn't been easy, and it has sometimes produced deeply ironic results. The Chinese Communists, for example, first met in Shanghai in a building now called the First Congress Meeting Hall. It's surrounded by ranks of barrack-like apartments—*shikumen*—where a family of three squeeze into 260 square feet. Into this neighborhood comes Vincent Lo of Hong Kong. He's building a private, 129-acre redevelopment project called Xintiandi, or New Heaven and Earth. It's only partly finished, but rather than put up modern towers, his architects—from Singapore and the United States—are renovating the old *shikumen*. They've opened pedestrian spaces and added fountains. The result is a welcome contrast to Shanghai's militantly bright developments. Xintiandi's many customers find a Starbucks and a McCafe, a French bakery and a restaurant owned by Jackie Chan.

The barrack dwellers are meanwhile told to buy apartments on the city's outskirts, where there are new and shiny commercial centers. Xujiahui, they're told, offers a Grand Gateway, a Metro Pacific, and an Oriental Department Store. The residents reply that their compensation won't cover the cost of a new apartment. A few have even dared to file lawsuits demanding more, but most know the Chinese government too well to bother. Across the city, 850,000 families have already moved to make way for the city's prodigious building program. Meanwhile, the old First Congress Meeting Hall has been swallowed up by Xintiandi, where it is carefully preserved as a monument to the Communist Revolution.

The shift to private investors reaches beyond the housing market to infrastructural systems that until now have almost always been a government undertaking. The United States may have turned perhaps 1% of its municipal water systems over to private management, for example, but privatization overseas is much further advanced. The biggest providers are Vivendi and Suez, both French. With a history going back to the canal, Suez now operates water systems in Santiago and Buenos Aires, Casablanca and Amman, New Delhi, Jakarta, and Manila. It has operations in 130 countries and serves 100 million customers. Bids have been let recently for private contractors to renovate the decaying water mains of Sofia, Rio de Janeiro, Panama City, and San Juan, Puerto Rico.

SUCCESS STORIES

Public and private investment can pay off spectacularly, not only with improved housing but with the equally or more important development of good jobs.

A *shikumen,* one of many of these apartment buildings that were popular when built and are still so, in a city as expensive as Shanghai.

Singapore's the obvious example. No American airport is as attractive or clean as Singapore's Changi—and it's a fair axiom that you can tell a great deal about a country from its airports. Singapore's streets are so clean that visiting American children for the first time in their lives think that their own hometowns are dirty. Its supermarkets have a greater variety of goods than you'll find in most American ones. Many outsiders deplore the regimentation of Singapore, but there are plenty of people around the world who would happily accept such controls if, in exchange, they could share in such prosperity. Now the government of Singapore is looking beyond electronics to biotechnology: It has set up what it calls Biopolis, a miniature city of research institutes and private companies. The big pharmaceutical companies are watching: Novartis has taken the plunge and begun setting up an Institute for Tropical Diseases.

Off to a late start, China is doing phenomenally well. Its commitment is symbolized by the fact that it now has 5 of the world's 10 tallest buildings. (The United States has only two—the Empire State Building and the Sears Tower; the other three are the twin Petronas Towers in Kuala Lumpur and the Taipei 101 tower, currently the world's tallest.) One major Chinese project is Hong Kong's Cyberport, which will offer 1 million square feet of office space, plus 2,700 apartments, retail space, and a hotel. It's big, but Hong Kong is only a small part of the action in South China. Shenzhen started the boom: It was a village in 1980 and now has 7 million people. There are other prospering cities around the Pearl River delta from Shenzhen to Guangzhou and Zhuhai, near Macao. They include big places foreigners have never heard of, like Huizhou, Dongguan, Foshan, Jiangmen, and Zhongshan. Taken as a whole, the delta produces a third of all Chinese exports and is the recipient of a third of all foreign investment in China.

Xintiandi: A former *shikumen* has been recycled into upscale retail space. It's very nice, so long as you don't ask too many questions about the people who had to move.

Farther north, the new district of Pudong, across the Huangpu from central Shanghai, must rank as one of, if not the most, dramatic cases of rapid urbanization anywhere in the world. The World Financial Center, planned to be 1,500 feet tall, was put on hold in 1998 but is still scheduled for completion in 2007. Asia's biggest shopping mall has already opened in Pudong, though: It's the Super Brand Mall, not far from the soaring Jin Mao tower, which for now is metropolitan Shanghai's tallest building. The Super Brand Mall looks a lot like Bangkok's shopping centers, which is to say like a fire marshal's nightmare. No surprise: Thailand's Charoen Group is the owner. Charoen plans to open 100 superstores in China. A few miles away, a new industrial park is emerging at Zhanjiang, linked to the city by a subway that speeds across what a decade ago were rice paddies.

All told, Shanghai in 2003 had 2,880 buildings 18 stories or higher. Almost all had been built since 1990—some as offices, many as apartments. GM is here. Sony is making televisions. NEC is building computer chips. Land in Pudong is so expensive that developers look an hour's drive west to Suzhou. In the past, Suzhou was famous for gardens and silk, but they've been overshadowed since Singapore in 1994 helped establish the Suzhou Industrial Park on the city's east or Shanghai side. Drive around the district and you'll see an amazing collection, including Alcatel, Bosch, Emerson, GlaxoSmithKline, Hitachi, Kraft, Kubota, Nokia, Philips, Samsung, and Selectron. Tidbit: In 2003 Logitech here assembled 20 million computer mice.

Similarly ambitious plans are being laid for Chongqing (Chungking), which was administratively separated from Sichuan in 1997. The government plans to invest $200 billion in the expanded muncipality—population 31 million—over the next decade. Ford has a plant here that will build 50,000 cars a year. Other factories make motorcycles, mobile phones, while cruise ships sail down the Yangtze and through the locks at the Three Gorges Dam.

If it's any consolation, China has spent a lot of money developing midsize cities only to find that investors sit on their hands. Beihei, on the west coast of Guangxi, is a good example; so are Changsha, the capital of Hunan; Nanchang, the capital of Jiangxi; Fuzhou, across from Taiwan; and Shantou, up the coast from Hong Kong. There's plenty of shiny infrastructure in these places but little economic growth. They're just too remote from the country's centers. As things stand, half of the $50 billion annually invested by foreigners in China goes to Guangzhou, Shanghai, and Shanghai's neighboring province, Jiangsu.

India, too, is in the game, with glitzy office buildings in every major city except Kolkata (Calcutta). In many respects India's seriously behind China. If you believe in the airport-indicator law, you'll suspect the worst the moment you land at Mumbai. A related indicator is the fact that India's four airlines (Air India, Indian Airlines, Jet Air, and Sahara) have a total of 140 jets, while Air China, China Southern, and China Eastern have 480. Less visibly, total foreign investment in India runs only about $5 billion annually, less than one-tenth that of China.

Still, India has a huge advantage in its widespread use of English. Starting with little in its favor besides an attractive climate, Bangalore has attracted Sony, Hitachi, and Siemens. A Singapore-funded International Tech Park stands 10 miles east of the city. Two hundred miles north, the old city of Hyderabad now relishes the nickname Cyberabad. There's a satellite town called Hi-Tec City, and though the name is cheesy the buildings themselves

A supermarket in the basement of Pudong's Super Brand Mall has a great selection of many things, including salad and cooking greens.

Two of the many buildings clustered at Hi-Tec City, a northwest suburb of Hyderabad. The city until 1948 was the capital of a princely state and, as such, was famously reactionary. No more.

would look sharp in Silicon Valley. On the other side of Hyderabad is New Oroville, created by Catalytic Software, which is controlled by ex-Microsoft people who plan to deliver just-in-time sophisticated software at bargain prices. Delhi, too, has high-end office districts: One is at Gurgaon, which was the site in the 1920s of the Gurgaon Experiment, a then famous but now just about forgotten colonial project in rural development.

India is beginning to compete even in businesses where English is not especially important. A good example is the Indica, a small car made by the Tata Group. Twenty thousand are to be exported annually to England, where they will be badged as Rovers. The volume is small but amazing to anyone who remembers the wretched state of Indian-made cars 20 years ago.

In the Middle East, Dubai among others is thriving, although it is also developing a worrisome tripartitite population, with Arab owners outnumbered by European visitors and Indian and Filipino workers. The Crown Prince has taken the lead on several megaprojects. The lavish Burj Al Arab hotel is one, standing offshore like a giant dhow. Another is a $5.5 billion project called Palm Island, where some 40,000 people will live on an artificial island built in the Persian Gulf. There will be almost 50 hotels plus 5,000 residences, built on a giant landfill in the shape of a palm tree, its fronds protected by an enclosing dike. Each of the 17 fronds will be designed with a different regional motif—Venice, Bali, Okinawa, even Art Nouveau Miami Beach. Among the newcomers is Kerzner International, which owns the Atlantis resort in the Bahamas and is planning another here, this time for Europeans. Back on the nearby beach, the Madinat Jumeirah is under construction with hotels, villas, and, according to one of the architects, a "fully authenticated" traditional market, or *suq*, rising from the sand. The Chinese have meanwhile announced the construction of the 60-acre Dubai Distribution Center for Chinese Commodities, a new Chinatown with residential as well as commercial space.

Japanese developers are doing megaprojects. One is Tokyo's Roppongi Hills, where Minoru Mori has a $4 billion investment in offices, apartments, a hotel, and shops on a 27-acre site. Rents are high, starting at $2,000 a month for an apartment. Mori says, "I want to change the character of the city." Critics say he's changed it all right: built a generic district that could be anywhere in the world.

Big developments—and successful ones, too—don't come only from Asia. Curitiba, the capital of Brazil's Paraná State, has grown from 150,000 people in 1950 to 2.5 million today. It could be a case study in squalor, but the city has a research- and urban-planning institute that has worked with the long-serving mayor, Jaime Lerner. There's an excellent bus system, with a route system of spokes and concentric belts. Triple-unit buses carry 300 people along restricted lanes as fast as subways but at a much lower cost. Above the bus terminals are clusters of services such as day care, job training, and small claims courts. Downtown shopping streets are closed to traffic. Scattered throughout the city, 50 Lighthouses of Knowledge provide libraries and internet access. Flooding used to be a problem, but an elaborate system of parks and lakes has reduced it. Teenagers are hired to keep the parks clean. Orphans and abandoned children, common in Brazil, are informally adopted by businesses that are urged to give them simple jobs in exchange for food and a little money. Street vendors are grouped into fairs that move through the city's neighborhoods. There's extensive trash recycling, with the city distributing food in exchange for sorted waste. Surveys show that almost all Curitiba's residents have no desire to live elsewhere in Brazil. Such success has attracted investment by Renault, Daimler, and Volkswagen.

Brazil's federal government has meanwhile announced plans to grant land titles to the residents of *favelas*. The hope is that these people will finally have the collateral against which they can borrow money to start or expand small businesses. It's risky for the borrowers, and the Peruvian Hernando de Soto warns that "the history of Latin America, including that of Brazil, is littered with property schemes for the poor that were not done correctly and ended with broken illusions and money thrown out the window." Still, the need is huge, and President Lula da Silva is determined to act. Very simply, he says: "I want to do a lot for those who live in shacks on stilts." He's Brazil's first president to have grown up in a *favela*.

PRESERVING THE URBAN HERITAGE

As fast as these cities are growing, their cores are often filled with decrepit buildings left over from the colonial era. Many of the world's biggest cities, after all, were established as European ports. The list of such places in Africa alone would include Casablanca, Dakar, Accra, Lagos, and Capetown. Africa's inland cities, too, are usually of European origin: Think of Johannesburg, Nairobi, and Kinshasa (Léopoldville). It's the same in Asia, where the European ports include Karachi and Mumbai (Bombay), Chennai (Madras) and Kolkata (Calcutta), Dhaka (Dacca), Yangon (Rangoon), Ho Chi Minh City (Saigon), Singapore, Jakarta, Manila, Shanghai, and Hong Kong. In both Africa and Asia, you almost have to hunt for cities with non-European roots. In Africa you'd come up with Cairo, Mombasa, and Dar es Salaam. In Asia, the list would include Lahore, Delhi (but not adjoining New Delhi), most of the cities of interior China, and the cities of Japan and Korea.

After World War II, most of these nations did what they could to clear away the remnants of their hated colonial past. What to replace them with? Despite their nationalism, the leaders of these countries chose to build symbols showing that they, too, could play the Baconian game. A good example is Kuala Lumpur, where Malaysia's prime minister, Mahathir Mohammad, poured money into the Petronas Twin Towers. "A country needs something to look up to," he explained.

In neighboring Singapore you have to hunt for signs of the colonial past, but most former colonies haven't been able to afford much more than carting statues away, renaming streets, and removing plaques. Kolkata (Calcutta) is a good example. So is Rangoon (Yangon), time-warped with its shabby old brick buildings, plastered and ornamented with the arches and columns favored by colonial builders. (Trained architects were involved only rarely in these cities, although there are exceptions like Edward Lutyens in New Delhi and Herbert Baker in Pretoria.) Fifty years on, Hanoi remains famous for its French atmosphere. Semarang, on Java, is a museum of Dutch colonial architecture, while nearby Bandung has a clutch of Art Deco buildings.

The "clear out the old mentality" is changing now, as governments discover that old buildings are a bankable asset. Once again, though they'd rather not hear this, they're copying Europe, which made the same discovery decades ago—even centuries ago in cases like Rome, Florence, and Venice.

Restoration efforts go back to the colonial powers themselves, who, as they began to modernize the landscapes of Asia in particular, realized also that they should preserve the great monuments in their care. The Dutch, for example, undertook the early restoration of Borobudur, a Buddhist prayer center near Yogyakarta on Java's south coast. The French

The Strand Warehouse, along the Hooghly River in Kolkata. Although Kolkata's port is now concentrated downstream, the warehouse is still crammed with baled textiles.

Rangoon until 1948 was in Burma but not of Burma. The British had surveyed it, and they ruled it; most of its residents, however, were Indians, economic migrants in today's terminology. They're gone now, more or less evicted after Burma became independent, but the apartments in which they lived still line most of the downtown streets.

undertook the forest clearance and renovation of the temples and palaces of Angkor, in what is now western Cambodia but which was then part of Indochina. The British perhaps did the most, undertaking early in the 20th century a raft of projects including the restoration of the Taj Mahal, the nearby abandoned city of Fatehpur Sikri, and the Mogul tombs of Jehangir near Lahore and Humayun near Delhi. It's hardly an exaggeration to say that the high points of most trips to India today are the monuments that the British rescued from long neglect.

The viceroy at the time, Lord Curzon, also spoke out against the habitual practice of carting off important archaeological relics to distant museums. In a speech accompanying the passage of the ancient monuments bill in 1904, he said: "the plan has hitherto been to snatch up any sculptured fragment in a Province or Presidency, and send it off to the Provincial Museum. This seemed to me, when I looked into, to be all wrong. Objects of archaeological interest can best be studied in relation and in close proximity to the group and style of buildings to which they belong. . . . If transferred elsewhere, they lose focus, and are apt to become meaningless." The result was the creation of local museums at places like Pagan in Burma and Peshawar, in what is now northwest Pakistan.

The British were also probably the first to undertake to preserve the historic core of a city. They had had a long, deep, and personal interest in the Holy Land; the important Palestine Exploration Fund was headquartered in London and for decades sponsored investigations that sought to determine the historical accuracy of the Bible. Then, with the retreat of the Turkish army in World War I, Palestine and Jerusalem in particular came under British administration. From Lord Allenby down, the British were determined to restore what the Turks had neglected or vandalized.

The Hall of Private Audience, built by Akbar at Fatehpur Sikri, abandoned for centuries until the British found it worth restoring as a historic monument.

The first British governor of Jerusalem, Sir Ronald Storrs, hired a disciple of William Morris, C. R. Ashbee, to take charge of the work. The walls were cleared of garbage and a walkway opened around the entire circuit, except for the corner occupied by the Dome of the Rock. The land rising from the Old City to the Mount of Olives was declared off-limits to builders, so the city at least to the east could retain its historic integrity and setting in biblical Judea. Within the Old City, construction materials and designs were sharply restricted to maintain the historic character of the place. Next to the Dome of the Rock, Ashbee restored the finest *suq* in the city, and he worked hard to reintroduce traditional crafts—tile, weaving, glass.

When the British hurriedly left Palestine in 1948, the last High Commissioner, General Alan Cunningham, wrote that the city for nearly 30 years had been a "sacred trust" and that he hoped the work of the British in Jerusalem would be "an example to the future generation in whose care our Holy City must rest." It probably has not been anything of the sort—neither the Jordanians nor the Israelis have been much inclined to thank the British—but the measures undertaken by the British in Palestine have now become staples of urban preservation around the world, including Israel and Jordan.

After 1948, the Old City became part of Jordan, which did little to maintain it. In 1967 the Israelis took charge. They completely reconstructed the Jewish Quarter but allowed the much larger Arab quarter to deteriorate into the worst slum in all of greater Jerusalem. The Palestinian Authority can do nothing about this decay—it has no control over Jerusalem. With donor help, however, the Authority has been very active rehabilitating Bethlehem and Hebron.

Given half a chance, these cities could show their European peers a thing or two. The historic core of Seville, for example, is a major asset of the Spanish tourist industry, but it has nothing on Hebron, whose core has been handsomely rebuilt by the Palestinians with

Saudi money and has achieved the remarkable result of renovating a historic district without displacing the poor residents. The Israelis have meanwhile restored ancient neighborhoods in Israel. There's Acre, with its Crusader foundations, and there's Old Jaffa, once an Arab city but now an Israeli artist's colony. Ironically, Jaffa's Palestinians live nearby in grim, modern apartment blocks. Even if they wanted to live in Old Jaffa among Israelis, they couldn't afford to.

City planners around the world see tourists flocking to Bruges, Granada, and dozens of other European cities. "Why not us?" they ask. And so tourists are now drawn to the colonial core of Bahia, the 17th-century capital of Brazil and the first slave market in the New World. They come to La Fortaleza, the Spanish fortress that once protected San Juan, Puerto Rico. Old Havana is now a UNESCO World Heritage Site, and a Cuban architect named Eusebio Leal has set up a company, Habanaguanex, that has raised over $200 million from Europeans who want to help renovate buildings in the district.

Americans are much more likely to visit the Zócalo, the square at the center of the decrepit colonial heart of Mexico City. The Mexico City government is finally putting money into restoration of the neighborhood's 3 square miles of largely abandoned colonial buildings. It'll be a big job, calling not only for architectural restoration but for new infrastructure, new businesses, and reductions in street crime. North of Mexico City, there's the nicknamed Sistine Chapel of Mexico. It's in Atotonilco, Guanajuato, and it's the Calvary Chapel of the Sanctuary of Jesus of Nazareth. Whatever you call it, it's been restored for about $800,000, of which about $300,000 came from an organization in Los Angeles known as The Friends of Heritage Preservation. A Dutch investor has meanwhile bought whole blocks of Curaçao and is restoring them as a commercial venture called Kura Hulanda.

A corner of Jerusalem's wall, last rebuilt by Suleiman the Magnificent but opened to curious tourists in the 1920s, during the British Mandate.

Courtyard houses in Hebron, carefully restored despite the city's tortured recent history.

From Mexico City to Buenos Aires, the Latin American city has an additional element wedged between the colonial core and modern developments. Buenos Aires doubled its population between 1850 and 1870, and many Latin American cities, including Buenos Aires, São Paulo, and Mexico City, saw their population triple between 1900 and 1930. During this period, Latin Americans looked almost without exception to France, especially the France of Napoleon III, for cultural leadership. That's why beyond their colonial grids all these cities have broad, tree-lined avenues. For Buenos Aires, that avenue is the Avenida de Mayo; for Mexico City, it's the Paseo de la Reforma. That's also why these avenues are bordered by five-story buildings with mansard roofs and Beaux-Arts decoration. Buenos Aires was especially ambitious. Calling itself the American Paris, it hoped to provide a civilized counterpoint to the crude United States. Those ambitions seem absurd today, but their physical expression remains.

Will this postcolonial but premodern zone in Latin American cities survive? A warning sign comes from Casablanca, which between the French occupation in 1907 and independence about 50 years later was a model of a sophisticated design fusion. The French began—it's almost too hackneyed to be true—by building a clock tower right next to the crowded native city. This was 1910, and the Moroccans were apparently going to be shown, first thing, how to organize time. The French then built a neighborhood, ironically named Liberté. It is full of elegant, architect-designed buildings—Art Nouveau at first, then Art Deco and Moderne. Seeking to integrate Moorish elements into European design, French architects developed a blend known as *Arabisance*. Their work is pretty much ignored now, and tourist guidebooks hurry readers along to the older cities of Fez and Marrakech; many elegant French buildings have already been razed. (For an illustrated history of this remarkably sophisticated city, see Jean-Louis Cohen and Monique Eleb's *Casablanca,* 2002.)

There are plans now to restore old Tbilisi with World Bank help, even though the buildings aren't particularly old—only the street pattern is. The World Bank is financing Eritrea's Cultural Assets Rehabilitation Project, which will help restore and preserve Asmara's outstanding collection of modernist buildings built by the Italians in the 1920s and 1930s. (For more, see *Asmara: Africa's Secret Modernist City,* 2004, by Edward Denison and others.) Farther east, the old palace of the Nawab of Dhaka has been restored from a grossly dilapidated condition.

Even Singapore is saving what survived the decades of modernizing with a vengeance. In 1985, rooms at the long-decayed Raffles hotel were cheap, but maintenance was so far gone that electrical outlets burst into flame if a guest plugged in a radio. It's a different story now: The lavishly renovated Raffles is the most expensive place in town. The other Chinese city of the Malay Peninsula is Georgetown, on Penang. Its colonial core is much more intact than Singapore's, and some residents want to capitalize on this heritage. In a time when governments insist on changing the colonial names of cities, Georgetown is still Georgetown. One architect who's influential in the preservation movement says, "We know who we are, and if we didn't, changing our name wouldn't help."

China is slowly coming round to see the value of preservation. The Beijing architect Liang Sicheng in the 1950s argued that Beijing should be preserved, while a new administrative capital was built nearby. Mao didn't have a preservationist bone in his body, however, and Beijing's outer wall, to take a single example, was destroyed to build a ring road. (Liang

The Singapore planners are honest enough to admit that the surviving shophouses have "a charm not found in much of our modern architecture."

is said to have wept.) To mark the 10th anniversary of the revolution, 10 supposedly great projects were undertaken, including the Great Hall of the People and the Museum of History and Revolution. Together, these two buildings frame Tiananman Square, the iconic heart of the nation. In an attempt to get away from gray blocks, Chinese architects added elements of traditional Chinese design, but the results were absurd, concrete blocks with bits of Chinese ornament.

Under the pen name Simon Leys, the architect Pierre Ryckmans wrote *Chinese Shadows* (1977), a blistering critique of this barbarism. China's leaders after Mao, however, were engineers, not architects, and to the slight extent they cared about architecture, they preferred modern buildings. Beijing's National Grand Theater is a good example. Designed by Paul Andreu, this is a 6,200-seat theater behind the Great Hall of the People. Critics say it is as Chinese as a hamburger, which it does resemble, but the theater is the legacy of President Jiang Zemin, and criticism has been muted. Besides, as one Chinese architect says, "in Beijing, people hate old things. The socialist upbringing did so much to destroy tradition, so people have no sense of it."

A similar attitude still prevails in China's smaller cities. Bulldozers in 2003 leveled 3 square kilometers of the oldest part of Kashgar, or Kashi. Perhaps it was because the old, labyrinthine neighborhoods were hard to police. The police are unlikely to be very busy in the new neighborhood, because its apartments are too expensive for the indigenous Uighurs. Says one Uighur cobbler, "If we say something against it, we will go to jail."

Despite all this, the Chinese have finally realized that there's money to be made from preservation, and not just from famous attractions like the Forbidden City. Shortly after the

A neighborhood of Beijing *siheyuan* yields to a phalanx of apartments; one of the old buildings is allowed to stand.

announcement that Beijing would be the site of the 2008 Summer Olympic Games, the Chinese announced that they would spend $200 million on the city's cultural preservation. That's not much compared to the $22 billion that they will spend on the games overall, but it's not chump change either. Most of the city's 6,000 alleys, or *hutong*, are gone, along with most of the courtyard houses, or *siheyuan*, that lined them until demolitions began in the 1960s. The entrances were always on the auspicious south side, while the courtyards were balanced along an east–west axis. Some of these neighborhoods survive, for example near the Drum Tower. Most have decayed into slums. The Chinese announced in 2003 that they would preserve 2,500 of then, but where will the crowded occupants go, and who will move into the restored homes? The city government has told McDonald's that the golden arches are inappropriate on 30 of the city's McDonald's restaurants. Down in Shanghai, the old mansions of the French Concession are meanwhile being restored as very expensive private residences. This comes after decades when they were chopped into decaying apartments.

Authentic restoration isn't easy. Many Palestinian buildings in Bethlehem and Hebron have been spoiled during renovation because concrete was used instead of limestone mortar. It's a mistake because concrete, unlike limestone, traps water behind rubble-filled walls, which soon crack. When UNESCO set out to help restore Luang Prabang, the precolonial capital of Laos, it had another mortar problem: Nobody remembered the traditional formula. Eventually, conservators found two old men who gave them the recipe. It included sand and straw, particular leaves and bark, tamarind seeds, sugarcane juice, and—*pièce de résistance*—a glue extracted by long boiling of buffalo hides. Now that the recipe has been recovered, UNESCO must persuade the people of Luang Prabang that traditional houses are good. It's no easy task: The people want concrete houses, like those they see in Bangkok.

Those aren't the only problems: With preservation comes tourists. UNESCO estimates that designating a place a World Heritage Site immediately increases the number of visitors by 40%. Tourists who formerly wished to visit the old Chinese caravan city of Lijiang had to drive for 2 days from Kunming. Now Lijiang is a designated heritage site, and tourists fly to the city's new airport. Do they notice that Han Chinese are displacing Lijiang's native Naxi people? Chinese, not Naxi, serve as guides and run tourist shops selling bogus goods.

CHOICES

Is it moral to spend money on heritage preservation when people are hungry? The question often arises. India's supreme court recently ordered small Delhi workshops to stop dumping their waste in the grossly polluted Yumuna (Jumna) River. The companies promptly went on strike, arguing that their survival was more important than the river's purity. There are some 90,000 such factories. Most are small, but if you figure 10 employees per factory, that's a lot of jobs. What comes first? The problem is that historic cities are like old-growth forest: Once you destroy them there's no going back. The best practical solution may be to base preservation on economic arguments, so the two goals are compatible, instead of conflicting.

CHAPTER 22

COUNTRYSIDES ABROAD

On the face of it, this last chapter is absurd: One chapter covering almost everything. The dark moist earth of the Nile Delta, the ragged sugarcane plantations of Trinidad, the lonely crofts of the Scottish Highlands: They're all here, along with a thousand other places. But there hasn't been a chapter yet whose subject couldn't fill a bookcase. Besides, I'm going to keep this last chapter short and focus on only two points.

The first is that we're looking here at population source areas, the places from which the world's cities are drawing migrants. From Mexico to Argentina, Morocco to South Africa, Egypt to the Philippines—almost wherever you look, people are leaving the land. Some migrants come from prime agricultural lands that are now being farmed with methods that require fewer and fewer hands. Others come from places so dry or hilly or infertile that they are being abandoned. Either way, people are leaving. "No jobs," says a taxi driver in Delhi who's explaining why he left his ancestral home in the Himalayan foothills.

Just possibly—and this is the second point—some of these marginal lands may become agricultural-heritage landscapes, the rural equivalent of protected urban areas. Call it our final look at the romantic impulse.

SOURCE AREAS

Look again at the demographics. China's rural population is growing at only 0.2% annually. Not 2%: two-tenths of 1%. The birthrate's higher than that, so the explanation is that people are calling it quits and moving to town. China's rural population constitutes a large part of the 400 million Chinese making less than $2 a day. They're a big part of the 30 million who fall beneath the Chinese government's own grinding definition of poverty, which is an income of less than $220 a year.

Think of a barefoot young Chinese farmer walking in deep mud behind an ox and occasionally grunting directional orders. "In the city, they have everything. Here, we have nothing." That's one Chinese peasant. Another adds, "If we had more electricity, we could use an electric stove, tape player, a rice cooker, a washing machine." These aren't the voices of happy, contented people.

Think of a Vietnamese farmer who's only 30 miles from Ho Chi Minh City (Saigon) but who says that nothing's changed in her lifetime. Think of a farm laborer in Bangladesh who uses a long-handled wooden mallet to break clods in a field sprinkled with thousands

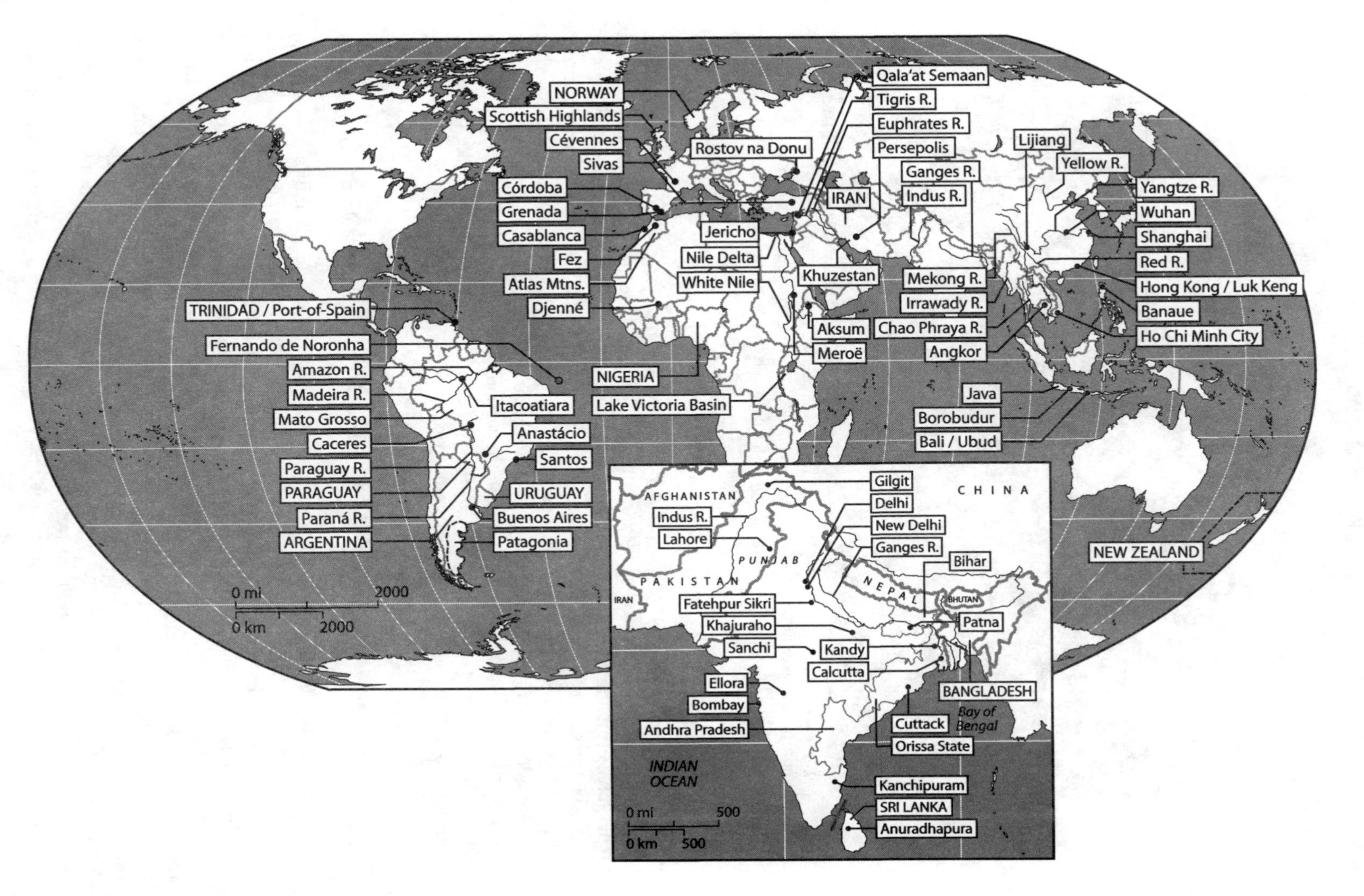

NORWAY
Scottish Highlands
Cévennes
Sivas
Córdoba
Grenada
Casablanca
Fez
Atlas Mtns.
Djenné
TRINIDAD / Port-of-Spain
Fernando de Noronha
Amazon R.
Madeira R.
Mato Grosso
Caceres
Paraguay R.
PARAGUAY
Paraná R.
ARGENTINA
Itacoatiara
Anastácio
Santos
URUGUAY
Buenos Aires
Patagonia
0 mi
2000
0 km
2000
Rostov na Donu
Jericho
Nile Delta
White Nile
NIGERIA
Lake Victoria Basin
Qala'at Semaan
Tigris R.
Euphrates R.
Persepolis
Ganges R.
Indus R.
IRAN
Khuzestan
Aksum
Meroë
Mekong R.
Irrawady R.
Chao Phraya R.
Angkor
Lijiang
Yellow R.
Yangtze R.
Wuhan
Shanghai
Red R.
Hong Kong / Luk Keng
Banaue
Ho Chi Minh City
Java
Borobudur
Bali / Ubud
NEW ZEALAND
AFGHANISTAN
CHINA
PAKISTAN
PUNJAB
IRAN
NEPAL
BHUTAN
Gilgit
Delhi
New Delhi
Ganges R.
Indus R.
Lahore
Bihar
Fatehpur Sikri
Khajuraho
Patna
Sanchi
Kandy
Calcutta
Ellora
Bombay
BANGLADESH
Bay of Bengal
Andhra Pradesh
Cuttack
Orissa State
INDIAN OCEAN
Kanchipuram
SRI LANKA
Anuradhapura
0 mi
500
0 km
500

A moment's pause in an otherwise endless task of clodbreaking near Kushtia, Bangladesh. A plow lies on the ground. A typical farmstead is in the background. In the summer, it will be surrounded by rice paddies.

more, all waiting. Think of an Egyptian peasant sitting next to a ditch and lifting a thin stream of water onto a cotton field. Gravity might bring water to the fields, but Egyptian ditches have been dug at a level that forces farmers to lift every drop. It's a deliberate way of forcing farmers to conserve water. Viewed more broadly, it's a strategy that reduces the waste of water by requiring the waste of labor and time. If this is the peasant life, perhaps we should do everything in our power to eradicate it, even if that means making the world look like the American Midwest. The Shah tried it. He imported American farmers to Khuzestan, in the southwest corner of Iran. For a while, the place looked like California's Central Valley, with huge fields of irrigated tomatoes.

PRODUCTION PROSPECTS FOR PRIME LANDS

It's already happening on some of the land best suited to farming. A million tons of soybeans are shipped annually down the Madeira River to the Amazon port of Itacoatiara. The beans are removed by giant vacuums and dumped in freighters heading to China. Other growers ship by truck and wait for the *Ferronorte*, a railroad scheduled to open in 2004 and connect Mato Grosso to the port of Santos. Another transportation system is in the works, too. It's the Hidrovia Scheme, which aims to canalize the Paraguay River and form a 10-foot-deep navigation channel running 2,000 miles south from Caceres in Mato Grosso through Paraguay to Buenos Aires. All told, Mato Grosso's prospects are bright enough that American soybean producers are now buying Brazilian land. Good land in Mato Grosso costs about $700 an acre, one-fifth the price in Illinois. Brazil is also looking now to beef exports. "This story is only just beginning," says the head of Independencia Alimentos, in

Anastacio, Mato Grosso do Sul. The company raises grass-fed beef and is already supplying beef slaughtered in accordance with both Muslim and Kosher law.

Then there's Argentina. During the 1990s George Soros invested heavily in Cresud, an Argentinian corporation that aims to make a fortune in crop and livestock production. In 1994, the company owned about 50,000 acres of the Pampas; by 2001, it owned more than 1 million. Soybean profits were good, especially after the collapse of the Argentine peso, because soybeans are sold abroad for dollars. The prospects for beef exports suddenly looked good, too. Since the 1930s, unprocessed Argentine beef has been excluded from world markets because of hoof-and-mouth disease, but in 2000 the country was declared disease-free. Argentina's ranchers and processors now hope to export beef to Korea and Japan, even though those Asian markets are a month away by sea, three times the shipping time from Australia and the United States. Argentine beef would be competitive in the United States and Europe, too, if quotas didn't nearly exclude it. It's a familiar story. One Argentinian dairy farmer, thinking of the tariffs levied on his products, says that the United States makes more money from his milk than he does. Unless the rich countries lower their tariffs, he continues, "they're condemning us to eternal recession and the debt spiral."

Modernization is coming more slowly to Russia, though its wheat exports are growing. In 2002–2003 they exceeded the previous record, which was set (unbelievably) in czarist Russia in 1913. There's a transport bottleneck at Russia's ports—the port of Rostov is especially important for wheat—but it could and probably will be fixed. Also, Russian farmers use very little fertilizer. If they applied it the way Americans do, Russia could easily rival the United States as a wheat exporter. For the time being, Russia exports 10 million tons annually, less than half the 25 million tons exported by the United States. Its customers are chiefly around the Mediterranean Basin, reached through the Black Sea. But farmers in South Dakota are already amazed: "We never thought Russia could come back," says one.

Keep an eye on the Ukraine, too. Its grain and beef could be very competitive if the Ukraine's institutional problems are resolved, if its product quality is satisfactory, and if health-conscious Americans keep eating beef.

Another more distant possibility is grain production from the plain of the White Nile in southern Sudan. Arab-funded plans for the development of this huge area of flat, fertile soil came to nothing when civil war broke out between the government and the Dinka, Nuer, and Shilluk. When peace comes, there will be another shock, because these tribes are immensely devoted to their cattle and may not welcome intensive farming.

Last and perhaps least, we should look for modernization on the prime lands of the North China and Indo-Gangetic plains. I say last not because these places are physically poor. They're not. When the British after 1900 established an agricultural-research institute in India with funds donated by a visiting Pittsburgh industrialist, they chose to locate it at Pusa, now an hour's drive north of Patna in the lower Ganges Valley. They did so because the local soil and climate could grow anything. Critics at the time argued that Pusa was too remote, and a devastating earthquake 30 years later provided a convenient excuse to move the establishment to Delhi, where the Pusa Institute still exists. (The name Pusa, by the way, is not the name of an Indian village. It's an acronym for Phipps U.S.A., in honor of Henry Phipps of Pittsburgh.)

I say we should look last to these areas because they are so densely populated that you can't modernize their agriculture without immense social dislocation. That kind of upheaval will and probably should eventually take place. Certainly the Chinese and Indian governments are willing to force millions of people to get out of the way of reservoirs. Still, those changes are almost trivial compared to the displacements necessary to modernize the agri-

Fruit for sale, Suzhou.

culture in the great plains of China and India. The chief executive of Cargill, a huge private company involved in the global marketing of food, puts it this way, with regard to China: "What are you going to do with 600 million people? They are going to move to the big cities, and that's the last thing China needs—more people heading for the big cities. . . . It's a social problem, not just an economic problem."

Still, the train's moving, at least in China. Chinese citrus growers fear that falling tariffs on imported citrus will put them out of business, because Chinese oranges have rough, thick, scarred skin. Sunkist representatives meanwhile visit Shanghai.

It's not all bad news for the Chinese. Their apple production rose from 4 million tons in 1990 to 23 million in 2000. The market is almost all domestic, but the best of the apples are export quality. You'd think you were in a Japanese market when you see how the best Chinese apples are individually wrapped in foam for sale in China's larger towns. Those same high-quality apples are already pushing American apples aside in Hong Kong and Singapore. Taiwan is next, now that China has entered the World Trade Organization. China could even export apples to the United States, but the Chinese remember 1994, when their garlic exports to the United States jumped from 3 million to 64 million pounds in one year. The next year the United States responded with a 376% duty on Chinese garlic. In defense, the American growers point out that Chinese farm workers make $2 a day. Their full-time California counterparts make $12 an hour, plus benefits.

How will it work out? The likeliest prospect is for Chinese farmers to move away from wheat, corn, rice, and cotton. Those staples will be increasingly imported, and the Chinese will concentrate on fruits, vegetables, and meat, which will increasingly be exported. This change is already happening: Chinese production of rice, corn, and wheat fell 18% between

1998 and 2003. The drop was so swift that the government grew alarmed and quickly offered subsidies to grain producers. The emergence of superhybrid rice may yet allow China to grow much of its rice requirement on a relatively small amount of land.

Meanwhile, the huge populations along the lower Ganges watch on television as the distant world changes. Ask why the plains around Patna, naturally so rich, fall so far short of what they could produce, even with the tiny holdings characteristic of them. Politics will be the answer: The Bihar government fails to support better farming methods, marketing mechanisms, and infrastructure like telephones and roads.

HALF FULL OR HALF EMPTY?

Go to a supermarket in Muscat, the long-shrouded capital of Oman, and you can buy contorted Omani carrots. But you can also buy Egyptian and South African oranges, Turkish grapefruit, Ugandan avocados and pineapples, Indian onions, Chinese garlic, and even American apples. Go to Sri Lanka and stop at the village of Tambuteggama, nearly 100 miles from Colombo and 15 shy of Anuradhapura. There's no supermarket here, but there's a roadside shack with local mangos, pineapples, coconuts, and bananas. There are local eggs. There are oranges. The shopkeeper says he has two kinds: one from Pakistan and the other from Australia. He has apples, too. From the highlands above Kandy? No: from China.

How do people here buy such things? It's part of a bigger question: Across the street an agricultural-equipment dealer sells hand tractors from China, along with bicycles and

Out of the way, but not out of the global flow: Tambuteggama, Sri Lanka.

motorcycles from India. Those tractors cost over $1,000, and they're replacing animals. People say it's because Sri Lanka's so crowded now that it's very difficult to find fodder for buffalo. Still, there's money to spend, on tractors as well as on Chinese apples, and there's a global market in which to shop.

For every man walking behind his plow ox in the Indian Punjab, you see another driving a tractor or even a combine. Those machines are harbingers of a future when many farm laborers in this part of India will be out of work. Punjab farms are already getting larger. They're also much more rationally organized than in other Indian states, because after independence the Punjab and neighboring Haryana carried out a thorough reparcellation. A checkerboard of tidy squares replaced the aboriginal maze in which a landowner owned this sliver and that bit. Reparcellation made the use of machinery feasible. The system is in jeopardy now because, as the reform generation dies, ownership of the squares is being divided among children. Partible inheritance is the law, but it's no way to make a living. In practice, one son stays to farm; the others leave. As more efficient methods are adopted, the temptation will be for smart operators to rent or buy neighboring squares. It will be a visible change, because the Punjab and neighboring Haryana are the only states where Indian farmers live, as American farmers do, on their land. Doing so didn't make sense when holdings were scattered in a dozen fragments; now it does, in these states. With enlargement, many of those isolated farmsteads could be empty or gone in a generation or two. The Biharis who often work as seasonal laborers here won't be needed.

We're looking now at the social cost of modernization. Indian farmers with a couple of scattered acres down south try making money with cotton, billed as white gold. They get hit with disease and buy pesticides. Crop prices fall, leaving the farmers deeply in debt. Each year, dozens commit suicide by drinking what's left in the pesticide container. In Andhra Pradesh alone, the government says that 3,000 farmers have killed themselves since 1997. Their widows become field laborers, paid 45 cents per day.

The Turkish villagers of Calli, near Sivas, got so far into debt to buy seed and fertilizer that in 2001 they put their entire village up for sale. Anyone who paid off the villagers' debts of about $1 million could have 120 homes, plus all the village land.

Over on the other side of the world, in Mexico, there are 1 million migrant families—*jornaleros*, day workers—who travel around the country picking vegetables, often for the American market. They are like the migrant workers who come to the United States, but they're paid even less, about $5 a day. Their children work too, in the hope that the family will be able in the summer to bring $1,500 back home, where the farm is too small to support everyone.

PRESERVATION OF MARGINAL LANDS

When you examine the world's better cropland there's a bright side to agricultural modernization; there isn't when you consider the world's marginal lands. The Sahel since the 1970s has become a byword for food insecurity and wholesale migration. A grim story could be told, too, about the so-called rainfed areas of India, where a failed monsoon means that sorghum and millet shrivel. A third bitter case is the newly cleared lands of Amazonia, where farmers once looked at the forest and assumed the soils were rich. Not so. The nutrients are cycled between the trees and the topsoil; clear the trees, and the nutrients ebb into the subsoil, beyond crop roots. Pioneer farms are engulfed by cattle ranches.

When urban opportunities present themselves—even when it's merely the illusion of opportunity—such land is abandoned, just as it was in the farms of New England a century ago. In 1989 the paddy fields behind Luk Keng, an outlying village in Hong Kong, were still in production, sustained by water from a newly lined irrigation ditch. A few years later, the ditch was dry, and the fields were full of weeds. It's not unique. Frustrated irrigation engineers talk about paddy farmers on Malaysia's Penang Island who walk away from their land and go to work in nearby factories. In Korea, too, paddy lands are interspersed with fields of weeds. It was such a sight, and the heartbreak associated with them, that led Lee Kyang Ja, a perennial protester at World Trade Organization meetings, to go to a meeting in Mexico in 2003 and fatally stab himself. A neighbor in Jangsu said, "Mr. Lee committed suicide to save the farmers. He sacrificed himself for farmers like me."

Forestry can be profitable on some of this land. International Paper has established fast-growing pine plantations on over 1 million acres of abandoned sheep pasture in New Zealand. Weyerhaeuser's doing the same thing not only in New Zealand but in Uruguay, where it has a 300,000-acre tree farm on old sheep ranches.

Rather than watch these places revert to forest, we could preserve some of them as traditional agrarian landscapes. These places, after all, are not just places of crushing poverty; they're also places of surpassing beauty. Some of the poems of Rabindranath Tagore (1861–1941) touch on this, particularly those in Tagore's Nobel prize-winning collection *Gitanjali*. But perhaps it's better to draw on a prosaic author, one whose style is literal and flat. Prafulla Mohanti's *My Village, My Life* (1973) includes a recollection of childhood in a village near Cuttack, in Orissa. Mohanti writes: "I wandered by myself for hours. I watched the palm trees swaying in the wind, the ripples on the river, the small fish darting in the water. I enjoyed the brilliant moonlit nights. I lay on a mat in the courtyard counting the stars and watching the moon play hide-and-seek with the floating clouds. I saw the paddy fields change through the seasons, from vivid green to gold, the mango trees blossom, the

Rice terraces in decay, along with a now useless irrigation ditch. Luk Keng, Hong Kong.

sweet-scented flowers open out in the evening, the multi-coloured birds, the brilliant sunsets and sunrises. It was magic. I felt peaceful and at one with Nature." That last sentence helps explain Mohanti's title, *My Village, My Life*. It's as though the Utilitarians had never set foot in India.

A risky but obvious way to preserve such places is to stimulate tourism. Residents of Banaue in the highland spine of northern Luzon are used to tourists coming to see their famously steep rice terraces. So are the people around Ubud, the cultural center of Bali. Chinese villages in Yunnan meanwhile lay claim to being Shangri-La. The claim is spurious—the place is fictional, from James Hilton's *Lost Horizon* (1933). Still, the Chinese government in 2002 allowed the villagers of Zhongdian, northwest of Lijiang in northwest Yunnan, to change their village's name to Xiang-ge-li-la. Shangri-La is also claimed by northern Pakistan, where Hunza villagers rent out stone cottages surrounded by apricot orchards irrigated by snowmelt from towering mountains. These are not the Blenheim apricots of American commerce. They're so tangy that the locally bottled juice burns the throat.

How should tourism be stimulated? An obvious first step is to do no wrong. It sounds simple, but there's a famous West Bank village named Artas. The Crusaders thought it was the site of Solomon's gardens, so they called it Hortus, or garden, which was corrupted over the centuries into Artas. It's just downstream from some Roman pools, misattributed to Solomon. The Crusaders made lots of such mistakes, but Artas attracted Europeans who came in the 19th century and built simple cottages in the Palestinian style, with massive masonry walls bearing a heavy masonry dome. They reintroduced fruit trees to what had become a chronically insecure area, plagued by Bedouin raiders. The trees were irrigated by an ancient village spring.

There are Palestinians today who would like to build a tourist center here—there's already a simple museum—but Artas is being visually destroyed. A very large Israeli settlement named Ephrata is steadily encroaching from the south, while on the north a Palestinian marble-cutting mill piles waste into the valley. In the very center of the gardens, Palestinians have erected plastic greenhouses.

A few miles away there's a lovely valley with a name that doesn't fare well in English. Still, Wadi Fukin (pronounced foo-KEEN) has an ancient spring, rock-cut irrigation channels built by the Romans, and maybe 50 acres of vegetables stretching down the valley from the settlement at its head, near the spring. What does a helpful American irrigation engineer do? Why, replace the open ditches with black plastic hoses that, for a while at least, don't leak. The engineer goes to a nearby village, Battir, where another spring has irrigated a narrow, steep valley since at least Roman times. He replaces the rock-cut channel with a buried pipe. Other consulting geniuses go down to Jericho and see water running down unlined ditches to orange orchards. "Waste!" they cry. It's a no-brainer for them: Water in pipes won't be lost to seepage. They're right. But the name of the game is income, not acre-feet, and tourists won't come to a village that engineers have improved in this way. There's no reason to stroll along a buried pipe. That's why the Acequia Madre in Santa Fe still burbles along.

Every now and then, engineers do get it right. The Aga Khan's rural support program in northern Pakistan often builds new irrigation channels, but they're left open and unlined. The reason has nothing to do with esthetics but with the simple fact that villagers can maintain and modify an open channel much more easily than they can work on a buried pipe or a concrete-lined ditch. So what? Take what victories you can.

An obvious second step is to link such tourism to architectural monuments. Many thousands of visitors come each year to Cambodia's Angkor and Java's Borobudur, both of

Battir, Palestine: the village spring is at the upper right, and the water is piped to the slide in which it descends to the storage pool, where it accumulates at night for use the next day. An Israeli road scars the distant hill.

which are located in agrarian environments and were built by people who understood exactly what Mohanti means when he writes about the magic of the countryside. There are other important Buddhist centers that, with better infrastructure, could accommodate many more visitors. Among them are Burma's Pagan, Sri Lanka's Anuradhapura, and India's Sanchi. India is awash with Hindu temples in rural settings: Khajuraho and Kanchipuram are two of the best known. It has Islamic monuments, too. A good example is Akbar's abandoned capital, Fatehpur Sikri, an hour's drive west of Agra.

Other Islamic monuments stretch all the way west to the spectacular adobe architecture of West Africa, best known from the much photographed grand mosque at Djenné, in Mali. If you want a sense of the market potential of these places, just visit Granada. There, the flimsy palace of a third-rate sultan—so poor that the fabulous ornamentation is in plaster, instead of the stone used earlier at Córdoba—has been a cultural magnet for over a century. Granada's economy today depends heavily on the thousands of visitors who storm the Alhambra daily except Christmas, but there's hardly a city from Fez to Lahore that doesn't have equally interesting buildings.

There are spectacular Christian monuments at Ethiopia's Aksum and Syria's Qala'at Semaan, which is the ruined church and monastery of Simeon Stylites. There's the Holy Land itself, not only the monuments of Jerusalem, Bethlehem, and Hebron but the hills around them. For the time being, politics make a mockery of any program to develop tourism here, but the potential is immense. Terraced olives and vines cover the moister slopes facing the Mediterranean, while rugged cliffs and ravines overlook the Jordan Valley and Dead Sea. Link these places to tourism development in Jerusalem, Bethlehem, Hebron, and the other places with biblical associations—places like Jericho, Beth-el, Shilo—and you could generate more money than will ever come from the high-tech activities that Palestinian planners usually advocate.

There's no guarantee that farmers in such places will continue farming, once they're surrounded by hotels, restaurants, and craft shops. On the contrary, tourism can lead to derelict fields as new, more profitable jobs become available. An alternative approach to preserve traditional rural landscapes is through direct government intervention. Heaven knows, the idea of subsidizing farmers shouldn't come as a surprise to Americans, Europeans, or the Japanese. One country that has actually acted to save traditional agricultural landscapes in this way is France. In 1970 it created a national park in the Cévennes, where population peaked before 1850. Farmers in the park are paid to farm as their grandfathers did and to keep buildings in the old style. If improvements are needed, and if new materials are cheaper than the old ones, then the farmer uses the old ones and the government pays him the difference. In Norway, meanwhile, hundreds of millions of dollars are spent on tunnels to remote islands with tiny populations that the nation is determined to support in a traditional setting.

Land trusts could play a role here in this work, as they do in the United States. Still another approach, on the user-fee principle, is to charge visitors a daily fee. Brazil does this for visitors to Fernando de Noronha, an island off its easternmost point, Cape São Roque. Each visitor pays about $8 a day, and with tourists limited to 500 at any time, the fees generate about $100,000 a month.

Besides stimulating tourism and subsidizing traditional farming, governments can protect the producers of regionally unique products. In the United States, for example, a wide range of cheese is legally sold as parmesan. Not so in Europe, where only cheese from Parmigiano–Reggiano can carry that label. The European Union now protects some 600 regional foods, including Kalamata olives—cheeses including feta, gorgonzola, roquefort, provolone, fontina, and asiago—Parma ham, Orkney beef and lamb, champagne, Kentish ale, and items for dessert like Cornish clotted cream, Lübecker marzipan, and the Wachauer marille, a special apricot. It's a program that can be traced back to 1935, when France instituted its AOC (*appellation d'origine controlée*) system of protecting wines from particular locales.

In 2003 the European Union formally asked the WTO's TRIPS Council (the Trade-Related Aspects of Intellectual Property Rights Council) to establish a global register of 41 of these products. Wisconsin cheesemakers fought back. They didn't want to have to label their feta as Mediterranean white cheese. The original producers of the products, however, argue that their economy depends on this protection. They may get help from the makers of Napa Valley wine and Canadian rye whisky. Maybe California's asparagus growers will join up: They've seen their production slide from 75,000 acres in the 1950s to 24,000 acres today. Mexico and Peru are shipping more and more of this very labor-intensive crop to the United States, and no California grower wants to think about China, which already grows 85% of the world total. Asparagus will probably survive in California only if farmers can persuade consumers to spend more for a local product.

The same strategy could help the producers of many non-European crops. There's Antigua coffee from Guatemala. There's India's Darjeeling tea. You can try Sri Lankan tea, too: Uda Pussellawa is particularly recommended with fish or chicken. Sri Lanka has another specialty in the king coconut. Even a staple like rice has a thousand flavors—Japan alone is said to grow 300 varieties. Basmati rice from the Himalaya foothills has lately become well-known in North America, but there are many other Indian types. Nobody who has tasted Konkan rice, from the west coast of peninsular India, will ever again think that rice has to be doctored with sauce.

Many countries, including Venezuela, Cuba, Nigeria, Morocco, and India, have expressed an interest in participating in this program. You can see why if you take a stroll in a Dubai supermarket and see mangoes from the Philippines and Australia, Alphonso mangoes from India, and two other varieties from Kenya. You'll find many varieties of dates, too, including American ones. There's no reason to restrict protection to foods, either. The Czech Republic already protects Bohemia crystal and Vamberk lace. Slovakia protects Modranská majolica, a hand-painted pottery. For many countries, craft protection could help save traditional economies and, in this way, landscapes.

How will it play out? On a purely technical level, such a program could sink under the weight of its complexities. The French experience with the AOC designations is a warning here, because with 466 designations now, consumers are overwhelmed, baffled. Fearful that they will choose wines from other countries, French vintners are now trying to revise their system and make it more comprehensible.

The same question can be asked and answered more broadly, though. On the one hand, we're irresistibly drawn to modernization. On the other, we're emotionally attached to agrarian landscapes. Why else would Filipinos at Banaue continue with what they themselves call sentimental agriculture, rice terraces maintained as a cultural rather than economic activity? Why else have hundreds of massive old olive trees been stolen from Spanish hillsides? They're trucked to Belgium and Switzerland and sold, unbelievably, for over $100,000 apiece to homeowners seriously eager for roots. It's an upscale version of American cactus rustlers, who swipe a 100-year-old saguaro and sell it for $5,000. Once again there's a contest, and it's plain to see, spread out before us in the world's cultural landscapes.

SOURCES OF QUOTATIONS

Abbreviations: BW, *Business Week*; DMN, *Dallas Morning News*; F, *Fortune*; FEER, *Far Eastern Economic Review*; FT, *Financial Times*; G, *Guardian*; GM, *Globe and Mail*; H, *Ha'aretz*; IHT, *International Herald Tribune*; LAT, *Los Angeles Times*; NYR, *New York Review of Books*; NYT, *New York Times*; SCMP, *South China Morning Post*; WP, *Washington Post*; WSJ, *Wall Street Journal.*

CHAPTER 2. HUMAN EVOLUTION, DIFFUSION, AND CULTURE

"a practice that causes so much pain": David Blunkett in Press Association, "Tougher Penalties for Genital Mutilation," G, 3.3.04.

"the hardest problem": Morten H. Christiansen and Simon Kirby, "Language Evolution: The Hardest Problem in Science?" in Christiansen and Kirby, eds., *Language Evolution*, 2003.

"of selfhood, of aesthetics": Sally McBrearty in Robert Lee Hotz, "With Ancient Jewelry, It's the Thought That Counts," LAT, 4.16.04.

CHAPTER 3. FORAGERS

"the rest of the human race as their enemies": the historic quotations regarding the Jarawa come from *http://www.andaman.org/book/originals/Jarawa/art-jarawa.htm*, sampled 10/11/03; for the Lamphere letter, see *http://www.aaanet.org/committees/cfhr/ltrjar1.htm*, sampled 10/11/03.

"nasty, brutish, and short": Thomas Hobbes, *Leviathan*, Chapter 13, paragraph 9.

"free as nature first made me": John Dryden, "The Conquest of Granada," I, i, 208–210.

"what can be more revolting": Sharon Morgan, *Land Settlement in Early Tasmania*, 1992, p. 150.

"*exterminate the brutes*": Bronislaw Malinowski, *A Diary in the Strict Sense of the Word,* 1967, p. 69.

"the slaughter-house of their own firesides": Charles Darwin, *The Voyage of the Beagle*, Chapter 10, paragraph 18.

"Arise! Arise! They come, they come!": Arthur Grimble, *We Chose the Islands*, 1952, p. 189

CHAPTER 4. DOMESTICATION

"somewhere among the forest-clad mountains": Wilhelm G. Solheim II, "An Earlier Agricultural Revolution," *Scientific American*, April 1972, p. 34.

"a more cautious interpretation": Charles Higham, *Early Cultures of Mainland Southeast Asia*, 2002, p. 53.
"entirely new relationship of subordination": Jacques Cauvin, *The Birth of the Gods and the Origins of Agriculture*, 2000, p. 69.
"a different conception of the world": *ibid.*, p. 220.

CHAPTER 5. THE DIFFUSION AND EARLY DEVELOPMENT OF AGRICULTURE

"our present knowledge does not permit": Václav Blažek, "Elam: A Bridge Between Ancient Near East and Dravidian India?" in Roger Blench and Matthew Spriggs, eds., *Archaeology and Language IV: Language Change and Cultural Transformation*, 1999, p. 54.
"values his freedom": Jacques Cauvin, *The Birth of the Gods and the Origins of Agriculture*, 2000, p. 197.
"a fugitive and a vagabond": Genesis 4:12.
"in the sweat of thy face": Genesis 3:19.
"gross degenerative disease": T. I. Molleson, "The People of Abu Hureyra," in A. M. T. Moore, G. C. Hillman, and A. J. Legge, *Village on the Euphrates: From Foraging to Farming at Abu Hureyra*, 2000, p. 356.
"binary oppositions": Julian Thomas, *Rethinking the Neolithic*, 1991, p. 182 and 28.

CHAPTER 6. THE EMERGENCE OF CIVILIZATION

"that children should command old men, fools wise men": Jean Jacques Rousseau, "A Discourse Upon the Origin and the Foundation of the Inequality among Mankind," last paragraph.
"to be governed is to be": Pierre-Joseph Proudhon in Adam T. Smith, *The Political Landscape: Constellations of Authority in Early Complex Polities*, 2003, p. 182.
"primordial protection racket": *ibid.*, p. 91.
"acquiring the status of a divine king": Aidan Southall, *The City in Time and Space*, 1998, p. 27.
"The earth was a wilderness": Adam T. Smith, *op. cit.*, p. 160.
"indirectly accentuate the power of government": Patricia Leigh Brown, "Ancient Dunes vs. Exotic Trees," NYT, 3.9.03.

CHAPTER 7. CHINA

"supposed to act like children": Jiao Guobiao in Joseph Kahn, "Let Freedom Ring? Not So Fast. China's Still China," NYT, 5.3.04.
"all under Heaven": from the Shi Jing (Book of Odes), quoted in Yinong Xu, *The Chinese City in Space and Time: The Development of Urban Form in Suzhou*, 2000, p. 63.
"the emperor was commissioned": Jeffrey F. Meyer, *The Dragons of Tiananmen: Beijing as a Sacred City*, 1991, p. 61.
"where earth and sky meet": quoted in Yinong Xu, *The Chinese City in Space and Time: The Development of Urban Form in Suzhou*, 2000, pp. 32–33.
"demarcated it as a square": *ibid.*, p. 34.
"designed to awe and affirm" " Rhoads Murphey, "The City as a Mirror of Society: China, Tradition and Transformation," in John A. Agnew, John Mercer, and David E. Sopher, eds., *The City in Cultural Context*, 1984, p. 189. Quoted in Yinong Xu, *The Chinese City in Space and Time: The Development of Urban Form in Suzhou*, 2000, p. 87.
"nine meridianal": *ibid.*, p. 34.

"in the front is the Imperial Court": *ibid.*, p. 36.
"commercial and residential activities": *ibid.*, p. 71.
"round on top": from the Bai Hu Tang, quoted by Victor F. S. Sit, *Beijing: The Nature and Planning of a Chinese Capital City*, 1995, p. 16.
"Evolved to become the basic concept": *ibid.*, p. 18.

CHAPTER 8. INDIA

"Who shed rivers of blood": Edward Luce and Demetri Savastopulo, "Blood and Money," FT, 2.20.03.
"How can I show my face": Atal Behari Vajpayee in Edward Luce, "Master of Ambiguity, India's Inscrutable Prime Minister . . ., " FT, 4.3.04.
"better to live in Bihar": Edward Luce, "Faith, Caste, and Poverty," FT, 7.4.03.
"now it is our turn to take control": *ibid.*.

CHAPTER 9. A TECHNOLOGICAL CIVILIZATION

"reject vain speculations": Francis Bacon, *The Advancement of Learning*, Chapter 5, paragraph 11.
"the effecting of all things possible": the *New Atlantis* has no subdivisions for ease of citation; the quoted material here appears on p. 288 of the edition published by Oxford University Press in 1966 as *The Advancement of Learning and New Atlantis.*
"cultivated by many centuries": *Discourse on Method*, Part One, paragraph 12.
"dominion": *Genesis* 1:26.
"the aim of Platonic philosophy": "Francis Bacon," *Edinburgh Review*, July 1837, pp. 80–81.
"wit, reason, memory like an immortal god": Kenneth Clark, *Civilization: A Personal View*, 1969, p. 89.
"in apprehension how like a God": *Hamlet*, II, ii, 315.
"this cruel, harsh, brutal agglomeration": quoted in Robert Twombley, ed., *Louis Sullivan: The Public Papers* (1988), p. 105; the original paper, "The Tall Office Building Artistically Considered," appeared in *Lippincott's Magazine*, March 1896, pp. 403–409.
"House-Machine": Le Corbusier, *Towards a New Architecture*, 1980 (1927), p. 210.
"avoiding all romantic embellishment": Robert Hughes, *The Shock of the New*, 1980, p. 195.
"my architecture is almost nothing": Tom Wolfe, *From Bauhaus to Our House*, 1981, p. 74.
"rising tier on tier": Commonwealth of Australia, Department of Home Affairs, *The Federal Capital: Report Explanatory on the Preliminary General Plan*, no date (1913), p. 3.
"Death is assumed to be the end of life": Ruth Gledhill, "Christianity Almost Beaten Says Cardinal," *The Times* [London], 9.6.01.
"Europe is no longer Christian": David Cornick, in June/July 2003 issue of *Inside Out*; quoted in Frank Bruni, "Faith Fades Where It Once Burned Strong," NYT, 10.13.03.

CHAPTER 10. GLOBALIZATION

"Arms are my theme": Camões, *The Lusiads* (1997), translated by Landeg White, p. 3.
"this execrable crew of butchers": Jonathan Swift, *Gulliver's Travels*, Part IV, Chapter 12.
"The conquest of the earth": Joseph Conrad, *Heart of Darkness*, 1902, Part I, paragraph 12.
"Whatever happens, we have got": Hilaire Belloc, *The Modern Traveller*, 1898, p. 41.
"promiscuously scattered about the beach": Charles C. Mann, "1491," *The Atlantic Monthly*, March 2002, sampled online at *www.theatlantic.com* on 7.5.03.

"the Rock of Doom": Thomas Raleigh, ed., *Lord Curzon in India: Being a Selection from his Speeches as Viceroy and Governor-General of India, 1898-1905*, 1906, p. 46.

"the treasures of India": *The Times* [London], February 4, 1847, p. 4.

"signifies stagnation": quoted in Betty Gosling, *Old Luang Prabang*, Oxford University Press, Kuala Lumpur, 1996, p. 57.

"about as much as a big retriever dog": Edward Cecil, *The Leisure of an Egyptian Official*, 1921, p. 11.

"nauseating": Guy Arnold, "Romance Raised a Global Row," GM, 6.5.02.

"the wind of change is blowing through this continent": the phrase comes from a speech that Harold Macmillan gave in South Africa at the end of an African tour in 1960; it can be heard on the BBC's sound archive at news.bbc.co.uk/hi/english/static/in_depth/uk_politics/ 2001/open_politics/foreign_policy/timeline.stm.

"our culture, our cosmology": Virginia Haoa in Héctor Tobar, "Keeping a Culture Afloat," LAT, 1.28.04.

"the Russians destroyed our way of life": Klavdia Demko in Jeanne Whalen, "In Siberia, A Race to Write a Dictionary Before the Subject Dies," WSJ, 3.26.04.

"locked into a monolingual view": Raymond Cohen, "Speaking the Same Language: The Benefits and Pitfalls of English," in Stefania Panebianco, ed., *A New Euro-Mediterranean Cultural Identity*, 2003, p. 162.

"to revisit the country she hates": Maud Gonne, "The Famine Queen," *United Irishman*, 7.4.00, contained in Karen Steele, ed., *Maud Gonne's Irish Nationalist Writings: 1895–1946*, 2004, p. 55.

"those university-educated intellectuals": Bernard Lewis, *A Middle East Mosaic*, 2000, p. 263, excerpted from Ayatollah Khomeini, *Islam and Revolution*, translated by Hamid Algar, 1981.

"We did not have interesting lives": Bharati Mukherjee, "On Being an American," *http://164.109.48.86/products/pubs/writers/mukherjee.htm* sampled 12.7.02.

"they like to wage war": Michael Vatikiotis, "Mahathir's Parting Shot at the West," FEER, 7.3.03.

"Comfort and his brother, Respectability": "Confession of a Young Bengal" (1872), in Amit Chaudhuri, ed., *The Picador Book of Modern Indian Literature*, 2001, pp. 19–21.

"We hate these straight lines": quoted in Arthur Gaitskell, *Gezira: A Story of Development in the Sahara*, 1959, p. 202.

"a clock ticking and time rushing by": Ron Suskind, "A Plunge into the Present," NYT, 12.2.01.

"Ah, what avails the sceptred race!": "Rose Aylmer," *Complete Works of Walter Savage Landor*, edited by T. Earle Welby (1927-36), volume 15, p. 399. Welby says that the plaque was attached to the tomb only in 1906.

"marriage and me will never mix": Nisha Kalro in Joanna Slater, "For India's Youth, New Money Fuels a Revolution," WSJ, 1.27.04.

CHAPTER 11. RESOURCE PRODUCTION

"borders on communism": Peter de Savary in Thomas Wagner, "Farmers May Reclaim the Highlands," WP, 4.20.03.

"the rolling, coffee-clad hills": William H. Ukers, "A Trip to Brazil," pamphlet copyright 1924 by the Tea and Coffee Trade Journal Co., New York, p. 31.

"An empire of agriculture": Frederick Adams, *Conquest of the Tropics*, 1914, p. 122.

"fruit laden and foul": Pablo Neruda, "The United Fruit Co." in *Selected Poems of Pablo Neruda*, translated by Ben Belitt, 1961, pp. 149–151.

"it may be you tomorrow": William Naggaga in Felicity Lawrence, John Vidal, and Steven Morris, "Unfair Trade Winds," G, 5.17.03.

"forced to live like rats": Salifou Kabore in Alan Cowell, "War Inflates Cocoa Prices but Leaves Africans Poor," NYT, 10.31.02.

"corrupted world trading system": John Howard in Edward Alden and Neil Buckley, "Sweet Deals:

'Big Sugar' Fights Threats from Free Trade and a Global Drive to Limit Consumption," FT, 2.27.04.
"everybody else gets killed": Luther Markwart in *ibid.*.
"we were digging holes in the ground": Hugh Grant in David Barboza, "Monsanto Struggles Even as It Dominates," NYT, 5.31.03.
"We've bet the farm": Hugh Grant in Stefan Stem, "We've bet the Farm," FT, 6.12/13.04.
"dishonest or irresponsibly naïve": Keith Buchanan, *The Transformation of the Chinese Earth*, 1970, p. 312.
"all hell breaks loose": John Aluma in Roger Thurow, Brandon Mitchener, and Scott Kilman, "Tinkering with Banana Genes Could Save Ugandan Staple, but the Seeds Stay in a Lab," WSJ, 12.26.02.
"if we miss the GM revolution": Patrick Rubaihayo in *ibid.*.
"as much diversity as a carpark": Chrissie Hynde in David Fickling, "Tasmanian Boycott Urged Over Threat to Forests," G, 3.2.04.
"Americans are like terrorists to us" Victor Omunu in Norimitsu Onishi, "Flow of Wealth Skirts Nigerian Village," NYT, 12.22.02.
"One of the most richly endowed": Diepreye Alamieyeseigha, Governor of Bayelsa State, in David White, "Shell Tries to Repair Troubled Delta Relations," FT, 2.24.04.
"a web of opaque offshore accounts": quoted in Ken Silverstein, "Gusher to a Few, Trickle to the Rest," LAT, 5.12.04.
"willful, systemic": Kathleen Day, "Record Fine Levied for Riggs Bank Violations," WP, 5.13.04.
"one party, one ruler": Elizabeth Rubin, "The Jihadi Who Kept Asking Why," NYT, 3.7.04.
"drink camel's milk": *ibid.*
"We would, one thousand times over, have preferred Shell": Lelis Rivera in James Grimaldi, "Texas Firms Line Up U.S. Aid in Peru," WP, 11.20.02.

CHAPTER 12. MANUFACTURING

"just as one pin is like another pin": Peter Hall, *Cities in Civilization*, 1998, p. 406, quoting Allan Nevins, *Ford: The Times, The Man, The Company*, 1954, p. 276.
"the most stupid man": National Geographic Society, *Historical Atlas of the United States*, 1988, p. 154.
"the crunching wheels of machinery": Peter Hall, *Cities in Civilization*, 1998) p. 310, quoting Alexis de Tocqueville, *Journeys to England and Ireland*, 1958, pp. 107–108.
"some of this plane and some of that plane": Yan Zhiqing in Ben Dolven, "Dogfight Over China," FEER, 2.6.03.
"terrible dilemma": Mike Turner in Mark Odell and Peter Spiegel, "Boeing 'Too Late' to Catch Rival Airbus," FT, 2.27.04.
"Boy, we needed to get that kind of organization": Jack Smith in Alex Taylor III, "GM Gets Its Act Together. Finally, How America's No. 1 Car Company Changed Its Ways and Started Looking like . . . Toyota," F, 3.21.04.
"the fast that eat the slow": Assif Shameen, "You Want It, We'll Make It," FEER, 3.20.03.
"It's not what the consumer wants": David Kolpak in Amy Tsao, "Campbell Soup Is Leaving Investors Cold," BW, 3.11.03.
"no end to the red ink": Rick Scully in Gabriel Kahn, Dan Bilefsky, and Christopher Lawton, "Burned Once, Brewers Return to China—with Pint-Size Goals," WSJ, 3.10.04.
"not a national market": Wai Kee Tan in *ibid.*
"It tore our hearts out": William Kilgallen in Joseph Kahn, "An Ohio Town Is Hard Hit as Leading Industry Moves to China," NYT, 7.7.03

"Those days are gone": Norman Allard in Jeffrey Zaslow and Gregory L. White, "For GM's Retirees, It Feels Less Like 'Generous Motors,' " WSJ, 2.21.03
"not an issue we have any business being in": James Pagano in Ginger Thompson, "Behind Roses' Beauty, Poor and Ill Workers," NYT, 2.13.03
"intentionally waste time": Joseph Kahn, "The World's Sweatshop—the Etch A Sketch Connection," NYT, 12.7.03.
"the pay stays the same": Isabel Reyes in Nancy Cleeland, Evelyn Iritani and Tyler Marshall, "Scouring the Globe to Give Shoppers an $8.63 Polo Shirt," LAT, 11.24.03
"We just have a big problem": Igusti Made Arka in Rebecca Buckman and Trish Saywell, "For Asia's Maids, Years of Abuse Spill Into the Open," WSJ, 2.19.04
"Comparing Chinese and Japanese engineers on a cost-performance basis": Hiroshi Matsuo in Ken Belson, "Japanese Capital and Jobs Flowing to China," NYT, 2.27.04.
"we try to take advantage of it": Gary Meyers in Peter Wonacott, "Behind China's Export Boom, Heated Battle Among Factories," WSJ, 11.13.2003

CHAPTER 13. SERVICES

"respective McGraves": Sherri Day, "After Years at Top, McDonald's Strives to Regain Ground," NYT, 3.3.03.
"not only are people not cooking": Doug H. Brooks in Karen Robinson-Jacobs and Victor Godinez, "Brinker Has Plenty of Options on Menu," DMN, 2.22.04.
"1,500 businesses": Bob Nardelli in Neil Buckley and Betty Liu, "Fixer Puts the Final Touches to a DIY Refit," FT, 7.8.03.
"the world has never known a company": the Boston Consulting Group in Anthony Bianco and Wendy Zellner, "Is Wal-Mart Too Powerful?," BW, 10.6.03.
"now we are raisins": Angel Burgos in "Wal Around the World," *Economist*, 12.13.01.
"I'm going to do what I have to do": Carl Krauss in Abigail Goldman and Nancy Cleeland, "An Empire Built on Bargains Remakes the Working World," LAT, 11.23.03.
"If we don't win at Wal-Mart": Joe Galli in Matthew Boyle, "Newell Rubbermaid," F, 12.17.02.
"It's very pure": Steven Scheyer in Jerry Useem, "One Nation Under Wal-Mart," F, 2.18.03.
"Public Enemy No. 1": Dennis Gannon in Fran Spielman, "Unions Say Wal-Mart Snookered City, Vow to Block S. Side Site" *Chicago Sun-Times*, 3.27.04.
"L.A. County is not Arkansas": Robert McAdam in Jessica Garrison, Abigail Goldman, and David Pierson, "Wal-Mart to Push Southland Agenda," LAT, 4.8.04.
"We are not going to get pushed around": *ibid.*.
"the middle class is becoming the lower class": Barbara Harrison in Charlie LeDuff and Steven Greenhouse, "Grocery Workers Relieved, if Not Happy, at Strike's End," NYT, 2.28.04.
"A good life with the income": Georgene Hanbenreisser in Ronald D. White, "This Family's Getting Out of Grocery Work," LAT, 3.2.04.
"you bet we'll lose jobs": Larree Renda in Abigail Goldman and Nancy Cleeland, "An Empire Built on Bargains Remakes the Working World," LAT, 11.12.03.
"the union is a business": Mona Williams in Nancy Cleeland and Abigail Goldman, "Unions Stage Protests Against Wal-Mart Stores," LAT, 11.22.02.
"two years to get to nothing": Tom Fisher in Patricia Sellers, "Gap's New Guy Upstairs," F, 4.13.03.
"merchandising is a combat sport": Hirotake Yano in David Kruger and Ichiko Fuyuno, "King of the Mall," FEER, 8.30.01.
"salesmanship in print": www.ciadvertising.org/SA/fall_02/adv382j/ hae382/project2/advertising3.htm sampled 1.19.04.
"sell gum to tourists": Victor Velazquez in Ricardo Sandoval, "Mexico Has Grand Plan for Baja Tourism," DMN, 11.20.02.

"our entire ecology has been 'concessionized' ": Aniseto Caamal Colón in Carol J. Williams, "An Ugly Fight at Pretty Site," LAT, 1.20.03.

CHAPTER 14. TRANSPORTATION AND COMMUNICATION

"pick up trade all along the route": Ewart S. Grogan and Arthur H. Sharp, *From the Cape to Cairo: the First Traverse of Africa from South to North*, 1900, p. viii.
"should like to have the spray of the water": *ibid.*.
"time, high time": Michael Myers Shoemaker, *The Great Siberian Railway* (1904), p. 3.
"the only mode of travelling is by post" *Handbook for Travellers in Russia, Poland, and Finland, Including the Crimea, Caucasus, Siberia, and Central Asia*, John Murray, 1875, pp. 413–415.
"not one single American transcontinental line": Frederick C. Giffin, "An American Railroad Man East of the Urals," *The Historian*, Summer 1998, p. 812.
"icebergs drifting south": Michael E. Levine in Eric Torbenson, "'Legacy' Airlines Facing Doom," DMN, 4.7.04.
"just feel sorry for them": Michael O'Leary in Kerry Capell, "Ryanair Rising," BW, 6.2.03.
"Why don't you take me to England": Felicity Lawrence, "Growers' Market," G, 5.17.03.
"within a minute of Cape Town": Ewart S. Grogan and Arthur H. Sharp, *From the Cape to Cairo: the First Traverse of Africa from South to North* (1900), p. 89.

CHAPTER 15. ROMANTIC RESPONSES

"we murder to dissect": William Wordsworth, "An Evening Scene, on the Same Subject" in William Wordsworth and Samuel Taylor Coleridge, *Lyrical Ballads*, Section 8.
"the heart follows its own path": Blaise Pascal, *Pensées sur la religion* (1670), fragment 423, p. 241; this is a literal translation, courtesy of Dr. Pamela Genova, of the sentence familiarly known in English as "the heart has reasons of which the mind knows nothing."
"this quintessence of dust": *Hamlet*, II, ii, 315.
"a tale told by an idiot": *Macbeth*, V, v, 19.
"getting and spending": William Wordsworth, "The World Is Too Much With Us," *Miscellaneous Sonnets*, 1807.
"more socialist than democratic": Fil Hearn, *Ideas That Shaped Buildings*, 2003, p. 212.
"you've never designed a tall building": Larry Silverstein in David Leonard, "Tower Struggle," F, 1.12.04.

CHAPTER 16. POLITICAL REACTIONS

"there's bad and there's worse": Orna Coussin, "The Poverty Line Runs Through the Malls," H, 6.20.01.
"when you're married with kids": Jerry Stuchly in Terry Box, "Arlington GM plant Insists on Quality as Line Speeds Along," DMN, 10.27.02.
"have to make enough to hold my house": Katherine Boo, "Letter from South Texas," *New Yorker*, 3.25.04.
"a monstrous killing": Mirjana Kontevska in Salmon Masood, "Passage Out of Poverty is Cut Short by Antiterror Snare, NYT, 5.10.04.
"the opium of the people": from the introduction to Karl Marx's "Contribution to the Critique of Hegel's Philosophy of Right," *Deutsch-Franzosische Jahrbucher*, February 1844.
"eat dirt": Muhammad Mehdi in *ibid.*.

"worse than being exploited in Fresno": Eduardo Porter, "Immigrants Keep Arriving at Record Pace," WSJ, 3.11.03.

"we have to ask ourselves what we have to offer": Heinrich von Pierer in Matthew Karnitschnig, "Vaulted German Engineers Face Competition from China," WSJ, 6.15.04.

"necessary and correct": Gerhard Schröder in Luke Harding, "Schröder Vows No Turning Back on Reform," G, 3.33.04.

"to end starvation wages": Rick Wartzman, "How Support Eroded for a Program Intended to 'End Starvation Wages,' " WSJ, 7.19.01.

"than anything since the Great Depression": *ibid.*.

"racial animosity": Alberto Alesina, Elward Glaeser, and Bruce Sacerdote, "Why Doesn't the U.S. Have a European-Style Welfare State?" Harvard Institute of Economic Research, Discussion Paper 1933, 2001, http://post.economics.Harvard.edu/hier/2001papers/2001list.html, sampled 2.28.04, p. 2.

"pull primitive peoples": Paula Weideger, "Re-Reading the Diaries of War-Time," FT, 2.21.04.

"ferociously competitive": James Kynge, "The Chinese Boom is Bound to End in Tears. But It Might Not End For Another 10 Years... With Bumps Along the Way," FT, 3.24.04.

"so rich": Alexandra Zelenskaya in David Holley, "Missing the Good Old Soviet Days," LAT, 3.7.04.

"everything was stolen": Anatoly Grachyov in *ibid.*.

"How can people live like this?": Roza Khazuyeva in Seth Mydans, "Chechens Rebuild Lives One Window at a Time," NYT, 10.12.03.

"the urban underbelly": Gloria Macapagal Arroyo in John Burton, "Poverty Stalks the Iron Lady of Asia", FT, 12.27.01.

"It's not hatred, but it's what God says": Widi bin Hasbi in Timothy Mapes, "Indonesian School Has Chilling Roster of Terrorist Alumni," WSJ, 9.3.03.

"a famous Mohammadan theological school": *The Traveller's Companion, Containing a Brief Description of Places of Pilgrimage and Important Towns in India,* compiled by Abdur Rasheed under the orders of the Railway Board, Calcutta, 1911, p. 85.

"Why must Punjab be my destiny?": Khair Bakhsh Marri in Barry Bearak, "Pakistan is," NYT, 12.7.03.

"they are a defeated people": Moshe Ya'alon in Henry Siegman, "Israel Is an Occupier With a Duty to Protect," FT, 4.22.03.

"Children Throwing Stones": Nizar Qabbani, *On Entering the Sea*, 1996, p. 147.

"the jewel of Africa": quoted in Doris Lessing, "The Jewel of Africa," NYR, 4.10.03.

"people are just worn out": an unnamed diplomat in Craig S. Smith, "Algerian President Overwhelmingly Wins Re-election," NYT, 4.10.04.

"when two elephants meet": Mayar Mayar Kuethpiny in Abdalla Hassan, "When Two Elephants Meet: Sudanese Refugees in Cairo," in *World Press Review Online*, 7.31.01.

"I would literally be up to my waistline in bodies": Romeo Dallaire in Guy Lawson, "The Rwanda Witness," NYT, 4.2.04.

"Why don't we have any aunties": Claude Hope in Emily Wax, "Rwandans Are Struggling to Love Children of Hate," WP, 2.27.04.

"unreasonably opposed": Nicol Degli Innocenti, "A Determination to Right Past Wrongs," FT, 4.13.04.

"I only care about whether we can eat": Jeanne Bazard in Tim Weiner, "Life in Hard and Short in Haiti's Bleak Villages," NYT, 3.23.04.

"Perhaps I was too tolerant": Marjorie Valbrun, "Exile in France Takes Toll on Ex-Tyrant 'Baby Doc,' " WSJ, 4.16.03.

"we've known something better": Raul Quiemalinos in Hector Tobar, "The Good Life Is No More for Argentina," LAT, 2.18.03.

"We never thought this could happen": Teresa Acuna in Larry Rohter, "Once Secure, Argentines Now Lack Food and Hope," NYT, 3.2.03.

"the first time in my life the government has done something": Maria Lopez in T. Christian Miller, "Poor's Hopes Take Root Under Chavez," LAT, 9.17.03.
"our people are being shot": William Ospina in James Wilson, "Community Peace Initiatives Rub Up Against Colombian Paramilitaries," FT, 11.21.01.
"the violence found her": T. Christian Miller, "A War Without End Roils Colombia," LAT, 2.21.03.
"Globalization is just another name for submission": Nicanor Apaza in Larry Rohter, "Bolivia's Poor Proclaim Abiding Distrust of Globalization," LAT, 10.17.03.
"we want to be our own masters": *ibid.*.
"Indians no longer want to watch Bolivia from the mountains": Manfredo Kempff in *La Razon*, quoted in *Latin Trade*, 3.04, p. 22.
"who in their right mind is going to be willing to invest": *ibid.*.
"we're sick of them killing us": Bonita Bulger in Jodi Wilgoren, "Fatal Police Chase Ignites Rioting in Michigan Town," NYT, 6.19.03.
"the tribes never lose": Dan Walters in Fred Dickey, "Who's Watching the Casinos?," LAT, 2.16.03.
"When we try to organize against the casino": Paul Muller in Rone Tempest, "Indian Casino Has Rural Residents Up in Arms," LAT, 3.28.04.
"this canyon and everything in it": Louis Sahagun, "Tribes Buying Back Ancestral Lands," LAT, 10.20.03.
"I did what was possible": Luiz da Silva in "Brazil Pledges Land to 400,000 Families," Reuters in NYT, 11.23.03.
"distributing money is not easy": Luiz da Silva in "Poor Man's Burden," NYT, 6.27.04.
"I want America to be finished": Shaheen in David Rohde, "A Dead End for Afghan Children Adrift in Pakistan," NYT, 3.7.03.
"They kill our people": Muhammad in *ibid.*.
"I will poison his pizza": Adam Davidson, "Loves Microsoft, Hates America," NYT, 3.9.03.
"the source of evil on planet earth": Ahmed Kamal Aboulmagd in Susan Sachs, "Intellectual Speaks of the Arab World's Despair," NYT, 4.8.03.
"the more poverty you have, the more fundamentalism": Turki Hamad in Kim Murphy, "Saudi's Quicksand of Poverty," LAT, 5.16.03.
"the pursuit of peace is likely to be evasive": Faisal Islam, "An Audience with the King of World Banking," *Observer*, 3.16.03.
"I saw my son change before my eyes": Safia Damir in Daniel Ben Simon, "Morocco Crosses Its Rubicon," H, 9.26.03.
"experiment with taking away subsidies": Kwame Amezah in Scott Kilman and Roger Thurow, "Africa Could Feed Itself But Many Ask: Should It?," WSJ, 12.3.02.
"people like me are looking to religion": Kemal in Michael Vatikiotis, "The Struggle for Islam," FEER, 12.15.03.
"weary of this life of humiliation and restriction": Anthony Shadid, *Legacy of the Prophet*, 2001, p. 52.
"if all the world were America": *ibid.*, p. 57.
"the emptiness people feel" Rashid Khashana in Roula Khalaf, "Resentful of Repression, Tunisians Turn to Islam," FT, 3.3-4.04.
"With an iron fist": David Remnick, "Seasons in Hell," *The New Yorker*, 4.14.03.
"God is our purpose": http://weekly.ahram.org.eg/archives/parties/muslimb/polgod.htm; sampled 1.19.04.
"Go burn a school": Mahammed Azghar in Hamida Ghafour, "Afghan Villages Locked in Grip of Taliban Forces," NYT, 3.5.04.
"the great poison": Lakhdar Brahimi in Warren Hoge, "U.N. Distances Itself from an Envoy's Rebuke of Israel and the U.S.," NYT, 4.24.04.
"the most preposterous, idealistic statement": Kim Murphy, "Saudis Take the Slow Road," LAT, 4.9.03.
"what is almost impossible": James Traub, "Making Sense of the Mission," NYT, 4.16.04.
"I am neutral and I will vote for no one": Abdramane Ben Essayouti in Yaroslav Trofimov, "Islamic Democracy? Mali Finds a Way to Make It Work," WSJ, 6.22.04.

"those who find militant Islam terrifying": MacGregor Knox, "America Is Fighting a Global Civil War," FT, 4.14.03.
"Germans are a truly chosen people": Jason Epstein, "Leviathan," in NYR, 5.1.03.
"to Americanize the world": ibid.
"We cannot leave one nation": José Zapatero in Leslie Crawford, "Do Not Fail Us," FT, 7.17-18.04.

CHAPTER 17. CONSERVATION, NATURAL RESOURCES, AND POPULATION

"a skeleton of a body wasted by disease": Plato, *Critias*, section 3 (Stephanus 111).
"no": Robert Graves, *Good-Bye to All That*, 1929, p. 300.
"more than required by international standards": Kim Creak in Richard C. Paddock, "Tasmania's Mammoth Trees Don't Fall Quietly," LAT, 2.29.04.
"Nubia, my homeland": Ali Idris, *Dongola*, 1998, p. 58.
"Support the big home": Ching-Ching Ni, "Holdouts in Yangtze Deluge," LAT, 8.13.01.
"people know you have a great nation": Lu Yenshang in Sam Howe Verhovek, "A Reservoir of Pride, Dread," LAT, 6.2.03.
"if the government wants to go ahead": Hu Huashen in Jim Yardley, "Dam Building Threatens China's Grand Canyon" NYT, 3.10.04.
"like monkeys or ape-man": Wufan in ibid.
"I don't think I've ever heard": He Daming in Jim Yardley, "Beijing Suspends Plan for Large Dam," IHT, 4.8.04.
"ominously evident": Conference of the Governors of the United States, *Proceedings* (1908), pp. 7–9.
"This is a hill we're prepared to die on": Andy Horne in Tony Perry, "Farmers Oppose Call to Idle Land," LAT, 6.17.02.
"without water, the Imperial Valley is nothing": Stella Mendoza in Tony Perry, "Inland Water Sale Rejected; Coastal Cutback Threatened," LAT, 12.10.02.
"Wow, how generous": Stella Mendoza in Tony Perry, "Imperial Farmers Should Get Less Water, U.S. Report Says," LAT, 7.4.03.
"kangaroo court": Andy Horne, in Tony Perry, "Imperial Farmers."
"the deck is stacked against us": John Pierre Menvielle, in Tony Perry, "Imperial Farmers."
"a smooth, closely shaven surface of green": quoted in Preston Lerner, "Whither the Lawn," LAT, 5.4.03.
"two tons of gravel": J.C. Davis in Bettina Boxall, "Water Conservationists Step on the Grass," LAT, 9.6.03.
"living in unbelievable misery": Kenzo Oshima in David Finkel, "The Road of Last Resort," WP, 3.18.01.
"everything is lost": Vicente Vasquez in T. Christian Miller, "Severe Drought Imperils Poor of Central America," LAT, 8.17.01.
"destroy the meters": Richard Makolo in Ginger Thompson, "Water Tap Often Shut to South Africa Poor," NYT, 5.29.03.
"the power of population is indefinitely greater": E. A. Wrigley and David Souden, *Works of Thomas Robert Malthus*, 1986, volume 1, pp. 3–10.
"undue stress on fundamental rights": Amy Waldman, "States in India Take New Steps to Limit Births," NYT, 11.7.03.

CHAPTER 18. POLLUTION, BIODIVERSITY, AND CLIMATE CHANGE

"the most significant noncompliance pattern": Sylvia Lowrance in Bruce Barcott, "Changing All the Rules," NYT, 4.2.04.
"such a huge loophole": Frank O'Donnell in *ibid.*.

"the most harmful and illegal air-pollution initiative": *ibid.*

"an ignominious end": Seema Mehta, "Christmas Trees That Appeal to the High-Minded," LAT, 12.14.02.

"treat their pets better than human beings": Meeka Mike in Tom Cohen, "Inuits Defend Their Seal-Hunting Traditions," LAT, 3.23.03.

"everything is going": Frank V. Paladino in Kenneth R. Weiss, "Sea Turtle Is Losing the Race," LAT, 4.21.03.

"We ate them": Ransom A. Myers in Kenneth R. Weiss, "Seas Being Stripped of Big Fish, Study Finds," LAT, 5.15.03.

"if you get enough caviar, you sell it": Yuri Slobodchikov in Kim Murphy, "A Loss to Man and Nature," LAT, 11.8.03.

"we had it all at our fingertips": Henry Scow in Clifford Krauss, "Canadian Indians Challenge Fish Farms in Court," NYT, 9.14.03.

"they will destroy the Great Lakes": Richard Daley in Jeremy Grant, "Great Lakes Beware, the Asian Carp Are Here," FT, 7.17/18.04.

"They scare me": Louie Galeano in Joshua Partlow and David A. Fahrenthold, "Potomac Fishermen Enlisted to Expunge Snakeheads," WP, 5.14.04.

"We just don't know when": Rone Tempest, "Showdown at the Blair Ranch," LAT, 2.9.03.

"if you have animals in the zoo": Simon Levin in Kenneth R. Weiss, "Action to Protect Salmon Urged; Scientists Say Their Advice Was Dropped From a Report to the U.S. Fisheries Service," LAT, 3.25.04.

"the Grand Canyon river corridor is getting nuked": David Haskell in Bettina Boxall, "A River Losing Its Soul," LAT, 5.10.04.

"to incorporate conservation science": Bruce Nelson in Amy Yee, "Office Depot in Forest Conservation Alliance," FT, 3.23.04.

"a war of extermination against our indigenous communities": Tim Weiner, "Growing Poverty Is Shrinking Mexico's Rain Forest," NYT, 12.8.02.

"will have the tough job of explaining what a fish is": Klaus Topfer in "Protecting the Oceans," IHT, 9.9.03.

"oceans are in trouble": James D. Watkins in Kenneth R. Weiss, "Panel Presses New Ocean Safeguards," LAT, 4.20.04.

"most of the warming observed over the last 50 years": Intergovernmental Panel on Climate Change, *Climate Change 2001: Synthesis Report, Summary for Policymakers*, www.grid/no/climate/ipcc-tar/vol4/english/pdf/spm.pdf, sampled 7.20.04.

"seems determined to weaken the rules": Eric V. Schaeffer in Christopher Drew and Richard A. Oppel Jr., "How Industry Won the Battle of Pollution Control at E.P.A.," NYT, 3.6.04.

CHAPTER 19. AMERICAN CITIES

"low & convenient, and the streets light and airy": quoted by Roger K. Lewis in "Testing the Upper Limits of D.C. Building Height Act," WP, 3.16.03.

"cold, harsh, hard to get to": David Childs in David Leonard, "Tower Struggle," F, 1.12.04.

"skyscrapers chill communication": Kevin Helliker, "Sprint to Build Own Metropolis with Banks, Dry Cleaners, Zip Code," WSJ, 10.21.98.

"highways as forbidding as moats": Ross Miller, *Here's the Deal: The Making and Breaking of a Great American City*, 2003, p. 15.

"marvels of dullness": Jane Jacobs, *The Death and Life of Great American Cities*, 1972, p. 4.

"seems natural to want something that feels a little old": Paul Sternberg in Simon Romero, "SoHo-Inspired Lofts With Views of Houston," NYT, 8.9.03.

"defined by two theme parks": Carol Schatz in Allison B. Cohen, "Buying into L.A.," LAT, 4.13.03.

"the opportunity to grow": Stacey Dwyer in Sharon O'Malley, "The Dominators," *The Builder*, May 2002, *www.builderonline.com* sampled 2.29.04.
"buying and entitling": Ian McCarthy in Alison Rice, "Endangered Species," *The Builder*, May 2002, *www.builderonline.com* sampled 2.29.04.
"stuck with the banks": John Stanley in *ibid*..
"This is Minnesota": Jodi Shafer in Eric Slater, "10-Year-Old Mall of America Discounts Doubters," LAT, 9.15.02.
"This is not going to be a Rodeo Drive": Anne Davidoff in Steve Quinn, "Willow Bend Finding Its Stride," DMN, 12.22.02.
"This was his deal": Bob Warren in Scott Farwell and Lee Powell, "Stonebriar Centre Lands Frisco in Big Leagues," DMN, 3.31.03.
"the only place we like to walk around": Alexander Aledia in Valli Herman-Cohen, "World Class, Next Door," LAT, 11.30.01.
"it sends a message": Robert Bobb in Carol Pogash, "Idea Aims to Slam Door on Urban Blight in Oakland," LAT, 3.30.03.
"there isn't a single person in this room who could afford": Rep. Farr in Dan Baum, "The Battle of Ft. Ord," LAT, 4.27.03.
"being pushed elsewhere is their lot": *ibid*..
"treated like a crop on the plantation": Camille J. Strachan in Constance L. Hays, "For Wal-Mart, New Orleans is Hardly the Big Easy," NYT, 4.27.03.

CHAPTER 20. RURAL AMERICA

"one vast city-line along the Atlantic": Patrick Geddes, *Cities in Evolution*, 1968 (1915), p. 49.
"to reduce costs and improve operating margins": Sykes Enterprises annual report for 2002, http://media.corporate-ir.net/media_files/NSD/SYKE/reports/sykes_compiled2.pdf sampled 3.27.04.
"We plan to stay": Jim Marsicano in David Streigfeld, "A Town's Future Is Leaving the Country," LAT, 3.27.04.
"I was watching a TV program": John Clay Stanley in *ibid*..
"the working class will simply be ruined" Lewis Loflin in *ibid*..
"not quite as bad as being a nuclear-waste dump site": John Clay Stanley in *ibid*..
"don't call it a dump": James Latham in Andrew Jacobs, "In One Small Town, Radioactive Waste Is a Welcome Sight," WSJ, 3.28.04.
"we're happy to have this stuff": Tim Moore in *ibid*.
"the best job there is": Eddie Green in *ibid*..
"your average county garbage dump": Bob Guild in *ibid*..
"drinking the water all my life": Tim Moore in *ibid*..
"I have striven in vain": Josiah Gregg, *Commerce of the Prairie*, 1844; there are many editions of this work, but all the quotes here come from the last three pages of "Preparation for Returning Home," which is Chapter 8 of the second volume.
"Visit Biltmore Estate": *http://www.biltmore.com/visit/index.html* sampled 1.15.04.
"people who vote against this today": David Firestone, "Drilling in Alaska, a Priority for Bush, Fails in the Senate," NYT, 3.20.03.
"there's something out there more powerful than any senator": *ibid*.
"Gucci sprawl": Peter Whoriskey, "Density Limits Only Add to Sprawl," WP, 3.9.03.
"ambitious, controversial, and failed": D'Vera Cohn and Michael Laris, "Loudoun Leads Nation in Growth," WP, 4.8.04.
"left-wing extreme enviro member": *http://www.sierratimes.com/pf.php* sampled 1.15.04. The member in question was Lois Capps.
"not a feasible addition": National Park Service news release of 4.8.03 at *www.nps.gov/pwro/gaviota*, sampled 3.1.04.

CHAPTER 21. CITIES ABROAD

"Stop flying toilets": Davan Maharaj, "Kenyan Slum's Rent Battles Reflect Plight of Migrant Workers," LAT, 12.10.01.
"as long as we're spared drugs": Maria de Lourdes Luna in Miriam Jordan, "A Rio de Janeiro Slum Credits Shadowy Vigilantes for Safety," WSJ, 5.6.03.
"We are human beings": Kartini in Chris Brummitt, "Indonesian Squatters Forced to Make Way for Development," LAT, 1.11.04.
"How would anyone know if I died?": Chan Bak-choy in Sherry Lee, "Home Truths," SCMP, 1.20.03.
"I want to change the character of the city": Joel Dreyfuss and Desmond Hutton, "Japan Real Estate Mogul Makes a $2.2 Billion Bet," IHT, 10.27.03.
"littered with property schemes": Larry Rohter, "Brazil to Let Squatters Own Homes," NYT, 4.19.03.
"I want to do a lot for those who live in shacks": Larry Rohter, "Brazil to Let Squatters Own Homes," LAT, 4.19.03.
"a country needs something to look up to": Chen May Yee, "Monuments to Mahathir," WSJ, 9.10.01.
"The plan has hitherto been": *Speeches by Lord Curzon of Kedleston*, 1900, vol. 1, p. 214.
"sacred trust": Henry Kendall, *Jerusalem: The City Plan; Preservation and Development During the British Mandate, 1918-1948*, 1948, p. v.
"people hate old things": Zhang Xin in "First Out of the Blocks to Modernise," FT, 3.20.04.
"if we say something against it, we will go to jail": Umer in Ashley Gilbertson, "In Kashgar's Dust," FEER, 11.27.03.

CHAPTER 22. COUNTRYSIDES ABROAD

"in the city they have everything": Tashi Phongtsok in Peter S. Goodman, "Out of the Dark in Rural China," WP, 6.25.04.
"if we had more electricity": Tsering Lhazom in *ibid.*
"this story is only beginning": Antonio Tusso in Matt Moffett, "How a Brazilian Cattle Baron Shakes Up World's Beef Trade," WSJ, 6.22.04.
"eternal recession and the debt spiral": Gonzalez Fraga in Tony Smith, "Farm Exports Boom in Argentina," NYT, 3.26.03.
1"we never thought the Russians would come back": Greg Grenz in Roger Thurow, Scott Kilman, and Gregory L. White, "New Farm Powers Sow the Seeds of America's Agricultural Woes," WSJ, 6.18.04.
"that's the last thing China needs": Warren Staley in Caroline Daniel, "The Cargill Approach: We Don't Lobby. We Go and Share the Information," FT, 2.26.04.
"Mr. Lee committed suicide to save the farmers": An Sung Hyun in James Brooke, "Farming Is Korean's Life and He Ends It in Despair," NYT, 9.16.03.
"I wandered by myself for hours": Prafulla Mohanti, *My Village, My Life*, 1975, p. 15.

BOOKS AND ARTICLES CITED OR QUOTED

Frederick Adams, *The Conquest of the Tropics*, New York: Doubleday, 1914.

James Agee, *Let Us Now Praise Famous Men*, Boston: Houghton Mifflin, 1941.

Guillermo Algaze, *The Uruk World System*, Chicago: University of Chicago Press, 1993.

Idris Ali, *Dongola*, English translation by Peter Theroux, Fayetteville: University of Arkansas Press, 1998.

William Allan, *The African Husbandman*, New York: Barnes & Noble, 1965.

Robert Allshouser, *Photographs for the Czar*, New York: Dial, 1980.

Francis Bacon, *The Advancement of Learning* (1605) and the *New Atlantis* (1626), Oxford: Oxford University Press, 1966.

Jacques Barzun, *From Dawn to Decadence*, New York: HarperCollins, 2000.

Edward Bellamy, *Looking Backward, 2000–1887*, New York: Ticknor, 1888.

Hillaire Belloc, *The Modern Traveller* (1898), London: Duckworth, 1972.

Ruth Benedict, *Patterns of Culture*, Boston: Houghton Mifflin, 1934.

Edward Blakely and Mary Gail Snyder, *Fortress America*, Washington, DC: Brookings Institution, 1999.

Willem Blaue, *Le Grand Atlas*, (c. 1667, variants published as *Le Théâtre du Monde*), excerpts ed. by John Goss published as *Blaue's The Great Atlas*, London: Studio, 1997.

Roger Blench and Matthew Spriggs, eds., *Archaeology and Language IV: Language Change and Cultural Transformation*, New York: Routledge, 1999.

Karen Blixen, *Out of Africa*, London: Putnam, 1937.

Peter Bogucki, *The Origins of Human Society*, Malden, UK: Blackwell, 1999.

Stewart Brand, *How Buildings Learn*, New York: Penguin, 1994.

Henri Breuil, *Four Hundred Centuries of Cave Art*, Dordogne, France: Sapho, 1952.

Keith Buchanan, *The Transformation of the Chinese Earth*, London: Bell, 1970.

Samuel Butler, *Erewhon*, London: Trubner, 1872.

Luís de Camões, *The Lusiads* (1572), tr. by Landeg White, New York: Oxford, 2001.

Alphonse de Candolle, *The Origin of Cultivated Plants*, London: Kegan Paul, Trench, 1884.

Robert Carneiro, "A Theory of the Origin of the State," *Science*, vol. 169, 1970, pp. 733–738.

Robert Caro, *The Power Broker*, New York: Knopf, 1974.

Rachel Carson, *Silent Spring*, Boston: Houghton Mifflin, 1962.

Jacques Cauvin, *The Birth of the Gods and the Origins of Agriculture*, Cambridge: Cambridge University Press, 2000.

Edward Cecil, *The Leisure of an Egyptian Official*, London: Hodder and Stoughton, 1921.

Napoleon Chagnon, *Yanomamo: The Fierce People*, New York: Holt, Rinehart & Winston, 1968.

Amit Chaudhuri, ed., *The Picador Book of Modern Indian Literature*, London: Picador, 2001.
V. Gordon Childe, *New Light on the Most Ancient East*, London: Kegan Paul, Trench, Trubner, 1934.
Morten H. Christiansen and Simon Kirby, eds., *Language Evolution*, New York: Oxford, 2003.
Kenneth Clark, *Civilization: A Personal View*, New York: Harper & Row, 1969.
Jean-Louis Cohen and Monique Eleb, *Casablanca*, New York: Monacelli, 2002.
Harold Conklin, *Ethnographic Atlas of Ifugao*, New Haven: Yale University Press, 1980.
Diego Fernandez Correa, *The Three Voyages of Vasco Da Gama and His Viceroyalty, from the Lendas da India of Gaspar Correa*, Hakluyt Society, First Series, No. 42, London: The Society, 1869.
George Nathaniel Curzon, *Speeches by Lord Curzon of Kedleston*, two volumes, Calcutta: Superintendent of Government Printing, 1900.
Charles Darwin, *Journal . . . during the Voyage of the H. M. S. Beagle*, London: John Murray, 1860.
Gurcharan Das, *India Unbound*, New York: Knopf, 2001.
Daniel Defoe, *Robinson Crusoe*, London: W. Taylor, 1719.
Edward Denison, et al., *Asmara: Africa's Secret Modernist City*, New York: Merrell, 2004.
Jared Diamond, *Guns, Germs, and Steel*, New York: Norton, 1997.
Jared Diamond and Peter Bellwood, "Farmers and Their Languages: The First Expansions," *Science*, vol. 300, 2003, pp. 597–603.
Denis Diderot, *A Diderot Pictorial Encyclopedia of Trades and Industry* (1751–1772), ed. by Charles C. Gillespie, New York: Dover, 1959.
Roxanne Dunbar-Ortiz, *Red Dirt: Growing Up Okie*, New York: Verso, 1997.
[Lord] Ernle (Rowland Edmund Prothero), *English Farming: Past and Present*, London: Longmans, Green, 1912.
Brian Fagan, ed., *The Oxford Companion to Archaeology*, New York: Oxford University Press, 1996.
Hugh Ferris, *The Metropolis of Tomorrow*, New York: Ives Washburn, 1929.
Larry Ford, *America's New Downtowns*, Baltimore: Johns Hopkins University Press, 2003.
James G. Frazer, *The Golden Bough*, London: Macmillan, 1890.
Thomas Friedman, *The Lexus and the Olive Tree*, New York: Farrar, Straus and Giroux, 1999.
Arthur Gaitskell, *Gezira: a Story of Development in the Sahara*, London: Faber and Faber, 1959.
Joel Garreau, *Edge City: Life on the New Frontier*, New York: Doubleday, 1991.
Patrick Geddes, *Cities in Evolution*, London: Williams and Norgate, 1915.
Siegfried Giedion, *Mechanization Takes Command*, New York: Oxford University Press, 1948.
Gideon S. Golany, *Urban Design Ethics in Ancient China*, Lewiston: Mellon, 2001.
Goode's World Atlas, ed. by John C. Hudson, Skokie, IL: Rand McNally, 2000.
Jean Gottman, *Megalopolis*, New York: Twentieth Century Fund, 1961.
Robert Graves, *Good-Bye to All That*, London: Jonathan Cape, 1929.
Howard Gray, *English Field Systems*, Cambridge: Harvard University Press, 1915.
Josiah Gregg, *Commerce of the Prairies*, New York: Langley, 1844.
Arthur Grimble, *We Chose the Islands*, New York: Morrow, 1952.
Ewart S. Grogan and Arthur H. Sharp, *From the Cape to Cairo: The First Traverse of Africa from South to North*, London: Hurst and Blackett, 1900.
Peter Hall, *Cities in Civilization*, New York: Pantheon, 1998.
Richard Hakluyt, *Principal Navigations . . . of the English Nation*, London: George Bishop, 1599–1600.
Jack Harlan, *The Living Fields*, Cambridge: Cambridge University Press, 1990.

Fil Hearn, *Ideas That Shaped Buildings*, Cambridge: MIT Press, 2003.
Charles Higham, *Early Cultures of Mainland Southeast Asia*, London: Thames and Hudson, 2002.
Emmeline W. Hill, Mark A. Jobling, Daniel G. Bradley, "Y-Chromosome Variation and Irish Origins," *Nature*, vol. 404, 2000, p. 351.
James Hilton, *Lost Horizon*, New York: Grosset & Dunlap, 1933.
Henry-Russell Hitchcock and Philip Johnson, *The International Style: Architecture Since 1922*, New York: Norton, 1932.
Thomas Hobbes, *Leviathan*, London: Andrew Crooke, 1651.
Adam Hochshild, *King Leopold's Ghost*, Boston: Houghton Mifflin, 1998.
Erik Hornung and Betsy M. Bryan, eds., *The Quest for Immortality: Treasures of Ancient Egypt*, Washington, DC: National Gallery, 2002.
Ebenezer Howard, *Garden Cities of To-morrow*, London: Sonnenschein, 1902.
Robert Hughes, *The Shock of the New*, New York: Knopf, 1981.
Jane Jacobs, *The Death and Life of Great American Cities*, New York: Random House, 1961.
Osa Johnson, *I Married Adventure*, Philadelphia: Lippincott, 1940.
Henry Kendall, *Jerusalem: The City Plan; Preservation and Development During the British Mandate, 1918–1948*, London: HMSO, 1948.
Julius Klein, *The Mesta: A Study in Spanish Economic History, 1273–1836*, Cambridge: Harvard University Press, 1920.
Walter Savage Landor, *Complete Works*, ed. by T. Earle Welby, London: Chapman and Hall, 1927–1936.
J. H. G. Lebon, *Land-Use in Sudan*, Bude, UK: Geographical Publications Ltd., 1965.
Le Corbusier (pseud. of Charles Édouard Jeanneret), *Towards a New Architecture*, London: John Rodker, 1927.
Richard B. Lee and Irven Devore, eds., *Man the Hunter*, Chicago: Aldine, 1968.
Richard B. Lee and Richard Daly, eds., *The Cambridge Encyclopedia of Hunters and Gatherers*, Cambridge: Cambridge University Press, 1999.
Claude Levi-Strauss, *The Savage Mind*, Chicago: University of Chicago Press, 1966.
Bernard Lewis, *A Middle East Mosaic*, New York: Random House, 2000.
Simon Leys (pseud. of Pierre Ryckmans), *Chinese Shadows*, New York: Viking, 1977.
Luigi Luca and Francesco Cavalli-Sforza, *The Great Human Diasporas*, Reading: Addison-Wesley, 1996.
Norman Maclean, *A River Runs Through It*, Chicago: University of Chicago Press, 1976.
Ross MacPhee, ed., *Extinctions in Near Time: Causes, Contexts, Consequences*, New York: Kluwer Academic, 1999.
James C. Malin, *The Grassland of North America: Prolegomena to Its History,* Gloucester, MA: Peter Smith, 1947.
Bronislaw Malinowski, *A Diary in the Strict Sense of the Term*, New York: Harcourt Brace, 1967.
George Perkins Marsh, *Man and Nature*, New York: Scribner, 1864.
Carey McWilliams, *Factories in the Fields*, Boston: Little Brown, 1939.
W. J. McGee, ed., *Conference of Governors of the United States, Proceedings,* Washington, DC: U.S. Government Printing Office, 1908.
Aubrey Menen, *Dead Man in the Silver Market*, New York: Scribner's, 1953.
Merriam-Webster's Encyclopedia of World Religions, ed. by Wendy Doniger, Springfield, MA: Merriam-Webster, 2000.

Jeffrey Meyer, *The Dragons of Tiananmen: Beijing as a Sacred City*, Columbia: University of South Carolina, 1991.

Ross Miller, *Here's the Deal: The Buying and Selling of a Great American City*, New York: Knopf, 1996.

Marvin Miracle, *Agriculture in the Congo Basin*, Madison: University of Wisconsin Press, 1967.

Prafulla Mohanti, *My Village, My Life*, New York: Praeger, 1973.

A. M. T. Moore, G. C. Hillman, and A. J. Legge, *Village on the Euphrates: From Foraging to Farming at Abu Hureyra*, New York: Oxford University Press, 2000.

Lewis Mumford, *The Myth of the Machine*, New York: Harcourt Brace, 1967.

Martha Mundy and Basil Musallam, eds., *The Transformation of Nomadic Society in the Arab East*, Cambridge: Cambridge University Press, 2000.

Sebastian Münster, *Universalis Cosmographia*, Basel, Switzerland: Heinrichum Petri, 1550.

[Murray's] *Handbook for Travellers in Russia, Poland, and Finland, Including the Crimea, Caucasus, Siberia, and Central Asia*, London: John Murray, 1875.

National Geographic Society, *Historical Atlas of the United States*, Washington, DC: National Geographic Society, 1988.

Rodney Needham, *Science and Civilization in China*, vol. 4, part 3, *Civil Engineering and Nautics*, Cambridge: Cambridge University Press, 1971..

John G. Neihardt, *Black Elk Speaks*, New York: Morrow, 1932.

Pablo Neruda, *Selected Poems of Pablo Neruda*, trans. by Ben Belitt, New York: Grove, 1961.

George Orwell, *The Road to Wigan Pier*, London: Gollancz, 1937.

Stefania Panebianco, ed., *A New Euro-Mediterranean Cultural Identify*, London: Frank Cass, 2003.

Blaise Pascal, *Pensées sur la religion* (1670), Paris: Editions du Luxembourg, ed. by Luis Lafuma, 1951.

Peter Perdue, *Exhausting the Earth: State and Peasant in Hunan, 1500–1850*, Cambridge: Harvard University Press, 1987.

Gifford Pinchot, *Breaking New Ground*, New York, Harcourt, Brace, 1947.

Susan Pollock, *Ancient Mesopotamia*, New York: Cambridge University Press, 1999.

Stephen A. D. Puter, *Looters of the Public Domain*, Portland, OR: Portland Printing House, 1908.

Nizar Qabbani, *On Entering the Sea*, New York: Interlink, 1996.

Thomas Raleigh, ed., *[Lord] Curzon in India: Being a Selection from his Speeches as Viceroy and Governor-General of India, 1898–1905*, London: Macmillan, 1906.

Abdur Rasheed, comp., *The Traveller's Companion, Containing a Brief Description of Places of Pilgrimage and Important Towns in India*, Calcutta: Superintendent of Government Printing, 1911.

Jacob Riis, *How the Other Half Lives*, New York: Scribner's, 1890.

Kevin Roberts, *Lovemarks: The Future Beyond Brands*, New York: Power House, 2004.

Jean Jacques Rousseau, *Discourse on the Origin of Inequality* (French original 1754), London: R. and J. Dodsley, 1761.

Marshall Sahlins, *Stone-Age Economics*, Hawthorne: Aldine, 1972.

Antoine de Saint-Exupéry, *Night Flight*, London: Desmond Harmsworth, 1921.

Carl O. Sauer, *Agricultural Origins and Dispersals*, New York: American Geographical Society, 1952.

Franz Schwanitz, *The Origin of Cultivated Plants*, Cambridge: Harvard University Press, 1967.

Frank Scott, *The Art of Beautifying the Suburban Home Grounds of Small Extent*, New York: D. Appleton, 1870.

Elman Service, *Primitive Social Organization: An Evolutionary Perspective*, New York: Random House, 1962.
Anthony Shadid, *Legacy of the Prophet*, Boulder: Westview, 2002.
Pat Shipman, *The Man Who Found the Missing Link*, New York: Simon and Schuster, 2001.
Michael Myers Shoemaker, *The Great Siberian Railway*, New York: G.P. Putnam's, 1903.
Alyse Simpson, *The Land that Never Was*, London: Selwyn and Blount, 1937.
Adam T. Smith, *The Political Landscape: Constellations of Authority in Early Complex Polities*, Berkeley, University of California Press, 2003.
Anthony Smith, *Blind White Fish in Persia*, New York: Dutton, 1953.
Bruce Smith, *The Emergence of Agriculture*, New York: Scientific American Library, 1998.
David Sneath, *Changing Inner Mongolia: Pastoral Mongolian Society and the Chinese State*, New York: Oxford University Press, 2000.
Wilhelm G. Solheim II, "An Earlier Agricultural Revolution," *Scientific American*, vol. 226, 1972, pp. 34–41.
O. H. K. Spate, *India*, London: Methuen, 1954.
Craig Stanford, *Upright: The Evolutionary Key to Becoming Human*, Boston: Houghton Mifflin, 2003.
Chris Stringer, *African Exodus: the Origins of Modern Humanity*, New York: Henry Holt, 1997.
Jonathan Swift, *Gulliver's Travels*, London: Benj. Motte, 1726.
Rabindranath Tagore, *Gitanjali*, London: Chiswick, 1912.
Julian Thomas, *Rethinking the Neolithic*, New York: Cambridge University Press, 1991.
J. Walter Thompson, *Population and its Characteristics*, New York: J. Walter Thompson Company, 1912, and later dates.
Henry David Thoreau, *Walden*, Boston: Ticknor and Fields, 1854.
Mark Twain, *Roughing It*, Hartford, CT: American Publishing Company, 1872.
Robert Twombley, ed., *Louis Sullivan: The Public Papers*, Chicago: University of Chicago Press, 1988.
Edward Tylor, *Primitive Culture*, London: John Murray, 1871.
Urban Land Institute, *Business and Industrial Park Development Handbook*, Washington, DC: The Institute, 1988.
Urban Land Institute, *Shopping Center Development Handbook*, 3rd edition, Washington, DC: The Institute, 1999.
Laurens van der Post, *Lost World of the Kalahari*, New York: Morrow, 1958.
Nikolai Vavilov, *The Origin and Geography of Cultivated Plants*, Cambridge: Cambridge University Press, 1992.
Paul Wheatley, *The Pivot of the Four Quarters*, Chicago: Aldine, 1971.
William Willcocks, *Egyptian Irrigation*, London: Spon and Chamberlain, 1899.
E. O. Wilson, *Naturalist*, Washington, DC: Island Press, 1994.
Tom Wolfe, *From Bauhaus to Our House*, New York: Farrar, Straus and Giroux, 1981.
Leonard Woolf, *Growing: An Autobiography of the Years 1904–1911*, London: Hogarth, 1961.
William Wordsworth and Samuel Taylor Coleridge, *Lyrical Ballads*, London: Longman and Rees, 1798.
Yinong Xu, *The Chinese City in Space and Time: The Development of Urban Form in Suzhou*, Honolulu: University of Hawaii Press, 2000.

INDEX

B

D

E

I

J

K

L

M

N

O

ABOUT THE AUTHOR

Bret Wallach (*www.geosciences.ou.edu/~bwallach*) teaches geography at the University of Oklahoma. A MacArthur Fellow, he has previously published *At Odds with Progress: Conservation and Americans* (1991) and *Losing Asia: Modernization and the Culture of Development* (1996). He is presently working on a book about the rural landscapes of Eurasia.